Yuanzhi Xu, Jia Yao

Electron Magnetic Resonance Principles

Also of Interest

Atomic Emission Spectrometry.
AES – Spark, Arc, Laser Excitation
Golloch, Joosten, Killewald, Flock, 2019
ISBN 978-3-11-052768-1, e-ISBN 978-3-11-052969-2

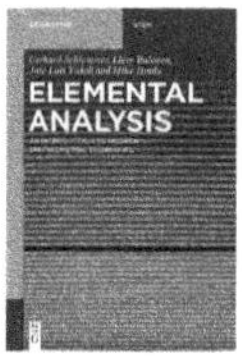

Elemental Analysis.
An Introduction to Modern Spectrometric Techniques
Schlemmer, Balcaen, Todolí, Hinds, 2018
ISBN 978-3-11-050107-0, e-ISBN 978-3-11-050108-7

Elastic Light Scattering Spectrometry.
Huang, Ling, Wang, 2018
ISBN 978-3-11-057310-7, e-ISBN 978-3-11-057313-8

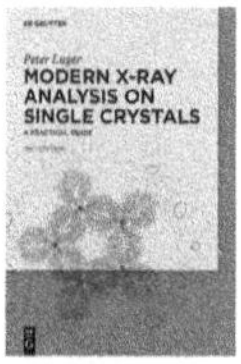

Modern X-Ray Analysis on Single Crystals.
A Practical Guide
Luger, 2014
ISBN 978-3-11-030823-5, e-ISBN 978-3-11-030828-0

Yuanzhi Xu, Jia Yao

Electron Magnetic Resonance Principles

DE GRUYTER

清華大學出版社
TSINGHUA UNIVERSITY PRESS

Authors
Dr. Yuanzhi Xu
Department of Chemistry
Zhejiang University
Hangzhou
P. R. China

Dr. Jia Yao
Department of Chemistry
Zhejiang University
Hangzhou
P. R. China

ISBN 978-3-11-052800-8
e-ISBN (PDF) 978-3-11-056857-8
e-ISBN (EPUB) 978-3-11-062010-8

Library of Congress Control Number: 2018951868

Bibliographic information published by the Deutsche Nationalbibliothek
The Deutsche Nationalbibliothek lists this publication in the Deutsche Nationalbibliografie; detailed bibliographic data are available on the Internet at http://dnb.dnb.de.

Typesetting: Integra Software Services Pvt. Ltd.
Printing and binding: CPI books GmbH, Leck
Cover image: Jakarin2521/ iStock / Getty Images Plus

www.degruyter.com

To

My Dear Friend
Professor Zuwen Qiu

Foreword

This book more specifically stresses on the aspect of foundational principle of EMR. The contents include the following: Introduction, theoretical basics, $\boldsymbol{g}$-tensor theory, isotropical hyperfine structure of spectrum, anisotropic hyperfine structure of spectrum, fine structure of spectrum, relaxation theory and lineshape, linewidths, quantitative determination, paramagnetic species in gas phase and lineshape, linewidths, quantitative determination, paramagnetic species in gas phase and inorganic radicals, spectra of transition metal ions and their complexes, and so on.

Quantitative determination of spectrum, discussed in Chapter 8, is one of the most difficult problems in EMR. Usually, the samples of EMR are in liquid or solid state. Chapter 9 discusses paramagnetic species in gas phase, and inorganic radical specially. Of the 107 elements in the periodic table, there are 57 transition metal (including rare earth) group elements, and their EMR spectra have some special characteristics. This is discussed in Chapter 10.

ENDOR, ELDOR, Pulse-EMR, and EMRI have been included in Appendix 1 for readers to refer to. Appendix 2 "Mathematic Preparation" and Appendix 3 "Angular Momentum and Stable State Perturbation Theory in Quantum Mechanics" help readers to replenish the basics of mathematics and physics.

Important aspects of modern EMR methods are considered in Appendix 1. This is pulse technique. In EMR, these methods have been developed in the last decade. It can be expected that these methods will be described in detail in the main part of the book in future, when the theory and practice of pulsed EMR will receive their completed development.

Foundational principle of EMR is the distinguishing feature of this book. It is an advanced specialzed book of science and technology. This book contains materials for further study on the basic theory of EMR for researchers and technicians who work in the field of EMR. This book will also serve as educational material for advanced study for graduate students and young teachers of related specialist fields. In all aspects, I recommend this book for study and use.

Professor of Physical Chemistry
Yu. D. Tsvetkov

https://doi.org/10.1515/9783110568578-201

Author's Preface

"Electron Magnetic Resonance (EMR)" instructive training course was conducted in the Chemistry Department of Tsinghua University during autumn of 1983. I was invited as a chief lecturer. The draft of my lecture then was an original blank of this book.

In the fall of 1984, I moved to Department of Chemistry, Zhejiang University, as a professor. I also served for two terms in the EMR training course in 1985 and 1986. My lecture drafts at Tsinghua training course in 1983 were printed by mimeograph and were provided to the students to use as teaching material.

The JEOL Company of Japan printed 200 copies of this material and provided to its users.

Till my retirement in 1998, graduate students had been provided EMR course every year. It is the main teaching material. Some modifications and compliments to the material have been received from time to time.

Ten years after my retirement (2008), this draft as a monograph of "Applied Electron Magnetic Resonance Spectroscopy" has been published formally by Science Press (Beijing), funded by the publishing foundation of the Department of Science & Technology.

And seven years thereon (2015), after receiving opinion of readers, the book has passed through considerable revision and modification, and has been renamed as "Principle of Electron Magnetic Resonance" (Chinese edition), published by Tsinghua University Press.

In order to satisfy the requirements of international exchange as well as students studying EMR abroad, the English edition of "Electron Magnetic Resonance Principles" is published now.

The author expresses heartfelt gratitude to Professor Yu. D. Tsvetkov for writing Foreword of this book.

Professor of Physical Chemistry
Yuanzhi Xu

https://doi.org/10.1515/9783110568578-202

Contents

1 Introduction

Abstract: This chapter discusses four topics: (1) Origin of EMR; (2) experimental apparatus; (3) target of research; (4) prospects for future.

1.1 Origin of EMR

If a magnetic moment (vector) of the electron is placed into an external magnetic field, it would be divided into two different levels: one parallel to the applied field (high level), and another antiparallel to the field (low level). The magnetic moments of low level transit to the high level when they absorb an appropriate frequency of electromagnetic wave under certain conditions. Such transition is called *magnetic resonance transition*. When the course of the magnetic moment is from the electron, it is called "electronic magnetic resonance" (*EMR*) *transition*; when the course is from the nucleus, it is called "nuclear magnetic resonance" (*NMR*) *transition*.

The "electron magnetic resonance" (EMR) phenomenon was first observed in the laboratory successfully by Zavoisky [1], a physicist from the former Soviet Union. He used $CuCl_2{\cdot}2H_2O$ as sample in the external magnetic field of 4.76 mT, applied alternating electromagnetic wave irradiation of 133 MHz frequency on the sample, and detected the resonance signal of the absorbed electromagnetic wave. Frenkel [2] explained the experimental results of Zavoisky. This experiment was successfully performed in 1945 and it has been declared as the official birth of EMR.

The spectral line splitting phenomenon resulting from the action of the magnetic field was found by Zeeman in 1895 (the so-called Zeeman effect [3]). This incubation period passed through half a century. In the meantime, the experiment of Gerlach and Stern [4] proved that with the application of external magnetic field, the orientation of electron magnetic moment becomes discrete; subsequently, Uhlenbeck and Goudsmit [5] proposed that the electron magnetic moment is mainly caused by the "spin motion," while the magnetic moment caused by "spin motion" is "inherent magnetic moment" or "eigen moment." Breit and Rabi [6] have described hydrogen atom under the effect of external magnetic field – the electron magnetic moment orientation is divided into two different states: "parallel" and "antiparallel" to the direction of the external magnetic field, with the energy corresponding to two levels is $\Delta E = h\nu$, where h is the Planck constant, ν is the frequency of electromagnetic wave, and ΔE is proportional to the strength of the external magnetic field H. In 1938, Rabi et al. [7] discovered that the transitions between different levels could be induced by alternating magnetic field (electromagnetic waves), and the conditions of EMR transitions were also predicted theoretically; in other words, the birth of EMR was predicted. Since then, a large number of scientists had been trying to observe the resonance phenomenon experimentally. Seven years later, it was first observed in the

https://doi.org/10.1515/9783110568578-001

laboratory of Zavoisky, which he named it as "electron paramagnetic resonance" (EPR). Owing to this achievement, Zavoisky is considered the father of EMR.

For more than a decade, research on EMR was vigorously carried out in the United States [8] and the United Kingdom [9]. Physicists of the Oxford University, namely, Abragam, Bleaney, Pryce, and Van Vleck, have made a significant contribution to the theoretical analysis of the magnetic resonance spectroscopy. At that time (i.e., the 1950s), the main objective of EPR research was radical, and since the magnetic moment of the electron in free radical is mainly contributed by the motion of electron spin, the physicists thought that it should be renamed as "electron spin resonance" (*ESR*). In 1958, J. E. D. Ingram of the University of Southampton in the United Kingdom changed all the names of EPR to ESR in his monograph *Free Radicals as Studied by Electron Spin Resonance*; he said: "It will be used all of the 'ESR' in future." Since then, most of literatures have preferred using ESR. By the end of 1980s, it was found that the use of expression ESR in many of the research samples was not appropriate. Accordingly, the expression ESR was reverted to EPR; in fact, both the EPR and ESR are used today. International society set up in 1989 is named International EPR/ESR Society. Since the 1990s, some literatures have used the name *Electron Magnetic Resonance*. In our opinion, this name is more exact, as it corresponds to NMR, so EMR has been used in this monograph. However, for historical reasons and in reference to the previous literature, at some places we have made use of ESR and EPR. For the early history of EMR, a review by Ramsey [10] could be referred.

In 1946, Purcell [11] of Harvard University and Bloch [12] of Stanford University independently observed the phenomenon of NMR in their respective laboratory.

Before 1950, the research on magnetic resonance was mainly undertaken by physicists. The first EMR spectroscopy of organic free radical was reported [13] in 1952. Since then, as a new experimental technique and research method, magnetic resonance has aroused wide interest among chemists, biologists, and medical scientists.

Since 1950s, important progress has been made in two aspects: one is the theoretical foundation of magnetic resonance laid by Abragam, Bleaney, Pryce, and van Vleck, as well as by Bloch, Purcell, Pound, Bloembergen, and others; another is the continuous improvement in the instrument performance and experimental techniques. A commercial magnetic resonance spectrometer was placed on the market by Varian Company in the early 1950s, thus enabling for a wide range of applications in chemistry, biology, medicine, and other fields.

During the early 1960s, a vigorous application of magnetic resonance in various fields was observed. Until the end of 1960s, the EMR spectrum of the stable organic free radical was exclusively studied. Due to the advances in experimental techniques, such as the application of cryogenic technology, in situ detection, and spin trapping, unstable free radicals were also studied and the research on the EMR was extended to the diamagnetic material by using spin labeling technique.

Usually, the EMR signals are absorption lines; however, in some chemical reactions (especially thermal reaction or photoreaction), there are emission lines and enhanced absorption lines. This is the so-called chemically induced dynamic electron polarization (CIDEP). The first CIDEP phenomenon was observed by Fessenden and Schuler [14] in the reaction of liquid methane at ~100 K, using 2.8 MeV electron pulse. The corresponding CIDNP is the "chemically induced dynamic nuclear polarization." In 1967, Bargon and Fischer [15] as well as Ward and Lawer [16] independently observed the phenomenon of CIDNP for the first time. The discovery of CIDEP and CIDNP provides a powerful research tool for the rapid reaction kinetics, especially for the study of photochemical and radiation chemical reaction kinetics.

By the end of 1970s, owing to the development of computer technology, magnetic resonance spectrometer reached a new level. In the 1970s (especially in the late 1970s), most of the spectrometers had embedded computers, which enabled the weak signal to be observed through Fourier transform enhancement; and the overlapped spectral line could be resolved by the computer simulation. After the 1990s, almost all the spectrometers were equipped with computers.

After the 1970s, on the basis of EMR, a number of related technologies were developed, such as double resonance (including electron nuclear double resonance [ENDOR] and electro-electron double resonance [ELDOR]) and electron magnetic resonance imaging (EMRI); from the continuous wave electron magnetic resonance (CW-EMR) to pulsed electron magnetic resonance (pulsed-EMR); from the electron spin echo (ESE) to electron spin echo envelope modulation (ESEEM) and saturation recovery (SR); and so on. The pulsed excitation sequence of EMR has not only been widely applied in kinetics of ultrahigh speed reaction, but it has also become a powerful tool in the study of the structure of biological macromolecules and in the development of a single channel and multiple channels, with frequency from hundreds MHz to 95 GHz or even higher, in order to meet the needs of various applications, from one-dimensional spectrum to multidimensional spectrum, from frequency domain to time domain. In the early 1990s, a remarkable achievement was the observation of the single-electron EMR signal in the electromagnetic field (Penning) [17].

1.2 Experimental apparatus

Modern EMR spectrometer has made epoch-making progress than the equipment of that time used by Zavoisky. However, not much changes have been observed in the principal structure of spectrometer. Since this book focuses on the basic principles, a schematic diagram of the apparatus is only introduced here. The reader who is interested in details of the spectrometer may refer to the relevant literature [18,19] and references therein.

The principle block diagram of the EMR device is shown in Figure 1.1.

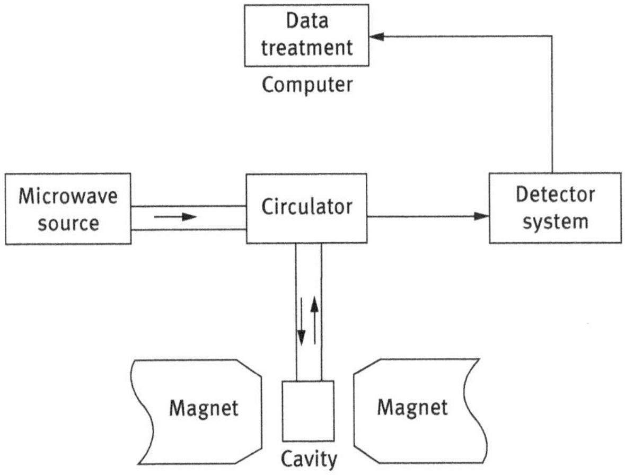

Figure 1.1: The principle block diagram of EMR spectrometer.

1.2.1 Microwave source

The microwave frequencies are provided in the range of $1 \le \nu \le 100$ GHz generally, by using a variety of special oscillation tubes, including backward-wave oscillating tube, special diode and triode tubes (e.g., the Gunn diode), klystron, magnetron, traveling wave tube, and so on. The klystron is the best and the most common used microwave source in these oscillation tubes. Compared to klystron, the Gunn diode is cheaper, and hence in order to reduce the cost, the manufacturers have gradually replaced the klystron with the Gunn diode.

The klystron is a vacuum tube producing microwave oscillation. Its oscillation frequency is concentrated in a very narrow range, and the output power is function of the *frequency for Klystron mode*. The resonance frequency established in a *metal cavity* (this is a very sensitive place for temperature) is very stable. A klystron has several *modes*, which can be displayed on the oscilloscope. Usually, the mode working on the maximum of output power is chosen.

The microwave frequency of output could be obtained by means of mechanical tuning of the klystron cavity and also by adjusting the voltage of the klystron. These adjustments could only be carried out in a selected mode, by changing slightly the frequency to quite a limited range. The output microwave frequency of the klystron could be kept stable, and the temperature and voltage fluctuation of the klystron as well as the mechanical vibration should be suppressed to the minimum. In order to detect the spectral parameters accurately, the work frequency must be accurate at least to 1 MHz, and would be better if kept accurate to 1 kHz.

Despite the extensive studies made on the solid-state devices producing microwave, their frequency stability and their inherent noise are no better than the klystron as of now. However, for the pulse EMR spectrometer, which requires a short-period operation, the solid-state devices as microwave source would be a better choice.

Microwave transmission can be effectively carried out by using the catheter, including coaxial cable, narrow flat wire, and waveguide. Although all these three kinds are used in modern microwave systems, the waveguide is most universally applied to the microwave transmission in/out of the resonant cavity [20]. The waveguide has a strict size requirements; the waveguide used in the most common "X-band" spectrometer is made of a rectangular copper tube with external size 12.7 × 25.4 mm, and it can transmit microwave of 8.2–10.9 GHz with a low loss. The frequency range corresponding to the traditional microwave band and the field strength corresponding to $g = 2$ are listed in Table 1.1.

Table 1.1: The frequency range corresponding to the traditional microwave band, and the field strength corresponds to $g = 2$.

Band	Frequency range (GHz)	Typical frequency (GHz)	Typical field (mT)
L	0.390–1.550	1.5	54
S	1.550–3.900	3.0	110
C	3.900–6.200	6.0	220
X	6.200–10.900	9.5	340
K	10.900–36.000	23	820
Q	36.000–46.000	36	1300
V	46.000–56.000	50	1800
W	56.000–100.000	95	3400

1.2.2 Resonant cavity and coupling system

The microwave is transmitted from the microwave source and coupled to the sample shelf in the resonant cavity. Three important devices, namely, *isolator, attenuator*, and *circulator*, connected through waveguide, constitute a coupling transmission system.

The role of *isolator* is to enable the microwave to be transmitted through it only and to ensure that it is not reflected back, just like the unidirectional valve of water pipe. The isolator thus protects the klystron and ensures that the microwave frequency is highly stable. Even a little bit of disturbance in the reflected microwave would cause instability in the microwave frequency emitting from klystron.

The role of *attenuator* is to adjust the microwave energy transmitting into the sample chamber.

The role of *circulator* (shown in Figure 1.2) is similar to that of isolator in that it is also unidirectional, but is more complex than the isolator. Microwave passes through attenuator and enters into the circulator via ①. Then, the microwave enters the resonant cavity as long as possible via ② to be resonantly absorbed by the sample. It is then exported via ③ into the detector system. It is not possible that the microwave entering circulator has not passed through resonant cavity and be exported via ③ to the detector system directly. But it is possible that some of the microwave reflected back via ③ from detector system, which has to be absorbed at the end ④ with microwave absorption materials.

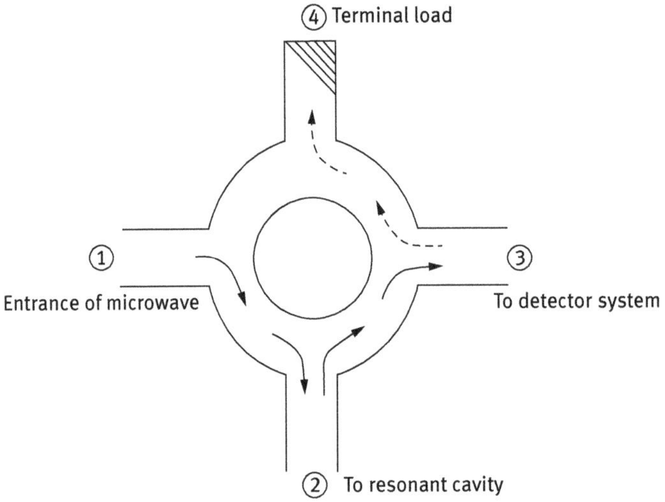

Figure 1.2: Illustrated diagram of four terminal microwave circulator.

Cavity. People often analogize the resonance cavity as the heart of the EMR spectrometer. The tested sample is inserted here. In fact, obtaining the resonance signal in the EMR experiment is not limited to using resonant cavity. Some other resonance devices such as helical resonator and loop-gap resonator have also been used to detect EMR signals. For very high-frequency spectrometer, Fabry–Pérot resonator [21] is used. However, a vast majority of EMR spectrometers use the resonant cavity as the resonant device. This is because it can achieve rather high Q value, and the sample insertion and removal are very convenient.

An effective resistance could be caused by the skin effect [22] of microwave on the metal conductor. The resonant cavity size has a proportional wavelength. The microwave will oscillate due to the multiple reflective in the cavity and will form standing waves, the mode of oscillation depends on the size and shape of the cavity as well as the excitation form.

To match the terminal load in the waveguide, both transverse electric field and transverse magnetic field in the waveguide are transmitted along the longitudinal direction, and reach to the maximum at the same time. However, the electromagnetic field due to the formation of standing wave in the cavity, the maximum of the transverse electric field, and the maximum of the transverse magnetic field along the longitudinal direction are separated by 1/4 wavelength, and the electric field and the magnetic field have phase difference of 90°, that is, when the electric field reaches maximum, the magnetic field reaches minimum, and vice versa. Therefore, the resonant cavity should be optimized: (1) It should have the highest energy density; (2) the position of samples in the resonant cavity should be with the maximum magnetic field strength H_1 (the minimum electric field strength E_1); (3) H_1 is required to be perpendicular to the external magnetic field H; (4) Q value should be as high as possible.

Two common forms of the resonant cavity are the rectangular cavity TE_{102} and the cylindrical cavity TE_{011}. These two kinds of resonant cavity are herein briefly introduced.

The most commonly used rectangular cavity is the TE_{102} type (see Figure 1.3). TE stands for transverse electric, the subscript 102 means the size of *a* direction is 1 "half wavelength," *c* direction is 2 "half wavelength"; *b* direction is 0 [20, 23]. Figure 1.3(b) shows the distribution of electric field in the *xz* plane, in the *x* (*c*) direction, the size should be strictly equal to two half wavelength, and *z* (*b*) is not the strict size required, but it must be less than a half wavelength; Figure 1.3(c) shows the

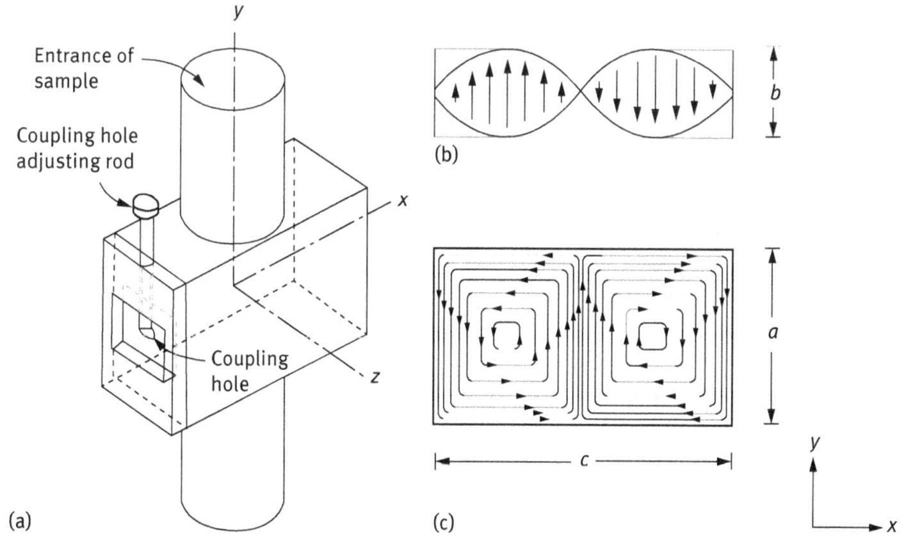

Figure 1.3: Illustrated diagram of rectangular cavity TE_{102}.

distribution of the magnetic field in the xy plane, the size of y (a) direction is 1 half wavelength. This kind of cavity can be inserted by large size and low dielectric loss samples, without causing sharp decrease of energy density. Especially, it is suitable for the liquid sample, and the sample can fill the whole height of cavity.

The most common cylindrical cavity is the TE_{011} type. Figure 1.4(b) shows the distribution of electric field in the cylindrical section; Figure 1.4(c) shows the distribution of magnetic field in the cylindrical diameter profile. It has the advantage that the Q value can be reached more than 20,000, which is three times higher than the rectangular cavity TE_{102}. It is especially suitable for gas samples, and the diameter of the sample tube can be as large as 25 mm.

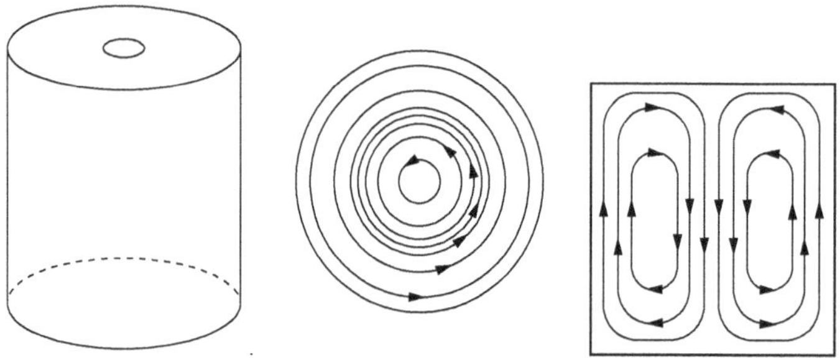

(a) (b) (c)

Figure 1.4: Illustrated diagram of cylindrical cavity TE_{011}.

A very important index for evaluating the resonant cavity is the Q value. Its definition is as follows:

$$Q = \omega \frac{\text{Energy stored in the cavity}}{\text{Average power loss}} \tag{1.1}$$

The loss of cavity inside is caused by the thermal loss produced from the surface current density J in the skin effect resistance R_s, and the ohmic power loss P_L is

$$P_L = \frac{R_s}{2} \int (H_{tm})^2 ds \tag{1.2}$$

The H_{tm} is the maximum tangential direction magnetic field of the surface, and the integral range includes the whole cavity wall area. H_{tm} is numerically equal to J, which is perpendicular to the direction of J, but is parallel to the surface. In addition, there are other losses, such as dielectric loss (especially water) and the loss of paramagnetic samples with large loss angle, as well as the loss of microwave through coupling holes, and so on.

The Q value of the resonant cavity without loading, which is the Q value having solely considered for the ohmic loss of the cavity wall, is expressed as Q_o:

$$Q_o = \frac{\omega\mu}{R_s}\frac{\int |H_m|^2 dV}{\int |H_{tm}|^2 dV} \tag{1.3}$$

The Q value of the resonant cavity with loading is expressed as Q_L:

$$\frac{1}{Q_L} = \frac{1}{Q_o} + \frac{1}{Q_\varepsilon} + \frac{1}{Q_r} \tag{1.4}$$

Here Q_ε is the Q value of the dielectric loss, Q_r is the loss of microwave passing through coupling holes.

$$Q_\varepsilon = \frac{2\pi(\text{energy stored in the cavity})}{\text{Energy loss through the medium per cycle}} = \frac{\mu \int |H_m|^2 dV}{\int \varepsilon'' |E_m|^2 dV} \tag{1.5}$$

Here ε'' is the imaginary part of the dielectric constant, and E_m is the maximum value of the electric field.

$$Q_r = \frac{2\pi(\text{energy stored in the cavity})}{\text{Energy loss through the coupling hole per cycle}} \tag{1.6}$$

The microwave from the coupling system that enters into the resonant cavity should be passed through a "coupling hole" (Iris), see Figure 11.4(a). It is directly proportional to the size and frequency of the cavity as well as to the Q value (sensitivity) [20, 24]. The upside of coupling hole has a spiral Tuner Iris, which is used to adjust the size of the coupling hole in order to get the best match.

The loop-gap resonator [25–27] was developed in the 1980s. There is a series of gaps in the inside of two concentric cylindrical metal rings. A comparison with the operating performance of the TE_{102} resonator shows that the three-loop – two-gap is suitable for the X-band [28]. Loop-gap devices have excellent filling factors, and thus are especially suitable for nonsaturated samples, and in the case of pulse work. The loop-gap resonator is also suitable for the detection of spin on the surface of the sample, such as biological living body in the EMR 1 GHz test [29].

The *dielectric resonator* is completely made of the diamagnetic materials such as quartz. It can concentrate the microwave excitation field H_1 to the whole sample space to enhance the fill factor.

The *microwave spire (slow wave) resonator* although has slightly poor sensitivity (lower Q values), it is useful for inhibiting sharp frequency response (screaming) [20].

There are some special resonators, such as high-temperature and low-temperature resonator, high-voltage resonator, light (ray) irradiation sample cavity, dual mode resonator, and double sample cavity. Please refer to the References section.

1.2.3 Magnet

The magnet is required to provide the stable and uniform magnetic field, as well as the field strength in accord with demands. The so-called stability and uniformity are required in time and space (around the sample), which do not change more than $\pm 1\mu T$, in order to distinguish the narrow spectral lines of EMR and get the correct line shape [30]. The position of pole direction is not important, although the direction of the radio frequency field H_1 should be perpendicular to the field direction $\boldsymbol{H}$ of the magnet. The direction of sample insert to the cavity has to be vertical to the direction of H_1 and H mutually.

In order to get high stability of the magnetic field, we must make the power supply of the electromagnet highly stable. Owing to the existence of hysteresis phenomenon, the relationship between current intensity and the magnetic field strength must be nonlinear. Therefore, the magnetic field strength must be detected timely and the current of magnet could be controlled for any type of EMR spectrometer. The easiest way is to fix a Hall probe in the polar head of the magnetic field. The output voltage of the Hall probe is proportional to the field strength of $\boldsymbol{H}$. Also, it is temperature adjustable or temperature compensation function, preventing the magnetic field drift resulting from the probe temperature change. There is a feedback system connected to the current of magnet to maintain a stable constant voltage of the Hall probe. In order to maintain strict linear magnetic field scan (sweep) in the whole EMR absorption range, the control system of Hall probe must take measures in advance. Field scanning must have good repeatability and linearity to guarantee high accuracy of the measured spectral parameters.

The accurate determination of the magnetic field intensity during the sample's resonance absorption: The proton NMR probe is placed at the edge of the resonant cavity; the magnetic field strength can be calculated accurately from the NMR signal and microwave frequency data. Commodity name: *NMR magnetic field measuring instrument.*

The magnets of high strength field are still being developed. The DC magnet will be able to produce electromagnet ~30 T, and the pulse electromagnet will be able to produce ~100 T [31]. They are required for a very high power supply, as well as for high-frequency microwave source [31–33].

Modulation of Field. The magnetic field modulation EMR spectrometer uses the technology of high-frequency small-field modulation to reduce the output signal noise [34], and also the technology of *phase-sensitive detection* to filter the noise in the vicinity of the modulation frequency. The modulation frequency often used is 100 kHz. The magnetic field modulation method involves the following: On both sides of the sample chamber, a pair of Helmholtz coils are installed, and the 100 kHz current flowing in the coil generates high-frequency magnetic field, which can pass through the sample and in the same direction of static magnetic field. The amplitude of the high-frequency modulation should be less than the linewidth of the EMR signal.

The more smaller the modulation amplitude, the more the approach to the first derivative of the resonance absorption curve of output signal; however, the amplitude of the output signal is also smaller, reducing the sensitivity as well. In order to keep the spectral line away from distortion, as well as taking into account the sensitivity, the modulation amplitude is usually chosen about 1/10 of the signal linewidth. The principle of modulation is shown in Figure 1.5. The point at which the first derivative curve of the signal crosses the horizontal axis always corresponds to the summit of the symmetric absorption peak. However, in order to improve the sensitivity, as long as there is no obvious distortion of the line shape, the amplitude of modulation will be increased to more than 1/10. The range of modulation amplitude may be chosen from 5×10^{-4} to 4.0 mT, depending on the concrete condition of the signal linewidth and sensitivity – the users can make their own choice. There is another design of a field sweep coil in the electromagnet, which carries out the linear field sweep slowly by a DC current with slow change. If the sweep speed is too fast, it

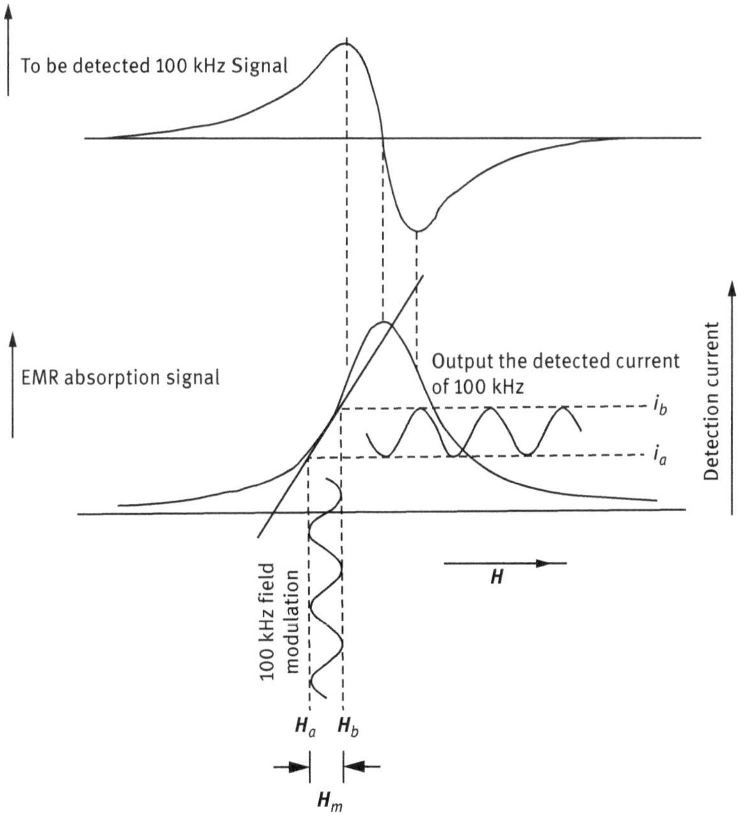

Figure 1.5: Illustrated diagram of the signal by smaller amplitude modulation with high frequency.

may cause spectral line distortion; and if it is too slow, it may affect the efficiency of the work. Commodity EMR spectrometers set several gear, such as the sweep time of 1, 2, 5, 10, 20 min for the operator to choose.

1.2.4 Detection system

When the external magnetic field slowly sweeps through the resonance point, the microwave power is modulated by the current of 100 kHz. After resonant absorption by sample from the resonant cavity, the microwave power was modulated, loading the signal and passing through the circulator. After the detection of crystal detector, it entered into a narrowband amplifier of 100 kHz. Then, it enters into the phase-sensitive detector, where the signal modulated by 100 kHz is combined with a reference signal. This reference signal is phase-adjustable equiamplitude wave, which is generated from the 100 kHz oscillation generator of modulating field. Here, it is just detected whether the frequency and phase of received signal agree with the reference signal [35]. Owing to the fact that phase-sensitive detector has a characteristic sensitivity for the frequency and phase, all the hybrid signal whose frequency and phase disagree with reference signal and the noise could be removed. In addition, the phase-sensitive detector can also play the role of the compression signal bandwidth, thus helping to reduce the final bandwidth of noise and enhancing the signal-to-noise ratio. The output signal of the phase-sensitive detector after being amplified is fed into a data treating system (e.g., a computer, pen recorder, or an oscilloscope).

The EMR spectrometer of high-frequency small modulation field is characterized by high sensitivity, and the structure is also not too complex. But the shortcoming is the spectral line broadening, which may be caused by the magnetic field modulation. Due to the 100 kHz magnetic field modulation, ±100 kHz frequency conversion (corresponding to $\pm 3.6\,\mu T$) may be caused at the midpoint of the spectrum line, which would affect the resolution of the narrow line ($< 5.0\,\mu\text{T}$). It is possible to improve the resolution by suitable lower modulation frequency, with not much impact on the sensitivity of spectrometer. Usually the commodity EMR spectrometer is designed with several modulation frequencies as per user's choice.

1.2.5 Data treatment system

The data treatment system of old spectrometer is an oscilloscope or a recorder. The oscilloscope and the recorder were replaced by the computer that was introduced into the spectrometer. The computer could manage and control the majority operations as well as the final data treatment. Each instrument company has developed many more software combined with the spectrometer. It is not discussed here.

1.3 Target of research

E. Zavoisky was the first to give the name electron paramagnetic resonance. As its name implies, the objective of the research is limited to paramagnetic substances, which are classified into the following categories:

1. **Free radicals** Including organic, inorganic, solid, liquid, gaseous; in other words, all the atoms, molecules, or ions that contain an unpaired electron are free radicals.
2. **Transition (containing lanthanide and actinide) ions** They usually contain multiple unpaired electrons.
3. **Point defect** The negative ion vacancies in the lattice of a point defect crystal or glass, capturing an electron, usually called the *F* center. In addition, the electron-deficient vacancies (positive electron holes) are also paramagnetic.
4. **The systems of more than one unpaired electrons** They do not include the following transition (as lanthanide and actinide) ions:
   (a) **Triplets** It contains two unpaired electrons, having a strong interaction between them. Most of triplets are in the excited state and not stable, but the ground-state triplet is stable.
   (b) **Biradicals** It also contains two unpaired electrons, but having sufficient long distance between them and the interaction is rather weak.
   (c) **Multiple radicals** It is a system containing more than three unpaired electrons.
5. **Contain unpaired conduct electron system** Such as some metals and semiconductors.

The above-mentioned five types of substances are only scarcely found in Nature, the majority of materials are diamagnetic, which is beyond the scope of this study. These systems can however become the objective of EMR study with the aid of spin label technology [36, 37].

The reaction process, especially radiation chemical reaction (including light) that produced unstable, short-lived free radical, need to be studied using *spin trapping* technology [38, 39] or *time-domain EMR spectrometer* [40, 41].

1.4 Prospects for future

Since its inception in 1945, EMR is becoming perfect day by day in theory, its technology has been continuously developed, and its field of application is expanding increasingly. In fact, the development of theory and technology and the expansion of application field are supplementary and complementary to each other.

Experimental technique. The spectrometer with CW-EMR has made great progress in the past several decades. In particular, computer is enhancing the intellectual ability of the spectrometer. The microwave frequency has been expanded from the X-band (9–10 GHz in the 1950s) and the L-band (0.4–1.5 GHz) to the W-band (56–100 GHz), to be a total of eight bands. Actually, the frequency of the commodity spectrometer in common use is still fixed in the X-band, because owing to its wide adaptability, it can achieve high sensitivity and also high resolution.

Owing to the low frequency, the resonant cavity size is relatively larger (can accommodate small animals), and the nonresonance absorption is relatively smaller for the samples of high dielectric loss (e.g., water); L-band spectrometer in biomedical applications has made greater advancement since the 1980s. In particular, the in vivo test provides the technical support for the future clinical application of EMR. It can be expected that in the near future the EMR will be used in the hospitals as commonly as the NMR is used today.

For W-band (the so-called *high-frequency EMR*), when it deals with $S \geq 1$ system, especially for the system with negative zero field splitting parameter, a full fine structure spectra can be given, not only giving the value of the zero field splitting parameter but also determining its symbol. At present it is used as a specific tool to measure the magneton and magnetic anisotropy of molecules of the nanomaterials. With the development of nanomaterials, HF-EMR should receive more attention.

The introducing of pulse technology into the EMR experimental equipment not only improves the sensitivity and resolution but also detects the spectral line of the weak hyperfine interaction, which is difficult to detect in CW-EMR. Since the 1990s, with the commercialization of the pulse spectrometer, the price is declining day by day. The application thus gradually expanded from the frequency domain to the time domain, and multidimensional correlation spectrum and multidimensional exchange spectrum are developed. Pulse-EMR is in the ascendant.

In addition, the special type and portable small spectrometer will also have a certain developing prospect.

Application field. The application of EMR in the field of physics, chemistry, materials science, and other traditional disciplines will continue to deepen and extend; at the same time, due to the great progress in pulse technology and relaxation theory, EMR will be widely applied in the field of medicine and biology – especially, its application in clinical medicine displays a glorious prospect.

In conclusion, due to the progress of technology and the introducing of computer, the EMR in related fields (especially in the field of medicine and biology) has always made new achievements and will continue to grow in the next few decades.

References

[1] Zavoisky E. *J. Phys. USSR*. 1945, **9**: 211; *ibid.*, **10**:170.
[2] Frenkel J. *J. Phys. USSR*. 1945, **9**: 299.

[3] Zeeman P. *Nobel Lectures (Physics 1901–1921)*, Elsevier Publishing Company, Amsterdam, 1967.
[4] Gerlach W, Stern O. *Z. Phys.* 1921, **8**: 110; *ibid.* 1922, **9**: 353.
[5] Uhlenbeck G. E., Goudsmit S. *Naturwissenschaften*. 1925, **13**: 953.
[6] Breit G., Rabi I. I. *Phys. Rev.* 1831, **38**: 2982.
[7] Rabi I. I., Zacharias J. R., Millman S., Kusch P. *Phys. Rev.* 1938, **53**: 318.
[8] Cummerow R. L, Halliday D. *Phys. Rev.* 1946, **70**: 433.
[9] Bagguley D. M. S., Griffiths J. H. E. *Nature (Lond.)*. 1947, **160**: 532.
[10] Ramsey N. F. *Bull. Magn. Reson.* 1985, **7**: 94.
[11] Purcell E. M., et al. *Phys. Rev.* 1946, **69**: 37.
[12] Bloch F., et al. *Phys. Rev.* 1946, **69**: 127.
[13] Hutchison C. A., Jr., Postor R. C., Kowalsky A. G. *J. Chem. Phys.* 1952, **28**: 534.
[14] Fessenden R. W., Schuler R. H. *J. Chem. Phys.* 1963, **39**: 2147.
[15] Bargon J., Fischer H. *Z. Natarforsch.* 1967, **22a**: 1551.
[16] Ward H. R., Lawer R. *J. Am. Chem. Soc.* 1967, **89**: 5517.
[17] Dehmelt H. *Am. J. Phys.* 1990, **58**, 17.
[18] Xu Y Z. Applied Electron Magnetic Resonance Spectroscopy. Science Press, Beijing. 2008, Chap.11.
[19] Lu J F. Modern Electron Paramagnetic Resonance Spectroscopy and Its Applications. Peking University Medical Press, Beijing. 2012, Chap.3 and 4.
[20] Bowman M. K. "Fourier Transform Electron Spin Resonance", in *Modern Pulsed and Continuous-wave Electron Spin Resonance*, Kevan L., Bowman M. K., eds., John Wiley & Sons, Inc., New York, NY, 1990, Chapter 1.
[21] Belford R. L., Clarkson R. B., Cornelius J. B., et. al. "EPR over Three Decades of Frequency, Radiofrequency to Infrared", in *Electronic Magnetic Resonance of the Solid State*, Weil J. A., ed., Canadian Society for Chemistry, Ottawa, ON, Canada, 1987.
[22] Smith G. S. *Am. J. Phys.* 1990, **58**: 996.
[23] Liao S. Y. *Microwave Devices and Circuits*, 3rd edn, Prentice Hall, Englewood Cliffs, NJ, 1990, Chapter 4.
[24] Feher F. *Bell Syst. Tech. J.* 1957, **36**: 449.
[25] Hyde J. S., Froncisz W. "Loop-gap Resonators," in *Advanced EPR: Applications in Biology & Biochemistry*, Hoff A. J., ed., Elsevier, Amsterdam, The Netherlands, 1989, Chapter 7.
[26] Pfenninger S., Forrer J., Schweiger A., Weiland Th. *Rev. Sci. Instrum.* 1988, **59**: 752.
[27] Hardy W. N., Whitehead L. A. *Rev. Sci. Instrum.* 1981, **52**: 213.
[28] Hyde J. S., Froncisz W., Oles T. *J. Magn. Reson.* 1989, **82**: 233.
[29] Nilges M. J., Walczak T., Swartz H. M. *Phys. Med.* 1989, **2–4**: 195.
[30] Burt J. A. *J. Magn. Reson.* 1980, **37**: 129.
[31] Witters J., Herlach F. *Bull. Magn. Reson.* 1987, **9**: 132.
[32] Date M. *J. Phys. Soc. Jpn.* 1975, **39**: 892.
[33] Date M., Motokawa M., Seki A., Mollymoto H. *J. Phys. Soc. Jpn.* 1975, **39**: 898.
[34] Hyde J. S., Sczaniecki P. W., Froncisz W. *J. Chem. Soc. Faraday Trans. I* 1989, **85**: 3901.
[35] Slichter C. P. *Principles of Magnetic Resonance*, 3rd ed., Springe, Berlin, Germany, 1990, p. 184.
[36] Berliner L. J. (ed.). *Spin Labeling I: Theory and Application*, Academic Press Inc., New York, 1979, Chapters 2 and 3.
[37] Zhang J. Z., Zhao B. L., Zhang Q. G. The Principle and Application of Spin Labeled ESR Spectroscopy. Science Press, Beijing. 1987, Chap.3 and 6.
[38] Zhang J. Z., Zhao B. L. Chemistry (Huaxue Tongbao). 1988. 1, 27–33.
[39] Angelo M., Singel D. J., Stamler J. S. *Proc. Natl. Acad. Sci. USA* 2006, **103**: 8366.
[40] McLauchlan K. A., Stevens D. G. *Acc. Chem. Rev.* 1988, **21**: 54–59.
[41] Forbes M.D.E. *Photochem. Photobiol.* 1997. **65**: 73–81.

2 Theoretical basics

Abstract: This chapter discusses the energy-level splitting of the electron angular momentum and its magnetic moment in an external magnetic field, the resonance absorption and energy level transition produced by the interaction with radio frequency field, and the spectral line hyperfine splitting under the action of the nuclear magnetic moments. The chapter enables the readers to have a preliminary and fundamental understanding of the basic principles of EMR.

2.1 Phenomenal description of EMR

The simplest energy-level diagram of a particle with spin quantum number $s = 1/2$ in a variable magnetic field H is shown in Figure 2.1. When the external magnetic field increases gradually from 0, the electron spin energy level of the particle splits into two levels. The higher level is marked as α, with a magnetic quantum number $m_s = +\frac{1}{2}$, and the value of energy level is

$$E_\alpha = +\frac{1}{2}g_e\beta_e H \tag{2.1}$$

Here g_e is the Landé factor of electrons (which will be thematically discussed in Chapter 3) and β_e is Bohr magneton of electrons. The lower level is marked as β, with a magnetic quantum number $m_s = -\frac{1}{2}$, and the value of energy level is

$$E_\beta = -\frac{1}{2}g_e\beta_e H \tag{2.2}$$

Between two levels, the difference is ΔE:

$$\Delta E = E_\alpha - E_\beta = g_e\beta_e H \tag{2.3}$$

ΔE increases with the strength of the applied magnetic field:

$$\Delta E = h\nu = g_e\beta_e H \tag{2.4}$$

When a radio frequency field H_1 with the center frequency ν is added to the external magnetic field in perpendicular direction, and satisfied Eq. (2.4), the particles in the E_β level will absorb the energy of the radio frequency field and transit to the E_α level. This is called *resonance transition*. To produce resonance transition when the external magnetic field strength H is around 0.3 T, the frequency ν of about 9 GHz of the radio frequency field is required. In practical operation, the variation of the external magnetic field intensity should be linear.

https://doi.org/10.1515/9783110568578-002

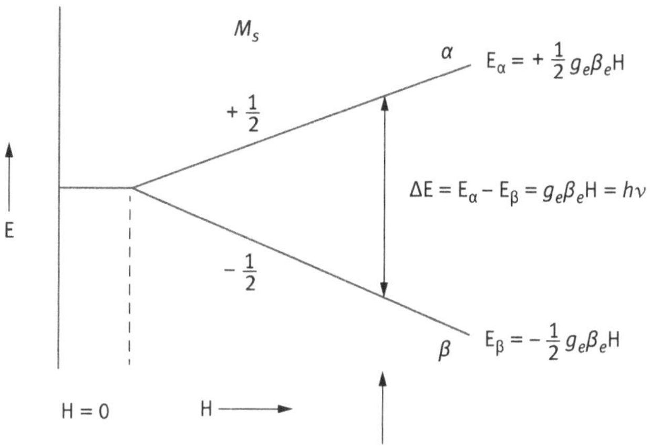

Figure 2.1: Diagram of the level variation with strength of the applied field $s = 1/2$ spin system.

The energy-level splitting of complex systems is more complicated. It will be discussed in detail in subsequent relevant chapters.

2.2 Angular momentum and magnetic moment

2.2.1 Orbital motion of electron and its magnetic moment

A particle with mass m and charge e in the xy plane makes a circular motion along an elliptical orbit in time period T (as shown in Figure 2.2). The current i generated is

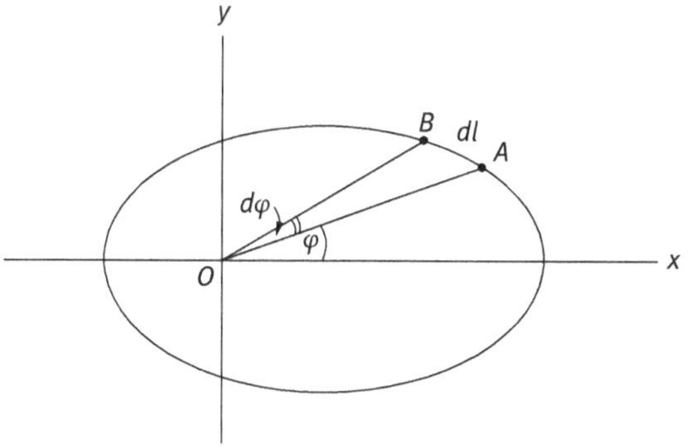

Figure 2.2: Classical phenomenal description of electron makes orbital moment.

$$i = \frac{e}{cT} \tag{2.5}$$

Here c is the speed of light. The resulting magnetic moment μ_l is

$$\mu_l = i \cdot \mathscr{A} = \frac{e}{cT} \cdot \mathscr{A} \tag{2.6}$$

Here $\mathscr{A}$ is the area surrounded by the orbit. In Figure 2.2, the distance from the original point O to the point A of orbit is r, and φ is the angle between r and x-axis.

$$\mathscr{A} = \frac{1}{2}\int_0^{2\pi} r^2 \mathrm{d}\varphi \tag{2.7}$$

$\mathrm{d}\varphi$ is the angle faced by the displacement $\mathrm{d}l$ (from A to B) on the orbit that the electron moves on in time $\mathrm{d}t$. The electron has mass, and the angular momentum P_φ is generated when it makes a circular motion:

$$P_\varphi = mv \cdot r \tag{2.8}$$

Here v is the tangent velocity of the electron making circular motion.

$$v = \frac{\mathrm{d}l}{\mathrm{d}t} \tag{2.9}$$

When $\mathrm{d}\varphi \to 0$,

$$\mathrm{d}l \approx r\mathrm{d}\varphi \tag{2.10}$$

then

$$v = r\frac{\mathrm{d}\varphi}{\mathrm{d}t} \tag{2.11}$$

Substituting Eq. (2.11) into Eqs. (2.9) and (2.8),

$$P_\varphi = mr^2\frac{\mathrm{d}\varphi}{\mathrm{d}t} \tag{2.12}$$

Moving term,

$$\frac{P_\varphi}{m}\mathrm{d}t = r^2\mathrm{d}\varphi \tag{2.13}$$

Multiplying by $\frac{1}{2}$ and integrating both sides,

$$\frac{1}{2}\int_0^T \frac{P_\varphi}{m}\mathrm{d}t = \frac{1}{2}\int_0^{2\pi} r^2\mathrm{d}\varphi$$

Integrating the left side,

$$\frac{P_\varphi}{2m}\int_0^T \mathrm{d}t = \frac{P_\varphi T}{2m} \tag{2.14}$$

Comparing with Eq. (2.7), the integral of the right side is the area $\mathscr{A}$.

$$\frac{1}{2}\int_0^{2\pi} r^2 \mathrm{d}\varphi = \mathscr{A} \tag{2.15}$$

Thus,

$$\mathscr{A} = \frac{P_\varphi T}{2m}$$

Substituting into Eq. (2.6),

$$\mu_l = i \cdot \mathscr{A} = \frac{e}{cT} \cdot \mathscr{A} = \frac{e}{cT} \cdot \frac{P_\varphi T}{2m} = \frac{e}{2mc} P_\varphi \tag{2.16}$$

Moving term,

$$\frac{\mu_l}{P_\varphi} = \frac{e}{2mc} \tag{2.17}$$

Let

$$\gamma = \frac{e}{2mc} \tag{2.18}$$

γ is the ratio of the orbital moment μ_l to the orbital angular momentum P_φ, which is called *magnetogyric ratio*. At this moment, the results are entirely derived from the concept of classical mechanics.

Considering the old quantum theory, that the angular momentum generated by the orbital motion of electrons is quantization, it means

$$P_\varphi = l\hbar \tag{2.19}$$

Here $\hbar = h/2\pi$, which is a basic unit of measuring angular momentum; h is the Planck constant; and l is the angular quantum number, $l = 1, 2, 3, \ldots, n$; here n is the principal quantum number. It can also be obtained that the magnetic moment μ_l, generated by the orbital motion, is also quantization. Substituting Eq. (2.19) into Eq. (2.17), we get

$$\mu_l = \frac{e}{2mc} l\hbar = l\frac{e\hbar}{2mc} \tag{2.20}$$

Let

$$\frac{e\hbar}{2mc} = \beta \tag{2.21}$$

Then Eqs. (2.16) and (2.20) can be written as

$$\mu_l = l\beta \tag{2.22}$$

Here β is the basic unit of measuring magnetic moment, called Bohr magneton. So, it is evident that the magnetic moment is also quantization. It is an integer times of the

Bohr magneton. Until this point, the old quantum theory has made a very satisfactory description of the orbital angular momentum and the orbital magnetic moment of the electron.

2.2.2 Eigen motion of electrons and its magnetic moment

At the end of the nineteenth century, it was confirmed that the mass and charge are the basic properties of the electron. The fine structure of atomic spectra is an obvious explanation of the basic properties of the electrons: that the spin motion is the eigen motion of electron. However, this includes some assumptions:

(1) There are only two orientations of the angular momentum vector producing spin motion in the external magnetic field: parallel or antiparallel to the direction of the applied magnetic field.
(2) The projection of the angular momentum vectors of the spin motion in the direction of the outer magnetic field must be equal in magnitude but opposite in direction. That is,

$$s_z = m_s \hbar \tag{2.23}$$

Here, s_z is the projection of the angular momentum vector s in the z direction (or the direction of external magnetic field) of the spin motion and m_s is the magnetic quantum number. For a single electron,

$$m_s = \pm \frac{1}{2}$$

Substituting into Eq. (2.23), we get,

$$s_z = \pm \frac{1}{2}\hbar = \pm \hbar/2 \tag{2.24}$$

That is, the projection of the electron spin angular momentum s in the direction of the outer magnetic field (z) should be m_s times of $\hbar$, with $\hbar$ being the half integer.

Spin motion is the intrinsic motion of electron, and the electron is a charged particle. When electron makes spin motion, the magnetic moment is generated. Thus, *spin magnetic moment* is the intrinsic magnetic moment of electron. From the above-mentioned two fundamental assumptions of the spin motion, the following can be derived:

(1) In the external magnetic field, the electron spin magnetic moment has only two possible orientations: parallel or antiparallel to the direction of the magnetic field.
(2) The direction of the projection in the external magnetic field of the spin magnetic moment must be opposite and the value should be same.

The basic unit of electron eigen magnetic moment is *Bohr magneton*. For a single electron, the projection of eigen magnetic moment along the external magnetic field (z), μ_{s_z}, is given by

$$\mu_{s_z} = \pm\beta = \pm\frac{e\hbar}{2mc} = \pm\frac{e}{mc}\frac{\hbar}{2} \tag{2.25}$$

Substituting Eq. (2.24) into Eq. (2.25) gives

$$\mu_{s_z} = \pm\frac{e}{mc}s_z$$

$$\frac{\mu_{s_z}}{s_z} = \pm\frac{e}{mc} \tag{2.26}$$

$\frac{\mu_{s_z}}{s_z}$ is the ratio of electron spin magnetic moment to its angular momentum (i.e. *magnetogyric ratio* of electron spin motion) denoted by γ_s:

$$\gamma_s = \frac{\mu_{s_z}}{s_z} = \pm\frac{e}{mc} \tag{2.27}$$

We have got the magnetogyric ratio of electron orbit motion discussed before, Eq. (2.18), and now we use γ_l to designate it:

$$\gamma_l = \frac{e}{2mc} \tag{2.28}$$

Comparing Eqs. (2.27) and (2.28) gives

$$\gamma_s = 2\gamma_l \tag{2.29}$$

The above discussion of the electron spin motion and its magnetic moment is based on the old quantum theory. It still has not shaken off the trammels of classical mechanics concepts. Using "classical" explanation for this new basic attribute of electron has however the following three insurmountable difficulties:

(1) According to the classical mechanics, how to calculate magnetic moment should be given assumptions of the electron structure (shape, size, distribution of charge, etc.).
(2) Why the angular momentum vector of the electron spin motion in the external magnetic field only have two possible orientations? It could not be explained by classical mechanics.
(3) Why do the relationship $\gamma_s = 2\gamma_l$ exist? It could even not be explained by classical mechanics (including old quantum theory).

Until the early 1930s, after the establishment of relativistic quantum mechanics by Dirac, above conclusion could be obtained and they do not need any assumption.

Dirac pointed out clearly that the classic explanation of the spin motion like a small ball in rotation is unacceptable in principle.

The famous experiment of Stern–Gerlach [1] proves the existence of spin motion and spin magnetic moment of the electron. However, the experiment itself cannot distinguish between orbit magnetic moment and spin magnetic moment clearly.

Afterward, Bohr pointed out too: "Any attempt to measure the magnetic moment of free electron should be doomed to failure." It is decided by the *uncertainty principle*. The problem just exactly lies in that the electron spin magnetic moment brings the kinematics attribute of quantum mechanics. Therefore, it is impossible to distinguish exactly between the electron spin magnetic moment and the magnetic effect of related translational motion of a charged particle. Any attempt to determine the spin magnetic moment will inevitably introduce uncertainty in the numerical value of the electron angular momentum.

From the general theorems of quantum mechanics, it could be derived that the absolute value of the projection of the spin angular momentum along the external magnetic field (z) direction $|s_z|$ is given by

$$|s_z| = \sqrt{s(s+1)}\hbar \tag{2.30}$$

When $s = \frac{1}{2}$, it becomes

$$|s_z| = \frac{\sqrt{3}}{2}\hbar = \sqrt{3}\frac{\hbar}{2}$$

The absolute value of the projection of the z-axis $|\mu_{s_z}|$ is

$$|\mu_{s_z}| = \frac{e}{mc}\sqrt{s(s+1)}\hbar = \sqrt{s(s+1)}\beta \tag{2.31}$$

When $s = \frac{1}{2}$, it becomes

$$|\mu_{s_z}| = \frac{\sqrt{3}}{2}\beta \tag{2.32}$$

Refer back to the orbital angular momentum P_φ, is it not equal to $l\hbar$ in quantum mechanics as

$$P_\varphi \neq l\hbar$$

But

$$P_\varphi = \sqrt{l(l+1)}\hbar \tag{2.33}$$

here $l = 0, 1, 2, \ldots, (n-1)$.

Likewise, for the orbital magnetic moment μ_l, is it not equal to $l\beta$ as

$$\mu_l \neq l\beta$$

But

$$\mu_l = \sqrt{l(l+1)}\beta \tag{2.34}$$

When $l = 0$, $\mu_l = 0$, however, $s \neq 0$ thus $\mu_s \neq 0$. It means that the existence of $l = 0$ state is reasonable (i.e., s state).

In order to solve the problem of $\gamma_s = 2\gamma_l$, we still define the gyromagnetic ratio as

$$\gamma = \frac{e}{2mc} \tag{2.18}$$

Introducing a factor g,

$$\frac{\mu}{P} = g\gamma = g\frac{e}{2mc} \tag{2.35}$$

When the magnetic moments are contributed by the pure orbital motion,

$$g = 1; \quad \gamma = \gamma_l$$

When the magnetic moments are contributed by the pure spin motion,

$$g = 2; \quad \gamma = \gamma_s$$

This g-factor, also called Landé factor, is very important in electron magnetic resonance! It not only indicates the contribution ratio of the orbital motion to the spin motion but also reflects the position of EMR spectrum line. Hence, g-value in practice is not limited between 1 and 2. (It will be discussed in detail in Chapter 3.)

2.2.3 Spin angular momentum and magnetic moment of atomic nucleus

If we do not consider the other elementary particles temporary, we can simply think that a nucleus consists of proton and neutron. The proton is charged positive, while the neutron is uncharged. If we regard the neutron as a proton composed of an electron, we can think the mass of a proton is approximately the same as a neutron.

The atomic nucleus is has positive charge, and its charge number is the same as the number of protons, that is, the atomic number represented by Z; the mass number is the number of protons plus the number of neutrons, represented by A; using X as representing some element, it can be notated as ${}^{A}X_{Z}$. For example, ${}^{1}H_{1}$ expresses the hydrogen element, with mass number as 1 and atomic number also as 1. Also, ${}^{13}C_{6}$ expresses a carbon element with the mass number as 13 and atomic number as 6.

To obtain spin motion of an atomic nucleus, it ought to have spin angular momentum P_N, which should also be quantized:

$$P_N = \sqrt{I(I+1)}\hbar \tag{2.36}$$

where I is the quantum number of nuclear spin, which can be 0, integer, or half-integer, depending on the collocation of the mass number A with the order number Z. For instance, when the mass number A is an odd number, the I always is a half-integer, irrespective of whether the atomic order number Z is odd or even. If the mass number A is odd, the atomic number Z is also odd, such as $I = \frac{1}{2}, \frac{3}{2}, \frac{5}{2}, \frac{7}{2}$ for $^{1}H_1$, $^{7}Li_3$, $^{55}Mn_{25}$, $^{51}V_{23}$, respectively. If the mass number A is odd, the atomic number Z is even, such as $I = \frac{1}{2}, \frac{3}{2}, \frac{5}{2}, \frac{7}{2}$ for $^{13}C_6$, $^{9}Be_4$, $^{17}O_8$, $^{143}Nd_{60}$, respectively. If mass number A is even and atomic number Z is odd, then I must be integer, such as $I = 1$ for $^{14}N_7$, $I = 3$ for $^{10}B_5$, $I = 6$ for $^{50}V_{23}$, and so on. When both the mass number and atomic number are even (i.e. nonmagnetic nucleus), then I is zero, such as for $^{12}C_6$, $^{16}O_8$, and $^{32}S_{16}$.

According to the principles of quantum mechanics, P_{N_z}, the component along the external magnetic field (z) direction of the nuclear spin angular momentum (P_N), should be also quantized:

$$P_{N_z} = m_I\hbar \tag{2.37}$$

where m_I is the magnetic quantum number of nucleus:

$$m_I = I,\ I-1,\ I-2, \ldots \quad -I+2,\ -I+1,\ -I$$

It may be seen that the maximum of the component along the external magnetic field (z) direction of the nuclear spin angular momentum is given by

$$(P_{N_z})_{\max} = I\hbar \tag{2.38}$$

where $I = 0$, $(P_{N_z})_{\max} = 0$.

When the nucleus $I \neq 0$, spin motion should be generated by the nuclear magnetic moment μ_N:

$$\mu_N = \gamma_N P_N \tag{2.39}$$

where γ_N is the gyromagnetic ratio of the nucleus:

$$\gamma_N = \frac{e_N}{2m_N c} \tag{2.40}$$

where e_N is the dynamic charge of nucleus, m_N is the dynamic mass of nucleus, and c is the speed of light. The γ_N value can either be positive or negative, such as

$$\gamma_N = 26.7519 \times 10^7 \text{rad} \cdot \text{T}^{-1}\text{s}^{-1} \quad \text{for} \quad {}^1\text{H}$$
$$\gamma_N = -2.712 \times 10^7 \text{rad} \cdot \text{T}^{-1}\text{s}^{-1} \quad \text{for} \quad {}^{15}\text{N}.$$

μ_{Nz}, the component along the external magnetic field (z) direction of the nuclear magnetic moment (P_N), is given by

$$\mu_{Nz} = \gamma_N P_{N_z} = \gamma_N m_I \hbar = m_I \beta_N \tag{2.41}$$

$(\mu_{N_z})_{\max}$, the maximum of the component along the external magnetic field (z) direction of the nuclear spin angular momentum (μ_N), is given by

$$(\mu_{N_z})_{\max} = \gamma_N I \hbar = I \beta_N \tag{2.42}$$

where β_N is the nuclear magneton, which is the smallest unit of nuclear magnetic moment:

$$\beta_N = \gamma_N \hbar = \frac{e_N}{2m_{NC}} \hbar \tag{2.43}$$

The absolute value of nuclear magnetic moment is given by

$$|\mu_N| = \sqrt{I(I+1)}\beta_N \tag{2.44}$$

where $I = 0$, $|\mu_N| = 0$; hence, the nuclei with $I = 0$ is also called nonmagnetic nuclei.

2.2.4 Electric quadrupole moment of atomic nucleus

As known from the experimental results, the charge distribution is spherical symmetry for nuclei with $I = \frac{1}{2}$. The electric quadrupole moment of these nuclei $Q = 0$ is easy to get for high-resolution NMR spectra as well as for the EMR spectrum with hyperfine structure of good resolution.

The nuclei with $I > \frac{1}{2}$ not only have the nuclear magnetic moment but also the electric quadrupole moment (i.e., $Q \neq 0$). The charge distribution of these nuclei are ellipsoidal spherically symmetry, and nuclear spin axis is parallel to the direction of the nuclear magnetic moment.

Designating $\rho(x, y, z)$ as the charge distribution function in the nucleus, the nuclear quadrupole moment Q is defined as

$$Q = \int \rho(x, y, z)(3z^2 - r^2)\text{d}x\text{d}y\text{d}z \tag{2.45}$$

By definition, the nuclear magnetic moment is parallel to the direction of the external magnetic field (z), r is the distance from the point (x, y, z) to origin, and a is the radius of the spin axis, the radius of the equatorial plane is b, as shown in Figure 2.3.

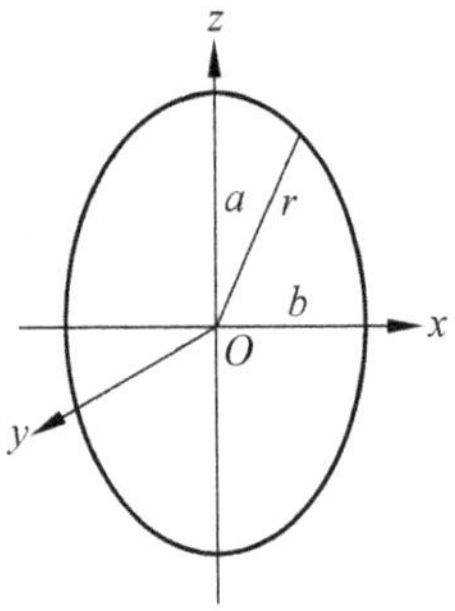

Figure 2.3: The distribution illustrating diagram of the electric quadrupole moment.

The electric quadrupole moment and the atomic number Z has the relationship as follows:

$$Q = \frac{2}{5} Z(a^2 - b^2) \tag{2.46}$$

When $a = b$, $Q = 0$; when $a > b$, $Q > 0$; when $a < b$, $Q < 0$. It can be seen from Eq. (2.46) that the dimension of the nuclear quadrupole moment Q is quadratic of length. The dimension of Q in the CGS system is cm^2, and the Q value increases with increasing atomic number Z. Since the nuclear radius is about 10^{-12} cm, the magnitude order of Q is about 10^{-24} cm^2.

Q value and the nuclear spin quantum number I have the relationship as follows:

$$Q = CI(2I - 1) \tag{2.47}$$

where C is the constant; when $I = 0$, $I = \frac{1}{2}$ and $Q = 0$; when $I > \frac{1}{2}$, the value of Q is not equal to 0. The interaction of the electric field gradient with the nuclear electric quadrupole causes additional complexity. In many cases, this interaction is very strong, only weaker than the Zeeman interaction of the nuclear magnetic moment with the external magnetic field.

Thus, we divide all nuclei into four types as follows:

$I = 0$	$Q = 0$	^{16}O, ^{12}C, ^{32}S, ...
$I = \frac{1}{2}$	$Q = 0$	^{1}H, ^{13}C, ^{15}N, ^{19}F, ^{31}P, ...
$I \geq 1$	$Q > 0$	^{2}H, ^{10}B, ^{14}N, ...
$I \leq 1$	$Q < 0$	^{7}Li, ^{17}O, ^{33}S, ^{35}Cl, ...

2.3 Unit of magnetic field

The magnetic field can be expressed in two ways: magnetic field strength H and the magnetic-induced density B. The relation between both is as follows:

$$H = B/\mu_m \tag{2.48}$$

Where μ_m is the magnetic permeability of the medium:

$$\mu_m = \kappa_m \mu_o \tag{2.49}$$

Here μ_o is the magnetic permeability in vacuum; κ_m is the dimensionless parameter, $\kappa_m = 1$ in vacuum; $\kappa_m = \mu_m/\mu_o$ in medium. The value of κ_m is different in variable medium.

Here

$$\mu_o = 4\pi \times 10^{-7}\,\mathrm{J \cdot C^{-2} \cdot S^2 \cdot m^{-1}(T^2 \cdot J^{-1} \cdot m^3)}$$

Of course, it is a universal constant; here J is Joule, its unit is $\mathrm{kg \cdot m^2 \cdot s^{-2}}$; C is Coulomb, the unit of the quantity of electric charge; the unit of electric current is ampere, meaning $\mathrm{C \cdot s^{-1}}$. Obviously, the magnetic-induced density B is different from the magnetic field strength H either in dimension or in unit. Nevertheless, the magnetic field strength H is universally used for magnetic resonance. The unit of H is Tesla (T) and its dimension is $\mathrm{kg \cdot s^{-1} \cdot C^{-1}}$ or $\mathrm{J \cdot C^{-1} \cdot m^{-2} \cdot s}$:

$$1\mathrm{T\,(Tesla)} = 1 \times 10^4\mathrm{G\,(Gauss)} \tag{2.50}$$

Another very important physical quantity is the magnetic dipole moment μ, and its unit is $\mathrm{J \cdot T^{-1}}$. There are N magnetic dipole moments in a given volume (V), the macroscopic magnetic moment is given by

$$M = \frac{1}{V}\sum_{i}^{N} \mu_i \tag{2.51}$$

$\boldsymbol{M}$ is the magnetization, which has the unit of $\mathrm{J \cdot T^{-1} \cdot m^{-3}}$. $\boldsymbol{M}$ is the net magnetic moment per unit volume.

Since the magnetic moments of nuclei, atoms, and molecules are proportional to the angular moment of these species, we introduce the product of a dimensionless g factor and a dimensioned factor (an aggregate of physical constants) called the *magneton*.

$$\boldsymbol{\mu} = \alpha g \beta \boldsymbol{J} \tag{2.52}$$

Here $\boldsymbol{\beta}$ has the same unit with vector $\boldsymbol{\mu}$; the magnitude order of g is same with the Zeeman factor of electron. It is dimensionless. When using $\boldsymbol{\hbar}$ as the unit, and merged with $\boldsymbol{\beta}$ into one, the unit is J s; $\alpha = \pm 1$.

For free electron (i.e., a single electron in vacuum), $\alpha_e = -1$; operator $\hat{\boldsymbol{J}}$ should be the electron spin operator $\hat{\boldsymbol{S}}$, hence β becomes[3]

$$\beta_e = \frac{e\hbar}{2m_e} = 9.2740154(31) \times 10^{-24}\,\mathrm{J \cdot T^{-1}} \tag{2.53}$$

Here, β_e is the *Bohr magneton*; e is the electron charge, $2\pi\hbar = h$ is Planck's constant, and m_e is the mass of the electron. The Zeeman splitting factor for the free electron is

$$g_e = 2.002319304386(20) \tag{2.54}$$

It is one of the most accurate physical constants known. When the electron interact with other particles, $g \neq g_e$.

For nuclei, $\hat{J}$ is the nuclear spin operator $\hat{I}$, $\alpha = +1$; nuclear magneton β_n is given by

$$\beta_n = \frac{e\hbar}{2m_P} = 5.0507866(17) \times 10^{-27} \mathrm{J} \cdot \mathrm{T}^{-1} \tag{2.55}$$

Here m_p is the mass of the proton (^1H); the values of nuclear g factors g_n is 5.585564 ± 0.000017.

2.4 The interaction between external fields and magnetic moment

Now, we consider the projection of magnetic moment μ along z (taken along external magnetic field) in a external magnetic field **H**. μ_z is generally defined as follows:

$$\mu_z = -\left.\frac{\partial E}{\partial H}\right|_{H=0} \tag{2.56}$$

Here $\boldsymbol{E(H)}$ is the energy of a magnetic dipole of magnetic moment $\boldsymbol{\mu}$ in magnetic field $\boldsymbol{H}$, which is the function of $\boldsymbol{H}$. Under most situation, the energy $\boldsymbol{E}$ of the interaction between magnetic moment and magnetic field can be described in terms of scalar product of $\boldsymbol{\mu}$ and $\boldsymbol{H}$:

$$E = -\boldsymbol{\mu} \cdot \boldsymbol{H} = -|\mu H| \cos(\boldsymbol{\mu}, \boldsymbol{H}) \tag{2.57}$$

Here $(\boldsymbol{\mu}, \boldsymbol{H})$ is the angle between vector $\boldsymbol{\mu}$ and $\boldsymbol{H}$. when $(\boldsymbol{\mu}, \boldsymbol{H}) = 0$, $\boldsymbol{E}$ is the minimum, equal to $-|\mu H|$; when $(\boldsymbol{\mu}, \boldsymbol{H}) = \pi$, $\boldsymbol{E}$ is the maximum, equal to $+|\mu H|$. When the angle lies between 0 and $\boldsymbol{\pi}$, the $\boldsymbol{E}$ value lies between $-|\mu H|$ and $+|\mu H|$ extremes.

Now consider the total effect of the classical magnetic moment $\boldsymbol{\mu}$ that has not interacted with the external magnetic field. If the interaction energy $\boldsymbol{\mu} \cdot \boldsymbol{H}$ is larger than the thermal energy kT, the macroscopic magnetization $\boldsymbol{M}$ would be approximately equal to $N_v\boldsymbol{\mu}$, where N_v is the number of the dipoles per unit volume. However, in almost all cases, the orientation of the magnetic moment is disordered, so in most cases $|\mu H/kT| \ll 1$. Therefore, the macroscopic magnetization $\boldsymbol{M}$ is several orders of magnitude smaller than $N_v\boldsymbol{\mu}$.

Macroscopic magnetization could be related to external magnetic field by a dimensionless proportional factor χ_m ; it is called *volume susceptibility*, which can be obtained by measure of the sample endurable force in heterogeneous static magnetic field [2,3]. In the simplest (isotropic) case, a set of noninteraction magnetic dipoles contributes to χ_m as follows:

$$\boldsymbol{M} = -\alpha[g/|g|]\chi_m \boldsymbol{H} \tag{2.58}$$

For electron, $\alpha = -1$; $g > 0$, then

$$\chi_m = \frac{\boldsymbol{M}}{\boldsymbol{H}} = \frac{\boldsymbol{M}}{B/\kappa_m \mu_0} \tag{2.59}$$

With assumption under the case of equilibrium (according to Boltzmann distribution) [2], it can be rewritten as follows:

$$\chi_m = \frac{N_v \mu^2}{3k_b T} \kappa_m \mu_o = \frac{C}{T} \geq 0 \tag{2.60}$$

where $\mu^2 = g^2 \beta_e^2 S(S+1)$ and C is the Curie "constant" under common situation $\chi_m \approx 10^{-6}$. The relative magnetic permeability (related to free space) κ_m is equal to $(1+\chi_m)$. In addition, there is also a diamagnetic, negative, and almost independent of temperature, contribution to χ_m, they arise from the motion and distribution of all electrons. Define the material of $\chi_m > 0$ as paramagnetics, and the material of $\chi_m < 0$ as diamagnetic.

The simplest paramagnetic substances are the radicals with only one unpaired electron. Their orbital angular momentum is almost zero. The experimental determination of χ_m obtained the product of $N_v \mu$. In order to obtain μ, N_v should be determined from other data. However, N_v and μ can be determined separately by EMR measurements.

The magnetic energy $\boldsymbol{E}$ is proportional to magnetic moment μ (Eq. (2.57)). If the direction of external field $\boldsymbol{H}$ is z-direction, and the system only has spin, then $E = -\mu_z H$. Substitution of $g_e \beta_e M_S$ for μ_z gives the energy

$$E = g_e \beta_e M_S H \tag{2.61}$$

For an unpaired electron, the possible value of M_S are $+\frac{1}{2}$ and $-\frac{1}{2}$; Hence, μ_z also has only two possible values $\mp \frac{1}{2} g_e \beta_e$, and the energy is given by

$$E = \pm \frac{1}{2} g_e \beta_e H \tag{2.62}$$

$$\Delta E = E_+ - E_-$$

$$\Delta E = g_e \beta_e H = -\gamma_e \hbar H \tag{2.63}$$

It correspond to $|\Delta M_S| = 1$. As shown in Figure 2.1, ΔE increases linearly with the magnetic field. Briefly, the energy level splitting is the result of the interaction between magnetic moment and external magnetic field. The splitting energy ΔE is a linear function of external magnetic field H.

The states of the magnetic system, as already mentioned, are finite number generally. If all of these states have same energy, they are called *degenerated state.* Every state has a corresponding quantum number. Thus, the corresponding quantum number is M_S for one unpaired electron system. As we shall see in the later chapter, The *Dirac mark* $|M_S>$ or $<M_S|$ is often used to express the states. For a single electron, $M_S = +\frac{1}{2}$ or $-\frac{1}{2}$, then the Dirac marks are $|+\frac{1}{2}>$ and $|-\frac{1}{2}>$ or $|\alpha>$ and $|\beta>$. When there are several spin-relating particles in a same magnetic space, the spin state of the system need to be described by the quantum numbers of each particles. In atomic and molecular systems, there are no more than two electrons to occupy the same given space orbit. Based on the *Pauli exclusion principle*, the sign of spin quantum numbers M_S of the two electrons occupying the same orbit must be opposite. So the magnetic moments produced from their spin movement should be of same size and the sign opposite, counteracting each other. An EMR signal could be observed only when the atomic or molecular orbit is in the semioccupied situation.

2.5 Interaction of magnetic moment with electromagnetic field in the external magnetic field

Particles on the two Zeeman energy levels of electron could be induced by an electromagnetic field H_1 with an appropriate frequency v causing transition. This *appropriate* frequency v should satisfy the equation, $\Delta E = hv$. Then:

$$\Delta E = hv = g_e \beta_e H \tag{2.64}$$

where H is the external magnetic field satisfying the resonance condition. The number of splitting energy levels is more than 2 for the spin system of $S > 1/2$. Which transitions between energy levels are allowed? Only those transitions that satisfy the selection rule of $|\Delta M_S| = 1$.

Now, let us consider briefly single photon to be absorbed and emitted by an unpaired electron system. The photon has its spin angular momentum component ($\pm\hbar$) *along* or *opposite* to its motion direction [4]. It corresponds to *rotational polarization* toward right and left. However, the photon has no magnetic moment. It depends on the direction of the axis of the electron spin for absorption, which can transmit the energy hv and angular momentum (as σ-type photon) or only energy (as π-type photon).

For the energy requirement of Eq. (2.64), some photons can be worked in coordination. In order to meet the condition of total (photon + electron) angular momentum conservation, only σ-type is suitable. As shown in Figure 2.4, EMR experiment of double frequency is not common, but it has been carried out indeed,

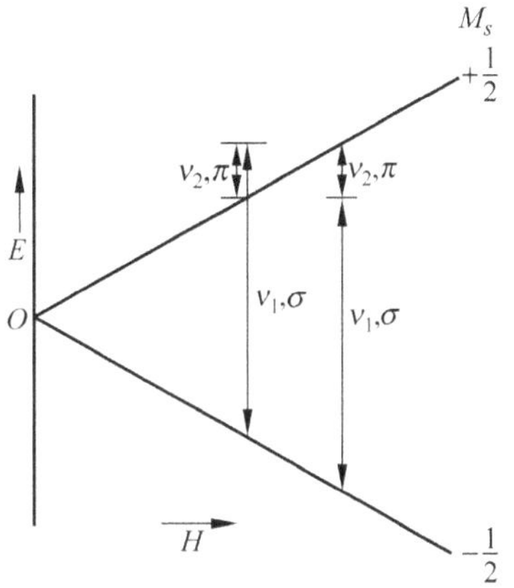

Figure 2.4: Diagram of photon transition in the presence of two excitation fields.

using stable organic radical DPPH [5]. In the most EMR experiments, only the single photon (σ-type) plays an effective role in the transition excitation. Let us not consider this kind of transition for the moment now. In the EMR experiments of recent years, the importance of multi-quantum phenomena has become even more obvious. These effects (e.g., developing new EMR lines), such as demand of excitation field H_1, have increased clearly.

The transitions between Zeeman energy levels require change in the orientation of the electron magnetic moment, because the transition can only be generated under electromagnetic radiation inducing reorientation (oriented reverse) of the electron magnetic moment. To make the transition possible, the electromagnetic radiation must be polarized and the alternative magnetic field should be perpendicular to the external magnetic field. With suitable alternative magnetic field (σ-type photon) and the frequency of the microwave band, it could be absorbed by the magnetic dipoles on lower energy level and transit to the higher energy level; this is called the *resonance transition.* If the orientation of the electromagnetic radiation field H_1 is parallel to the external magnetic field H, the radiation effect causes merely an energy oscillation with frequency ν on the Zeeman energy level. And if it did not happen, the inverse orientation of electron magnetic moment will not have the transition of particles between Zeeman energy levels and thus there will be no resonance absorption.

We can fix the intensity of external magnetic field H and adjust the microwave frequency ν to satisfy Eq. (2.64). Conversely, we can also adjust the intensity of external magnetic field H at fixed radiation frequency ν to satisfy Eq. (2.64). The most extant EMR spectrometer selects the latter mode, as the adjustable range of the microwave frequency produced by the oscillating tube (Klystron or Gunn diode) is very small.

The transition rule of the electron spin system (photon and electron) already discussed are suitable to nuclear spin system. Analogous to Eq. (2.61), the nuclear Zeeman energy is given by

$$E = -g_n \beta_n M_I H \tag{2.65}$$

where g_n is the nuclear Zeeman splitting factor g, β_n is the nuclear magneton, and M_I is the z-component of the nuclear spin angular momentum vector along the direction of external magnetic field. Similar to electron spin system, dipolar transition is allowed only in the condition of $|\Delta M_I| = 1$. Hence, it is analogous to Eq. (2.64):

$$\Delta E = hv = g_n \beta_n H \tag{2.66}$$

This is the most basic equation of nuclear magnetic resonance (NMR).

Nuclear spin and nuclear magnetic moment are very important in EMR studies. The interaction between unpaired electron with magnetic nuclei give rise to hyperfine splitting in EMR spectrum, supplying extremely abundant information for EMR research.

It should be noted that the spin species to actually affecting the magnetic field are not only the magnetic field H_{ext} applied to the spin system from outside of sample but also the field existing within the spin system itself (*local field*) H_{local}. This can be expressed as follows:

$$H_{eff} = H + H_{local} \tag{2.67}$$

where H is H_{ext} exactly. H_{local} is composed of two parts: one is induced by the external magnetic field, thus the magnitude is dependent on H; the other is spin system itself, thus the magnitude is independent of H, only related to the orientation of the external magnetic field.

If we consider merely the former factor, that is the induction contributing to H_{local}, H in Eq. (3.32) could be substituted by H_{eff} in principle. Actually, it is more suitable to retain the "external magnetic field" H. Then g_e must be replaced by effective g factor, and Eq. (2.67) would be rewritten as

$$H_{eff} = (1 - \sigma)H = (g/g_e)H \tag{2.68}$$

where σ corresponds to the "chemical shift" parameter (shielding coefficient) used in the NMR spectroscopy, and the g factor is called the *effective Zeeman factor* by EPR spectroscopists. Many free radicals and some transition metal ions have $g \approx g_e$, but there are many spin systems, especially most of the transition metal ions, have the effective g factor deviated far from g_e. The effective g factor of some rare earth ions are even negative.

We noted from Eqs. (2.35) and (2.55) that the g factor in broad sense contains magnetic moment. Practically, these local magnetic fields often are contributed by the orbit motion of the unpaired electron. Hence, g factor should have a variable value. If the EMR spectrum merely consists of a singlet line with g=g_e, it would not be interesting. It is exactly due to the variation of the g factor, the adjacent varying dipoles caused the variation of the local magnetic field and produced various spectra lines, making the EMR spectra provide abundant and multi-color microscopic information for us.

2.6 Interaction of nuclear magnetic moment with electron magnetic moment in the external magnetic field

Generally, the electron always depend on nucleus and mutually coexist in the substance. There are three kinds of interactions between magnetic moments: (1) the interaction between electron magnetic moment and magnetic moment of magnetic nucleus; (2) the interaction between electron magnetic moments, it will be discussed in detail in Appendix 1; (3) the interaction between nuclear magnetic moments, which is the topic of nuclear magnetic resonance and beyond the scope of this monograph.

As described earlier, the nuclei could be divided roughly into magnetic and nonmagnetic nuclei. The spin quantum number of magnetic nuclei is $I \neq 0$. The value I of these nuclei could be 1/2, 1, 3/2, 2, and so on, and have corresponding $2I+1$ nuclear spin states. For electric magnetic moment, the magnetic nuclei can be considered as an additional magnetic field. In the external magnetic field, the additional field makes the resonance absorption lines of electrons producing re-split, that is, *hyperfine splitting.*

The simplest case is hydrogen atom. The proton has a spin $I = 1/2$, and nuclear spin angular momentum has merely two possible values $M_I = \pm 1/2$. Only one unpaired electron ($S = 1/2$) out of the nucleus, for this electron, the magnetic moment is caused by spin motion of nucleus, is analogous to an additional small magnetic field (local magnetic field H_{local}). Hence, the resonant magnetic field H_{r} is given by

$$H_r = H + H_{\text{local}} \tag{2.69}$$

$$H_{\text{local}} = a_{\text{o}} M_I \tag{2.70}$$

where a_{o} is hyperfine splitting constant. When $a_{\text{o}} = 0, H_{\text{local}} = 0$, and $H_{\text{r}} = H$, the resonant magnetic field is equal to external magnetic field. For hydrogen atom, $a_{\text{o}} = 50.684\,\text{mT}, M_I = \pm 1/2;$; Eq. (2.69) can be rewritten as

$$H_r = H - a_{\text{o}} M_I = H \pm a_{\text{o}}/2 \tag{2.71}$$

There are thus two resonant magnetic fields, that is, two spectral lines: $(H_{\text{r}})_1 = H - a_{\text{o}}/2$ and $(H_{\text{r}})_2 = H + a_{\text{o}}/2$. The distance between the two lines is about equal to $\sim a_{\text{o}}$. Experimental result tells us that the space of the hyperfine doublet is 50.970 mT at 9.5 GHz of the radiation frequency. An unpaired electron interact with a magnetic nucleus (I = 1/2) producing splitting of energy level, as shown in Figure 2.5.

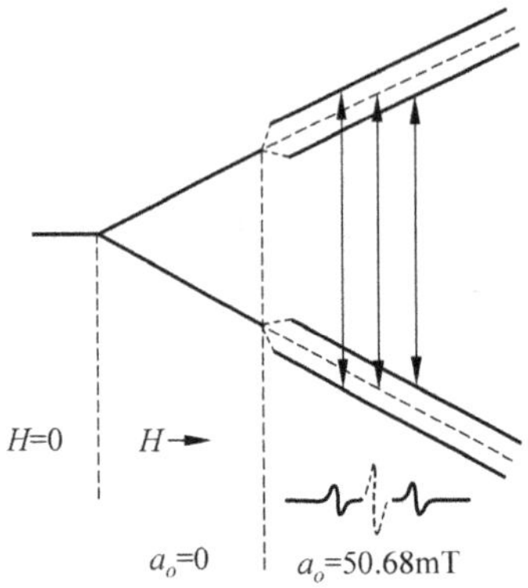

Figure 2.5: The hyperfine splitting of the energy level of free hydrogen atom.

For the system containing more than one unpaired electron, one magnetic nucleus, or $I > 1/2$, the situation would be much more complicated, and will be discussed in Chapters 4 and 5 in detail.

References

[1] Gerlach W, Stern O. *Z. Phys.* 1921, **8**: 110; *ibid.* 1922, **9**: 349, 353.
[2] Gerloch M. *Magnetism and Ligand-Field Analysis*, Cambridge University Press, Cambridge, UK, 1983.
[3] Drago R. S. *Physical Methods in Chemistry*, Saunders, Philadelphia, PA, 1977, Chapter 11.
[4] French A. P., Taylor E. F. *An Introduction to Quantum Physics*, W. W. Norton, New York, NY, 1978, Section 14.8.
[5] Berget J., Odchnal M., Petricek V., Sacha J., Trlfaj L. *Czech. J. Phys.* 1961, **B11**: 719.

Further reading

[1] Carrington A., McLachlan A. D. *Introduction to Magnetic Resonance*, Harper & Row, New York, NY, 1967.
[2] Alger R. S. *Electron Paramagnetic Resonance: Techniques and Applications*, John Wiley & Sons, Inc., New York, NY, 1967.
[3] Abragam A., Bleaney B. *Electron Paramagnetic Resonance of Transition Ions*, Oxford University Press. Oxford, UK, 1970.
[4] Talpe J. *Theory of Experiments in EPR*, Pergamon Press, Oxford, UK, 1971.
[5] Pake G. E., Estle T. L. *The Physical Principles of Electron Paramagnetic Resonance*, 2nd ed., Benjamin, Reading, MA, 1973.
[6] Weltner, W., Jr. *Magnetic Atoms and Molecules*, Van Nostrand Reinhold, New York, NY, 1983.
[7] Slichter C. P. *Principles of Magnetic Resonance*, 3rd ed. Springer, New York, NY, 1989.
[8] Pople J. A., Schneider W. G., Bernstein H. J. *High-Resolution Nuclear Magnetic Resonance*, McGraw-Hill, New York, NY, 1959.
[9] Van Vleck J. H. *The Theory of Electric and Magnetic Susceptibilities*, Oxford University Press. London, UK, 1932.

3 g-Tensor theory

Abstract: This chapter introduces the relationship between the Landé factor and the angular momentum vector; how to get the g-tensor and determine its principal axis coordinate in order orientation system; and how to determine the value of the g-tensor in disorder (random) oriented system.

3.1 Landé factor

The Landé factor g is isotropic in the free ions. The magnetic moments of orbital motion $\boldsymbol{\mu}_L$ and spin motion $\boldsymbol{\mu}_S$ are given by

$$\boldsymbol{\mu}_L = -g_L\beta\boldsymbol{L}(g_L = 1) \tag{3.1}$$

$$\boldsymbol{\mu}_S = -g_S\beta\boldsymbol{S}\,(g_S = 2) \tag{3.2}$$

It is known that the total angular momentum vector $\boldsymbol{J}$ is the sum of the orbital angular momentum vector $\boldsymbol{L}$ and the spin angular momentum vector $\boldsymbol{S}$. Then the sum of vector $\boldsymbol{\mu}_L$ and $\boldsymbol{\mu}_S$ should be total magnetic moment $\boldsymbol{\mu}_J$:

$$\boldsymbol{\mu}_J = -g_J\beta\boldsymbol{J} \tag{3.3}$$

We have many complex methods to derive g_J rigorously. Here we introduce a relatively simple method and the physical concept in a rather clear way: *Landé* method.

Since the coupling effect between $\boldsymbol{L}$ and $\boldsymbol{S}$ is very strong, the vectors $\boldsymbol{L}$ and $\boldsymbol{S}$ about vector $\boldsymbol{J}$ make quick precession motion and are more faster than the precession motion of vector $\boldsymbol{J}$ about the external magnetic field $\boldsymbol{H}$. Hence, the components of $\boldsymbol{L}$ and $\boldsymbol{S}$ perpendicular to $\boldsymbol{J}$ are averaged, and the projections in the direction of $\boldsymbol{J}$ are only considered (Figure 3.1).

According to Eqs. (3.1) and (3.2), the relationship of $\boldsymbol{\mu}_L$ and $\boldsymbol{L}$ is 1:1 and of $\boldsymbol{\mu}_S$ and $\boldsymbol{S}$ is 2:1. Else has the relationship as follows:

$$\boldsymbol{J} = \boldsymbol{L} + \boldsymbol{S} \tag{3.4}$$

Thus, the ratio relation between the total magnetic moment $\boldsymbol{\mu}_J$ and the total angular momentum $\boldsymbol{J}$ is not simple. It has an angler difference between them. According to *Landé*,

$$\boldsymbol{\mu}_J = (\boldsymbol{\mu}_L)_{\text{av}} + (\boldsymbol{\mu}_S)_{\text{av}} \tag{3.5}$$

Here, $(\boldsymbol{\mu}_L)_{\text{av}}$ and $(\boldsymbol{\mu}_S)_{\text{av}}$ are the average values of $\boldsymbol{\mu}_L$ and $\boldsymbol{\mu}_S$, that is, the projection of the direction in the vector $\boldsymbol{J}$. Hence,

https://doi.org/10.1515/9783110568578-003

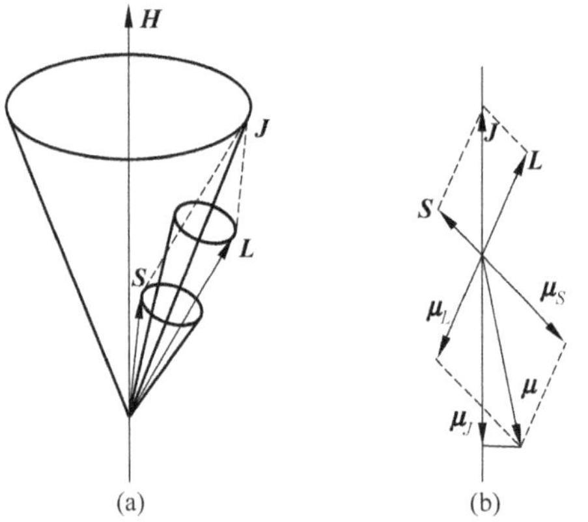

Figure 3.1: (a) The precession of S and L about J, and the precession of J about H. (b) The relationship between S, L, and J, and the relationship between μ_S, μ_L, and μ_J.

$$g_J\beta \boldsymbol{J} = g_L\beta \boldsymbol{L}_{av} + g_S\beta \boldsymbol{S}_{av} \tag{3.6}$$

$$g_J\boldsymbol{J} = 1\times\left(\boldsymbol{L}\frac{J}{|J|}\right)\frac{J}{|J|} + 2\times\left(\boldsymbol{S}\frac{J}{|J|}\right)\frac{J}{|J|}$$
$$g_J = \frac{LJ}{J^2} + 2\frac{SJ}{J^2} \tag{3.7}$$

Since $\boldsymbol{L} = \boldsymbol{J} - \boldsymbol{S}$, $\quad \boldsymbol{L}\cdot\boldsymbol{L} = J^2 + S^2 - 2J\cdot S$

Therefore, $\quad \boldsymbol{J}\cdot\boldsymbol{S} = \frac{1}{2}(J^2 + S^2 - L^2)$

Since $\boldsymbol{S} = \boldsymbol{J} - \boldsymbol{L}$, $\quad \boldsymbol{S}\cdot\boldsymbol{S} = J^2 + L^2 - 2J\cdot L$

Therefore, $\quad \boldsymbol{J}\cdot\boldsymbol{L} = \frac{1}{2}(J^2 + L^2 - S^2)$

Substituting Eq. (3.7), we obtain

$$g_J = \frac{3J^2 + S^2 - L^2}{2J^2} \tag{3.8}$$

Using operator express angular momentum in quantum mechanics, and

$$\langle J^2\rangle = J(J+1);\ \ \langle L^2\rangle = L(L+1);\ \ \langle S^2\rangle = S(S+1)$$

Hence, Eq. (3.8) could be rewritten as follows:

$$g_J = \frac{3}{2} + \frac{1}{2}\left[\frac{S(S+1) - L(L+1)}{J(J+1)}\right] \tag{3.9}$$

$J \neq 0$; or

$$g_J = 1 + \frac{J(J+1) + S(S+1) - L(L+1)}{2J(J+1)} \tag{3.10}$$

From Eq. (3.10), we can see that $J = L$, $g_J = 1$, when $S = 0$; $J = S$, $g_J = 2$, when $L = 0$. Above discussion is in case of free ions.

However, the free ions can exist only as gas state. In solution or solid state, the ions are always surrounded by solvent or coordinated with ligands forming complexes. Under the action of the ligand field, the energy level produce splitting and the ground state is nondegenerate, bringing about *orbital quenching*, and the *g*-factor approximate to 2. It can be seen from the susceptibility measurement that the *g*-factor of the transition metal ion is relatively close to 2, but the *g*-value of rare earth metal ions are more large deviating from 2. Because the 4f orbit of rare earth metal ions areshielded by 5s and 5p orbits, so the effect of ligand field is rather smaller. In fact, *g*-value of some transition metal ions is also deviated more than 2. The reason is that some excited states are mixed to the ground state via the "spin–orbit coupling," thus orbital motion of the ground state cannot be "quenched" completely. If the symmetry of ligand field is not too low, such as octahedron or tetrahedron symmetry, the ground state cannot be "orbital nondegenerate." If it also has a contribution of the orbital motion, their *g*-factor will be more deviate from 2.

If the ligand field of lower symmetry also has contribution of orbital motion, its *g*-factor will show a strong anisotropy, and the value of g_x, g_y, g_z, express more different from each other; therefore, the *g*-factor should be represented by tensor.

Solid samples are divided into order orientation (single crystalline, mainly doping and point defects, etc.) and disorder orientation (including polycrystalline, microcrystalline powder, amorphous, etc.).

3.2 Matrix presentation of *g*-tensor

In the crystal, under the situation of the magnetic dipoles arranged close together, owing to dipole–dipole interaction causing spectral line broading, the spectrum cannot provide any useful information. Therefore, to take the magnetic dipoles as impurities mixed into the nonmagnetic single-crystal samples or in the samples with point defects is interesting for studying EMR object.

3.2.1 *g*-Tensor of colour center (cubic symmetry and uniaxial symmetry system)

Colour center are divided into two kinds: one is *F* center represented by NaCl crystals (Figure 3.2a). It lacks a Cl^- ion in its center but captures an electron e^-. The charge is balanced, although this is an unpaired electron. The other kind is *V* center represented by MgO (Figure 3.2b). A Mg^{2+} ion is lost in its center. From the viewpoint of charge balance, this center corresponds to e^+ electron.

The *g*-factor of the unpaired electrons of these centers is strictly a scalar quantity. Its Hamiltonian operator has the form as follows:

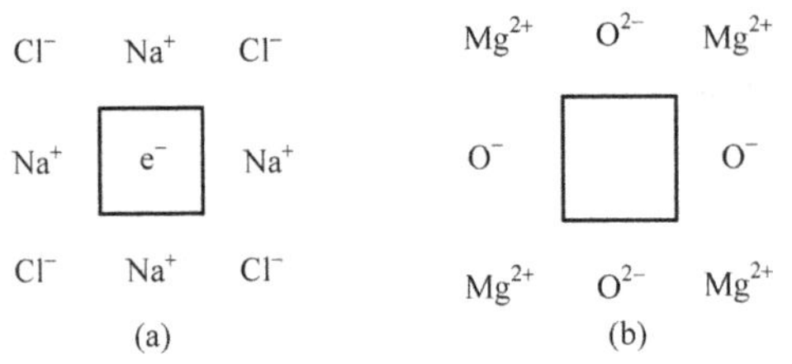

Figure 3.2: (a) Model of the F center in NaCl (cubic symmetry). (b) Model of the V2 Center in MgO (Tetragonal Symmetry).

$$\hat{\mathcal{H}} = g\beta_e(H_x\hat{S}_x + H_y\hat{S}_y + H_z\hat{S}_z) \tag{3.11}$$

For the F center in cubic crystal system (negative ion vacancy), EMR spectrum is isotropic, and the g-factor is not dependent on the orientation of the sample in a magnetic field. On the other hand, V center (positive ion vacancy) provides such an example when the octahedron is squeezed by external force along any one direction of three axes, and the symmetry reduced to tetrahedron, the V^- center (in the past called V_1 center) in MgO or CaO (NaCl-type; point group O) has an unpaired electron. In an ideal crystal, Mg^{2+} and O^{2-} ions are all on the octahedral symmetry positions [1–3]. After the X-ray irradiation at low temperatures, when an electron is taken off from any of six O^{2-} by the adjacent (preexisting) magnesium ion vacancies, the O^{2-} ion become O^- ion, as shown in Figure 3.2b. The O^- ion has an unpaired electron in the p-orbit, and has uniaxial symmetry, that is, tetrahedron symmetry. If the external magnetic field H is parallel to the Z-axis and $\nu = 9.0650$ GHz, the EMR spectrum is observed at 323.31 mT. When MgO crystal is rotated, corresponding to the external magnetic field on the YZ plane, from Z-direction rotating to Y-direction, the spectrum line of the V center from 323.31 mT shifts to 317.71 mT. A variety of line positions with orientation is shown in Figure 3.3. We can obtain the parameters from

$$g_\perp = \frac{h\nu}{\beta_e H_\perp} = \frac{6.62608\times10^{-34}\text{Js}\times9.0650\times10^9\,\text{s}^{-1}}{9.27402\times10^{-24}\text{JT}^{-1}\times0.31771\text{T}} = 2.0386 \tag{3.12a}$$

$$g_\parallel = \frac{h\nu}{\beta_e H_\parallel} = 2.0033 \tag{3.12b}$$

where $g_\perp$ and $g_\parallel$ are the g-facors corresponding to $H_\perp$ and $H_\parallel$, respectively.

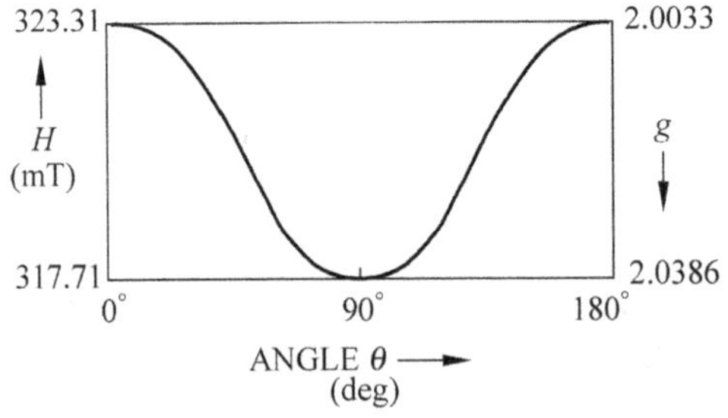

Figure 3.3: The angular dependence of EMR spectrum of the V^- center in MgO, for $H \perp X[100]$. Angle 0° indicates $H \parallel Z\,[0\ 0\ 1]$. Microwave frequency at 9.0650 GHz.

In addition, transition metal ions with electron configuration $nd^1 (s = \frac{1}{2})$ doped in orthophosphate tetrahedral also belong to the situation of uniaxial symmetry. It is meseared by the EMR spectral parameters, for instance, Ti^{3+} doped in $ScPO_4$ single crystal; Zr^{3+} doped in $LuPO_4$ single crystal; Hf^{3+} doped in YPO_4 single crystal; there are very similar g-factors at 77 K [4] – such as Ti^{3+} in $ScPO_4$ single crystal, $g_\perp = 1.961$; $g_\parallel = 1.913$. The effective g-factor is taken as the positive of the square root of the following formula:

$$g^2 = g_\perp^2 \sin^2\theta + g_\parallel^2 \cos^2\theta \tag{3.13}$$

where θ is the angle between the external magnetic field and the symmetry axis of the defect.

3.2.2 The g-tensor of nonaxisymmetric (lower than uniaxial symmetry) system

For the lower symmetric system, without considering nuclear hyperfine interactions and nuclear Zeeman interaction, the spin Hamiltonian expresses as follows:

$$\hat{\mathscr{H}} = \beta \hat{\boldsymbol{S}} \cdot g \cdot \boldsymbol{H} \tag{3.14}$$

Let

$$\boldsymbol{H_{eq}} = g_e^{-1} \cdot g \cdot \boldsymbol{H} = \frac{g \cdot \boldsymbol{H}}{g_e} \tag{3.15}$$

then Eq. (3.14) can be rewritten as

$$\hat{\mathscr{H}} = -(-g_e \beta \hat{\boldsymbol{S}}) \cdot \boldsymbol{H_{eq}} = g_e \beta \boldsymbol{H_{eq}} \hat{\boldsymbol{S}}_{\boldsymbol{H_{eq}}} \tag{3.16}$$

where $\boldsymbol{H}_{eq}$ is the equivalent magnetic field, $\hat{\boldsymbol{S}}_{H_{eq}}$ is the projection of the spin operator along the $\boldsymbol{H}_{eq}$. Set the eigenfunctions of $\hat{\boldsymbol{S}}_{H_{eq}}$ as $|\pm\rangle$, and the corresponding eigenvalue is $\pm\frac{1}{2}$:

$$\hat{\boldsymbol{S}}_{H_{eq}} |\pm\rangle = \pm \frac{1}{2} |\pm\rangle \tag{3.17}$$

then

$$E_\pm = \langle \pm | \hat{\mathscr{H}} | \pm \rangle = \pm \frac{1}{2} g_e \beta \boldsymbol{H}_{eq} \tag{3.18}$$

The condition for producing the resonance transition between two energy levels is

$$h\nu = E_+ - E_- = g_e \beta \boldsymbol{H}_{eq} \tag{3.19}$$

Equations (3.19) and (2.64) are very similar in form, only difference is that of $\boldsymbol{H}_{\text{eq}}$ and $\boldsymbol{H}$. $\boldsymbol{H}_{\text{eq}}$ is an equivalent magnetic field whose magnitude and direction are dependent on the orientation of crystal axis relating to the external magnetic field. In general, the directions of $\boldsymbol{H}_{\text{eq}}$ and $\boldsymbol{H}$ are different.

It should be pointed out that Eq. (3.19) is simplified only in format. Actually, it is still very complicated. From Eq. (3.19), one can obtain

$$(\Delta E)^2 = (g_e \beta \boldsymbol{H}_{\text{eq}})^2 = \beta^2 (g \boldsymbol{H}_{\text{eq}}) \cdot (g \boldsymbol{H}_{\text{eq}}) = \beta^2 (\boldsymbol{H} \cdot g \cdot g \cdot \boldsymbol{H}) \tag{3.20}$$

3.2.2.1 Determine the matrix elements of *g*-tensor in the selected orthogonal coordinate system (x, y, z)

Now we want to express Eq. (3.20) in a matrix form in selected (x, y, z) coordinates. In principle, choosing a set of orthogonal coordinate system does not have any restriction, but usually, according to outward appearance of crystal, always select one or two (monoclinic crystal has two) orthogonal crystal axes as the axes of the orthogonal coordinate system. Then, the third coordinate axis is selected on the basis of requirements of right-hand orthogonal system. It is for the experimental convenience entirely. After choosing the orthogonal coordinate system, $\hat{\mathbf{S}}$, g, $\boldsymbol{H}$ can be expressed as follows:

$$\left.\begin{aligned} \hat{\mathbf{S}} &= \sum_i \hat{S}_i \boldsymbol{e}_i \\ g &= \sum_{jk} g_{jk} \boldsymbol{e}_j \cdot \boldsymbol{e}_k \\ \boldsymbol{H} &= \sum_p \boldsymbol{H}_p \boldsymbol{e}_p \end{aligned}\right\} \tag{3.21}$$

where $i, j, k, p = 1, 2, 3$; $\boldsymbol{e}_1$, $\boldsymbol{e}_2$, $\boldsymbol{e}_3$ are the unit vectors along x-, y-, z-coordinate axes, respectively; then Eq. (3.14) can be written as follows:

$$\hat{\mathscr{H}} = \beta \hat{\mathbf{S}} \cdot g \cdot \boldsymbol{H} = \beta \sum_i \sum_j \sum_k \sum_p \hat{S}_i \, g_{jk} H_p (\boldsymbol{e}_i \cdot \boldsymbol{e}_j)(\boldsymbol{e}_k \cdot \boldsymbol{e}_p) \tag{3.22}$$

According to orthonormality:

$$\boldsymbol{e}_i \cdot \boldsymbol{e}_j = \delta_{ij}$$

$$\boldsymbol{e}_k \cdot \boldsymbol{e}_p = \delta_{kp}$$

where δ_{ij} and δ_{kp} are δ-Kronecker signs, when $i = j$, $\delta_{ij} = 1$; when $i \neq j$, $\delta_{ij} = 0$, Hence, Eq. (3.22) can be written as follows:

$$\hat{\mathscr{H}} = \beta \sum_i \sum_k \hat{S}_i g_{ik} \boldsymbol{H}_k \tag{3.23}$$

$$\hat{\mathscr{H}} = \beta(\hat{S}_x\hat{S}_y\hat{S}_z)\begin{pmatrix} g_{xx} & g_{xy} & g_{xz} \\ g_{yx} & g_{yy} & g_{yz} \\ g_{zx} & g_{zy} & g_{zz} \end{pmatrix}\begin{pmatrix} H_x \\ H_y \\ H_z \end{pmatrix} \tag{3.24}$$

Equation (3.24) is the matrix expression of Eq. (3.14) in selected orthogonal coordinate system (x, y, z). To note, the vector inner product is defined as a row vector multiplied by a column vector. Then,

$$(\Delta E)^2 = \beta^2(g_e\boldsymbol{H}_{eq}) \cdot (g_e\boldsymbol{H}_{eq}) = \beta^2(g \cdot \boldsymbol{H})_T \cdot (g \cdot \boldsymbol{H}) = \beta^2 \cdot \boldsymbol{H}_T \cdot g_T \cdot g \cdot \boldsymbol{H} \tag{3.25}$$

where $\boldsymbol{H_T}$ is the transport matrix of $\boldsymbol{H}$ matrix; $g_{\boldsymbol{T}}$ is the transport matrix of g matrix. Then the matrix expression of Eq. (3.25) is given as follows:

$$(\Delta E)^2 = \beta^2(H_x \quad H_y \quad H_z)\begin{pmatrix} g_{xx} & g_{yx} & g_{zx} \\ g_{xy} & g_{yy} & g_{zy} \\ g_{xz} & g_{yz} & g_{zz} \end{pmatrix}\begin{pmatrix} g_{xx} & g_{xy} & g_{xz} \\ g_{yx} & g_{yy} & g_{yz} \\ g_{zx} & g_{zy} & g_{zz} \end{pmatrix}\begin{pmatrix} H_x \\ H_y \\ H_z \end{pmatrix} \tag{3.26}$$

H_x, H_y, H_z are the projection of external magnetic field $\boldsymbol{H}$ along the x-, y-, z-coordinate axes, respectively. ℓ_x, ℓ_y, ℓ_z are defined as the cosine of the angle between the external magnetic field vectors with x-, y-, z-coordinate axes.

Let

$$(H_x \quad H_y \quad H_z) = \boldsymbol{H}\ (\ell_x \quad \ell_y \quad \ell_z) \tag{3.27}$$

$$(\Delta E)^2 = \beta^2 g_{eq}^2 \boldsymbol{H}^2 \tag{3.28}$$

Then, we obtain

$$g_{eq}^2 = (\ell_x \quad \ell_y \quad \ell_z)\begin{pmatrix} g_{xx}^2 & g_{xy}^2 & g_{xz}^2 \\ g_{yx}^2 & g_{yy}^2 & g_{yz}^2 \\ g_{zx}^2 & g_{zy}^2 & g_{zz}^2 \end{pmatrix}\begin{pmatrix} \ell_x \\ \ell_y \\ \ell_z \end{pmatrix} \tag{3.29}$$

g_{eq}^2 represents the square of equivalent g-tensor. g^2 ought to be $(g)_{\boldsymbol{T}} \cdot (g)$. And $(g)_{\boldsymbol{T}}$ is the transport matrix of (g). It should be noted: g^2 matrix is always symmetric matrix, that is, $(g^2)_{ij} = (g^2)_{ji}$. So g^2 only has six independent matrix elements.

How to get these matrix elements $(g^2)_{ij}$ from experiments: Let $\boldsymbol{H}$ be rotational on the xy, yz, zx planes, respetively. The usual operation is fixing the direction of the external magnetic field and rotating the selected orthogonal coordinate system on the crystal. If the $\boldsymbol{H}$ rotates in the zx plane, and θ is the angle between $\boldsymbol{H}$ and z-axis, then

$$(\ell_x \quad \ell_y \quad \ell_z) = (sin\theta, \quad 0, \quad cos\theta)$$

Hence,

$$g_{eq}^2(\theta) = (\sin\theta,\quad 0,\quad \cos\theta)\begin{pmatrix} g_{xx}^2 & g_{xy}^2 & g_{xz}^2 \\ g_{yx}^2 & g_{yy}^2 & g_{yz}^2 \\ g_{zx}^2 & g_{zy}^2 & g_{zz}^2 \end{pmatrix}\begin{pmatrix} \sin\theta \\ 0 \\ \cos\theta \end{pmatrix} \tag{3.30}$$

$$= g_{xx}^2 \sin^2\theta + 2g_{zx}^2 \sin\theta\cos\theta + g_{zz}^2\cos^2\theta$$

When $\theta = 0$, $g_{eq}^2(0) = g_{zz}^2$; when $\theta = \pi/2$, $g_{eq}^2\left(\frac{\pi}{2}\right) = g_{zx}^2$.

$$g_{eq}^2\left(\frac{\pi}{4}\right) = \frac{1}{2}[g_{xx}^2 + g_{zz}^2] + g_{xz}^2 \tag{3.31}$$

$$g_{eq}^2\left(\frac{3\pi}{4}\right) = \frac{1}{2}[g_{xx}^2 + g_{zz}^2] - g_{xz}^2 \tag{3.32}$$

Equation (3.31) minus Eq. (3.32) gives $g_{xz}^2 = \frac{1}{2}\left[g_{eq}^2\left(\frac{\pi}{4}\right) - g_{eq}^2\left(\frac{3\pi}{4}\right)\right]$

Using the same method, let ***H*** rotate in the *xy* plane, and θ be the angle between ***H*** and *x*-axis. Then:

$$g_{eq}^2(\theta) = g_{xx}^2\cos^2\theta + 2g_{xy}^2 \sin\theta\cos\theta + g_{yy}^2\cos^2\theta \tag{3.33}$$

g_{xx}^2, g_{yy}^2, and g_{xy}^2 can be obtained by the same method.

Again, let ***H*** rotate in the *yz* plane, and θ be the angle between ***H*** and *y*-axis. Then,

$$g_{eq}^2(\theta) = g_{yy}^2\cos^2\theta + 2g_{yz}^2 \sin\theta\cos\theta + g_{zz}^2\cos^2\theta \tag{3.34}$$

Similarly, we can also get g_{yy}^2, g_{zz}^2, and g_{yz}^2. So far we have got two g_{xx}^2, two g_{yy}^2, and two g_{zz}^2, one g_{xy}^2, one g_{yz}^2, and one g_{zz}^2. Each of g_{yy}^2 and g_{zz}^2 has two values; they are not equal usually, and ought to take average value.

3.2.2.2 Determination of the principal axis coordinates

For tensor, there must exist a set of orthogonal coordinate system (*X*, *Y*, *Z*) called the principal axis coordinate system of tensor. In this system, the matrix elements of tensor are diagonalized, that is, the diagonal matrix elements have a nonzero value, or say are zero. The coordinate system selected previously is not the principal axis coordinate of the sample's *g*-tensor. In order to get the principal values of *g*-tensor, we should first find out the principal axis coordinates. It means that former choice of coordinate system should be converted to the principal axis coordinate system of sample's *g*-tensor.

Between the experimental selected coordinate system (*x*, *y*, *z*) and the principal axis coordinate system (*X*, *Y*, *Z*) of sample's *g*-tensor exist the relationship as shown in Figure 3.4.

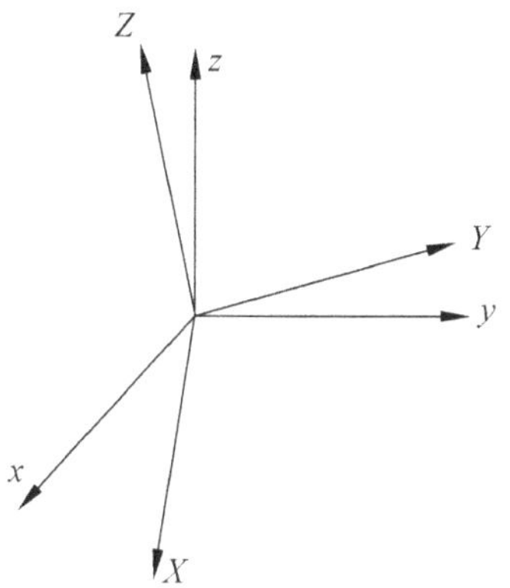

Figure 3.4: The relationship between experimental coordinate system (x, y, z) and principal axis coordinate system (X, Y, Z).

Let $\boldsymbol{e}_x$, $\boldsymbol{e}_y$, $\boldsymbol{e}_z$ be the unit vectors of the experimental coordinate system (x, y, z) and $\boldsymbol{e}_X$, $\boldsymbol{e}_Y$, $\boldsymbol{e}_Z$ be the unit vectors of the principal axis coordinate system (X, Y, Z) of sample's g-tensor. Then

$$\boldsymbol{e}_X = \boldsymbol{e}_x\,\ell_{Xx} + \boldsymbol{e}_y\,\ell_{Xy} + \boldsymbol{e}_z\,\ell_{Xz} \tag{3.35}$$

$$\boldsymbol{e}_Y = \boldsymbol{e}_x\,\ell_{Yx} + \boldsymbol{e}_y\,\ell_{Yy} + \boldsymbol{e}_z\,\ell_{Yz} \tag{3.36}$$

$$\boldsymbol{e}_Z = \boldsymbol{e}_x\,\ell_{Zx} + \boldsymbol{e}_y\,\ell_{Zy} + \boldsymbol{e}_z\,\ell_{Zz} \tag{3.37}$$

where $i = X, Y, Z$, representing the principal axis coordinates; $j = x, y, z$, representing the experimental coordinates. ℓ_{ij} is the cosine of angle between the axes of the two coordinate systems: $\ell_{Xx} = \cos(X, x)$. Combining Eqs. (3.35)–(3.37), the matrix form may be expressed as follows:

$$\begin{pmatrix} \boldsymbol{e}_X \\ \boldsymbol{e}_Y \\ \boldsymbol{e}_Z \end{pmatrix} = \begin{pmatrix} \ell_{Xx} & \ell_{Xy} & \ell_{Xz} \\ \ell_{Yx} & \ell_{Yy} & \ell_{Yz} \\ \ell_{Zx} & \ell_{Zy} & \ell_{Zz} \end{pmatrix} \begin{pmatrix} \boldsymbol{e}_x \\ \boldsymbol{e}_y \\ \boldsymbol{e}_z \end{pmatrix} = \mathscr{L} \begin{pmatrix} \boldsymbol{e}_x \\ \boldsymbol{e}_y \\ \boldsymbol{e}_z \end{pmatrix} \tag{3.38}$$

where $\mathscr{L}$ is composed of nine direction cosines, called $\mathscr{L}$ matrix. To look for the principal axis coordinate system (X, Y, Z), actually, is to seek the problem of these direction cosines. Since the direction cosine is satisfied, the normalization condition should be a unitary matrix, and its Hermitian matrix $\mathscr{L}^{\dagger}$ has to be equal to its inverse matrix, namely, $\mathscr{L}^{\dagger} = \mathscr{L}^{-1}$. It can be seen that the matrix representation of g^2 tensor in the principal coordinate system (X, Y, Z) is given by

$$g^2 = \sum_{i=X,Y,Z} (g_{ii}^2)\ \boldsymbol{e}_i\ \boldsymbol{e}_i = (\boldsymbol{e}_X \quad \boldsymbol{e}_Y \quad \boldsymbol{e}_Z) \begin{pmatrix} g_{XX}^2 & 0 & 0 \\ 0 & g_{YY}^2 & 0 \\ 0 & 0 & g_{ZZ}^2 \end{pmatrix} \begin{pmatrix} \boldsymbol{e}_X \\ \boldsymbol{e}_Y \\ \boldsymbol{e}_Z \end{pmatrix} \tag{3.39}$$

However, the matrix representation of g^2 tensor in our selected experimental coordinate system (x, y, z) is given by

$$g^2 = \sum_{i,j=x,y,z} (g_{ij}^2)\ \boldsymbol{e}_i\ \boldsymbol{e}_j = (\boldsymbol{e}_x \quad \boldsymbol{e}_y \quad \boldsymbol{e}_z) \begin{pmatrix} g_{xx}^2 & g_{xy}^2 & g_{xz}^2 \\ g_{yx}^2 & g_{yy}^2 & g_{yz}^2 \\ g_{zx}^2 & g_{zy}^2 & g_{zz}^2 \end{pmatrix} \begin{pmatrix} \boldsymbol{e}_x \\ \boldsymbol{e}_y \\ \boldsymbol{e}_z \end{pmatrix} \tag{3.40}$$

Substituting Eq. (3.38) into Eq. (3.39), we get

$$g^2 = (\boldsymbol{e}_x \quad \boldsymbol{e}_y \quad \boldsymbol{e}_z) \mathscr{L}^\dagger \begin{pmatrix} g_{XX}^2 & 0 & 0 \\ 0 & g_{YY}^2 & 0 \\ 0 & 0 & g_{ZZ}^2 \end{pmatrix} \mathscr{L} \begin{pmatrix} \boldsymbol{e}_x \\ \boldsymbol{e}_y \\ \boldsymbol{e}_z \end{pmatrix} \tag{3.41}$$

Comparing Eqs. (3.40) and (3.41), we get

$$\begin{pmatrix} g_{xx}^2 & g_{xy}^2 & g_{xz}^2 \\ g_{yx}^2 & g_{yy}^2 & g_{yz}^2 \\ g_{zx}^2 & g_{zy}^2 & g_{zz}^2 \end{pmatrix} = \mathscr{L}^\dagger \begin{pmatrix} g_{XX}^2 & 0 & 0 \\ 0 & g_{YY}^2 & 0 \\ 0 & 0 & g_{ZZ}^2 \end{pmatrix} \mathscr{L} \tag{3.42}$$

Let $(g^2) = \begin{pmatrix} g_{xx}^2 & g_{xy}^2 & g_{xz}^2 \\ g_{yx}^2 & g_{yy}^2 & g_{yz}^2 \\ g_{zx}^2 & g_{zy}^2 & g_{zz}^2 \end{pmatrix}$; $^d(g^2) = \begin{pmatrix} g_{XX}^2 & 0 & 0 \\ 0 & g_{YY}^2 & 0 \\ 0 & 0 & g_{ZZ}^2 \end{pmatrix}$

Equation (3.42) can be written as follows:

$$(g^2) = \mathscr{L}^{\dagger d}(g^2)\mathscr{L} \tag{3.43}$$

$$^d(g^2) = \mathscr{L}(g^2)\mathscr{L}^\dagger \tag{3.44}$$

Multiplying the former by $\mathscr{L}^{-1}$ on both sides and using the property of $\mathscr{L}^\dagger = \mathscr{L}^{-1}$ gives

$$\begin{aligned} \mathscr{L}^{-1}\ \mathscr{L}(g^2)\mathscr{L}^\dagger &= \mathscr{L}^{\dagger d}(g^2) \\ (g^2)\mathscr{L}^\dagger &= \mathscr{L}^{\dagger d}(g^2) \end{aligned} \tag{3.45}$$

$$\mathscr{L} = \begin{pmatrix} \ell_{Xx} & \ell_{Xy} & \ell_{Xz} \\ \ell_{Yx} & \ell_{Yy} & \ell_{Yz} \\ \ell_{Zx} & \ell_{Zy} & \ell_{Zz} \end{pmatrix} \qquad \mathscr{L}^\dagger = \begin{pmatrix} \ell_{Xx} & \ell_{Yx} & \ell_{Zx} \\ \ell_{Xy} & \ell_{Yy} & \ell_{Zy} \\ \ell_{Xz} & \ell_{Yz} & \ell_{Zz} \end{pmatrix}$$

Then, Eq. (3.45) can be rewritten as follows:

$$\begin{pmatrix} g_{xx}^2 & g_{xy}^2 & g_{xz}^2 \\ g_{yx}^2 & g_{yy}^2 & g_{yz}^2 \\ g_{zx}^2 & g_{zy}^2 & g_{zz}^2 \end{pmatrix} \begin{pmatrix} \ell_{Xx} & \ell_{Yx} & \ell_{Zx} \\ \ell_{Xy} & \ell_{Yy} & \ell_{Zy} \\ \ell_{Xz} & \ell_{Yz} & \ell_{Zz} \end{pmatrix} = \begin{pmatrix} \ell_{Xx} & \ell_{Yx} & \ell_{Zx} \\ \ell_{Xy} & \ell_{Yy} & \ell_{Zy} \\ \ell_{Xz} & \ell_{Yz} & \ell_{Zz} \end{pmatrix} \begin{pmatrix} g_{XX}^2 & 0 & 0 \\ 0 & g_{YY}^2 & 0 \\ 0 & 0 & g_{ZZ}^2 \end{pmatrix}$$
$$= \begin{pmatrix} g_{XX}^2\ell_{Xx} & g_{YY}^2\ell_{Yx} & g_{ZZ}^2\ell_{Zx} \\ g_{XX}^2\ell_{Xy} & g_{YY}^2\ell_{Yy} & g_{ZZ}^2\ell_{Zy} \\ g_{XX}^2\ell_{Xz} & g_{YY}^2\ell_{Yz} & g_{ZZ}^2\ell_{Zz} \end{pmatrix} \tag{3.46}$$

If the $\mathscr{L}^{\dagger}$ matrix consists of the following three column matrixes (column vectors),

$$\begin{pmatrix} \ell_{Xx} \\ \ell_{Xy} \\ \ell_{Xz} \end{pmatrix}, \quad \begin{pmatrix} \ell_{Yx} \\ \ell_{Yy} \\ \ell_{Yz} \end{pmatrix}, \quad \textit{and} \quad \begin{pmatrix} \ell_{Zx} \\ \ell_{Zy} \\ \ell_{Zz} \end{pmatrix} \tag{3.47}$$

then these three column vectors become the *eigenfunctions* of the (g^2) matrix operator, and g_{XX}^2, g_{YY}^2 and g_{ZZ}^2 are the *eigenvalues* of their corresponding three eigenfunctions.

Therefore, after determining the matrix representation of tensor (g^2) in the selected experimental coordinate system (x, y, z), the next step is to find out the *principal value* of matrix (g^2) and the relationship between *principal axis coordinate system (X, Y, Z) and experimental coordinate system (x, y, z)*. This process is to find out the eigenfunction and eigenvalue of (g^2) matrix in mathematics.

The problem of finding out the eigenvalue is a very common problem in quantum mechanics. It relates to the problem of solving secular determinant.

Suppose λ is an eigenvalue; then the secular determinant is given by

$$\begin{vmatrix} (g_{xx}^2 - \lambda) & g_{xy}^2 & g_{xz}^2 \\ g_{yx}^2 & (g_{yy}^2 - \lambda) & g_{yz}^2 \\ g_{zx}^2 & g_{zy}^2 & (g_{zz}^2 - \lambda) \end{vmatrix} = 0 \tag{3.48}$$

This is a cubic equation of λ; its solution gives out three roots λ_1, λ_2, and λ_3, which are g_{XX}^2, g_{YY}^2, and g_{ZZ}^2, respectively. However, it is still unknown how to determine the λ_1, λ_2, λ_3 corresponding relation with g_{XX}^2, g_{YY}^2, g_{ZZ}^2. Let us slove the eigenfunction first! Solving eigenfunction is solving the following equation group:

$$\left.\begin{aligned} (g_{xx}^2 - \lambda_i)\ell_{ix} + g_{yx}^2\ell_{iy} + g_{zx}^2\ell_{iz} &= 0 \\ g_{xy}^2\ell_{ix} + (g_{yy}^2 - \lambda_i)\ell_{iy} + g_{zy}^2\ell_{iz} &= 0 \\ g_{xz}^2\ell_{ix} + g_{yz}^2\ell_{iy} + (g_{zz}^2 - \lambda_i)\ell_{iz} &= 0 \\ \ell_{ix}^2 + \ell_{iy}^2 + \ell_{iz}^2 &= 1 \end{aligned}\right\} \tag{3.49}$$

Of these four equations, only three are independent; the three roots obtained earlier are substituted as the λ_i ($i = 1, 2, 3$) secular determinants to determine the matrix of direction cosine. Since the corresponding relationship between λ_1, λ_2, λ_3 and g^2_{XX}, g^2_{YY}, g^2_{ZZ} is not known, the method of guess can only be used. Fortunately, the row vector arrangement of direction cosine matrix cannot be wrong, and whether the arrangement of column vector order is right or not depends on whether the (g^2) matrix could be diagonalized. If it can be diagonalized, it is right. Otherwise, the order should be adjusted and try again. Not more than three trials need to be made, andthe right answer will be obtained. In fact, today there are readymade software for diagonalization to calculate these kinds of trivial problems quickly.

In brief, after obtaining the (g^2) matrix from the experimental coordinate system, it is diagonalized, and then the matrix elements on diagonal is the *principal value of g-tensor*. In order to diagonalize (g^2) matrix, we need to find $\mathscr{L}$ matrix; in fact, it is a process of seeking the eigenvector and eigenvalue.

3.3 *g*-Tensor of irregular orientation system

In our practical work, the vast majority of research samples are in random states, such as polycrystalline, powder, amorphous solid, and solution at low temperatures, frozen as glassy state, and so on. All of these belong to the irregular orientation system. Thus, it is more important and realistically significant to derive useful information from the EMR spectrum of irregular orientation system.

The small magnets in irregular orientation system can merely be distributed in accordance with a certain statistical regularity in magnetic field. The anisotropy of *g*-tensor is closely related with the geometrical configuration and symmetry of the molecule where the unpaired electron exists. If the paramagnetic particles have high symmetric molecular configuration (e.g., spherical, octahedral, n-cube), their *g*-tensor can be regarded as almost isotropic (i.e., $g_x = g_y = g_z$). For the slightly lower symmetry such as distorted octahedral, tetrahedral, and so on with C_{4v}, D_{4h} symmetry, we have $g_x = g_y \neq g_z$ (i.e., $g_x = g_y = g_\perp$; $g_z = g_\parallel$); if the symmetry is further reduced, such as C_{2v}, D_{2h}, and so on, to more lower symmetry, we have $g_x \neq g_y \neq g_z$. The latter two cases would be discussed here.

3.3.1 *g*-Tensor of axisymmetric system

In order to deal with the problem simply and distinctly, we choose the axial symmetric system with $S = \frac{1}{2}$; $I = 0$. Let (x, y, z) be the experimental orthogonal coordinates system, the *z*-direction as the direction of the external magnetic field, and the *x*-, *y*-axes perpendicular to the external magnetic field. The orientations of these small axisymmetric magnets in outer magnetic field are random. The g-tensor of the small magnets whose orientation of symmetry axes agree with the direction of outer magnetic field is defined as $g_P(g_z)$, and the tensor whose orientation of symmetry axes is perpendicular to the external magnetic field is defined as $g_\perp(g_x = g_y)$. Their corresponding resonant magnetic fields are given by

$$H_\| = \frac{h\nu_\circ}{g_\|\beta}; \quad H_\perp = \frac{h\nu_\circ}{g_\perp\beta} \tag{3.50}$$

In fact, the symmetric axes of the vast majority of the small magnets in system are neither parallel nor perpendicular to the z-axis. Their resonant magnetic field H_r have a probability distribution between $H_\|$ and $H_\perp$. We define the small angle between the magnetic symmetric axis and z-axis as θ; then

$$H_r = \frac{h\nu}{g_{eq}\beta} = \frac{h\nu}{\beta}(g_\|^2\cos^2\theta + g_\perp^2\sin^2\theta)^{-\frac{1}{2}} \tag{3.51}$$

$$g_{eq}^2 = g_\|^2\cos^2\theta + g_\perp^2\sin^2\theta = g_\perp^2 - (g_\perp^2 - g_\|^2)\cos^2\theta \tag{3.52}$$

Since all the orientation of the symmetric axes of the small magnets in space are of equal probability, we introduce a concept of stereo angle Ω:

$$\Omega = \frac{\mathscr{A}}{4\pi r^2} \tag{3.53}$$

where $4\pi r^2$ is the entire area of the spherical surface and $\mathscr{A}$ is the part of spherical surface area corresponding to the stereo angle Ω. Since every radius vector with the points on spherical surface corresponds to each other, the orientations of symmetric axes in a point space are all equivalent, that is, the number of symmetric axes in unit stereo angle is equal for all the regions of the sphere.

It has been stated that the direction of the external magnetic field H is parallel to the z-axis and the number of small magnets in the angle (from θ to $\theta + \mathrm{d}\theta$) between the symmetric axis and the external magnetic field is $\mathrm{d}N$. Their resonant magnetic field is between H_r and $H_r + dH_r$. The total number of small magnets in the entire sphere is N_o, the area of spherical surface is $4\pi r^2$, and the area of the ring on spherical surface (Figure 3.5) between θ and $\theta + \mathrm{d}\theta$ is given by

$$2\pi(r\sin\theta)r \cdot d\theta \tag{3.54}$$

Then the stereo angle has

$$\mathrm{d}\Omega = \frac{\mathrm{d}N}{N_\circ} = \frac{2\pi r^2\sin\theta\mathrm{d}\theta}{4\pi r^2} = \frac{1}{2}\sin\theta\mathrm{d}\theta$$

The resonance absorption intensity of system $\mathscr{I}$ is given by

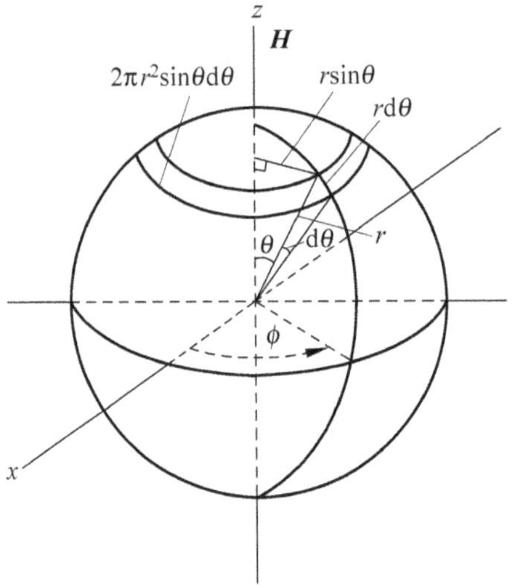

Figure 3.5: The area element on the spherical surface [5].

$$\mathcal{I} = \int_{H_{\parallel}}^{H_{\perp}} \mathcal{I}\,(H)\mathrm{d}H \tag{3.55}$$

The absorption intensity corresponding to the stereo angle is

$$\frac{\mathcal{I}\,(H)\mathrm{d}H}{\mathcal{I}} = \frac{\mathrm{d}N}{N_\circ} = \frac{1}{2}\sin\theta\mathrm{d}\theta \tag{3.56}$$

The absorption intensity $\mathcal{I}(H)dH$ corresponding to the stereo angle $d\Omega$ should be proportional to the probability of resonance transition ($P(H)\mathrm{d}H$) in H_r and $H_r + \mathrm{d}H_r$, that is,

$$\mathcal{I}\,(H)\mathrm{d}H \propto P(H)\mathrm{d}H \propto \sin\theta\mathrm{d}\theta \tag{3.57}$$

$$P(H) \propto \frac{\sin\theta}{\mathrm{d}H/\mathrm{d}\theta} \tag{3.58}$$

It can be seen from Eq. (3.58) that the numerator is $\sin\theta$, which is maximum at $\theta = \pi/2$. That is, the small magnets with symmetric axes perpendicular to the direction of the external magnetic field is majority in the tested sample. Then the signal at $H_{\perp}$ should be the strongest. Second, the more smaller the denominator in $\mathrm{d}H/\mathrm{d}\theta$, the larger the $P(H)$. This means that there is an extremum between $H_{\parallel}$ and $H_{\perp}$. Now, see the derivative of Eq. (3.51) with respect to $\mathrm{d}\theta$:

$$\begin{aligned}\mathrm{d}H &= \left(\frac{hv}{\beta}\right)\left(\frac{-1}{2}\right)(g_{\parallel}^2\cos^2\theta + g_{\perp}^2\sin^2\theta)^{-3/2}\,\mathrm{d}(g_{\parallel}^2\cos^2\theta + g_{\perp}^2\sin^2\theta)\\ &= \left(\frac{hv}{\beta}\right)(g_{\parallel}^2\cos^2\theta + g_{\perp}^2\sin^2\theta)^{-3/2}\,(g_{\parallel}^2 - g_{\perp}^2)\cos\theta\sin\theta\mathrm{d}\theta\end{aligned}$$

Then,

$$\frac{\mathrm{d}H}{\mathrm{d}\theta} = \left(\frac{hv}{\beta}\right)(g_{\|}^2\cos^2\theta + g_{\perp}^2\sin^2\theta)^{-3/2}(g_{\|}^2 - g_{\perp}^2)\cos\theta\sin\theta \tag{3.59}$$

$$\frac{\sin\theta}{\mathrm{d}H/\mathrm{d}\theta} = \left(\frac{\beta}{hv}\right)(g_{\|}^2\cos^2\theta + g_{\perp}^2\sin^2\theta)^{3/2}[(g_{\|}^2 - g_{\perp}^2)\cos\theta]^{-1} \tag{3.60}$$

$$P(H) \propto \left(\frac{\beta}{hv}\right)(g_{\|}^2\cos^2\theta + g_{\perp}^2\sin^2\theta)^{3/2}[(g_{\|}^2 - g_{\perp}^2)\cos\theta]^{-1} \tag{3.61}$$

From Eq. (3.51), we can obtain

$$H_{\mathrm{r}}^3 = \left(\frac{hv}{\beta}\right)^3(g_{\|}^2\cos^2\theta + g_{\perp}^2\sin^2\theta)^{-3/2} \tag{3.62}$$

Substituting Eq. (3.62) into Eq. (3.61) gives

$$P(H) \propto \left(\frac{hv}{\beta}\right)^2 \frac{1}{H_{\mathrm{r}}^3(g_{\|}^2 - g_{\perp}^2)\cos\theta} \tag{3.63}$$

When $\theta = 0°$, from Eq. (3.50),

$$\left(\frac{hv}{\beta}\right) = g_{\|} H_{\|}$$

Substituting into Eq. (3.63) gives

$$P(H) \propto \frac{1}{H_{\|}} \tag{3.64}$$

In an appointed system (sample), $H_{\|}$ is a fixed value. The absorption intensity corresponding to $H_{\|}$ (i.e., $\theta = 0°$) should also be a fixed value; when $\theta = \pi/2$, $P(H) \to \infty$. The absorption line in an ideal situation is shown in Figures 3.6a and 3.6b shows the computer-simulated absorption lines with different linewidths. When the linewidth is about 10 mT, the spectrum becomes a big envelope line and to identify $g_{\|}^2$ and $g_{\perp}^2$ is very difficult; Figure 3.6c is a first-order differential spectrum. If we get this kind of experimental spectra, we can determine the system as axial symmetry and can obtain $g_{\|}^2$ and $g_{\perp}^2$ from the spectra immediately.

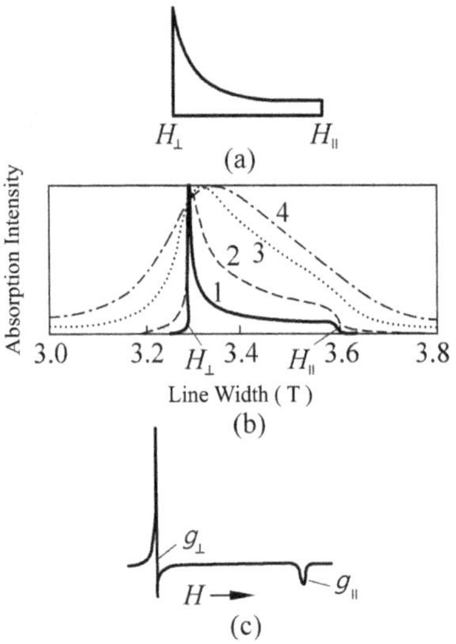

Figure 3.6: (a) Theoretical absorption spectra of polycrystalline system with axial symmetry and no hyperfine coupling. (b) Computer-simulated absorption lines with linewidths of 0.1, 1.0, 5.0, 10.0 mT. (c) First-order differential spectrum of powder sample with axial symmetry [6].

3.3.2 *g*-Tensor of nonaxisymmetric system

In order to discuss conveniently, we else choose the system of nonaxisymmetric polycrystalline with $S = 1/2$ and $I = 0$, without taking linewidth into account. Select a reference coordinate system and let the external magnetic field be parallel to the z-axis. When the external magnetic field is changed between H and $H + \mathrm{d}H$, there are $\mathrm{d}N$ particles generating resonance absorption. These $\mathrm{d}N$ particles are in the stereo angle from θ to $\theta + \mathrm{d}\theta$ and from ϕ to $\phi + d\phi$, where θ is the angle between g-tensor and z-aixs, ϕ is the angle between the projection of g-tensor in the xy plane and the x-axis. Then g-tensor is varied between $h\nu/\beta H$ and $h\nu/\beta(H + \mathrm{d}H)$. Here we introduce the concept of stereo angle $\mathrm{d}\Omega$:

$$\mathrm{d}\Omega = \frac{\mathrm{d}S}{r^2} = \frac{r\sin\theta(r\mathrm{d}\theta)\mathrm{d}\phi}{r^2} = \sin\theta\mathrm{d}\theta\mathrm{d}\phi \tag{3.65}$$

The transition probability of resonance transtion of $\mathrm{d}N$ particals in $\mathrm{d}\Omega$ is

$$P(H)\mathrm{d}H = \frac{\mathrm{d}N}{N_\circ} = \frac{\mathrm{d}\Omega}{4\pi} = \frac{\sin\theta\mathrm{d}\theta\mathrm{d}\phi}{4\pi} = \frac{\mathrm{d}(\cos\theta)\mathrm{d}\phi}{4\pi} \tag{3.66}$$

Now Eq. (3.51) can be rewritten as follows:

$$H_\mathrm{r} = \frac{h\nu}{\beta}(g_\mathrm{eq}^{-1}) = \frac{h\nu}{\beta}(g_x^2\sin^2\theta\cos^2\phi + g_y^2\sin^2\theta\sin^2\phi + g_z^2\cos^2\theta)^{-1/2} \tag{3.67}$$

Equation (3.52) can be written as follows:

$$g_{eq}^2 = g_x^2 \sin^2\theta \cos^2\phi + g_y^2 \sin^2\theta \sin^2\phi + g_z^2 \cos^2\theta \tag{3.68}$$

The resonant magnetic field H_r is a function of θ and ϕ, and should be written as $H_r(\theta, \phi)$, and the entire absorption line is a function of magnetic field, $f[H - H_r(\theta, \phi)]$:

$$\mathcal{I}\,(H) \propto \int_0^{4\pi} f[H - H_r(\theta,\ \phi)]d\Omega \tag{3.69}$$

Suppose the linear function is Lorentz-type, and $d\Omega = \sin\theta d\theta d\phi$, then Eq. (3.69) can be written as follows:

$$\mathcal{I}\,(\theta,\ \phi) \propto \int_{\phi=0}^{2\pi} \int_{\theta=0}^{\pi} \frac{\sin\theta d\theta d\phi}{[H - H_r(\theta,\ \phi)]^2 + \Delta H} \tag{3.70}$$

The integral of this equation is very complex; fortunately, it has already been solved by computer [7] as shown in Figure 3.7: (a), integral spectrum; (b) first-order differential spectrum; (c) the EMR experimental spectrum of CO_2^- on the surface of MgO powder [8], the left peak belongs to different centers.

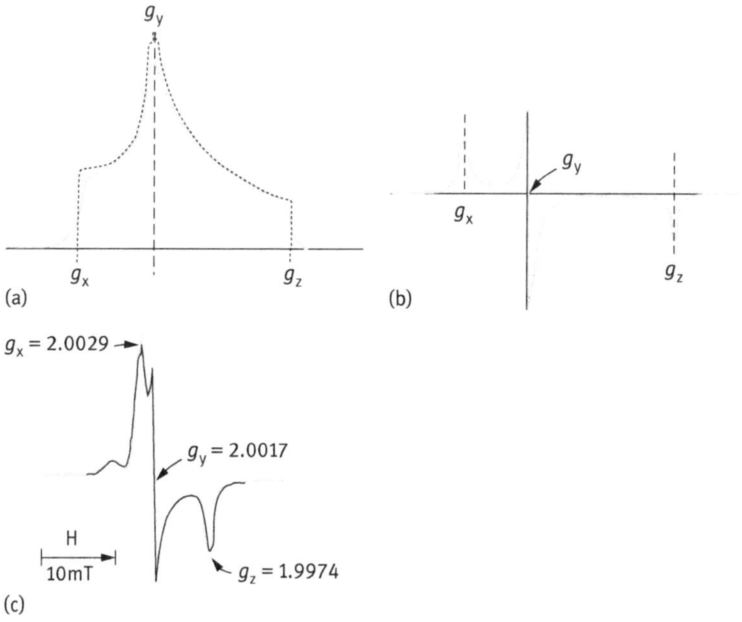

Figure 3.7: (a) Theoretical absorption spectrum of nonaxisymmetric powder sample. (b)First-order differential spectrum. (c) Experimental spectrum.

How to determine g_x, g_y, g_z ? Usually, we first determine the most strong among the three as g_y; from the tensor ellipsoid analysis, the probability maximum is also along the y-aixs. g_x and g_z are on both sides of g_y. Generally, the peak relatively far from the g_y is defined as g_z, and the near one as g_x. It could be seen from a large number of experimental results that the axial symmetric molecules are quite common in the samples of irregular orientation. Some molecules should be nonaxisymmetric strictly, but the g_x and g_y are very close in the spectrum, and very similar to the spectrum of axial symmetric system. That is why, the peak of near g_y is defined as the peak g_x [9].

A lot of literatures have reported about the computer simulations and related information of EMR spectra of powder samples [10–14]. With advances in computer technology and the popularity of graphical digitization, the spectral inherent parameters (e.g., g_x, g_y, g_z as well as line shape, linewidth, and line strength) have a special software package to be applied for routine operation [13,15].

References

[1] Wertz J. E., Auzins P., Griffiths J. H. E., Orton J. W. *Faraday Disc. Chem. Soc.* 1959, **28**: 136.
[2] Delbecq C. J., Hutchinson E., Schoemaker. D., Yasaitis E. L. Yuster P. H. *Phys. Rev.* 1969, **187**: 1103.
[3] Patten F. W., Keller F. J. *Phys. Rev.* 1969, **187**: 1120.
[4] Abraham M. M., Boatner L. A., Aronson M. A. *J. Chem. Phys.* 1986, **85**: 1.
[5] Barrow G. M. *Physical Chemistry*, 2nd edn, McGraw-Hill. New York, NY, 1966, p. 803.
[6] Ibers J. A., Swalen J. D. *Phys. Rev.* 1962, **127**: 1914.
[7] Swalen J. D., Gladney H. M. *IBM J.* 1964, **8**: 515.
[8] Lunsford J. H., Jayne J. P., *J. Phys. Chem.* 1965, **69**: 2182.
[9] Weil J. A., Hecht H. G., *J. Chem. Phys.* 1963, **38**: 281.
[10] Taylor P. C., Baugher J. F., Kriz H. M. *Chem. Rev.* 1975, **75**: 203.
[11] Van Veen G. *J. Magn. Reson.* 1978, **30**: 91.
[12] Siderer Y., Luz Z. *J. Magn. Reson.* 1980, **37**: 449.
[13] DeGray J .A., Reiger P. H. *Bull. Magn. Reson.* 1986, **8**: 95.
[14] Bernhard W. A., Fouse G. W. *J. Magn. Reson.* 1989, **82**: 156.
[15] She M., Chen X., Yu X. *Can. J. Chem.* 1989, **67**: 88.

Further reading

[1] Poole C. P., Jr., Farach H. A. *Theory of Magnetic Resonance*, John Wiley & Sons, Inc., New York, NY, 1987.
[2] Pryce M. H. L. *Paramagnetism in Crystals* (Lecture I, International School of Physics), *Nuovo Cimento*, 1957, **6** (Suppl.): 817.
[3] Pake G. E., Estle T. L. *The Physical Principles of Electron Paramagnetic Resonance*, 2nd ed., Benjamin, Reading, MA, 1973.

4 Isotropic hyperfine structure

Abstract: This chapter begins with the mechanism of hyperfine interaction and then discusses the interpretation of isotropic hyperfine spectrum, the calculation method of the hyperfine coupling constant of organic π-radical, and the number of spectral lines, splitting space, the intensity ratio, and their relationship with the probability density distribution of unpaired electron.

4.1 Theoretical exploration of hyperfine interaction

4.1.1 Dipole–dipole interaction

It has been mentioned briefly in Section 2.6 of Chapter 2 that the magnetic dipoles of unpaired electrons interact with the magnetic dipoles of those very near to its magnetic nuclei. This implies that dipole–dipole interaction generate hyperfine splitting.

If electron magnetic dipole and nuclear magnetic dipole are aligned along the external magnetic field H (// Z-axis) direction, the approximate expression of dipole–dipole interaction energy between them is as follows:

$$E_{\text{dipolar}} = -\frac{\mu_0}{4\pi}\frac{3\cos^2\theta - 1}{r^3}\mu_{N_z}\mu_{e_z} = -H_{\text{local}}\mu_{e_z} \tag{4.1}$$

where μ_{N_z} and μ_{e_z} are the components along the z-direction of nuclear dipole moment and electron dipole moment, respectively. As shown in Figure 4.1, r is the distance between these two dipoles, and θ is the angle between external magnetic field and the joining line of two dipoles. It can be seen from Eq. (4.1) and Figure 4.1 that the magnitude and direction of local magnetic field on the electron caused by nucleus are dependent on the angle θ and vector $\boldsymbol{r}$. The electronic magnetic dipoles on the various different positions of space, the magnitude and direction to be acted by the local magnetic field $\boldsymbol{H}_{\text{local}}$ that arise from nuclear magnetic dipoles, are all different. Since electron is not localized at the same fixed position in space, the effect value of H_{local} should be the average value of all appearing positions of electron in the space. For unpaired electron in s-orbital (as hydrogen atom), the appearance probability of all angle of θ are equal. Then $\cos^2\theta$ is the required average for a spherome:

$$\langle\cos^2\theta\rangle = \frac{\int_0^{2\pi}\int_0^{\pi}\cos^2\theta\sin\theta\,\mathrm{d}\theta\,\mathrm{d}\varphi}{\int_0^{2\pi}\int_0^{\pi}\sin\theta\,\mathrm{d}\theta\,\mathrm{d}\varphi} = \frac{1}{3} \tag{4.2}$$

Comparing with Eq. (4.1), if $H_{\text{local}} = 0$, then $E_{\text{dipolar}} \to 0$. It is not confirmed with experimental results, because the hyperfine splitting of EMR experimental spectrum of the hydrogen atom has 50.97 mT (when the microwave frequency is 9.5 GHz).

https://doi.org/10.1515/9783110568578-004

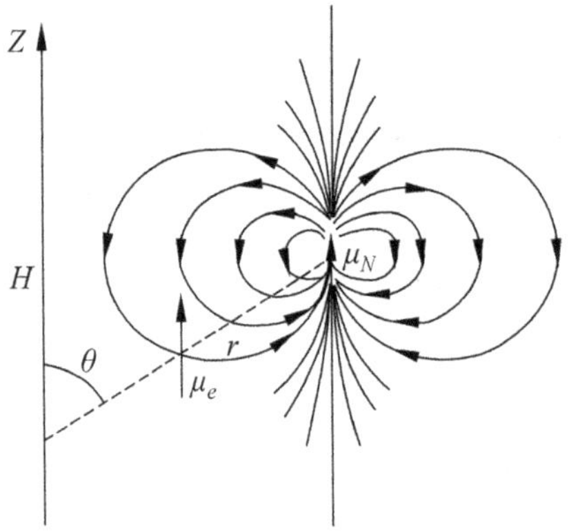

Figure 4.1: Electronic magnetic dipole μ_{e_z} and nuclear magnetic dipole μ_{N_z} in the external magnetic field.

That is to say, there must be some other interactions contributed to hyperfine splitting apart from dipole–dipole interaction. On the contrary, the dipole–dipole interaction is not equal to zero as long as the unpaired electron is not in s-orbital, because the probability density of unpaired electron appearance at nucleus is zero for the electron in p-, d-, f-orbitals. The hyperfine structure is mainly caused by the dipole–dipole interaction between unpaired electron and magnetic nucleus for the unpaired electron in the orbital nonspherical symmetric spin system.

4.1.2 Fermi contact interaction

From Eq. (4.1), when $r \to 0$, then $E_{\text{dipolar}} \to \infty$. For s-electron, the probability density at nucleus is not equal to 0; that is to say, the state of $r \to 0$ exists. However, $E_{\text{dipolar}} \to \infty$ is unreasonable obviously. Previously, interpretation of this problem was very knotty. This has been discussed by Skinner and Weil [1]. Another question that would emerged here is the same: Has some interaction other than dipole–dipole interaction contributes to the hyperfine structure?

Since the spatial distribution of s-electron is spherically symmetric as well as isotropic, Fermi has given an approximate expression for the isotropic magnetic interaction energy ($\boldsymbol{E}_{\text{iso}}$) between electron and nucleus [2]:

$$E_{\text{iso}} = -\frac{2\mu_0}{3}|\psi(0)|^2\mu_{N_z}\mu_{e_z} \tag{4.3}$$

where $\mu_0 = 4\pi \times 10^{-7}\,\text{J}\,\text{C}^{-2}\,\text{s}^2\,\text{m}^{-1}$, the vacuum permeability. When an external magnetic field is large enough, $\psi(0)$ is the electronic wave function at the nucleus of the center, and $|\psi(0)|^2$ is the probability density of electron at the nucleus. The ground-state wave function of hydrogen atom is given as

$$\psi_{1s}(r) = \left(\frac{1}{\pi r_b^3}\right)^{1/2} \exp\left(-\frac{r}{r_b}\right) \tag{4.4}$$

Here r_b is the first Bohr orbital radius (52.9×10^{-9} mm). By means of Eq. (4.3) with probability density $|\psi(0)|^2 = 1/\pi r_b^3$, the isotropic magnetic interaction energy E_{iso} could be obtained.

4.2 Energy operator of isotropic hyperfine interaction

4.2.1 Spin operator and hamiltonians

For an energy-level discrete system that could be described by definite quantum number, eigenequation can always be written. If $\hat{\Lambda}$ represents the energy operator, then the eigenequation of system is

$$\hat{\Lambda}\psi_k = \lambda_k \psi_k \tag{4.5}$$

Here, λ_k represents the eigenvalue of eigenfunction ψ_k of the state (k).

Since the spin angular momentum is quantized, for the system of electron spin $S = \frac{1}{2}$, there are two states characterized by the quantum numbers of $M_s = \pm\frac{1}{2}$. $\hat{S}_z$ is a component of the spin angular momentum operator $\hat{S}$ along the external magnetic field (Z); its eigenequation is

$$\hat{S}_z\phi_e = M_s\phi_e \tag{4.6}$$

Here M_s is the eigenvalue of the eigenfunction $\phi_e(M_s)$ of electron, acted upon by angular momentum operator $\hat{S}_z$.

Assuming $\alpha(e) = \phi_e(M_s = +\frac{1}{2})$ and $\beta(e) = \phi_e(M_s = -\frac{1}{2})$, we can obtain

$$\hat{S}_z\alpha(e) = +\frac{1}{2}\alpha(e) \tag{4.7a}$$

$$\hat{S}_z\beta(e) = -\frac{1}{2}\beta(e) \tag{4.7b}$$

Please note that the unit of angular momentum is $\hbar$.

For the system with nuclear spin $I = \frac{1}{2}$, there are two states characterized by quantum numbers $M_I = \pm\frac{1}{2}$. $\hat{I}_z$ is the component of nuclear spin angular momentum operator $\hat{I}$ along the external magnetic field (Z). Their eigenequations are

$$\hat{I}_z\alpha(n) = +\frac{1}{2}\alpha(n) \tag{4.8a}$$

$$\hat{I}_z\beta(n) = -\frac{1}{2}\beta(n) \tag{4.8b}$$

ψ_k is represented by Dirac mark $|k\rangle$; then Eqs. (4.7) and (4.8) can be written as follows:

$$\hat{S}_z|\alpha(e)\rangle = +\frac{1}{2}|\alpha(e)\rangle \tag{4.9a}$$

$$\hat{S}_z|\beta(e)\rangle = -\frac{1}{2}|\beta(e)\rangle \tag{4.9b}$$

and

$$\hat{I}_z|\alpha(n)\rangle = +\frac{1}{2}|\alpha(n)\rangle \tag{4.10a}$$

$$\hat{I}_z|\beta(n)\rangle = -\frac{1}{2}|\beta(n)\rangle \tag{4.10b}$$

The energy E_k of electron spin angular momentum M_s and nuclear spin angular momentum M_I of system can be calculated from the time-independent Schrödinger equation:

$$\mathscr{H}_e\phi_{ek} = E_{ek}\,\phi_{ek} \tag{4.11}$$

$$\mathscr{H}_n\phi_{nk} = E_{nk}\,\phi_{nk} \tag{4.12}$$

Here the Hamiltonian operators $\mathscr{H}_e$ and $\mathscr{H}_n$ are the energy operators of electron and nucleus (which can be replaced by $\hat{S}_z$ or $\hat{I}_z$). The subscript k represents any one of the eigenstates of the system. It is important that the component of the spin angular momentum operator along the direction z of external magnetic field in Eqs. (4.9) and (4.10), together with the energy operators of Eqs. (4.11) and (4.12), has common eigenfunction ϕ_k. Thus,

$$\mathscr{H}_e|\alpha(e)\rangle = E_{\alpha(e)}|\alpha(e)\rangle \tag{4.13a}$$

$$\mathscr{H}_e|\beta(e)\rangle = E_{\beta(e)}|\beta(e)\rangle \tag{4.13b}$$

and

$$\mathscr{H}_n|\alpha(n)\rangle = E_{\alpha(n)}|\alpha(n)\rangle \tag{4.14a}$$

$$\mathscr{H}_n|\beta(n)\rangle = E_{\beta(n)}|\beta(n)\rangle \tag{4.14b}$$

Generally, using a special and simple form, $\mathscr{H}$ is expressed. The Hamiltonian operator of system usually consists of two parts: one is the existing positions of whole particles (the spatial part) and magnetic moments, and another is their intrinsical angular momenta (the spin part). Since the Hamiltonian operator contains spin operators, it (in quantum mechanical state space) inevitably must be represented by the matrix.

4.2.2 Zeeman interaction of electrons and nuclei

For system with $S = \frac{1}{2}$ and $I = \frac{1}{2}$, the interaction between electron or nucleus and the external magnetic field is treated by means of spin operator method. Selecting the z-axis as the direction of the external magnetic field H, the electron magnetic moment operator $\hat{\mu}_{ez}$ is proportional to the electron spin operator $\hat{S}_z$. Similarly, the nuclear magnetic moment operator $\hat{\mu}_{nz}$ is proportional to the nuclear spin operator $\hat{I}_z$. Thus, referring to the related equations in Chapter 2, it gives

$$\hat{\mu}_{ez} = \gamma_e \hat{S}_z \hbar = -g\beta_e \hat{S}_z \tag{4.15}$$

$$\hat{\mu}_{nz} = \gamma_n \hat{I}_z \hbar = +g_n\beta_n \hat{I}_z \tag{4.16}$$

We can derive the electron and nuclear spin Hamiltonian operators from Eqs. (4.15) and (4.16):

$$\mathcal{H}_e = +g\beta_e H\hat{S}_z \tag{4.17}$$

$$\mathcal{H}_n = -g_n\beta_n H\hat{I}_z \tag{4.18}$$

Now application of the spin Hamiltonian operator in Eqs. (4.17) and (4.18) to the spin eigenfunction has the following results:

$$\mathcal{H}_e|\alpha(e)\rangle = +g\beta_e H\hat{S}_z|\alpha(e)\rangle = +\frac{1}{2}g\beta_e H|\alpha(e)\rangle \tag{4.19a}$$

and

$$\mathcal{H}_e|\beta(e)\rangle = +g\beta_e H\hat{S}_z|\beta(e)\rangle = -\frac{1}{2}g\beta_e H|\beta(e)\rangle \tag{4.19b}$$

Similarly,

$$\mathcal{H}_n|\alpha(n)\rangle = +g_n\beta_n H\hat{I}_z|\alpha(n)\rangle = -\frac{1}{2}g_n\beta_n H|\alpha(n)\rangle \tag{4.20a}$$

and

$$\mathcal{H}_n|\beta(n)\rangle = -g_n\beta_n H\hat{I}_z|\beta(n)\rangle = +\frac{1}{2}g_n\beta_n H|\beta(n)\rangle \tag{4.20b}$$

We can infer from Eqs. (4.19) and (4.20) that

$$E_{\alpha(e)} = +\frac{1}{2}g\beta_e H \tag{4.21a}$$

$$E_{\beta(e)} = -\frac{1}{2}g\beta_e H \tag{4.21b}$$

and

$$E_{\alpha(n)} = -\frac{1}{2}g_n\beta_n H \tag{4.22a}$$

$$E_{\beta(n)} = +\frac{1}{2}g_n\beta_n H \tag{4.22b}$$

Thus,

$$\Delta E_e = E_{\alpha(e)} - E_{\beta(e)} = g\beta_e H = h\nu_e \tag{4.23}$$

$$\Delta E_n = E_{\beta(n)} - E_{\alpha(n)} = g_n\beta_n H = h\nu_n \tag{4.24}$$

Resonance Eq. (4.23) corresponds to transition between the states $|\beta(e)\rangle$ and $|\alpha(e)\rangle$ (EMR transition), and resonance Eq. (4.24) (when $g_n > 0$) corresponds to transition between the states $|\alpha(n)\rangle$ and $|\beta(n)\rangle$ (NMR transition). Here, $h\nu_e$ and $h\nu_n$ are the photon energies of the stimulating electron and nuclear transition.

The procedure of determining energy E is given as follows: multiplication of both sides from left by the conjugate eigenfunction ϕ_k^* gives

$$\phi_k^*\mathcal{H}_e\phi_k = \phi_k^* E_k\phi_k$$

$$\phi_k^*\mathcal{H}_e\phi_k = E_k\phi_k^*\phi_k \tag{4.25}$$

Since here E_k is a constant, both sides are multiplied by $\mathrm{d}\tau$ and integrated over the variable τ (τ is a spatial variable), which give

$$\int_\tau \phi_k^*\mathcal{H}_e\phi_k d\tau = E_k\int_\tau \phi_k^*\phi_k \mathrm{d}\tau \tag{4.26}$$

Moving term,

$$E_k = \frac{\int_\tau \phi_k^*\mathcal{H}_e\phi_k \mathrm{d}\tau}{\int_\tau \phi_k^*\phi_k \mathrm{d}\tau} \tag{4.27}$$

If the function ϕ_k is normalized,

$$\int_\tau \phi_k^*\phi_k \mathrm{d}\tau = 1 \tag{4.28}$$

Thus,

$$E_k = \int_\tau \phi_k^* \mathscr{H}_e \phi_k \mathrm{d}\tau \tag{4.29}$$

It can be said that the expectation value of $\langle \mathscr{H} \rangle$ is the energy E_k of the system in its kth state.

Using the Dirac symbol of Eqs. (4.13) and (4.14) to rewrite Eqs. (4.25)–(4.29) and multiplying the left-hand side with ϕ_k^* and rewriting it as $\langle \phi_k|$, Dirac marked it as *bra*, with Dirac symbol *ket* $|\phi_k\rangle$ consisting of an integration $\langle \phi_k|\phi_k\rangle$ of full range, Eq. (4.26) can be written as

$$\langle \phi_k|\mathscr{H}|\phi_k\rangle = E_k\langle \phi_k|\phi_k\rangle \tag{4.30}$$

For normalized function Eq. (4.28),

$$\langle \phi_k|\phi_k\rangle = 1 \tag{4.31}$$

and Eq. (4.29) could be written as

$$E_k = \langle \phi_k|\mathscr{H}|\phi_k\rangle \tag{4.32}$$

We can see that the energy E_k is the kth diagonal element of matrix $\mathscr{H}$. The electronic and nuclear spin states can be written as

$$E_{\alpha(e)} = \langle \alpha(e|\mathscr{H}|\alpha(e)\rangle = +\frac{1}{2}g\beta_e H \tag{4.33a}$$

$$E_{\beta(e)} = \langle \beta(e)|\mathscr{H}|\beta(e)\rangle = -\frac{1}{2}g\beta_e H \tag{4.33b}$$

and

$$E_{\alpha(n)} = \langle \alpha(n)|\mathscr{H}|\alpha(n)\rangle = -\frac{1}{2}g_n\beta_n H \tag{4.34a}$$

$$E_{\beta(n)} = \langle \beta(n)|\mathscr{H}|\beta(n)\rangle = +\frac{1}{2}g_n\beta_n H \tag{4.34b}$$

4.2.3 Spin hamiltonian of isotropic hyperfine interaction

Let us now consider the effects of the isotropic hyperfine interaction. The anisotropic hyperfine interaction will be discussed in the next chapter. The isotropic spin Hamiltonian operator can be obtained by replacing the classical magnetic moment in Eq. (4.3) by the corresponding operator:

$$\mathscr{H}_{\mathbf{iso}} - \frac{2\mu_0}{3} g\beta_e g_n\beta_n|\psi(0)|^2\hat{S}_z\hat{I}_z \tag{4.35}$$

Setting a hyperfine coupling constants A_o,

$$A_o = \frac{2\mu_o}{3} g\beta_e g_n\beta_n|\psi(0)|^2 \tag{4.36}$$

It can be used to measure the magnetic interaction energy (Joule) between the electron and the nucleus. Hence, Eq. (4.35) becomes

$$\mathscr{H}_{\mathbf{iso}} = A_o\hat{S}_z\hat{I}_z \tag{4.37}$$

Hyperfine coupling constant can be expressed as A_o/h, and the common unit of frequency is MHz. It may also be expressed in magnetic field units and is called the hyperfine splitting constant $a_o = A_o/g_e\beta_e$. For the hydrogen atom and other isotropic systems with one electron and one magnetic nucleus $(I = \frac{1}{2})$, the spin Hamiltonian operator is obtained by adding Eqs. (4.17), (4.18), and (4.37):

$$\mathscr{H} = g\beta_e H\hat{S} - g_n\beta_n H\hat{I}_z + A_o\hat{S}_z\hat{I}_z \tag{4.38}$$

This equation is right when the external magnetic field is sufficiently large. If more than one magnetic nucleus interacts with electron, then

$$\mathscr{H} = g\beta_e H\hat{S} - \sum_i g_{ni}\beta_n\mathscr{H}\hat{I}_{zi} + A_{oi}\hat{S}_z\hat{I}_{zi} \tag{4.39}$$

When the hyperfine interaction term is relatively large (e.g., the hydrogen atom), the contribution of the nuclear Zeeman energy in the second term of Eqs. (4.38) and (4.39) is relatively small and even negligible. But in the case of anisotropic system it is not so.

4.3 Spectral isotropic hyperfine structure

In the study of hyperfine structure of EMR spectrum, the interesting problems are the number of lines, the intensity ratio of lines, the space between the lines (splitting space), and so on. In the case of lines not overlapped, the peaks should be of equal strength, and the splitting space should only be related with the hyperfine coupling parameters $\mathscr{A}$.

The number of spectral lines $(N = 2I + 1)$ is related to the spin quantum number I of the magnetic nucleus; the number of the magnetic nuclei n, and other factors; the intensity ratio of lines will be related to the number of lines and there are or no overlap between them; the splitting space is related to magnetic interactions between the nucleus and unpaired electron. These will be discussed separately as follows.

4.3.1 System with one magnetic nucleus and one unpaired electron

4.3.1.1 Hyperfine structure spectrum of system with $I = \frac{1}{2}$ and $S = \frac{1}{2}$

Hydrogen atom is the simplest model of hyperfine interaction; due to eigenvalue of the operator, $\hat{S}_z$, $\hat{I}_z$ are $M_S = \pm\frac{1}{2}$, $M_I = \pm\frac{1}{2}$, respectively. There are four possible combinations of their eigenfunctions, which can be expressed by *ket* vector as follows:

$$|\alpha(e)\alpha(n)\rangle \quad |\alpha(e)\beta(n)\rangle \quad |\beta(e)\alpha(n)\rangle \quad |\beta(e)\beta(n)\rangle$$

Using the spin operator, $\hat{S}_z$ and $\hat{I}_z$ act on these four eigenfunctions, respectively:

$$\hat{S}_z|\alpha(e)\alpha(n)\rangle = +\frac{1}{2}|\alpha(e)\alpha(n)\rangle \tag{4.40a}$$

$$\hat{I}_z|\alpha(e)\beta(n)\rangle = -\frac{1}{2}|\alpha(e)\beta(n)\rangle \tag{4.40b}$$

The remaining six are left to readers as exercises to solve by themselves.

The energies of these states according to Eqs. (4.33) and (4.34) can be written as follows:

$$\begin{aligned} E_{\alpha(e)\alpha(n)} &= \langle\alpha(e)\alpha(n)|\mathscr{H}|\alpha(e)\alpha(n)\rangle \\ &= \langle\alpha(e)\alpha(n)|g\beta_e H\hat{S}_z - g_n\beta_n H\hat{I}_z + A_0\hat{S}_z\hat{I}_z + \ldots + |\alpha(e)\alpha(n)\rangle \\ &= +\frac{1}{2}g\beta_e H - \frac{1}{2}g_n\beta_n H + \frac{1}{4}A_0 + \ldots \end{aligned} \tag{4.41a}$$

It could else be obtained as follows:

$$E_{\alpha(e)\beta(n)} = +\frac{1}{2}g\beta_e H + \frac{1}{2}g_n\beta_n H - \frac{1}{4}A_0 + \cdots \tag{4.41b}$$

$$E_{\beta(e)\alpha(n)} = -\frac{1}{2}g\beta_e H - \frac{1}{2}g_n\beta_n H - \frac{1}{4}A_0 + \cdots \tag{4.41c}$$

$$E_{\beta(e)\beta(n)} = -\frac{1}{2}g\beta_e H + \frac{1}{2}g_n\beta_n H + \frac{1}{4}A_0 + \cdots \tag{4.41d}$$

Above ellipses (...) represent the advanced items (it will be discussed later), the energy of these ignored advanced items is called the first-order energy. For a system $S = I = \frac{1}{2}$, the energy problem as a function of the field H has mathematically been solved by Breit and Rabi [3]. Equation (4.41) gives the solution of the first term that can be expanded as an infinite series. Figure 4.2a shows that the EMR transition can be observed at a moderately high intensity of magnetic field by scanning the frequency ν.

These quantitative expressions of energy levels are valid for sufficiently high external magnetic field H. We note that each nucleus has a doublet splitting with equal intensity. We also note that the subenergy levels (M_I) in the lower set is reversed as compared with the upper set. This is because the angular momentum of the electron should be changed with one unit ($\hbar$), but the angular momentum of the nucleus is not changed when a photon is absorbed. In the extreme cases, $H = 0$, the energy-level splitting caused by the hyperfine interaction still remains; so the zero

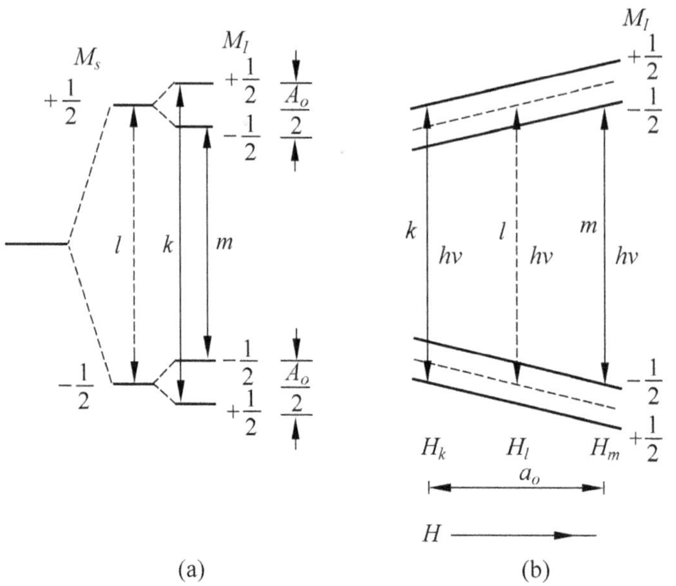

Figure 4.2: Energy levels of a system with a single unpaired electron and one nucleus with $I = 1/2$ (i.e., a hydrogen atom). (a) The dashed line would be transition corresponding to $hv = g\beta_e H$ without the hyperfine interaction (A_o) at a sufficiently high fixed magnetic field. The solid lines marked k and m correspond to the transitions of $hv = g\beta_e H \pm A_o/2$; here A_o is the isotropic hyperfine coupling constant. (b) As the function of external magnetic field, the dashed line corresponds to the transition with the assumption $A_o = 0$. The solid lines k and m refer to transitions induced by a constant microwave quantum of hv, which is the same energy for the transition l. Here values of the resonant magnetic field are given by $H = hv/g\beta_e \mp (g_e/g)a_o$. Hence, $(g_e/g)a_o$ is the hyperfine splitting constant, and it is almost given by $H_m - H_k$ (in mT). The above figures correspond to the system with positive g_n and A_o, such as hydrogen atom.

field transition at specific frequency with suitable radio frequency field H_1 could be observed. A notable example is the 1420 MHz emission of hydrogen atom in outer space.

The energy level differences of two EMR allowed transitions are as follows:

$$\Delta E_1 = E_{\alpha(e)\alpha(n)} - E_{\beta(e)\alpha(n)} = g\beta_e H + \frac{1}{2}A_o + \ldots \tag{4.42a}$$

$$\Delta E_2 = E_{\alpha(e)\beta(n)} - E_{\beta(e)\beta(n)} = g\beta_e H - \frac{1}{2}A_o + \ldots \tag{4.42b}$$

Note: The nuclear Zeeman terms are cancelled out.

Next we examine EMR transition under following two conditions:

1. *Keep the magnetic field H as a constant:* When $A_o = 0$ and the frequency is swept, a single transition occurs at $v = g\beta_e H/h$ (see the dashed line ℓ transition

in Figure 4.2a). For nonzero hyperfine interaction, transitions should occur at two frequencies: ν_k and ν_m (see transitions marked k and m solid lines in Figure 4.2a):

$$\nu_k = \left(g\beta_e H + \frac{1}{2}A_0 + \ldots\right)/h \quad \left(M_I = +\frac{1}{2}\right) \tag{4.43a}$$

$$\nu_m = \left(g\beta_e H - \frac{1}{2}A_0 + \ldots\right)/h \quad \left(M_I = -\frac{1}{2}\right) \tag{4.43b}$$

Note: Each of these two transitions occurs in the identical M_I. It implies that the selection rules for EMR absorption is $\Delta M_S = \pm 1$, $\Delta M_I = 0$.

2. *Keep the microwave frequency* ν *as a constant:* When $A_0 = 0$ and the magnetic field is swept, a single transition occurs at $H = h\nu/g\beta_e$ (see the dashed line in Figure 4.2b). If $A_0 \neq 0$, the EMR transitions occur at H_k and H_m (see the solid lines in Figure 4.2b).

$$H_k = h\nu/g\beta_e - A_0/2g\beta_e + \cdots \left(M_I = +\frac{1}{2}\right) \tag{4.44a}$$

$$H_m = h\nu/g\beta_e + A_0/2g\beta_e + \cdots \left(M_I = -\frac{1}{2}\right) \tag{4.44b}$$

The resonant equation becomes

$$h\nu = g\beta_e H + A_0 M_I + \cdots = g\beta_e[H + (g_e/g)a_0 M_I] + \cdots \tag{4.45}$$

where $a_0 = A_0/g_e\beta_e$ is called the hyperfine splitting constant (in magnetic field units); (g_e/g) represents the chemical shift correction. The first-order hyperfine splitting parameter is $(g_e/g)a_0$. For the vast majority of free radicals, g is very close to g_e, so (g_e/g) can be seen as 1 for radical systems.

In this type of chemical system, other magnetic resonance (i.e., NMR) transitions should also occur. There will be two pure NMR transition lines, if the electron spin orientation remains unchanged, but the nuclear spin orientation flips. This is very interesting for us. The electron-nuclear double-resonance (ENDOR) experiments can be realized if low appropriate excitation magnetic fields can be applied simultaneously. The biggest advantage of this technology is that the spectrum is simplified. For all the existing unpaired electronic system of nuclear spin, this technique can be easily applied to analyze the spectra and measure the hyperfine structure parameters (see Appendix A. 1.2).

4.3.1.2 Hyperfine structure spectrum of systems with $I = 1$ and $S = \frac{1}{2}$

^2H (deuterium) atom is the simplest example for system with $I = 1$ and $S = \frac{1}{2}$. As mentioned in the preceding section, the energy level can be calculated by Hamiltonian operator. There are six spin states represented by $|M_S M_I\rangle$ as follows:

$$|+1/2,-1\rangle \quad |-1/2,-1\rangle$$
$$|+1/2,0\rangle \quad |-1/2,0\rangle$$
$$|+1/2,-1\rangle \quad |-1/2,+1\rangle$$

Their first-order energies are written by the analogous method of Eq. (4.41):

$$\begin{aligned} E_{+1/2,+1} &= \tfrac{1}{2}g\beta_e H - g_n\beta_n H + \tfrac{1}{2}A_o, & E_{-1/2,-1} &= -\tfrac{1}{2}g\beta_e H - g_n\beta_n H + \tfrac{1}{2}A_o, \\ E_{+1/2,0} &= \tfrac{1}{2}g\beta_e H, & E_{-1/2,-1} &= -\tfrac{1}{2}g\beta_e H \\ E_{+1/2,-1} &= \tfrac{1}{2}g\beta_e H + g_n\beta_n H - \tfrac{1}{2}A_o, & E_{-1/2,+1} &= -\tfrac{1}{2}g\beta_e H - g_n\beta_n H - \tfrac{1}{2}A_o, \end{aligned} \tag{4.46}$$

According to the selection rules $\Delta M_s = \pm 1$ and $\Delta M_I = 0$, as described in Figure 4.3a, there are three allowed EMR transitions. With the increasing magnetic field, a typical first-derivative spectrum is shown in Figure 4.3b. The middle three lines as shown in Figure 4.4 are the spectrum of deuterium atom trapped in the crystalline quartz [4].

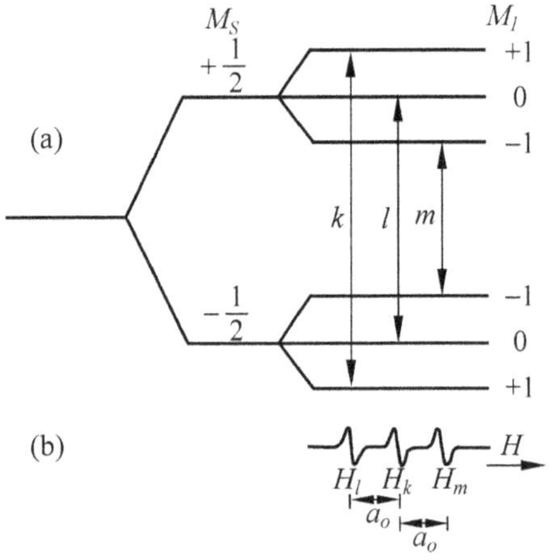

Figure 4.3: (a) Energy levels and allowed EMR transitions at constant magnetic field for an $S=\frac{1}{2}$, $I=1$ atom (e.g., deuterium); here $A_o > 0$. (b) Constant frequency spectrum.

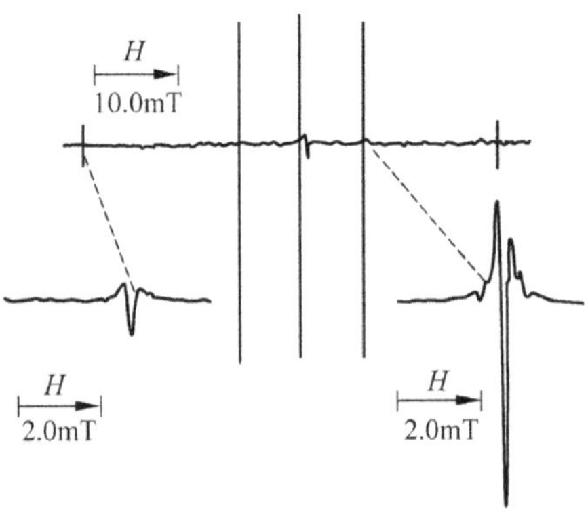

Figure 4.4: EMR spectrum of isotopically enriched atomic hydrogen trapped in X-irradiate α-quartz at 95 K. The outer lines arise from ^1H, and the three middle lines, from ^2H.

For $M_I = +1, 0, -1$, in the case of constant microwave frequency, the resonant magnetic fields for transitions in first approximation are as follows:

$$H_k = (hv/g\beta_e) - (g_e/g)a_o, \quad H_l = (hv/g\beta_e) \; H_m = (hv/g\beta_e) + (g_e/g)a_o \tag{4.47}$$

Since all states are nondegenerate and have no coincidence, these three lines are of equal intensity. It can easily be extended to system with $S = \frac{1}{2}$ and $I > 1$. For system with $I = \frac{3}{2}$, four lines of equal intensity are observed. In general, for the interaction between a single nucleus and a single unpaired electron, there are lines of equal intensity, and the spacing of the adjacent lines is a_o.

Here only the interaction between a single electron and one magnetic nucleus is involved. In most radicals, an electron usually interacts with several nuclei. This will be discussed in the next section.

4.3.1.3 The sign of isotropic hyperfine coupling parameter

The sign of the hyperfine coupling parameter determines the energy level order of the zero field. For $A_o > 0$ (e.g., atomic hydrogen), the triplet state lies above the singlet; but for $A_o < 0$, the situation is reversed. Here the EMR spectrum is unaffected by the sign of A_o. However, at the situation of sufficiently low temperature and magnetic field ($H > 0$), NMR spectra show the impact of the sign of A_o, since one of two lines would be of lower intensity.

For the simple case of one-electron atom, the sign of hyperfine coupling parameter A_o (as Eq. (4.36)) is determined by the sign of g_n. The sign of A_o implies that the arrangement of electronic and nuclear magnetic moments is parallel or antiparallel.

Note: $\mathbf{A}_o$ is an attribute of the spin system and does not depend on the direction and magnitude of external magnetic field.

For multielectron systems, the interaction between *unpaired* and *paired* electrons should also be taken into consideration. That is, the outer layer unpaired electron can cause *inner layer paired electrons* to exhibit *unpaired property*, the so-called *spin polarization*, making the inner electron parallel or antiparallel to the *outer unpaired electron*. In molecule, some polarization in local regions is possible. The *net electron spin polarization around any nucleus may be affected by the sign of* A_o.

4.3.2 Multimagnetic nuclei with one unpaired electron system

4.3.2.1 A set of equivalent magnetic nuclei with one unpaired electron system

A single unpaired electron ($S = \frac{1}{2}$) interacting with a proton ($I = \frac{1}{2}$) (as hydrogen atom) has two equal intensity lines, which has already been discussed previously, and the energy levels are shown in Figure 4.2. Here we will discuss the system with more than two equivalent protons. By equivalent, we mean that the magnitude of the energy levels causing splitting is equal (i.e., the hyperfine splitting parameter A of each proton is equal). *Note: A* is a very important spectral parameter determining the magnitude of hyperfine splitting (i.e., splitting space).

An unpaired electron interacting with two protons should split into four lines. Since these two protons are equivalent, *A* is equal (i.e., the splitting space are equivalent), and the middle two lines hence overlap according to the selection rules $\Delta M_s = \pm 1$; $\Delta M_I = 0$. Thus, only three lines appear, and the middle line intensity is twice the other two lines. The intensity ratio of these lines is 1:2:1, as shown in Figure 4.5b.

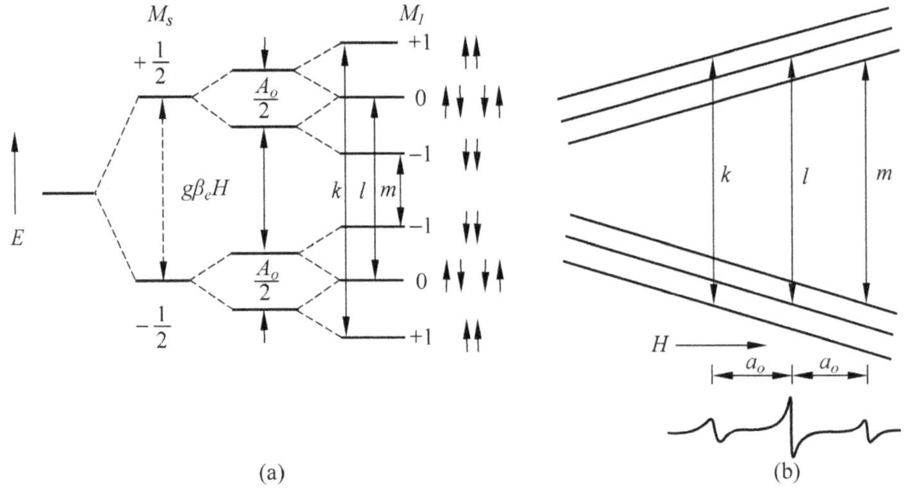

Figure 4.5: (a) Hyperfine splitting energy levels for a system with two equivalent protons. (b) EMR spectrum at constant frequency and swept magnetic field.

Similarly, the EMR spectrum of three equivalent protons consists of four hyperfine structural lines with equal spacing and their intensity ratio is 1:3:3:1. By analogy, the EMR spectrum of four equivalent protons should have five hyperfine structural lines with equal spacing and their intensity ratio should be 1:4:6:4:1. For *n* equivalent protons, there should be a set of $2nI + 1$ hyperfine lines with equal spacing. Their intensity ratio is the coefficient ratio of the binomial $(a + b)^n$ expansion, as shown in Table 4.1. Their simulated EMR spectra are shown in Figure 4.6.

Benzene anion radical has six equivalent protons. The X-band EMR experimental spectrum [5] of benzene anion radical in a solution of 2:1 tetrahydrofuran and dimethoxyethane at 173 K is shown in Figure 4.7. Spectral lines are due to the weak periphery without electrons and ^{13}C nuclear hyperfine interactions generated.

The very weak satellite lines are caused by the interaction between the unpaired electron and ^{13}C nucleus.

Table 4.1: Relation between the number of protons, hyperfine structural lines, and the intensity ratios.

Protons(n)	Lines $2nI + 1$	Intensity ratio
0	1	1
1	2	1 1
2	3	1 2 1
3	4	1 3 3 1
4	5	1 4 6 4 1
5	6	1 5 10 10 5 1
6	7	1 6 15 20 15 6 1
7	8	1 7 21 35 35 21 7 1
8	9	1 8 28 56 70 56 28 8 1
⋮	⋮	....

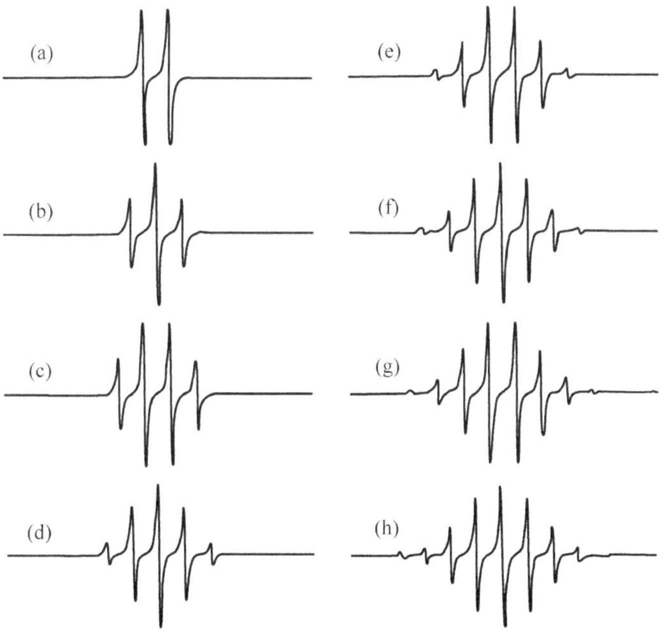

Figure 4.6: Computer-simulated EMR spectra for an unpaired electron interacting with (a) one, (b) two, (c) three, (d) four, (e) five, (f) six, (g) seven, and (h) eight equivalent nuclei. $v = 9.50$ GHz; $H_r = 339$ mT; $a_o = 0.5$ mT; and the Lorentzian peak-to-peak linewidths ΔH_{pp} are 0.05 mT.

4.3.2.2 Multiple sets of equivalent magnetic nuclei with an unpaired electron system

Multiple sets of equivalent protons imply that there are more than two groups of equivalent protons. The protons within a group are equivalent (there are same A),

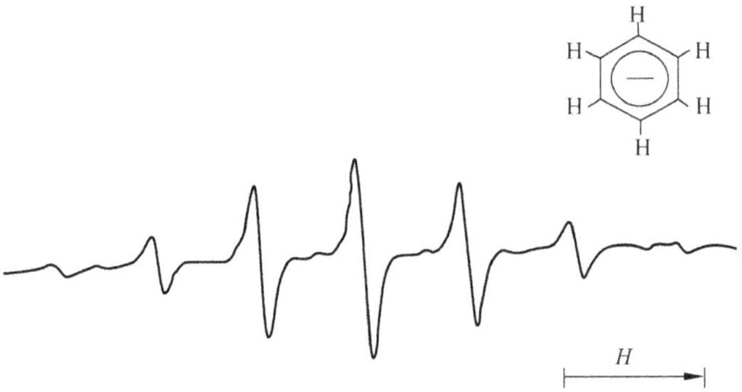

Figure 4.7: EMR spectrum of the benzene anion radical in solution at 173 K [5].

and the protons of different groups are not equivalent (i.e., $A_1 \neq A_2 \neq \cdots$). The simplest example is glycolic acid radical HOCHCOOH that has two sets of equivalent protons, and each set has only one proton. The unpaired electron is located on the orbital carbon atom of CH. The proton nearest to this C atom is the hydrogen atom of CH. The unpaired electron interacting with this proton causes the line split into two hyperfine structural lines with equivalent intensity. Then the proton on HOC causes these two lines split into four lines. Since this proton is separated by an oxygen atom with C atom, its hyperfine interaction energy is weaker than the proton of CH, and the splitting spacing is relatively smaller. So its g = 2.0038; $|a_{H(CH)}| = 1.725\,mT$; $|a_{H(HOC)}| = 0225\,mT$. The experimental spectrum of glycolic acid radical in aqueous solution with pH = 1.3 at 298 K is shown in Figure 4.8 [6].

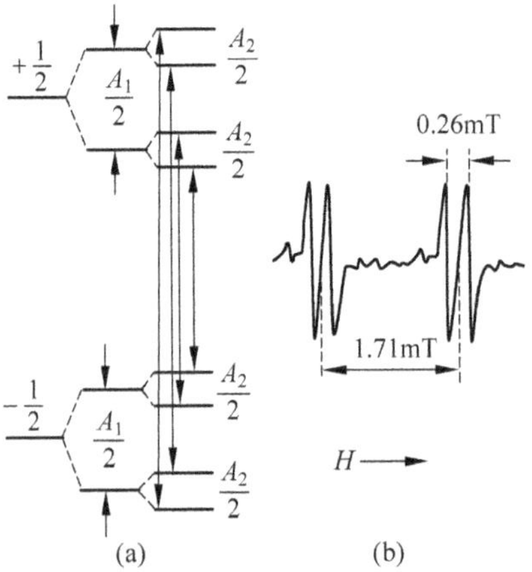

Figure 4.8: The EMR spectrum of glycolic acid radical [6].

Let us now consider methanol radical CH_2OH (generated by photolysis of methanol in H_2O_2 solution). It also has two sets of equivalent protons: one set has two equivalent protons on CH_2, while the other has only one proton on OH. The unpaired electron interacting with the two equivalent protons of CH_2 give rise to three lines with intensity ratio of 1:2:1; since the proton of OH with the unpaired electron causes hyperfine interaction, each of these three lines would split into two lines with equal intensity, as shown in Figure 4.9 [7]: $|a_{H(CH_2)}| = 1.738\,\text{mT}$; $|a_{H(OH)}| = 0.115\,\text{mT}$. In general, if there are two sets of equivalent protons in one molecule, one with m equivalent protons and the other with n equivalent protons, the maximum number of hyperfine lines should be $(m+1)(n+1)$. For N sets of equivalent proton systems, the maximum number of hyperfine lines is $\Pi_j(n_j+1)$, where $j = 1, 2, 3, \ldots, N$.

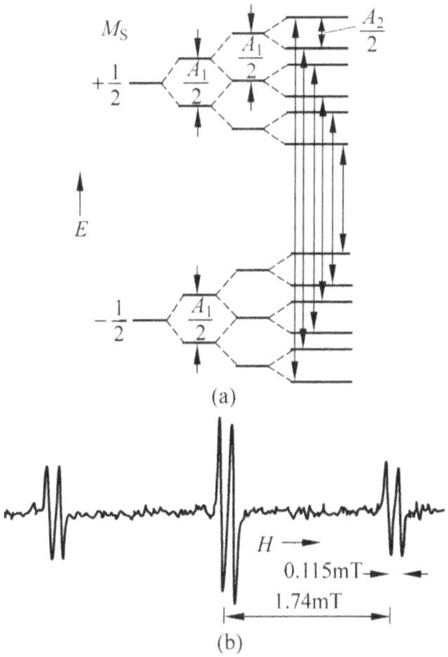

Figure 4.9: EMR spectrum of the methanol radical with an interaction between the unpaired electron and the two sets of equivalent protons [7].

Let us again see the 1,3-butadiene anion radical $(H_2C{=}CH{-}CH{=}CH_2)^{-}$. It has two sets of equivalent protons: one set has four equivalent protons, while the other set has two equivalent protons. The total number of its hyperfine lines should be $(4+1)(2+1) = 15$; four equivalent protons split into five hyperfine lines with intensity ratio of 1:4:6:4:1; and the other set of two equivalent protons split into three hyperfine lines with intensity ratio of 1:2:1. Its EMR spectrum [8] is shown in Figure 4.9. Figure 4.9a is the X-band EMR experimental spectrum of 1,3-butadiene anion radical obtained from electrolysis of 1,/3-butadiene in liquid ammonia at 195 K and Figure 4.9b is its theoretical stick spectrum. Here $|a_1| = |a_4| = 0.762\,\text{mT}$, where $|a_1| = 0.762\,\text{mT} > 2|a_2| = 0.558\,\text{mT}$, the 15 lines are represented very clearly. Under most situations, the lines cross over or even overlap.

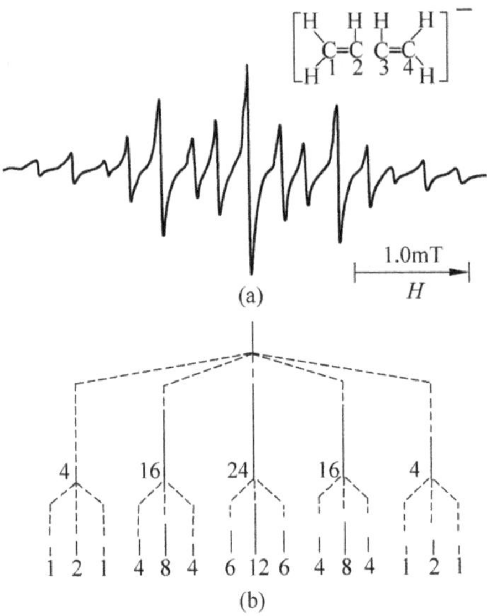

Figure 4.10: (a) *X*-band EMR spectrum of 1,3-butadiene anion radical at 195 K [8]. (b) Theoretical stick diagram.

Next we consider the naphthalene anion radical; it has two sets nonequivalent protons, and each set has four equivalent protons. There should be 25 hyperfine lines. Although it has a little overlap, the 25 lines still could be discerned. The EMR spectrum of naphthalene anion radical is generated by naphthalene reaction with metal potassium in dimethoxyethane solvent, and EMR spectrum [9] is measured at 298 K, as shown in Figure 4.11. It is the first EMR spectrum of radical for which the

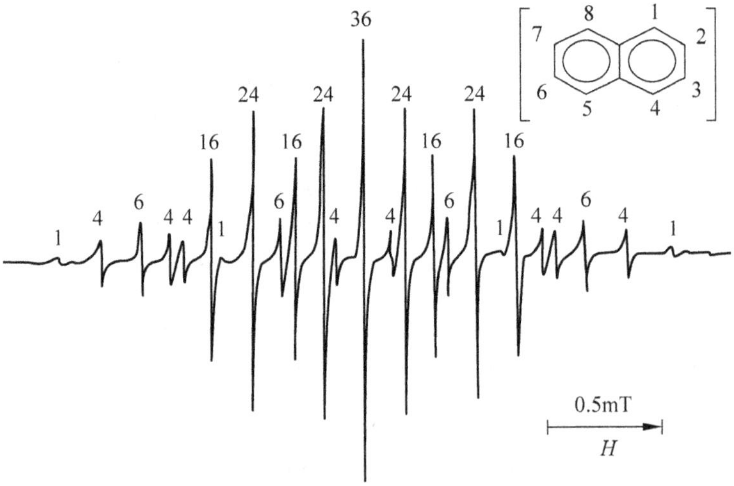

Figure 4.11: EMR spectrum of naphthalene radical anion at 298 K.

hyperfine splitting of proton has been observed. The absolute value of the hyperfine splitting parameters is $|a_1| = 0.495$ mT; $|a_2| = 0.187$ mT [10].

Anthracene anion radical is generated by anthracene reaction with metal potassium in the dimethoxyethane solvent. It has three sets of equivalent protons: 1, 4, 5, 8 is a set of (α proton); 2, 3, 6, 7 is a set of (β proton); and 9, 10 is a set of (γ proton). The total number of hyperfine lines should be (4+1) (4+1) (2+1) = 75. Here $|a_\alpha| = 0.273$ mT; $|a_\beta| = 0.151$ mT; $|a_\gamma| = 0.534$ mT. The EMR spectrum [11] is shown in Figure 4.12.

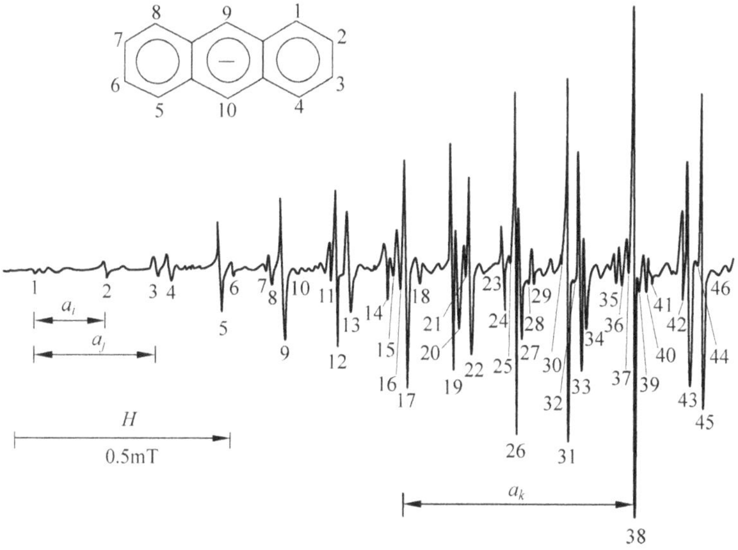

Figure 4.12: Low-field portion of the EMR spectrum of anthracene anion radical in dimethoxyethane at 295 K [11].

4.3.3 Hyperfine splitting arising from other magnetic nuclei

The magnetic nucleus with $I = \frac{1}{2}$ as hydrogen nucleus (proton) has been discussed previously. Here we would like to introduce the hyperfine splitting of non-proton magnetic nucleus with $I = \frac{1}{2}$ as well as magnetic nucleus with $I > \frac{1}{2}$.

4.3.3.1 Hyperfine splitting arising from the non-proton magnetic nucleus with $I = \frac{1}{2}$

The most common nuclei with $I = \frac{1}{2}$ in organic radical are ^{1}H, ^{13}C, ^{19}F, and ^{31}P. We have already made a detailed discussion about ^{1}H nucleus. The hyperfine splitting of ^{19}F and ^{31}P is usually difficult to distinguish from that of proton. In this case, the nuclear spin number and the number of hyperfine splitting lines to analyze spectrum are not enough. It should provide other evidences to identify the interacting nuclei. The methods mentioned previously for the analysis and reconstruction of spectra involving proton splitting are also applicable to nuclei of ^{19}F and ^{31}P. For nucleus ^{19}F, the cross-variation of hyperfine linewidths, in some instances, could be used to make an assignment [12].

The natural abundance of both ^{19}F and ^{31}P nuclei are 100%. With the element of more than one isotope and the existing number being large enough, the known nuclear spin amount and abundance can also be used generally to compare the strength of the hyperfine splitting of spectral lines to interpret spectrum.

Hyperfine splitting of ^{19}F nucleus has been observed in many organic radicals, such as perfluoro-*p*-benzosemiquinone [13]. CF_3 radical is a very interesting example [14], because it has been a controversial issue whether it is planar or pyramidal. The ^{13}C hyperfine splitting lines observed have been helpful in resolving this problem, and the pyramidal configuration is regarded as more reasonable. The EMR spectrum of CF_3 radical in liquid C_2F_6 at 110 K is shown in Figure 4.13, where the splitting of ^{19}F is 14.45 mT and the second-order hyperfine splitting is also observed. The middle of doublet arises from splitting of FO_2 [14].

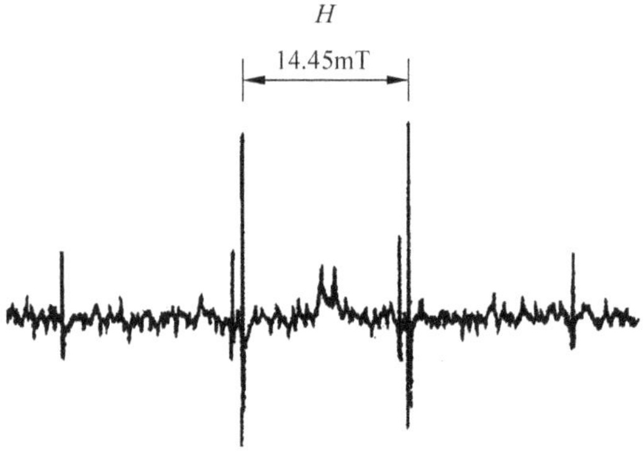

Figure 4.13: EMR spectrum of CF_3 radical in liquid C_2F_6 at 110 K [14].

PO_3 = radical is an example of revealing ^{31}P hyperfine splitting. This radical [15] has a very large isotropic hyperfine splitting (~60 mT). It expresses that the P atom of PO_3^- is sp^3 hybrid with a pyramid structure. If PO_3^- is a planar structure with sp^2 hybrid, it should show a smaller isotropic hyperfine splitting.

Hyperfine splitting of ^{13}C nucleus: The natural abundance of ^{13}C is only 1.11%, and ^{12}C (natural abundance 98.89%) is a nonmagnetic nucleus ($I = 0$). So the ^{13}C hyperfine splitting could only be measured by means of a spectrometer with high sensitivity. Because $^{12}CO_2^-$ radical has only one line while $^{13}CO_2^-$ has two lines, its intensity is only 100×(1.11/98.89)/2 = 0.561% of the $^{12}CO_2^-$ radical. For a molecule containing n equivalent carbon atoms, the intensity of satellite line caused by ^{13}C is relatively 0.00561n times the main line strength. As benzene radical anion (Figure 4.7), the intensity of satellite line caused by ^{13}C relative to that of the center line is 3.37%.

4.3.3.2 Hyperfine splitting arising from magnetic nuclei of $I > \frac{1}{2}$

^{2}H and ^{14}N are the most commonly encountered magnetic nuclei with $I = 1$. According to $N = 2I + 1$, each nucleus should be split into three lines of equal intensity. Frémy's salt $K_2NO(SO_3)_2$ is an example of $S = \frac{1}{2}$, $I = 1$. Solid Frémy's salt gives no EMR signal because the unpaired electrons are paired in the lattice. When it is dissolved in water (concentration ~0.005 M), three hyperfine splitting lines with equal intensity of ^{14}N nucleus at room temperature should occur; also, the low natural abundance of the isotopes ^{15}N, ^{33}S, ^{17}O produced weak satellite lines [16] as shown in Figure 4.14. It could be used as a good intensity standard for the sensitivity of EMR spectrometer and as a scale of calibration magnetic field. It has also aided the ENDOR techniques [17] to explore the mechanism of spin relaxation.

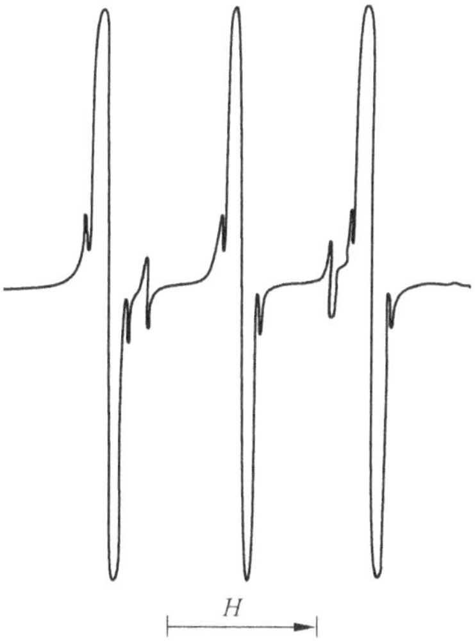

Figure 4.14: EMR spectrum of Frémy's salt in an aqueous solution with concentration of 0.005 M at room temperature [17].

For two equivalent nuclei with $I = 1$, there should be five hyperfine structural EMR lines with intensity ratio of 1:2:3:2:1. Nitronylnitroxide radical can be used as an example of hyperfine splitting for two equivalent ^{14}N nuclei. Figure 4.15 is the EMR spectrum of nitronylnitroxide in benzene solution at 295 K [18].

Another typical example is 2,2-diphenyl-1-picrylhydrazyl for short call DPPH. It is a stable organic radical, and usually used as a standard sample to calibrate the *g*-value of unknown samples and its radical concentration. The two N atoms of the hydrazino group can be regarded as equivalent. Therefore, in the solution, it also exhibits five lines hyperfine structure spectrum with 1:2:3:2:1 ratio of intensity distribution. In a very dilute deoxygenated solution, even the proton hyperfine splitting can be observed.

For two sets of different equivalent nuclei with equal number, such as naphthalene anion, there are four equivalent α-protons and four equivalent β-protons that

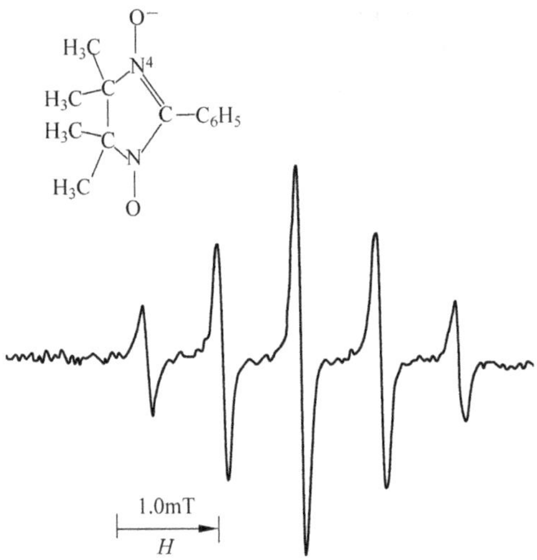

Figure 4.15: EMR spectrum of nitronylnitroxide radical in benzene solution at 295 K.

cannot be identified to analyze spectrum. If the α-protons is substituted partially by ^{2}H, it could be identified [19].

For the system of n equivalent nuclei with I = 1 interacting with an unpaired electron, the quantities and intensity ratios of their spectral lines of hyperfine structure are listed in Table 4.2.

Table 4.2: The number of equivalent nuclei with I = 1, the amount, and intensity ratio of the hyperfine structural lines.

Number of equivalent (n)	Amount of lines $2nI + 1$	Intensity ratio of the lines
0	1	1
1	3	1 1 1
2	5	1 2 3 2 1
3	7	1 3 6 7 6 3 1
4	9	1 4 10 16 19 16 10 4 1
5	11	1 5 15 30 45 51 45 30 15 5 1
⋮	⋮	⋮

In addition, there are many I = 3/2 nuclei, such as ^{7}Li, ^{11}B, ^{23}Na, ^{35}Cl, ^{37}Cl, ^{39}K, ^{53}Cr, ^{63}Cu, ^{65}Cu, and so on. Their hyperfine structural lines ought to have $2I$ + 1= 4 lines with equal intensity. Sometimes, the EMR spectra of anion radicals could exhibit hyperfine splitting of alkali–metal cations. A very interesting example is the pyrazine anion in dimethoxyethane solution; if it is reduced by the metal Na, its balance cation is $^{23}Na^+$, and then the EMR spectrum is composed of many equally intense quadruple

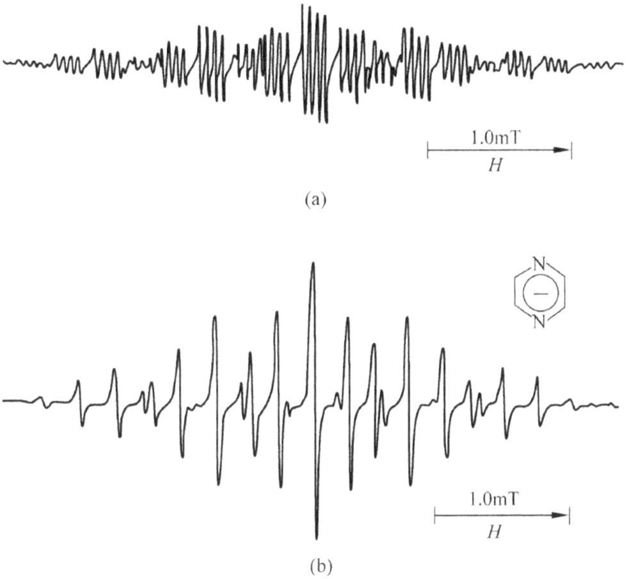

Figure 4.16: EMR spectrum of pyrazine anion radical in dimethoxyethane at 297 K.(a) Na^+ as the counterion [20]. (b) K^+ as the counterion [21].

lines [20], as shown in Figure 4.16a. If the reductant is metal K, its balance cation is $^{39}K^+$, and then the quadruple lines are replaced by 25 hyperfine structural lines caused by two equivalent ^{14}N nuclei and four equivalent protons, as shown in Figure 4.16b. Since the combination of Py–Na ion pair is more close than Py–K ion pair, the probability of the unpaired electron cloud appearing in the ^{23}Na nucleus is much larger than in ^{39}K nucleus, and the magnetic moment of ^{39}K nuclear is much smaller than that of the ^{23}Na nucleus.

4.3.4 Encountered problems in isotropic radical spectra

EMR spectra of the radicals mentioned in this chapter are all in liquid state or solution, the orientation of free radical is average to be isotropic due to the fast tumbling motion. In general, most of the free radicals (fragments) are asymmetrical, their g-factors and hyperfine splittings all are anisotropic. It should be known that hyperfine splitting parameter $(g_e/g)a$ is dependent on temperature and solvents.

Radical is closed usually in a hard "cage"; if the "cage" is a single crystal, we can get the greatest possible information from the EMR spectrum, since spectrum is a function of the orientation of single crystal in a magnetic field. If the radicals are located in a randomly oriented "cage," when the solution is frozen at low temperature, it is also possible to obtain sufficient information about their structure from the EMR spectrum.

Radicals in solvent of high viscosity still undergo some kind of reorientation, or some degree of rearrangement. The linewidths of whole spectrum would have a remarkable variation. We can obtain kinetic information from the analysis of such spectra.

4.4 Hyperfine structure of organic π-free radical spectrum

4.4.1 Hyperfine coupling constant of organic π-radical

We have already discussed that the spectra produced hyperfine splitting by the interaction of unpaired electrons with magnetic nuclei. We can predict the number and relative intensities of EMR spectral hyperfine structural lines, based on the symmetry of radical molecules and the number of magnetic nuclei. Then by comparing with experimental spectra, the value of hyperfine coupling constants a_i can be determined. In fact, this is only for a single nucleus or simple system. For complex multinuclei systems, it is not so simple. The first encountered problem relates to spectral line, such as naphthalene anion radical, which has four equivalent α-protons and four equivalent β-protons. We can measure $a_1 = 0.490$ mT, $a_2 = 0.183$ mT from the spectrum. However, we cannot decide which set of equivalent protons corresponds to a_1 or a_2 from the spectrum. Next, overlap between spectral lines often occurs in the complex system. Thus, the number of lines measured by experiment will be less than that of theoretical expectation. a_i value could not be measured from the experimental spectrum, and thus there is need to provide a method for calculating a_i theoretically.

The radicals could be divided into organic radicals, inorganic radicals, conductors and semiconductors, and so on. The organic radicals can also divide into π-radicals, σ-radical, biradical, triplet, and so on. The unpaired electrons lie on the π-orbital of the radical molecule and is called π-radicals, such as aromatic positive or negative ion radicals. They have complex hyperfine structure usually, since the unpaired electron in molecular π-electron orbital has a large delocalization, and with many magnetic nuclei in the molecule, hyperfine interactions can take place. Radicals with the unpaired electron in molecular σ-electron orbitals are called σ-radicals. This unpaired electron is commonly localized in the σ-orbital, which loses a hydrogen atom. It only has a hyperfine interaction with adjacent magnetic nuclei; thus, the hyperfine structure of its EMR spectrum would be relatively simple.

4.4.2 McConnell semiempirical formula

On the basis of a large number of experimental spectra of aromatic ion radicals, McConnell and Chesnut [22] put forward the relationship between the probability density (ρ_i) of the unpaired electron at the ith carbon atoms of π-radicals and the hyperfine splitting constant a_i of the proton in ith carbon atom as follows:

$$a_i = Q\rho_i \tag{4.48}$$

Here Q is the constant approximately under certain conditions, such as in the system of aromatic anion radicals.

$$Q = -2.25\,\text{mT or} - 63\text{MHz}$$

There are two basic assumptions for McConnell semiempirical formula:

(1) In π-radical system, the σ–π exchange interactions could be treated approximately by first-order perturbation.
(2) Energy of antibonding triplet state of C–H bond is much larger than the excitation energies of double and quadruple states of the π-electron.

Obtaining a_i from the EMR experimental spectra, this formula could be used to determine the probability density ρ_i of unpaired electron at the ith carbon atom. In fact, Q is a constant approximately, although, speaking strictly, it is not a constant. In the same series, there is a little difference between different molecules. Sometimes, the total width of spectrum can also be used as Q value. With regard to the Q value, it will be discussed in detail in Sections 4.4.5 and 4.5.4.

4.4.3 Hückel molecular orbital (HMO) theory

If the probability density ρ_i of unpaired electron at the ith carbon atom can be calculated theoretically, it could be related with the hyperfine splitting parameter of EMR spectrum and verified. In comparison to various different molecular orbital theories, Hückel molecular orbital theory (HMO) [23] is the simplest one.

Although the McConnell theory and HMO theory both are very impractical, considerable facts prove that they are still able to coincide with the objective reality. This is due to the fact that they have been able to assess the intrinsic nature and law of things. Therefore, they still play an important role in the interpretation of π-radical spectrum.

HMO theory can only be applied to planar conjugated molecules, that is, it can only be used to calculate the ρ_i value of π-radicals. The key points of HMO theory are as follows:

(1) The atomic orbitals 2s, $2p_x$, $2p_y$ of each carbon atom are hybridized to form three σ-bonds, with the 120 °angle mutually constituting planar molecular frame. The $2p_z$ orbital of carbon atom is perpendicular to the molecular plane, and the $2p_z$ orbitals of adjacent carbon atoms overlap to form Π bond. The electrons in Π bond are no longer localized on a fixed carbon atom; they rather distribute various carbon atoms of entire plane with a certain probability.

(2) Π bond has no relation to σ-bond, and the interaction between σ-electron and π-electron can be negligible.
(3) Molecular orbital ψ_i of π-electron location can be written as a linear combination of $2p_z$ atomic orbitals ϕ_j:

$$\psi_i = c_{i_1}\phi_1 + c_{i_2}\phi_2 + \cdots c_{in}\phi_n = \sum_{j=1}^{n} c_{ij}\phi_j \tag{4.49}$$

$$\sum_{j=1}^{n} |c_{ij}|^2 = 1 \tag{4.50}$$

$$\rho_i = |c_{ij}|^2 \tag{4.51}$$

(4) ψ_i is the eigenfunction of Hamiltonian operator $\mathcal{H}$ $\hat{\mathcal{H}}$, the corresponding eigenvalues are E_i, where $\mathcal{H}$ operator only considers the interaction between a single π-electron and the equivalent potential field established by nucleus and σ-electrons, and the interaction between π-electrons is excluded. The eigenequation is

$$\hat{\mathcal{H}}\psi_i = E_i\psi_i \tag{4.52}$$

(5) One can use the variational method for solving eigenequation to obtain energy of system. Using a trial function,

$$\psi = \sum_j c_j\phi_j \tag{4.53}$$

to calculate the system energy E,

$$E = \frac{\langle\psi|\hat{\mathcal{H}}|\psi\rangle}{\langle\psi|\psi\rangle} = \frac{\left\langle \sum_j c_j\phi_j \middle| \hat{\mathcal{H}} \middle| \sum_k c_k\phi_k \right\rangle}{\sum_j c_j\phi_j \middle| \sum_j c_k\phi_k} = \frac{\sum_j c_j^2\alpha_j + \sum_j\sum_k c_jc_k\beta_{jk}}{\sum_j c_j^2 + \sum_j\sum_k c_jc_kS_{jk}} \tag{4.54}$$

where $\alpha_j = \left\langle \phi_j \middle| \hat{\mathcal{H}} \middle| \phi_j \right\rangle$ is the coulomb integral;
$\beta_{jk} = \left\langle \phi_j \middle| \hat{\mathcal{H}} \middle| \phi_k \right\rangle$ is the exchange integral; and
$S_{jk} = \alpha_j = \langle\phi_j|\phi_j\rangle$ is the overlap integral, $S_{jk} = S_{kj}$.

If the trial function is the eigenfunction of $\hat{\mathcal{H}}$ indeed, the energy determined by Eq. (4.54) should be the lowest. Therefore, if we want to get the best approximate eigenfunction, their linear combination coefficients should be satisfied:

$$\frac{\partial E}{\partial c_i} = 0 \; (i = 1, 2, \ldots n) \tag{4.55}$$

(6) In order to further simplify the problem, Hückel assumes the following:
- All coulomb integrals are equal, that is, $\alpha_1 = \alpha_2 = \alpha_3 = \cdots = \alpha_n$.
- All exchange integrals are equal for adjacent atoms, that is, $\beta_1 = \beta_2 = \beta_3 = \cdots = \beta_n$.
- All exchange integrals of nonadjacent atoms are equal to zero, and all the overlap integrals are also zero.

4.4.4 Calculation of probability density distribution of unpaired electron

Taking naphthalene anion as an example, we try to calculate the molecular orbital of the unpaired electron and its energy using HMO theory. Let

$$\psi = \sum_{j=1}^{10} c_j \phi_j \tag{4.56}$$

$$E = \frac{\sum_{j=1}^{10} c_j^2 \alpha + 2\sum_{j<k}\sum_{k} c_j c_k \beta}{\sum_{j=1}^{10} c_j^2} \tag{4.57}$$

Thus,

$$E(c_1^2 + c_2^2 + \cdots c_{10}^2) = (c_1^2 + c_2^2 + \cdots c_{10}^2)\alpha + 2(c_1c_2 + c_2c_3 + \cdots c_{10}c_1)\beta \tag{4.58}$$

In order to make the trial function best approximate the true eigenfunction, the linear combination of coefficients must be satisfied with Eq. (4.55). Thus, 10 equations are obtained. Derivative of Eq. (4.55) with respect to c_1 on both sides gives

$$\frac{\partial E}{\partial c_1}(c_1^2 + c_2^2 + \cdots c_{10}^2) + 2Ec_1 = 2c_1\alpha + 2(c_2 + c_{10})\beta \tag{4.59}$$

Let $\frac{\partial E}{\partial c_1} = 0$. Thus,

$$c_{10}\beta + c_1(\alpha - E) + c_2\beta = 0 \tag{4.60}$$

Using Eq. (4.58) both sides with respect to $c_2, c_3, \ldots c_{10}$ for partial derivative and letting

$$\frac{\partial E}{\partial c_2} = \frac{\partial E}{\partial c_3} = \cdots = \frac{\partial E}{\partial c_{10}} = 0$$

we derive the following:

$$c_1\beta + c_2(\alpha - E) + c_3\beta = 0 \tag{4.61}$$

$$c_2\beta + c_3(\alpha - E) + c_4\beta = 0 \tag{4.62}$$

$$c_3\beta + c_4(\alpha - E) + c_5\beta = 0 \tag{4.63}$$

$$c_4\beta + c_5(\alpha - E) + c_6\beta = 0 \tag{4.64}$$

$$c_5\beta + c_6(\alpha - E) + c_7\beta = 0 \tag{4.65}$$

$$c_6\beta + c_7(\alpha - E) + c_8\beta = 0 \tag{4.66}$$

$$c_7\beta + c_8(\alpha - E) + c_9\beta = 0 \tag{4.67}$$

$$c_8\beta + c_9(\alpha - E) + c_{10}\beta = 0 \tag{4.68}$$

$$c_9\beta + c_{10}(\alpha - E) + c_1\beta = 0 \tag{4.69}$$

Equations (4.60)–(4.69) are a set of linear homogeneous equations. If $c_1, c_2, \ldots c_{10}$ were not at all zero solution, the essential condition would have been that the determinant consists of their coefficients that are equal to zero. Let

$$(\alpha - E)/\beta \equiv \omega \tag{4.70}$$

The diagonal elements of the determinant $d_{ii} = \omega$; nondiagonal elements with adjacent i and j $d_{ij} = 1$, nonadjacent i and j $d_{ij} = 0$. Using this rules, the secular equation of naphthalene is given by

$$\begin{vmatrix} \omega & 1 & & & & & & & 1 & \\ 1 & \omega & 1 & & & & & & & \\ & 1 & \omega & 1 & & & & & & \\ & & 1 & \omega & 0 & & & & & 1 \\ & & & 0 & \omega & 1 & & & & 1 \\ & & & & 1 & \omega & 1 & & & \\ & & & & & 1 & \omega & 1 & & \\ & & & & & & 1 & \omega & 1 & \\ 1 & & & & & & & 1 & \omega & 1 \\ & & & 1 & 1 & & & & 1 & \omega \end{vmatrix} = 0 \tag{4.71}$$

Expansion of Eq. (4.71) will give 10-degree equation of ω, and then 10 roots of ω can be obtained, $\omega_1, \omega_2, \ldots, \omega_{10}$. Corresponding to each ω_i, there should be a set of linear homogeneous equations and a ψ_i. If ψ_i satisfies the normalization condition,

$$\langle \psi_i \mid \psi_i \rangle = 1$$

then a set of coefficients $\{c_{ij}\}$ can be obtained. Thus,

$$\psi_i = \sum_{j=1}^{10} c_{ij}\phi_j$$

However, this is actually very cumbersome. We can simplify it by applying the symmetry principle. First, the naphthalene molecule is placed on a coordinate system, as shown in Figure 4.17.

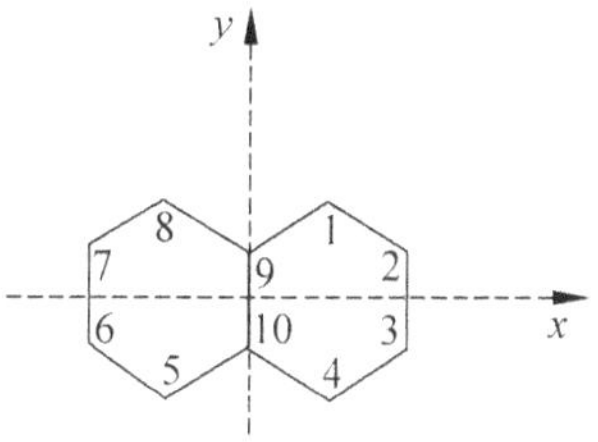

Figure 4.17: The naphthalene molecule is placed in the *xy* coordinate system.

According to the principle of symmetry, for the system with symmetry to *x*-axis, S_x, the coefficients should satisfy the following relationship:

$$c_1 = c_4;\ c_2 = c_3;\ c_5 = c_8;\ c_6 = c_7;\ c_9 = c_{10} \tag{4.72}$$

For the system with asymmetry to *x*-axis, A_x, the coefficients should satisfy the following relationship:

$$c_1 = -c_4;\ c_2 = -c_3;\ c_5 = -c_8;\ c_6 = -c_7;\ c_9 = -c_{10} \tag{4.73}$$

Similarly, for the system with symmetry to *y*-axis, S_y, the coefficients should satisfy the following relationship:

$$c_1 = c_8;\ c_2 = c_7;\ c_3 = c_6;\ c_4 = c_5;\ c_9 = c_{10} \tag{4.74}$$

For the system with asymmetry to *y*-axis, A_y, the coefficients should satisfy the following relationship:

$$c_1 = -c_8;\ c_2 = -c_7;\ c_3 = -c_6;\ c_4 = -c_5;\ c_9 = -c_9 \equiv 0;\ c_{10} = -c_{10} \equiv 0 \tag{4.75}$$

By combining these two kinds symmetries, we obtain the following:

$$S_xS_y\begin{cases} c_1 = c_4 = c_5 = c_8 \\ c_2 = c_3 = c_6 = c_7 \\ c_9 = c_{10} \end{cases} \qquad S_xA_y\begin{cases} c_1 = c_4 = -c_5 = -c_8 \\ c_2 = c_3 = -c_6 = -c_7 \\ c_9 = c_{10} = 0 \end{cases} \tag{4.76}$$

(c_1, c_2, c_9 are independent variables) (c_1, c_2 are independent variables)

$$A_xS_y\begin{cases} c_1 = -c_4 = -c_5 = c_8 \\ c_2 = -c_3 = -c_6 = c_7 \\ c_9 = -c_{10} \end{cases} \qquad A_xA_y\begin{cases} c_1 = -c_4 = -c_5 = -c_8 \\ c_2 = -c_3 = -c_6 = -c_7 \\ c_9 = -c_{10} = 0 \end{cases} \tag{4.77}$$

(c_1, c_2, c_9 are independent variables) (c_1, c_2 are independent variables)

Since energy level and wave function must belong separately to a certain symmetry, each of the three wave functions will belong to S_xS_y, A_xS_y symmetry, and each of the two wave functions will belong to S_xS_y, A_xA_y symmetry. The original 10 × 10 secular equation can be simplified into two 3 × 3 and two 2 × 2 subsecular equations to be solved easily. As for the S_xS_y, substitution of Eq. (4.76) into the original linear homogeneous equations provides three linear nonrelative equations:

$$\left.\begin{aligned} \omega c_1 + c_2 + c_9 = 0 \\ c_1 + (\omega + 1)c_2 = 0 \\ 2c_1 + (\omega + 1)c_9 = 0 \end{aligned}\right\} \tag{4.78}$$

If we want the solution of c_1, c_2, c_9 to be not at all zero, the coefficients of the determinant should be equal to zero. Thus,

$$\begin{vmatrix} \omega & 1 & 1 \\ 1 & \omega + 1 & 0 \\ 2 & 0 & \omega + 1 \end{vmatrix} = 0$$

Expanding gives

$$(\omega + 1)\,(\omega^2 + \omega - 3) = 0$$

$$\omega = -1; \quad \omega = \frac{-1 \pm \sqrt{13}}{2} = \begin{cases} -2.3028 \\ +1.3028 \end{cases} \tag{4.79}$$

Substituting into Eq. (4.70) gives

$$E = \alpha + \beta; \quad E = \alpha + 2.3028\beta; \quad E = \alpha - 1.3028\beta;$$

These ω values are substituted into Eq. (4.31), and under normalization condition,

$$4(c_1^2 + c_2^2) + 2c_9^2 = 1$$

$c_1, c_2, \ldots, c_9$ can be solved and the corresponding wave functions obtained. In the same way, E_i and ψ_i of other symmetry can be obtained and are all listed in Table 4.3.

From Table 4.3 we can see that they have the following two important properties:

(1) The energy levels have *pairing property*. If there is an energy level $\omega = -\omega_k$, there should also be an energy level $\omega = -\omega_k$. They are mutual mirror image, and the mirror face is $\omega = 0$. For example, ω_1 and ω_{10} are mutual *pairing energy levels*.
(2) The absolute value of the coefficient of linear combination of atomic obitals of their corresponding molecular obitals (HMO) ψ_k and ψ_{-k} of mutual pairing energy levels must be equal, and the sign of coefficient can be easily determined using symmetry.

Incidentally, a simple method has been introduced, by marking *to determine the sign of coefficients: For example, naphthalene molecules are marked * on No. 1

Table 4.3: Energy levels of naphthalene molecule and molecular orbitals ψ_i(HMO).

Symmetry	E_i	ψ_i(HMO)
S_xS_y	$E_1=\alpha+2.3028\beta$	$\psi_1=0.3005(\phi_1+\phi_4+\phi_5+\phi_8)+0.2307(\phi_2+\phi_3+\phi_6+\phi_7)+0.4614(\phi_9+\phi_{10})$
S_xA_y	$E_2=\alpha+1.6180\beta$	$\psi_2=0.2628(\phi_1+\phi_4-\phi_5-\phi_8)+0.4253(\phi_2+\phi_3-\phi_6-\phi_7)$
A_xS_y	$E_3=\alpha+1.3028\beta$	$\psi_3=0.3996(\phi_1-\phi_4-\phi_5+\phi_8)+0.1735(\phi_2-\phi_3-\phi_6+\phi_7)+0.3470(\phi_9-\phi_{10})$
S_xS_y	$E_4=\alpha+\beta$	$\psi_4=0.4083(\phi_1+\phi_4+\phi_5+\phi_8)+0.4083(\phi_9+\phi_{10})$
A_xA_y	$E_5=\alpha+0.6180\beta$	$\psi_5=0.4253(\phi_1-\phi_4+\phi_5-\phi_8)+0.2628(\phi_2-\phi_3+\phi_6-\phi_7)$
S_xA_y	$E_6=\alpha-0.6180\beta$	$\psi_6=0.4253(\phi_1+\phi_4-\phi_5-\phi_8)+0.2628(-\phi_2-\phi_3+\phi_6+\phi_7)$
A_xS_y	$E_7=\alpha-\beta$	$\psi_7=0.4083(\phi_1-\phi_4-\phi_5+\phi_8)+0.4083(\phi_9-\phi_{10})$
S_xS_y	$E_8=\alpha-1.3028\beta$	$\psi_8=0.3996(\phi_1+\phi_4+\phi_5+\phi_8)+0.1735(-\phi_2-\phi_3-\phi_6-\phi_7)+0.3470(-\phi_9-\phi_{10})$
A_xA_y	$E_9=\alpha-1.6180\beta$	$\psi_9=0.2628(\phi_1-\phi_4+\phi_5-\phi_8)+0.4253(-\phi_2+\phi_3-\phi_6+\phi_7)$
A_xS_y	$E_{10}=\alpha-2.3028\beta$	$\psi_{10}=0.3005(\phi_1-\phi_4-\phi_5+\phi_8)+0.2307(-\phi_2+\phi_3+\phi_6-\phi_7)+0.4614(-\phi_9+\phi_{10})$

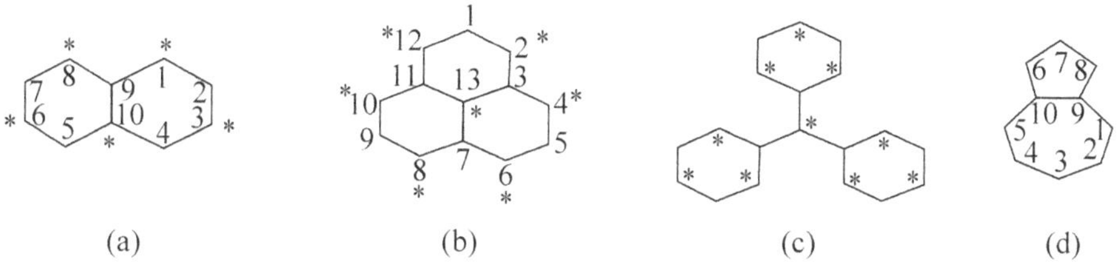

Figure 4.18: (a) Naphthalene. (b) Perinaphthene. (c) Triphenylmethyl. (d) Azulene.

carbon atom and not on the No. 2 carbon atom. At regular intervals the marking of * in the given order goes on, as shown in Figure 4.18a.

If the marked * carbon atom number and the nonmarked carbon atom number are equal, it is called *even alternant hydrocarbons*, such as naphthalene (Figure 4.18a). Conversely, if the marked * carbon number is not equal to the number of nonmarked carbon atom, it is called *odd alternant hydrocarbons*, such as perinaphthene (Figure 4.18b and triphenylmethyl (Figure 4.18c). There is also yet another kind that cannot be marked by *, such as azulene (Figure 4.18d), and this is called nonalternant hydrocarbons. For alternant hydrocarbons (whether odd or even), all can be solved very quickly by using symmetric principle.

Consider perinaphthene as an example. Perinaphthene is an odd alternant hydrocarbons, and the amount of its carbon atom is odd number, thus the energy level number also is odd. According to the pairing property of energy level, there should be a zero energy level (ω = 0). The corresponding molecular orbital ψ_o is called nonbonding orbital. This orbital can be obtained without solving secular equation.

For ψ_0, the coefficients of the unmarked carbon atomic orbitals c_i are zero. Thus,

$$c_1 = c_3 = c_5 = c_7 = c_9 = c_{11} = 0$$

In the linear homogeneous equations,

$$\omega c_1 + c_2 + c_{12} = 0$$

Since $c_1 = 0$, $c_2 = -c_{12}$; similarly, $c_4 = -c_6$; $c_8 = -c_{10}$, according to

$$c_2 + \omega c_3 + c_4 + c_{13} = 0$$

And since $c_3 = 0$,

$$c_2 + c_4 + c_{13} = 0$$
$$c_6 + c_8 + c_{13} = 0$$
$$c_{10} + c_{12} + c_{13} = 0$$

Summation of above three equations and applying $c_2 = -c_{12}$; $c_4 = -c_6$; $c_8 = -c_{10}$ gives $c_{13} = 0$, and

$$c_2 = -c_4 = c_6 = -c_8 = c_{10} = -c_{12}$$

Finally, using the normalization condition: $\sum_{i=1}^{13} c_i^2 = 1$, $c_2 = 1/\sqrt{6}$ is obtained, and thus

$$\psi_0 = (\phi_2 - \phi_4 + \phi_6 + \phi_8 + \phi_{10} - \phi_{12})/\sqrt{6} \tag{4.80}$$

Since the HMO is orthogonal normalized,

$$1 = \langle \psi_i | \psi_i \rangle = \sum_k \sum_s \langle c_{ik}\phi_k | c_{is}\phi_s \rangle = \sum_k \sum_s c_{ik}c_{is} \langle \phi_k | \phi_s \rangle = \sum_k \sum_s c_{ik}c_{is}\delta_{ks} = \sum_k c_{ik}^2 \tag{4.81}$$

It is thus clear that the physical meaning of c_{ik}^2 is the probability density or the density of electron cloud of the electron, which locate in ψ_i molecular orbital and appear at the kth carbon atom. In EMR, the focus is the molecular orbital that is occupied by unpaired electron, and this is called *frontier orbital*. Since the other bonding orbitals have all been filled with two electrons that spin in opposite directions according to Pauli principle, it has no contribution to EMR. Electron cloud density of frontier orbital is also called frontier π-electron density.

For even alternant hydrocarbons, the frontier orbitals are two adjacent HMO with $\omega = 0$; for example, the frontier orbital of naphthalene anion is ψ_6 and the frontier orbital of naphthalene cation is ψ_5 (see Table 4.1).

$$\psi_5 = 0.4253(\phi_1 - \phi_4 + \phi_5 - \phi_8) + 0.2628(\phi_2 - \phi_3 + \phi_6 + \phi_7)$$
$$\psi_6 = 0.4253(\phi_1 + \phi_4 + \phi_5 - \phi_8) + 0.2628(-\phi_2 - \phi_3 + \phi_6 + \phi_7)$$

For naphthalene anion, probability of the unpaired electron appearance at the first carbon atom is $c_{6,1}^2$. The value is

$$c_{6,1}^2 = (0.4253)^2 = 0.181$$

Similarly, probability of the unpaired electron appearance at the second carbon atom is $c_{6,2}^2$,

$$c_{6,2}^2 = (0.2628)^2 = 0.069$$

And probability of the unpaired electron appearance at the ninth carbon atom is $c_{6,9}^2$:

$$c_{6,9}^2 = (0)^2 = 0$$

With the frontier electron cloud density of HMO, ρ_i, we can use the McConnell formula to calculate the hyperfine coupling constant a_i. Taking naphthalene as an example, the experimental measures a_α and a_β are 0.495 and 0.187 mT, respectively; thus,

$$\frac{a_\alpha}{a_\beta} = \frac{0.495}{0.187} = 2.67$$

The ratio of electronic cloud density calculated from the HMO theory is

$$\frac{\rho_\alpha}{\rho_\beta} = \frac{0.1810}{0.0691} = 2.62$$

These two are very good coincidence indeed.

4.4.5 The Q value of the radical with fully symmetrical structure

The ρ_i value calculated by the HMO is able to reflect only approximately the objective reality. In order to test the reliability of the McConnell formula, it is necessary to avoid using the approximation of HMO. We have selected a number of molecules with fully symmetrical structures (see Table 4.4); their ρ_i can be obtained directly from the symmetry without any approximation, which has eliminated unreliability of ρ_i value. If the McConnell formula is accurate strictly, the Q value should be a constant.

From Table 4.4, we can see that the Q value is not a constant, and the deficiency of McConnell formula is also reflected. However, the Q values of two neutral radicals C_5H_5 and C_7H_7, as well as two anion radicals $C_6H_6^-$ and $C_8H_8^-$ are rather close. It means that the excess charge on the anion radical has effect on the Q value. Based on this concept, Colpa [30] and Bolton [31] and coworkers [32] proposed the amendments of McConnell

Table 4.4: Hyperfine coupling constant and *Q* value of radicals with fully symmetrical single ring structure.

Radical	Temperature (K)	a^H (mT)	*Q* value (mT)	References
C_5H_5	~200	0.600	3.00	[24]
$C_6H_6^-$	173	0.375	2.25	[25]
$C_6H_6^+$	298	0.428	2.57	[26]
C_7H_7	298	0.395	2.77	[27,28]
$C_8H_8^-$	~298	0.321	2.57	[29]

formula (Section 4.5.4). Despite this, through a vast amount of experimental facts trial, it has been proved that the McConnell formula still is a very useful formula.

4.4.6 Hyperfine coupling constant a value of the even alternant hydrocarbons

Combined with the application of HMO theory, the McConnell formula has achieved considerable success in the calculation of the hyperfine coupling constant ***a*** value. Since the frontier orbitals of the cations or anions of even alternant hydrocarbons are located on a couple orbital above or down the $\omega = 0$, according to the *pairing principle*, their linear combination coefficients are equal and the sign opposite. Hence, their electronic cloud density should be equal:

$$\rho_i^+ = \rho_i^- \tag{4.82}$$

Then, it can be conjectured that the EMR spectrum of cationic or anionic radicals of the even alternant hydrocarbon should be fully identical. The experimental results indicate that the two spectra are very similar, except for somewhat different Q value, caused by excess charge.

The total width of the spectral line should be the difference between the highest field H_{highest} and the lowest field H_{lowest}; thus,

$$H = H_0 - \sum_i a_i M_{Ii} \tag{4.83}$$

Then

$$\begin{aligned} H_{\text{highest}} &= H_0 + \tfrac{1}{2}\sum_i a_i \\ H_{\text{lowest}} &= H_0 - \tfrac{1}{2}\sum_i a_i \\ \left|H_{\text{highest}} - H_{\text{lowest}}\right| &= \sum_i a_i \end{aligned} \tag{4.84}$$

From the *McConnell formula*,

$$a_i = Q\rho_i$$

Summation of i on both sides gives

$$\sum_i a_i = \sum_i Q\rho_i = Q\sum_i \rho_i = Q$$

Because $\sum_i \rho_i = 1$,

$$Q = \sum_i a_i \tag{4.85}$$

So Q corresponds to the total width of the spectral line.

A large number of experimental results indicate that the total width of the anion radicals of aromatic hydrocarbons is always 2.8–3.0 mT. However, the total width of the spectral line of biphenyl anion radical is only 2.1 mT. The reason is the electron densities in 1 and 1′ are both 0.123 without proton and hyperfine splitting. Hence, the total width should be

$$2.8(1 - 2 \times 0.123) = 2.1\,\text{mT}$$

It should also coincide with the experimental results.

Now let us see the hyperfine coupling constant of benzene and its derivatives. EMR spectrum of benzene anion (reduction by metal potassium at −100 °C) has seven lines with intensity ratio of 1:6:15:20:15:6:1 and the total width of spectral lines is 2.25 mT. This indicates the six protons on benzene ring are equivalent, and the probability density of the unpaired electron on each carbon atom is 1/6. The energy levels E_i and molecular orbitals ψ_i calculated from HMO are listed in Table 4.5.

Table 4.5: The HMOs ψ_i and the corresponding energy levels E_i of benzene.

Symmetry	E_i	ψ_i
S_xS_y	$E_1 = \alpha + 2\beta$	$\psi_1 = \frac{1}{\sqrt{6}}(\phi_1 + \phi_2 + \phi_3 + \phi_4 + \phi_5 + \phi_6)$
A_xS_y	$E_2 = \alpha + \beta$	$\psi_2 = \frac{1}{\sqrt{12}}(2\phi_1 + \phi_2 - \phi_3 - 2\phi_4 - \phi_5 + \phi_6)$
S_xA_y	$E_3 = \alpha + \beta$	$\psi_3 = \frac{1}{\sqrt{4}}(\phi_2 + \phi_3 - \phi_5 - \phi_6)$
A_xA_y	$E_4 = \alpha - \beta$	$\psi_4 = \frac{1}{\sqrt{4}}(\phi_2 - \phi_3 + \phi_5 - \phi_6)$
S_xS_y	$E_5 = \alpha - \beta$	$\psi_5 = \frac{1}{\sqrt{12}}(2\phi_1 - \phi_2 - \phi_3 + 2\phi_4 - \phi_5 - \phi_6)$
A_xS_y	$E_6 = \alpha - 2\beta$	$\psi_6 = \frac{1}{\sqrt{6}}(\phi_1 - \phi_2 + \phi_3 - \phi_4 + \phi_5 - \phi_6)$

Its frontier orbital is degenerate with the energy equal to $\alpha - \beta$, the unpaired electron can be either in ψ_4 or in ψ_5:

$$\psi_4 = \frac{1}{\sqrt{4}}(\phi_2 - \phi_3 + \phi_5 - \phi_6) \qquad A_xA_y$$
$$\psi_5 = \frac{1}{\sqrt{12}}(2\phi_1 - \phi_2 - \phi_3 + 2\phi_4 - \phi_5 - \phi_6) \qquad S_xS_y$$

and the probability of two orbitals are equal. Since there is fast exchange between the S_xS_y and A_xA_y states, their average electron cloud density should be

$$\rho_1 = \frac{1}{2}\left(0 + \frac{4}{12}\right) = \frac{1}{6}$$

$$\rho_2 = \frac{1}{2}\left(\frac{1}{4} + \frac{1}{12}\right) = \frac{1}{6}$$

It is thus clear that the result is consistent with the electron cloud density calculated by the McConnell formula using experimental results.

Introducing of substituent groups can relieve degeneracy of the S_xS_y and A_xA_y states. For example, the hyperfine structure of the EMR spectrum of *p*-diethylbenzene radical has five lines with intensity ratio of 1:4:6:4:1 and a = 0.529 mT. Using McConnell formula, we have

$$a = 2.25 \times \frac{1}{4} = 0.56\ \text{mT}$$

These data are essentially in accord with experimental value.

Introducing not only a substituent group but also the hydrogen atom to be substituted by deuterium atom in benzene ring (C_6H_5D) can remove degeneracy [32]. However, since the chemical properties of H and D are very close, the degenerated E_4 and E_5 levels are still very near. Therefore, the unpaired electron is not always in the $\psi_4$orbital, while ψ_4 and ψ_5 orbitals are all occupied; the probability of unpaired electron in ψ_4 orbital is rather more than in ψ_5 orbital.

Suppose the probability of the unpaired electron occupying the ψ_5 orbital is x and that occupying ψ_4 orbital is $(1 - x)$. Hyperfine coupling constant measured from the experiments is a_4 = 0.341 mT; thus,

$$2.25\left[x \times \left\{(2)\left(\frac{1}{\sqrt{12}}\right)\right\}^2 + (1 - x) \times 0\right] = 0.341$$

Solving this equation gives $x = 0.45$, $(1-x) = 0.55$. Using this result we can calculate a_2:

$$a_2 = 2.25\left[0.45\left(\frac{1}{\sqrt{12}}\right)^2 + 0.55\left(\frac{1}{2}\right)^2\right] = 0.397$$

This is basically matched with experimental value of 0.392 mT.

The EMR spectrum of deuterated benzene-d_1 anion radical is shown in Figure 4.19. A set of seven lines with intensity ratio of 1:6:15:20:15:6:1 are contributed by the ^{13}C isotope.

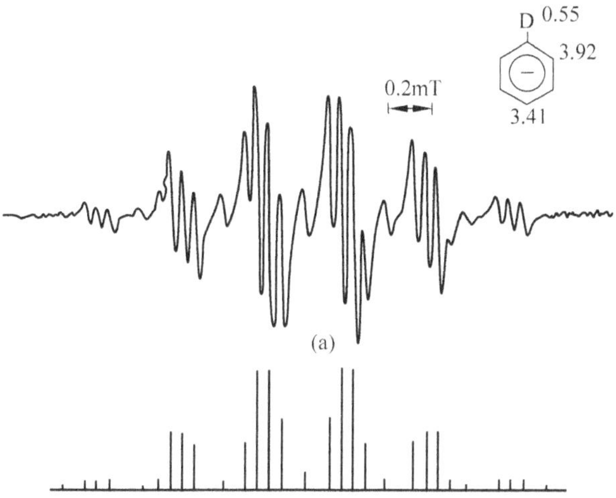

Figure 4.19: EMR spectrum of the deuterated benzene-d_1 anion radical $(C_6H_5D)^-$.

4.4.7 Hyperfine coupling constant a value of the even alternant heterocyclic hydrocarbons

Taking pyrazine as an example, let $\alpha_N = \alpha_c + \delta_N\beta$; here $\alpha_N = \langle\phi_N|\hat{\mathcal{H}}|\phi_N\rangle$; $\alpha_C = \langle\phi_C|\hat{\mathcal{H}}|\phi_C\rangle$. Due to the existence of induced effect, $\beta_{NC} = \langle\phi_N|\hat{\mathcal{H}}|\phi_C\rangle$ and $\beta_{CC} = \left\langle\phi_C'|\hat{\mathcal{H}}|\phi_C\right\rangle$ are different; we can assume

$$\beta_{NC} = (1+\varepsilon)\beta_{CC}$$

Furthermore, the induced effect also affects the coulombic integral of the carbon atom, which is adjacent to the nitrogen atom, α'_C.

$$\alpha_C' = \left\langle\phi_C'|\hat{\mathcal{H}}|\phi_C'\right\rangle$$

Let $\alpha'_C = \alpha_C + \delta'_C\beta$; thus, secular determinant of pyrazine is as follows:

$$\begin{vmatrix} \omega+\delta_N & 1+\varepsilon & 0 & 0 & 0 & 1+\varepsilon \\ 1+\varepsilon & \omega+\delta_C' & 1 & 0 & 0 & 0 \\ 0 & 1 & \omega+\delta_C' & 1+\varepsilon & 0 & 0 \\ 0 & 0 & 1+\varepsilon & \omega+\delta_N & 1+\varepsilon & 0 \\ 0 & 0 & 0 & 1+\varepsilon & \omega+\delta_C' & 1 \\ 1+\varepsilon & 0 & 0 & 0 & 1 & \omega+\delta_C' \end{vmatrix} = 0 \qquad (4.86)$$

The present problem is how to determine δ_N, δ'_C, and ε. From the analysis of experimental data, Carrington and McLachlan [33] obtained the following:

$$\delta_N = 0.75; \quad \delta'_C = 0; \quad \varepsilon = 0$$

So the secular determinant of pyrazine can be written as follows:

$$\begin{vmatrix} \omega+\delta_N & 1 & 0 & 0 & 0 & 1 \\ 1 & \omega & 1 & 0 & 0 & 0 \\ 0 & 1 & \omega & 1 & 0 & 0 \\ 0 & 0 & 1 & \omega+\delta_N & 1 & 0 \\ 0 & 0 & 0 & 1 & \omega & 1 \\ 1 & 0 & 0 & 0 & 1 & \omega \end{vmatrix} = 0 \tag{4.87}$$

Using the reductionism of symmetry, it is easy to obtain the energy levels and HMO. Its frontier orbital with $\omega = 0.545$ is

$$\psi_4 = 0.523(\phi_1 + \phi_4) + 0.338(\phi_2 + \phi_3 + \phi_5 + \phi_6)$$

Thus, $\rho_1 = 0.273; \quad \rho_2 = 0.114$

The experimental value of hyperfine coupling constants a_i and electron cloud density ρ_i value of HMO as well as the calculated values of a_i for some nitrogen heterocyclic compounds are listed in Table 4.6. It could be seen that the calculated results are coincident with the experimental values relatively.

4.4.8 Hyperfine coupling constant a value of the odd alternant and nonalternant hydrocarbons

The HMO theory and McConnell formula to be applied in even alternant hydrocarbons, including heterocyclic hydrocarbons, are still quite satisfactory despite some drawbacks. However, its application in the odd alternant hydrocarbons and nonalternant hydrocarbons seems unsatisfactory. The frontier orbital of the odd alternant hydrocarbon radical is ψ_o (nonbonding orbital). According to the HMO theory, the position with unmarked * should not have hyperfine splitting; however, the proton on the unmarked position of benzyl has a splitting of 0.175 mT. The proton on a_1 position of perinaphthene radical (Figure 4.18) has a 0.1833 mT splitting, with a total width of 4.31 mT far over from 3.0 mT. These cannot be solved only using the simple HMO theory, there should also be "electron correlation," which will be discussed in the next section.

Table 4.6: Hyperfine coupling constant of nitrogen heterocyclic compound[a].

Compound	Position	a_i (experiment) (mT)	HMO electron cloud density	Q-value (mT)	a_i (calculation) (mT)
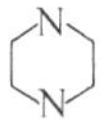	N	0.722	0.272	2.66	0.724
	C	0.266	0.114		0.258
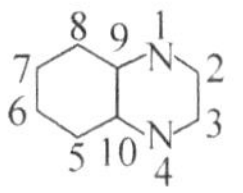	1	0.564	0.218	2.60	0.567
	2	0.332	0.128		0.333
	5	0.232	0.085		0.221
	6	0.100	0.085		0.516
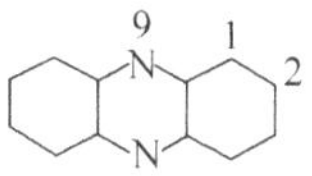	9	0.514	0.209	2.47	0.146
	1	0.193	0.059		0.119
	2	0.161	0.048		
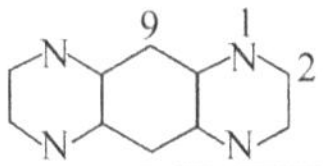	9	0.396	0.118		
	1	0.241	0.105		
	2	0.273	0.086		

[a]The data are from Table 2–6 on page 91 of *Electron Spin Resonance Spectroscopy*, edited by Qiu Zuwen, Science Press (1980). A correction of the calculation values of a_i has been done by the author.

4.5 Mechanism of hyperfine splitting in the spectrum of conjugated systems

Like McConnell formula mentioned previously, the hyperfine coupling constant of proton on planar conjugated radical is proportional to the unpaired π-electron density of the carbon atom, which is attached to the concerned proton (Eq. (4.48)). The necessary condition for proton hyperfine splitting is that the probability density of unpaired electron on the hydrogen nucleus is not zero. For radicals in solution, only the isotropic Fermi contact interaction can cause hyperfine splitting. In the π-radical, unpaired electron is in π-molecular orbital, which is linear combination of $2p_z$ atomic orbital of each carbon atom. Each $2p_z$ orbital has a node on the carbon atom, which is in the plane of the molecular skeleton. However, the C–H bond is formed by overlapping a sp^2 hybrid orbital of carbon atom and a s-orbital of the H atom. Here the question arises: How does the unpaired electron lying on the π molecular orbital appear at the hydrogen nucleus?

4.5.1 "Electronic correlation" effect

Referring to the HMO theory to be applied in π radicals, we assume that when the conjugated molecules get an electron, it becomes an anion radical; the other electrons in the molecule are not affected by this "new" electron. But in fact when the conjugated molecules get an electron, the other electrons in the molecule are affected. The original paired electrons of molecules display a slight "unpaired property". This is actually an "electronic correlation" effect. Hence, we must distinguish between the "electron cloud density" and the "spin density" of unpaired electron.

ρ_i is defined as the spin density of the ith region of a molecule, $p_i(\alpha)$ as the total spin density of the α spin electrons in the ith region, and $p_i(\beta)$ as the total spin density of β spin electrons in the ith region; thus,

$$p_i(\alpha) = p_{i,k}(\alpha); \quad p_i(\beta) = p_{i,k}(\beta); \tag{4.88}$$

Here $p_{i,k}(\alpha)$ is the spin density of the kth α spin electron in the ith region and $p_{i,k}(\beta)$ is the spin density of the kth β spin electron in the ith region. Hence,

$$\rho_i = \rho_{i,k}(\alpha) - p_{i,k}(\beta)$$

We take a fragment of C–H from the conjugated molecule (Figure 4.20). The C–H bond is composed of sp^2 hybrid orbital of the carbon atom and s-orbital of the hydrogen atom. When the unpaired electron has not drawn into the $2p_z$ orbital of the carbon atom, the existing probabilities of the configuration (a) and (b) in Figure 4.20 are equal. When an α spin electron is introduced into the $2p_z$ orbital of the carbon atom, the two paired electrons in C–H bond will be affected. According to the Hund rule, if two electrons occupy two different orbitals of the same carbon atom, the spin of two electrons should be as much parallel as possible. Therefore, the probability of the configuration shown in Figure 4.20a is larger than the configuration shown in Figure 4.20b for the ground state. From the viewpoint of proton, the electron of proton in configuration shown in Figure 4.20a is β spin (negative spin), and the electron of proton in configuration shown in Figure 4.20b is α spin (positive spin). Since the probability of configuration shown in Figure 4.20a is larger than the configuration shown in Figure 4.20b, the negative spin of electron on the proton is larger than that of the positive spin, and hence the net spin density of electron on the proton is "negative spin density." Negative spin density causes a negative proton hyperfine splitting, and the Q sign should also be negative. It

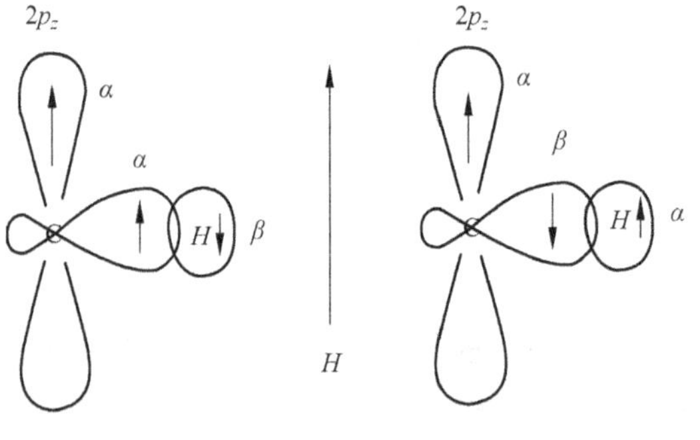

Figure 4.20: The possible configuration of electron spin in C–H fragment.

should be pointed out that the negative spin density appears on the proton corresponding to the positive spin density appearing on the carbon atom. This effect is called the "spin polarization effect."

4.5.2 The sign of proton hyperfine splitting constant

When the electron in the $2p_z$ orbital of carbon atom has a positive spin density, the spin density of the electron on proton should be negative, and the negative spin density causes a negative hyperfine splitting. Here we use proton magnetic resonance experiments to prove the existence of negative hyperfine splitting induced by negative spin density.

Measuring the shift of the proton magnetic resonance spectra and observing the direction of the paramagnetic chemical shift of the NMR spectra of the radical, the sign of the hyperfine splitting constant can be determined [34].

This experiment has certain requirements for the sample. Because of the existence of electron spin in the system, the spin relaxation of the proton will broaden the NMR spectral line and the degree of broadening is proportional to the square of the proton hyperfine splitting. Therefore, it is required that the linewidth of NMR must be very narrow, and the width of proton hyperfine splitting must not exceed 0.6 mT. The paramagnetic chemical shift of radicals in solution at room temperature is possible only when the proton hyperfine splitting is less than 0.6 mT [35,36]. According to this requirement, the anion radicals of biphenyl anion radical is a more suitable sample. Figure 4.21 shows the NMR spectrum at 60 MHz of 1.0 M solution of

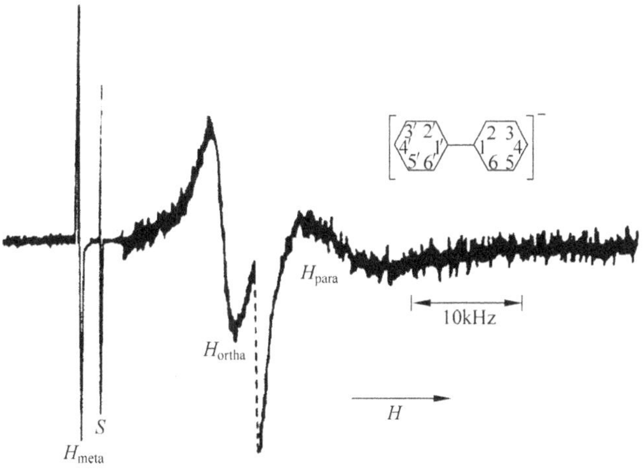

Figure 4.21: NMR spectrum at 60 MHz of a 1 M solution of the biphenyl anion in diethylene glycol dimethyl ether (diglyme) at room temperature [37].

the biphenyl anion radical in diethylene glycol dimethyl ether (diglyme) at room temperature [37].The spectral line marked with s is arising from the solvent.

If a_i is positive, the paramagnetic chemical shift should be negative, that is, shifting toward the lower field. The quantity of chemical shift is given by

$$\Delta H = H_i - H_0 = -\frac{gg_e\beta_e^2 H_0}{4g_p\beta_n k_b T} a_i \tag{4.89}$$

H_i is the resonance magnetic field after the chemical shift of the ith proton, and the H_0 corresponds to the magnetic field before the shift [38,39].

The first-order approximation of the spin Hamilton operator is given by

$$\hat{\mathcal{H}} = g\beta H\hat{S}_z + a\hat{S}_z\hat{I}_z - g_n\beta_n H\hat{I}_z$$

$$\hat{\mathcal{H}} = g\beta H\hat{S}_z - g_n\beta_n\left\{H - \frac{a\hat{S}}{g_n\beta_n}\right\}\hat{I}_z \tag{4.90}$$

From Eq. (4.90), we can see that the hyperfine interaction looks like the general chemical shift in form, but the value of shift is more large. It is called *contact shift* in NMR spectroscopy. It depends on the spin state of the electron. For $M_S = -\frac{1}{2}$,

$$\hat{\mathcal{H}} = -g_n\beta_n\left(H + \frac{a}{2g_n\beta_n}\right)\hat{I}_z - \frac{1}{2}g\beta H \tag{4.91}$$

The transition selection rule of NMR is $\Delta M_s = 0$; $\Delta M_I = 1$,

$$\Delta E_1 = g_n\beta_n\left(H + \frac{a}{2g_n\beta_n}\right) \tag{4.92}$$

Similarly, $M_S = +\frac{1}{2}$

$$\Delta E_2 = g_n\beta_n\left(H - \frac{a}{2g_n\beta_n}\right) \tag{4.93}$$

As the electron spin relaxation time is very short, the observed NMR spectra are not two lines, but single in the average frequency. The weighted average result is

$$\begin{aligned} h\bar{v} &= (\Delta E_1)\frac{N_\beta}{N_\alpha + N_\beta} + (\Delta E_2)\frac{N_\alpha}{N_\alpha + N_\beta} \\ &= g_n\beta_n H + \frac{a}{2}\frac{N_\beta - N_\alpha}{N_\beta + N_\alpha} \end{aligned} \tag{4.94}$$

where N_α and N_β are the number of electrons with α and β spin states, respectively. N_α and N_β could be considered to obey Boltzmann distribution, and under the condition of room temperature, $g\beta H/kT \ll 1$. Thus,

$$N_\alpha \approx \frac{N}{2}\left(1-\frac{g\beta H}{2kT}\right);\quad N_\beta \approx \frac{N}{2}\left(1+\frac{g\beta H}{2kT}\right)$$

and

$$\frac{N_\beta - N_\alpha}{N_\beta + N_\alpha} \approx \frac{g\beta H}{2kT} \tag{4.95}$$

Substituting Eq. (4.95) into Eq. (4.94) gives

$$h\bar{v} = g_n\beta_n H + \frac{a}{2}\frac{g\beta H}{2kT} = g_n\beta_n\left(H + \frac{g\beta H}{4g_n\beta_n kT}a\right) = g_n\beta_n(H+\delta) \tag{4.96}$$

where

$$\delta = \frac{g\beta H}{4g_n\beta_n kT}a \tag{4.97}$$

The unit of a is erg, while the unit of a_i is Tesla; here k is the Boltzmann constant k_b. Hence, Eq. (4.97) should be rewritten as follows:

$$\delta = \frac{gg_e\beta_e^2 H_o}{4g_p\beta_n k_b T}a_i \tag{4.98}$$

When $\delta = 0$, the magnetic field of NMR is $H = H_o$, that is, $h\bar{v} = g_n\beta_n H_o$; when $\delta \neq 0$, $H = H_i$:

$$h\bar{v} = g_n\beta_n(H_i + \delta) \tag{4.99}$$

$$H_i - H_o = -\delta = -\frac{gg_e\beta_e^2 H_o}{4g_p\beta_n k_b T}a_i$$

It is Eq. (4.89) that has been drawn before. From Eq. (4.89) it can be seen that when $a_i > 0$, the line should be shifted toward low field. Conversely, when $a_i < 0$, spectral line shifts toward high field. It coincides with the experiments. The spectral lines H_{ortho} and H_{para} of biphenyl anion radical also shift toward high field; it explains that the hyperfine coupling constants a_o and a_p of both proton *ortho*- and *para*-positions are negative. Since Q is negative, the spin densities of electrons on the *ortho*- and *para*-positions should be positive. However, H_{meta} is shifted toward low field, which implies that the spin density of unpaired electron on the *meta*-position is negative.

4.5.3 Negative spin density

The McConnell formula is still correct, but ρ_i should be seen as the *π-electron spin density* on the *i*th carbon atom rather than the *cloud density of* π-electron. If all the electron *spin densities* are positive, they are equal to the *electron cloud densities*. However, the electron spin density on the carbon atom sometimes is negative, and then both are not the same. In accordance with the requirements of the normalization condition, the algebraic sum of spin density should be equal to 1. If there is a negative spin density in some parts of the molecule, then to ensure that the algebraic sum is equal to 1, the other places must be more positive. Thus, the sum of the absolute values of the spin density will be greater than 1. Since the total width of the spectrum depends on the absolute value of the hyperfine splitting, the total width of the spectral line is greater than the *Q*-value for this case. This is the reason that the total width of perinaphthene free radical reaches to 4.31 mT.

The reason why the HMO theory application to the even alternant hydrocarbons could be successful is that their spin densities basically (the vast majority) are positive. While in the odd alternant hydrocarbons, due to the presence of negative spin density, the simple HMO theory faces difficulties, and we must apply the molecular orbital theory to be able to summarize *electron correlation effects*. See McLachlan [40] perturbation method, or else refer to Salem, L., *The Molecular Orbital Theory of Conjugated System*, Benjamin, New York, NY, Chapter 5 (1966).

4.5.4 About the Q value problem

The McConnell formula introduced in the previous section is concerned with the *Q* value problem. The experimental results explain that the *Q* value is not constant, and will also cause problems in the extensive application of McConnell formula. The application of McConnell equation to the even alternant hydrocarbon π-radical is quite satisfactory. Although the experimental results are in accordance with the principle of duality basically, the spin density of the anion or cation free radicals of the even alternant hydrocarbons should be the same. However, the experimental results indicate that the hyperfine splitting of the a_i value of cationic free radicals are always larger than the a_i value of anionic radicals [20], explaining the *Q* value between them are different slightly. The cause is that the excess charge of carbon atoms affect the *Q* value as mentioned above. Define ε_i as excess charge:

$$\varepsilon_i = 1 - q_i \tag{4.100}$$

where q_i is the total sum of π-electron density on the *i*th carbon atom. Colpa [30] and Bolton [31] proposed a modified McConnell formula based on the considering effect of excess charge:

$$a_i = [Q(0) + K\varepsilon_i]\rho_i \tag{4.101}$$

Here $Q(0)$ expresses the Q value of the neutral radical and K is a constant. With $Q(0) = -2.7$ mT and $K = -1.2$ mT, the results calculated by using Eq. (4.101) can better coincide with experimental results. For example, the experimental results $a_9^+ = 0.653$ mT (for anthracene cation), $a_9^- = 0.534$ mT (for anthracene anion), and the results calculated by Eq. (4.101), $\rho_9^+ = 0.224$, $\rho_9^- = 0.215$, both are relatively close. It must be pointed out that although the correction of excess charge can be better close to the experimental results, the deviation from the neutral radical is just only 15%. Thus, McConnell formula could still be used even without any correction.

Unlike the conjugated π-radical, alkyl radical is a σ-radical and the unpaired electron is localized on certain carbon atom. The carbon atom where the unpaired electron lies is called the α carbon atom, the closest one is called β carbon atom, and other carbon atoms in order of γ, δ, and so on. Their Q values also display some regularity. Hyperfine splitting a values and Q values of some alkyl radicals are listed in Table 4.7.

Table 4.7: Hyperfine splitting a_i values and Q values of some alkyl radicals [41].

Radical	ρ	a_α^H(mT)	Q_α(mT)	a_β^H(mT)	Q_β(mT)
CH_3	1.000	2.304	2.304	—	—
CH_3CH_2	0.919	2.238	2.435	2.687	2.925
$(CH_3)_2C$	0.844	2.211	2.620	2.468	2.925
$(CH_3)_3C$	0.776	—	—	2.272	2.930

From Table 4.7 it can be seen that with the increasing CH_3 substituents, the values of a_α and a_β are decreased, and a_α decreases more faster.

Chesnut [42] proposed an empirical equation to estimate the spin density on the α carbon atom:

$$\rho_\alpha = (1 - 0.081)^m = 0.919^m \tag{4.102}$$

where m is the number of methyl groups attached to α carbon atom. Assume that they also conform to the McConnell formula:

$$a_\alpha = Q_\alpha\rho_\alpha; \quad a_\beta = Q_\beta\rho_\beta \tag{4.103}$$

We can make use of the experimental measure a_α for estimating ρ_α by Eq. (4.102) and for calculating Q_α and Q_β by Eq. (4.103). It can be seen from Table 4.7 that Q_α is not a constant, while the value is very close to 2.7 mT, for the carbon atom with only one proton. It can be used to calculate the conjugated π-radicals by McConnell formula.

On the other hand, Q_β in Table 4.7 is very close to a constant. This important result inspired that the spin density on the β carbon atom ρ_β can be estimated using the experimentally measured a_β.

Fischer [43] has given the change of Q_α for $CH_3-\dot{C}H-X$ type radicals with the substituent groups, as shown in Figure 4.6. Q_α is calculated by the following equation:

$$Q_\alpha = \frac{a_\alpha Q_\beta}{a_\beta} = \frac{a_\alpha}{a_\beta} \times 29.25 \tag{4.104}$$

As seen from Table 4.8, Q_α is not a constant. The reason is not clear, may be it is caused by the inducing effect of substituting group. However, for neutral and structural similar compounds, the Q values are close to constant. The value of spin density can be estimated according to these empirical properties.

Table 4.8: The change of Q_α in the $CH_3-\dot{C}H-X$ type radicals.

X	a_α(mT)	a_β(mT)	Q_α(mT)
CH_3	2.211	2.468	2.62
H	2.238	2.687	2.44
$CO-CH_2-CH_3$	1.845	2.259	2.39
COOH	2.018	2.498	2.37
OH	1.504	2.261	1.95
$O-CH_2-CH_3$	1.396	2.228	1.83

Taking allyl as an example, for 1, 3 positions, Q_α is equal to 2.44 mT; for 2 position, Q_α is 2.62 mT; the measured data are $a_1 = a_3 = 1.438$ mT (average value) and $a_2 = 0.406$ mT. Substituting into McConnell formula,

$$\rho_1 = \rho_3 = \frac{1.438}{2.44} = 0.589$$

$$\rho_2 = \frac{-0.406}{2.62} = -0.155$$

$$\sum_{i-1}^{3} \rho_i = 0.589 - 0.155 + 0.589 = 1.023$$

This value is very close to 1, thus this rough estimate still has some practical value.

4.5.5 Hyperfine splitting and hyperconjugation effect of methyl protons

Experimental facts show that the hyperfine splitting a_i^H of methyl proton in methyl aromatic hydrocarbon radicals or semiquinone radicals is generally large, sometimes even larger than the ring protons. Unpaired electron is in π-orbital of the ring, the ring carbon atom and the carbon atom of methyl group are connected via σ-bond, as well as the methyl proton atom is also connected with the methyl carbon atom via σ-bond. Then how does the methyl proton couple with the unpaired π-electron? In addition, it is also found that in the σ-radical, the splitting of β proton of ethyl, isopropyl, and so on are also larger than the splitting of α proton. Let us see the experimental results. Hyperfine splitting values of 9,10-bimethyl anthracene cation or anion radical, ethyl, isopropyl, and *tert*-butyl are listed in Table 4.9.

Table 4.9: Hyperfine splitting values of 9,10-two methyl anthracene cation or anion radical, ethyl, isopropyl, and *tert*-butyl.

	a_1 (mT)	a_2 (mT)	a_{CH3} (mT)	a_α (mT)	a_β (mT)
9,10-Bimethyl anthracene cation	0.254	0.119	0.800		
9,10-Bimethyl anthracene anion	0.290	0.152	0.388		
$-CH_2CH_3$				2.238	2.687
$-CH(CH_3)_2$				2.211	2.468
$-C(CH_3)_3$					2.272

At first, the NMR experiments indicate that the sign should be positive [44,45]. Second, the hyperfine splitting of hydrogen in methyl of 9,10-bimethyl anthracene cation is 0.800 mT, while the hyperfine splitting of the corresponding anion, it is only 0.388 mT. Since the spin densities of the anion or cation in the 9,10-carbon atom are slightly different, the spin polarization cannot cause so large difference. It indicates that other mechanism of hyperfine coupling should exist. Because it does not exist in the conjugate system in the above discussion, we tentatively call it as *hyperconjugation effect*. Practically, in the above system, methyl hydrogen atom is produced by the hyperfine coupling interaction directly with the unpaired electron of the conjugated system.

We take the ethyl radical $\cdot C_\alpha H_2{-}C_\beta H_3$ as an example now to illustrate the mechanism of the hyperconjugation effect. The molecular orbital theory regarded that the C_β–H bond is not formed by the 1s orbital of hydrogen atom on the methyl group overlapping with the sp^3 hybrid orbitals of C_β atom. The three hydrogen atoms on CH_3 composed a "group orbitals" first. Thereafter, the three linear functions undergo linear combination, thus forming a new set of basic functions.

Let ϕ_1, ϕ_2, ϕ_3 represent the 1s orbitals of the three hydrogen atoms. Their linear combination function is

$$\psi = c_1\phi_1 + c_2\phi_2 + c_3\phi_3 \tag{4.105}$$

Let $c_1 = c_2 = c_3$. Using the normalization condition $\langle\psi|\psi\rangle = 1$, the following is obtained:

$$\psi_1 = \frac{1}{\sqrt{3}}(\phi_1 + \phi_2 + \phi_3) \tag{4.106}$$

Let c_1= 0. Using the orthogonal normalization conditions $\langle\psi_2|\psi_1\rangle = 0$, $\langle\psi_2|\psi_2\rangle = 1$, the following is obtained:

$$\psi_2 = \frac{1}{\sqrt{2}}(\phi_2 - \phi_3) \tag{4.107}$$

Finally, let $\psi_3 = c_1\phi_1 + c_2\phi_2 + c_3\phi_3$, and using the orthogonal normalization condition $\langle\psi_3|\psi_1\rangle = \langle\psi_3|\psi_2\rangle = 0$, $\langle\psi_3|\psi_3\rangle = 1$ is obtained:

$$\psi_3 = \frac{1}{\sqrt{6}}(2\phi_1 - \phi_2 - \phi_3) \tag{4.108}$$

Here ψ_1, ψ_2, ψ_3 are the recombinational three group orbitals. ψ_1 and ψ_2 are the σ symmetric group orbitals, which can only overlap with the 2s, $2p_x$, $2p_y$ atom orbitals on the C_β atom, and are irrelative with hyperfine coupling. Only ψ_3 has same symmetry with $2p_z$ orbitals of C_α, C_β, as shown in Figure 4.22, and can be composed a π molecular orbital:

$$\psi = aP_z + bp'_z + c\psi_3 \tag{4.109}$$

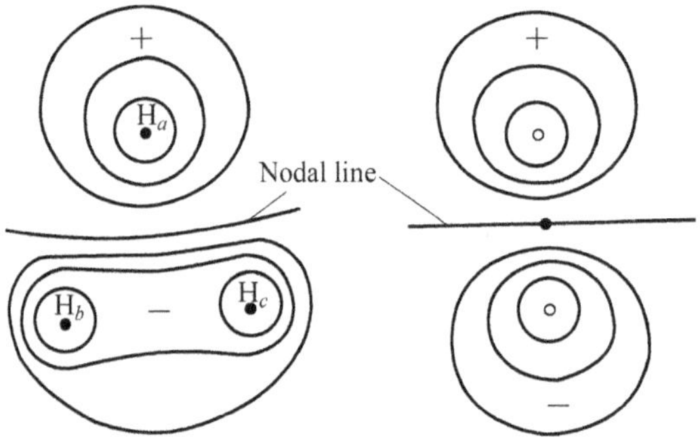

Figure 4.22: One of the three group orbitals composed of three hydrogen atoms of methyl, ψ_3, has the same symmetry with –π-orbital on carbon atom.

where p_z and p'_z are the $2p_z$ orbitals of C_α, C_β and *a*, *b*, *c* are the coefficients of linear combination. ψ_3 is called the pseudo π-orbital.

Since ψ_3 participates in the entire π-orbital system, the spin density of unpaired electron in the π system can directly couple to the methyl hydrogen atom producing a large value of hyperfine splitting. Since the spin density of the methyl proton is directly from the unpaired electron, the spin density of the proton should be positive. This is in coincidence with the NMR experimental result.

4.6 Hyperfine splitting of other (non-proton) nuclei

4.6.1 Hyperfine splitting of ^{13}C nucleus

All of the above discussion is about the hyperfine splitting of the ^1H nucleus. McConnell formula is used mainly in the splitting of ^1H nucleus of C–H. It is certainly not convenient to use in the hyperfine splitting of other magnetic nucleus. ^{13}C nucleus has $I = 1/2$, and the natural abundance is 1.108%. Although its natural abundance is not so large, but it is enough to be the subject of research. Its theory of isotropic hyperfine splitting is more complex than the ^1H nucleus. The splitting of ^{13}C nucleus and the π-electron spin density on carbon atoms is not a directly proportional relationship [46]. The splitting value of ^{13}C nucleus on methyl radical $\dot{C}H_3$ is 4.1 mT, while on benzene anion $C_6H_6^-$ it is only 0.28 mT. If there is a directly proportional relationship, the splitting value of ^{13}C nucleus on benzene anion should be (4.1/6) = 0.68 mT, which is far different from the value of experiment（0.28 mT）. Hence, the hyperfine splitting of ^{13}C nucleus on the ith carbon atom is related not only to π-electron spin density of the ith carbon atom but also to the π-electron spin density of the adjacent jth carbon atom.

As we have already discussed with the "electronic correlation," when the $2p_z$ orbital of C_i carbon atom is filled into an α spin electron, the C_i tip of C_i–H produces a positive spin density and the H end of C_i–H produces a negative spin density. It is called the "spin polarization" phenomenon. Similarly, on the $2p_z$ orbital of C_j carbon atom that is adjacent to the C_i carbon atom filled into an α spin electron, the C_j carbon atom causes a positive spin density and on the C_i carbon atom it causes a negative spin density.

As shown in Figure 4.23, $Q^C_{CC'}$ denotes that the spin polarization caused by the unpaired electron in the $2p_z$ orbital on C atom of the $C - C'$ bond contribute to the

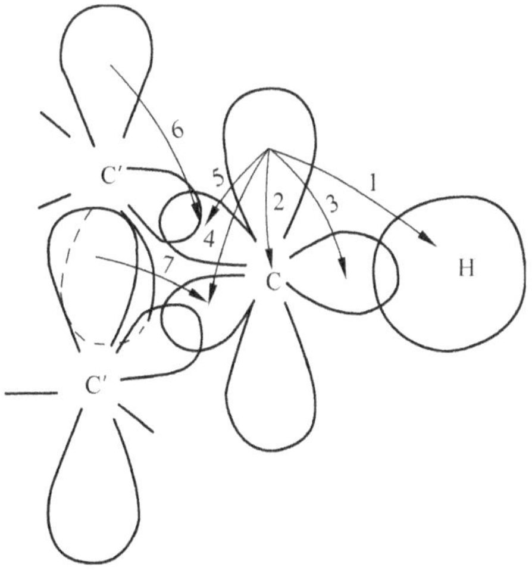

Figure 4.23: The contribution of spin polarization in $(C')_2 - C - H$ fragment to the hyperfine splitting of ^{13}C and ^1H nuclei. 1: Q^H_{CH}; 2: S^C; 3: Q^C_{CH}; 4 and 5: $Q^C_{CC'}$; 6 and 7: $Q^C_{C'}$.

hyperfine splitting of the ^{13}C nucleus on the C′ end. Similarly, $Q_{\mathrm{C'C}}^{\mathrm{C}}$ denotes the spin polarization caused by the unpaired electron in the $2p_z$ orbital on C′ atom of the C – C′ bond that contribute to the hyperfine splitting of the ^{13}C nucleus on the C end. Obviously, the $Q_{\mathrm{CC'}}^{\mathrm{C}}$ should be positive, and the $Q_{\mathrm{C'C}}^{\mathrm{C}}$ should be negative, since the unpaired electron in the $2p_z$ orbital of the C atom produces a positive spin density on the C atom and the unpaired electron in the $2p_z$ orbital on the C′ atom produces a negative spin density on the C atom – the reason being the same that $Q_{\mathrm{CH}}^{\mathrm{C}}$ is positive and $Q_{\mathrm{CH}}^{\mathrm{H}}$ is negative.

Let S^{C} represent the spin polarization caused by the unpaired electron in $2p_z$ orbital of the C atom on the 1s electron of C atom. Thus,

$$a_i^{\mathrm{C}} = \left(S^{\mathrm{C}} + \sum_{j=1}^{3} Q_{\mathrm{CX}_j}^{\mathrm{C}} \right) \rho_i + \sum_{j=1}^{3} Q_{\mathrm{X}_j\mathrm{C}} \rho_j \tag{4.110}$$

Quantitative calculation of spin polarization constants [46] are as follows:

$$S^{\mathrm{C}} = -1.27\,\mathrm{mT}, \quad Q_{\mathrm{CH}}^{\mathrm{C}} = +1.95\,\mathrm{mT}$$
$$Q_{\mathrm{CC'}}^{\mathrm{C}} = +1.44\,\mathrm{mT}, \quad Q_{\mathrm{C'C}}^{\mathrm{C}} = -1.39\,\mathrm{mT}$$

Substituting into Eq. (4.110), for $\mathrm{CC'_2H}$ fragment,

$$a_i^{\mathrm{C}} = (-1.27 + 1.44 + 1.44 + 1.95)\rho_i - 1.39 \sum_{j=1}^{2} \rho_j = 3.56\rho_i - 1.39 \sum_{j=1}^{2} \rho_j$$

$$a_i^{\mathrm{C}} = 3.56\rho_i - 1.39 \sum_{j=1}^{2} \rho_j \tag{4.111}$$

For $\mathrm{CC'_3}$ fragment,

$$a_i^{\mathrm{C}} = (-1.27 + 3[1.44])\rho_i - 1.39 \sum_{j=1}^{2} \rho_j \tag{4.112}$$

Calculating $\dot{\mathrm{C}}\mathrm{H}_3$ radical with Eq. (4.103) ($\rho = 1.0$),

$$a_i^{\mathrm{C}} = (-1.27 + 3[1.95]) \times 1.0 = 4.58\,\mathrm{mT}$$

The experimental result is 4.1 ± 0.3 mT.

Now, we use the hyperfine splitting value of proton and $Q_{\mathrm{CH}}^{\mathrm{H}}(0) = -2.70\,\mathrm{mT}$ to calculate ρ_i, and Eq. (4.103) to calculate a_i^{C}. It is found that the calculated results better coincide with the experimental results. Taking the anthracene cationic and anionic radicals as examples [47,48], ρ_i is calculated by taking average of the experimental value of a_i^{H} and ρ_δ can be obtained according to the normalization condition:

$$\rho_\delta = \frac{1}{4}[1 - (2\rho_\alpha + 4\rho_\beta + 4\rho_\gamma)] \tag{4.113}$$

After ρ_i is obtained, based on Eqs. (4.49) and (4.50) a_i^{C} can be calculated:

$$a_\alpha^{\mathrm{C}} = 3.56(0.220) - 1.39[-0.021 + (-0.021)] = 0.824\,\mathrm{mT}$$

The calculated results are listed in Table 4.10.

Table 4.10: The experimental and calculated values of hyperfine splittings of ^{1}H and ^{13}C nuclei of the anthracene cationic and anionic radicals [47,48].

	Position	Experimental a_i^{H} (mT)			ρ_i calculated value	a_i^{C} (mT) calcu-lated value	Experimental a_i^{C} (mT)	
		Anion	Cation	Average			Anion	Cation
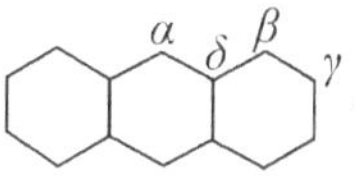	α	0.534	0.653	0.593	0.220	0.842	0.876	0.848
	β	0.274	0.306	0.290	0.107	0.337	+0.357	–
	γ	0.151	0.138	0.145	0.054	−0.033	−0.025	±0.037
	δ				−0.021	−0.490	−0.459	−0.450

From the data of Table 4.10, it can be seen that using the above calculated method, the values obtained of the hyperfine splittings of the ^{13}C of the anthracene cationic and anionic radicals are basically coincident with the experimental data. By comparison, the experimental results of its cationic radical are more close to the calculated results than that of the anionic radical.

4.6.2 Hyperfine splitting of ^{14}N nucleus

Organic radicals containing nitrogen are also common. The natural abundance of ^{14}N is 99.63% and nuclear spin $I = 1$. When there is unpaired π-electron on ^{14}N, it will present very beautiful hyperfine structure. Therefore, it is also one of the important research objects of EMR.

The relationship between the hyperfine splitting and spin density is different from ^{13}C nucleus. Experimental results indicate that the π-electron spin density of adjacent atoms slightly influences the hyperfine splitting – $Q_{\mathrm{C'N}}^{\mathrm{N}}$ value is very small, about ± 0.4 mT [47–51]. Therefore, hyperfine splitting value of the ^{14}N nucleus can be calculated by the following simple equation; for hydrogen atom attached on nitrogen atom,

$$a_i^N = Q_{N(\mathrm{C_2H})}^N \rho_i \tag{4.114}$$

For the situation of N atom does not attaching H atom, on the N atom has a lone pair electrons:

$$a_i^N = Q_{N(C_2P)}^N \rho_i \qquad (4.115)$$

The $Q_{N(C_2H)}^{N}$ value is about +2.7–3.0 mT; the $Q_{N(C_2P)}^{N}$ value is about +2.3–2.6 mT. Since the ^{14}N splitting is mainly from the spin density on the nitrogen atom, the Q value should be positive. It is in coincidence with the experimental results.

Using the hyperfine splitting a_i value of pyridine measured by the EMR experiment and letting $Q_{CH}^{H} = -2.7\,mT$, the ρ_i value of pyridine is calculated; letting $Q_{N(C_2P)}^{N} = -2.4\,mT$ the ρ_i value of pyrazine is calculated. The results are listed in Table 4.11.

Table 4.11: The calculated unpaired electron cloud density (ρ_i) of pyridine and pyrazine using the experimental values of hyperfine splittings (a_i) of pyridine and pyrazine.

	a_i(mT experiment)				ρ_i(calculation)				
	a_N	a_{ortho}	a_{meta}	a_{para}	ρ_N	ρ_{ortho}	ρ_{meta}	ρ_{para}	$\sum \rho_i$
Pyridine	0.628	0.355	0.082	0.970	0.262	0.132	0.030	0.359	0.944
Pyrazine	0.722	0.272			0.300	0.100			1.000

From the data of Table 4.11, it can be seen that this method is feasible.

4.6.3 Hyperfine splitting of ^{19}F nucleus

In the aromatic molecule, when H is replaced by F, the hyperfine splitting of ^{19}F is usually demonstrated. Its natural abundance with nuclear spin $I = \frac{1}{2}$ is 100%. Its hyperfine splitting value a_i^F and the π-electron spin density value of the ith carbon atom attaching F atom (ρ_i) have the following relationship:

$$a_i^F = Q_F \rho_i \qquad (4.116)$$

If the sign of the hyperfine splitting a_i^Fvalue of ^{19}F is positive, the π-electron spin density on the fluorine atom is positive. Since the nonbonding p electron on fluorine atom makes the F–C bond have partial double bond property, it participates in the whole conjugation system. Thus, unpaired electron density is partially transferred to the fluorine atom directly, making the fluorine has a net positive spin density. Hence, the hyperfine splitting of ^{19}F should be positive [52].

4.6.4 Hyperfine splittings of ^{17}O and ^{33}S nuclei

The natural abundance of ^{17}O is only 0.037% with I =5/2. EMR test was carried out using the ^{17}O isotope-enriched semiquinone radical and ketyl free radical; the obtained hyperfine splitting value a_i^O and spin density ρ_i of ^{17}O have the relationship of Eq. (4.110) [53], where $Q_{OC}^O = -4.45\,\text{mT}$ and $Q_{CO}^O = -1.43\,\text{mT}$. Because the natural abundance of $^{17}O$is very low and does not have sufficient number of experimental times, these values are still controversial.

The natural abundance of ^{33}S is 0.67% with nuclear spin $I = 3/2$. Using the ^{33}S isotope-enriched sulfur heterocyclic compounds, such as thianthrene [54], EMR spectrum experimental measurement was carried out, and the obtained hyperfine splitting value a^S and spin density ρ^S on sulfur atom have the following relationship:

$$a^S = Q_{S(C_2P)}^S \rho^S \tag{4.117}$$

where the $Q_{S(C_2P)}^S \approx 3.3\,\text{mT}$

References

[1] Skinner R., Weil J. A. *Am. J. Phys.* 1989, **57**: 777.
[2] Fermi E. *Z. Phys.* 1930, **60**: 320.
[3] Breit G., Rabi I. I. *Phys. Rev.* 1931, **38**: 2082.
[4] Isoya J., Weil J. A., Davis P. H. *J. Phys. Chem. Solids.* 1983, **44**: 335.
[5] Bolton J. R. *Mol. Phys.* 1963, **6**: 219.
[6] Dobbs A. J., Gilbert B. C., Norman O. C. *J. Chem. Soc.* 1972, 2053.
[7] Livingston R., Zeldes H. *J. Chem. Phys.* 1966, **44**:1245.
[8] Levy D. H., Zeldes H. *J. Chem. Phys.* 1964, **41**: 1062.
[9] Lipkin D., Paul D. E., Townsend. J., Weissman S. I. *Science.* 1953, **117**: 534.
[10] Atherton N. M., Weissman S. I. *J. Am. Chem. Soc.* 1961, **83**: 1330.
[11] Bolton J. R., Fraenkel G. K. *J. Chem. Phys.* 1964, **40**: 3307.
[12] Kaplan M., Bolton J. R., Fraenkel G. K. *J. Chem. Phys.* 1965, **42**: 955.
[13] Anderson D. H., Frank P. J, Gutowsky H. S. *J. Chem. Phys.* 1960, **32**: 196.
[14] Fessenden R. W., Schuler R. H. *J. Chem. Soc.* 1965, **43**: 2704.
[15] Horsfield A., Morton J. R., Whiffen D. H. *Mol. Phys.* 1961, **4**: 475.
[16] Windle J. J., Wiersema A. K. *J. Chem. Phys.* 1963, **39**: 1139.
[17] Atherton N. M., Brustolon M. *Mol. Phys.* 1976, **32**: 23.
[18] Osicki J. H., Ullman E. F. *J. Am. Chem. Soc.*, 1968, **90**: 1078.
[19] Tuttle T. R., Ward R. L., Weissman S. I. *J. Chem. Phys.* 1956, **25**: 189.
[20] dos Santos-Veiga J., Neiva-Correia A. F. *Mol. Phys.* 1965, **9**: 395.
[21] Carrington A, dos Santos-Veiga J. *Mol. Phys.* 1962, **5**: 21.
[22] McConnell H. M., Chesnut D. B. *J. Chem. Phys.* 1958, **28**: 107.
[23] Weissbluth M. *Atoms and Molecules*, Academic Press, New York, 1978.
[24] Fesseden R. W., Ogawa S. *J. Am. Chem. Soc.* 1964, **86**: 3591.
[25] Bolton J. R. *Mol. Phys.* 1963, **6**: 219.
[26] Carter M. K., Vincow G. *J. Chem. Phys.* 1967, **47**: 292.

[27] Carrington A., Smith C. P. *Mol. Phys.* 1963, 7: 99.
[28] Vincow G., Morrell W. V., Dauben H. J., Jr., Hunter F. R. *J. Am. Chem. Soc.* 1965, **87**: 3527.
[29] Katz T. J., Strauss H. L. *J. Chem. Phys.* 1960, **32**: 1837.
[30] Colpa J. P., Bolton J. R. *Mol. Phys.* 1963, **6**: 273.
[31] Bolton J. R. *J. Chem. Phys.* 1965, **43**: 309.
[32] Lawler R. G., Bolton J. R., Fraenkel G. K., Brown T. H. *J. Am. Chem. Soc.* 1964, **86**: 520.
[33] Carrington A., McLachlan A. D. *Introduction to Magnetic Resonance*, Harper & Row, New York, 1967.
[34] Drago R. S. *Physical Methods of Chemistry*, Saunders, Philadelphia, Chapter 12, 1977.
[35] de Boer E., MacLean C. *Mol. Phys.*1965, **9**: 191.
[36] Hausser K. H., Brunner H., Jochims J. C. 1966, *Mol. Phys.* 1966, **10**: 253.
[37] Canters G. W., de Boer E. *Mol. Phys.* 1967, **13**: 495.
[38] McConnell H. M., Holm C. H. *J. Chem. Phys.* 1957, **27**: 314.
[39] Eaton D. R., Phillips W. D. "Nuclear magnetic resonance of paramagnetic molecules" in *Advances in Magnetic Resonance*, Ed. J. S. Waugh, Vol. 1, Academic Press, New York, 1965.
[40] McLachlan A D. *Mol. Phys.* 1960, **3**: 233.
[41] Qiu Z W. Electron Spin Resonance Spectroscopy. Science Press, Beijing. 1980: 83, 84.
[42] Chesnut D. B. *J. Chem. Phys.* 1958, **29**: 43.
[43] Fischer H. *Z. Naturforsch.* 1965, **20A**: 428.
[44] Forman A., Murell J. N., Orgel L. E. *J. Chem. Phys.* 1959, **31**: 1129.
[45] Lazdins D., Karplus M. *J. Am. Chem. Soc.* 1965, **87**: 920.
[46] Karplus M., Fraenkel G. K. *J. Chem. Phys.* 1961, **35**: 1312.
[47] Ward R. L. *J. Am. Chem. Sco.* 1962, **84**: 332.
[48] Talcott C. L, Myers R. J. *Mol. Phys.* 1967, **12**: 549.
[49] Henning J. C. M., de Waard. *Phys. Lett.* 1962, **3**: 139.
[50] Geske D. H., Padmanabhan G. R. *J. Am. Chem. Sco.* 1965, **87**: 1651.
[51] Henning J. C. M. *J. Chem. Phys.* 1966, **44**: 2139.
[52] Eaton D. R., Josey A. D., Phillips W. D., Benson R. E. *Mol. Phys.* 1962, **5**: 407.
[53] Broze M., Luz Z., Silver B. L. *J. Chem. Phys.* 1967, **46**: 4891.
[54] Sullivan P. D. *J. Am. Chem. Sco.* 1968, **90**: 3618.

Further readings

[1] Kaiser E. T., Kevan L. Eds. *Radical Ions*, Wiley-Interscience, New York, 1968.
[2] Memory J. D. *Quantum Theory of Magnetic Resonance Parameters*, McGraw-Hill, New York, 1968.
[3] Rado G. T. Simple derivation of the electron–nucleus contact hyperfine interaction. *Am. J. Phys.* 1962, **30**: 716.
[4] Griffiths D. J. Hyperfine splitting in the ground state of hydrogen. *Am. J. Phys.* 1982, **50**: 698.
[5] Poole C. P., Jr., Farach H. A. *The Theory of Magnetic Resonance*, Wiley-Interscience, New York, 1972.
[6] Carrington A., McLachlan A. D. *Introduction to Magnetic Resonance*, Harper & Row, New York, 1967.

5 Anisotropic hyperfine structure

Abstract: In solid-state samples, either single crystal or powder, the hyperfine structure appears anisotropic. The spectra are more complex and more difficult to analyze, as well as the information they provide. This chapter focuses on the relationship between the hyperfine coupling tensor and the molecular structure, as well as on how to get the A tensor from the experimental spectra.

The experiments show that the amount, intensity, and composition of the hyperfine splitting should be changed when the orientation of solid samples is to be altered. For the samples with sufficiently large anisotropy, if the single crystal is rotated by only a very small angle, the features of spectrum can be changed largely. In order to make the problem as simple as possible, only the hyperfine coupling tensor $\boldsymbol{A}$ angle needs to be changed. And the changes accompanying the g tensor is ignored temporarily; at the same time, the electron spin of the system is restricted to $S = \frac{1}{2}$, and the hyperfine splitting arises from only a single nucleus.

A very simple example is the V_{OH} center [1,2] shown in Figure 5.1. It has a very strong anisotropic hyperfine interaction, but its g is almost isotropic. This center in MgO crystal is composed of linear defects of $^{-}$O□HO^{-} in which □ separates a paramagnetic O^{-} ion and the proton of a hydroxide impurity ion (by ~0.32 nm). If the crystal is rotated on the 100 $^{\infty}$ plane, taking the angle between the defect axis and the external magnetic field H as θ, the hyperfine coupling constants $\boldsymbol{A}(\theta)$ of hydrogen are given as follows:

$$A = A_o + \delta A\,(3\cos^2\theta - 1) \tag{5.1}$$

In this situation, using Eq. (3.4) and the relative data, we obtain the experimental form of the above equation as follows:

$$A/g_e\beta_e = 0.0016{+}0.08475(3cos^2\theta - 1)\cdot(mT)\cdot \tag{5.2}$$

The 0.1711 mT of $\theta = 0°$ becomes as 0 mT when $\cos^2\theta = (1 - 0.0016/0.08475)/3$ and 20.08315 mT of $\theta = 90°$. The doublet splitting is small enough (when microwave frequency is 9–10 GHz) and its magnitude is nearly equal to the order of $A/g_e\beta_e$. From Eq. (5.2), it can be seen that the proton hyperfine splitting is almost pure anisotropic. In most systems, this relates to the isotropic splitting term A_{o}, and the practical magnitude is of the same order with δA.

5.1 Anisotropic hyperfine interaction

In the previous chapter, we discussed the interaction of isotropic hyperfine interaction, which mainly originated from the interaction of Fermi. In addition, due to the rapid

https://doi.org/10.1515/9783110568578-005

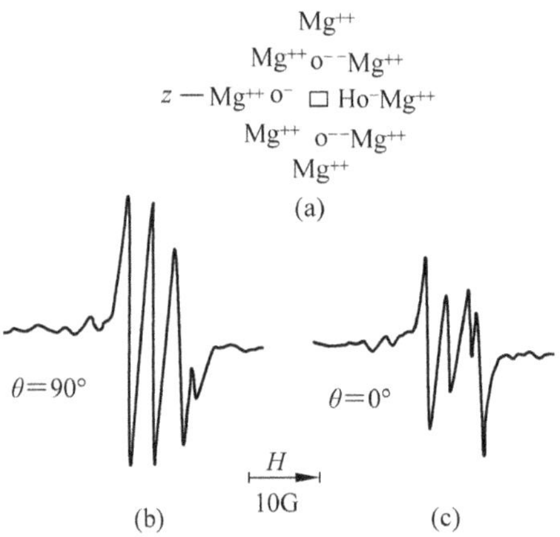

Figure 5.1: EPR spectra of the V_{OH} center in MgO. (a) Structure of the defect. The symmetry axis of the defect (tetragonal crystal axis) is labeled Z. (b) External magnetic field H is perpendicular to Z. (c) External magnetic field H is parallel to Z.

tumbling of magnetic particles in the low-viscosity liquid, the time-average value of anisotropic hyperfine interaction between the magnetic moment of the unpaired electron and the nuclear dipole approach to zero. However, in the rigid system, there is also a dipole–dipole interaction besides the Fermi contact interaction. This interaction strongly depends on orientation and exhibits a strict anisotropy. Therefore, it is a tensor. The classical theoretical expression of the dipolar interaction energy between the electron and the nucleus, separated by a distance r, is given by [3–5]

$$E_{dipolar}(\boldsymbol{r}) = \frac{\mu_e \cdot \mu_N}{r^3} - \frac{3(\mu_e \cdot r)(\mu_N \cdot r)}{r^5} \tag{5.3}$$

Here, $\boldsymbol{r}$ is the vector denoting the distance between the unpaired electron and the nucleus (see Figure 4.1). Vectors μ_e and μ_N are the electronic and nuclear magnetic moments, respectively. Equation (5.3) shows that the magnetic dipolar interactive energy between electron and nucleus is related to r^{-3} and is independent of the sign of $\boldsymbol{r}$. The dipolar interaction exist irrespective of whether the external magnetic field is applied or not.

For a system of quantum mechanics, Eq. (5.3) should be expressed by the corresponding operators. In order to make the problem simple, we shall temporarily ignore the anisotropy of g. It means g and g_n are seen as isotropic. Substituting $\mu_N = -g\beta\hat{S}$ and $\mu_N = +g_n\beta_n\hat{I}$ into Eq. (5.3), we obtain the Hamiltonian equation as

$$\hat{\mathscr{H}}_{dipolar}(\boldsymbol{r}) = -g\beta g_n\beta_n\left[\frac{\hat{S}\cdot\hat{I}}{r^3} - \frac{3(\hat{S}\cdot\boldsymbol{r})(\hat{I}\cdot\boldsymbol{r})}{r^5}\right] \tag{5.4}$$

Here $\hat{\mathscr{H}}_{dipolar}(\boldsymbol{r}) =$ is the energy of an anisotropic hyperfine dipolar interaction by vector. We put the magnetic nucleus at the original point of coordinate axes (x, y, z) of crystal sample, thus splitting Eq. (5.4) as follows:

$$\hat{\mathscr{H}}_{dipolar}(\boldsymbol{r}) = -g\beta g_n\beta\left[\left\langle\frac{r^2-3x^2}{r^5}\right\rangle\hat{S}_x\hat{I}_x+\left\langle\frac{r^2-3y^2}{r^5}\right\rangle\hat{S}_y\hat{I}_y+\left\langle\frac{r^2-3z^2}{r^5}\right\rangle\hat{S}_z\hat{I}_z\right.$$
$$\left.-\left\langle\frac{3xy}{r^5}\right\rangle(\hat{S}_x\hat{I}_y+\hat{S}_y\hat{I}_x)-\left\langle\frac{3yz}{r^5}\right\rangle(\hat{S}_y\hat{I}_z+\hat{S}_z\hat{I}_y)-\left\langle\frac{3zx}{r^5}\right\rangle(\hat{S}_z\hat{I}_x+\hat{S}_x\hat{I}_z)\right] \tag{5.5}$$

The value within angle brackets is the average value for the wave function of electron on the whole spatial distribution. Thus, Eq. (5.5) can be written as follows:

$$\hat{\mathscr{H}}_{dipolar}(r) = [\hat{S}_x\cdot\hat{S}_y\cdot\hat{S}_z]\begin{bmatrix} T_{xx} & T_{xy} & T_{xz} \\ T_{yx} & T_{yy} & T_{yz} \\ T_{zx} & T_{zy} & T_{zz}\end{bmatrix}\begin{bmatrix}\hat{I}_x \\ \hat{I}_y \\ \hat{I}_z\end{bmatrix} \tag{5.6a}$$

$$\hat{\mathscr{H}}_{dipolar} = \hat{\boldsymbol{S}}\cdot\boldsymbol{T}\cdot\hat{\boldsymbol{I}} \tag{5.6b}$$

Matrix element

$$T_{ij} = -g\beta g_n\beta_n\left\langle\frac{r^2\delta_{ij}-3ij}{r^5}\right\rangle\cdots(i,j=x,y,z)\cdots\cdot \tag{5.7}$$

$\boldsymbol{T}$ in Eq. (5.6b) is traceless matrix, that is, $T_r(\boldsymbol{T}) = T_{xx}+T_{yy}+T_{zz} = 0$. Since $\boldsymbol{T}$ is a traceless matrix, the rapid tumble of radical in solution is averaged, which is $\frac{1}{3}T_r(\boldsymbol{T}) = 0$.

The complete spin Hamiltonian should include the isotropic hyperfine term A_o and the electronic as well as nuclear Zeeman terms. Thus,

$$\hat{\mathscr{H}} = g\beta_e\boldsymbol{H}\cdot\hat{\boldsymbol{S}}+\hat{\boldsymbol{S}}\cdot\boldsymbol{A}\cdot\hat{\boldsymbol{I}}-g_n\beta_n\boldsymbol{H}\cdot\hat{\boldsymbol{I}} \tag{5.8}$$

where the hyperfine parameter should be (3×3) matrix; here $\boldsymbol{A}$ is

$$\boldsymbol{A} = A_o\boldsymbol{1}_3+\vec{\boldsymbol{T}} \tag{5.9}$$

Here A_o is the isotropic hyperfine coupling; $\boldsymbol{1}_3$ is the (3×3) unit matrix.

The general form of the Hamiltonian operator of hyperfine interaction is

$$\hat{\mathscr{H}}_{\text{hf}} = \hat{\boldsymbol{S}}\cdot\boldsymbol{A}\cdot\hat{\boldsymbol{I}} \tag{5.10}$$

For the multi-nuclear system,

$$\hat{\mathscr{H}}_{\text{hf}} = \sum_i\hat{\boldsymbol{S}}\times\boldsymbol{A}_i\times\hat{\boldsymbol{I}}_i \tag{5.11}$$

5.2 Matrix interpretation of anisotropic hyperfine interaction

Here, we only introduce the electronic Zeeman energy dominant system, and $g = g_e\boldsymbol{n}$ (anisotropy of g is very weak). $\hat{\boldsymbol{S}}$ is quantized along the direction of external magnetic field $\boldsymbol{H}$, that is, $\boldsymbol{H} = H\boldsymbol{n}$; thus, $\hat{S} = \hat{S}_H\boldsymbol{n}$. Here $\boldsymbol{n}$ is the unit vector along the direction $\boldsymbol{H}$; then Eq. (5.8) can be written as follows:

$$\hat{\mathscr{H}} = g\beta_e H\hat{S}_H - g_n\beta_n\left(\frac{-\hat{S}_H}{g_n\beta_n}\boldsymbol{A}^T\boldsymbol{n}+\boldsymbol{H}\right)\cdot\hat{\boldsymbol{I}}$$

$$= g\beta_e HMs - g_n\beta_n\left(\frac{-M_S}{g_n\beta_n}\boldsymbol{A}^T\boldsymbol{n}+\boldsymbol{H}\right)\cdot\hat{\boldsymbol{I}} \tag{5.12}$$

Or

$$\hat{\mathscr{H}} = \beta_e H\hat{S}_H - g_n\beta_n\boldsymbol{H}_{eff}\cdot\hat{\boldsymbol{I}} \tag{5.13}$$

where $|M_s\rangle$ is the eigenfunction of $\hat{\boldsymbol{S}}_H$, the corresponding eigenvalue $Ms = \pm\frac{1}{2}\boldsymbol{H}_{\text{eff}}$ is the effective magnetic field:

$$\boldsymbol{H}_{\text{eff}} = \boldsymbol{H} + \boldsymbol{H}_{\text{hf}} \tag{5.14}$$

$$\boldsymbol{H}_{\text{hf}} = \frac{-M_S}{g_n\beta_n}\boldsymbol{A}^T\cdot\boldsymbol{n} \tag{5.15}$$

$$\boldsymbol{H}_{\text{hf}}^T = \frac{-M_S}{g_n\beta_n}\boldsymbol{n}^T\boldsymbol{A}$$

The physical meaning of $\boldsymbol{H}_{\text{hf}}$ is hyperfine splitting magnetic field at the nucleus, which can be very large, such as the proton hyperfine coupling constant is 100° MHz; thus, $|\boldsymbol{H}_{\text{hf}}|$ can be 1.17 T.

It should be pointed out that the hyperfine magnetic field on the nucleus $\boldsymbol{H}_{\text{hf}}$ should not be confused with the hyperfine magnetic field of nucleus established on the electron $\frac{1}{2}|\Delta\boldsymbol{H}_{\text{hf}}|$

$$|\boldsymbol{H}_{\text{hf}}| = \left|\frac{-M_S\boldsymbol{1}\cdot\boldsymbol{A}}{\frac{g_n\beta_n}{h}}\right| = \frac{\frac{1}{2}\times 100\times 10^6\text{s}^{-1}}{\frac{5.585\times 5.051\times 10^{-20}erg/T}{6.626\times 10^{-27}erg\cdot\text{s}}} = 1.1744\,T$$

and

$$\frac{1}{2}|\Delta\boldsymbol{H}_{\text{hf}}| = \left|\frac{-M_S 1\cdot\boldsymbol{A}}{\frac{g_e\beta}{h}}\right| = \frac{\frac{1}{2}\times 100\times 10^6\text{s}^{-1}}{\frac{2.0023\times 9.274\times 10^{-17}erg/\text{T}}{6.626\times 10^{-27}\text{erg}\cdot\text{s}}} = 1.784\,\text{mT}$$

The physical meaning of $|\Delta\boldsymbol{H}_{\text{hf}}|$ is the splitting between the two hyperfine splitting lines, its half is the hyperfine magnetic field of the nucleus established on the electron. Here 1 is the unit vector.

The relationship between the vector of external magnetic field $\boldsymbol{H}$ and the hyperfine magnetic field of the nucleus $\boldsymbol{H}_{\text{hf}}$ is shown in Figure 5.2. For system $S = I = \frac{1}{2}$ superscripts α and β represent $Ms = +\frac{1}{2}$ and $Ms = -\frac{1}{2}$; we define $\boldsymbol{H}^{\alpha}_{\text{hf}} = -\boldsymbol{H}^{\beta}_{\text{hf}}$, and it can be concluded that (a)$|\boldsymbol{H}| < |\boldsymbol{H}|$; (b)$|\boldsymbol{H}| \approx |\boldsymbol{H}_{\text{hf}}|$; (c)$|\boldsymbol{H}| > |\boldsymbol{H}_{\text{hf}}|$.

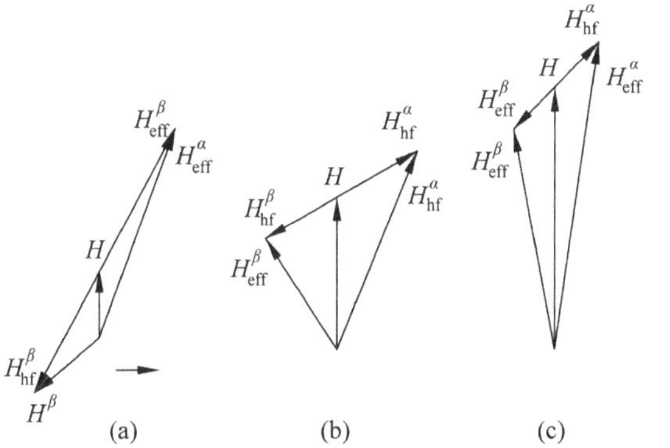

Figure 5.2: Vector relationship between external magnetic field H and hyperfine magnetic field $H_{\text{hf.}}$

Case 1:

$|\boldsymbol{H}| \gg |\boldsymbol{H}_{\text{hf}}|$ ($|\boldsymbol{A}| < 1.0mT$). This case is very rare. Under this case, $\hat{\boldsymbol{I}}$ is quantized along $\boldsymbol{H}$, that is, $\hat{\boldsymbol{I}} = \hat{I}_H\boldsymbol{1}$. Hence, if the eigenfunction of $\hat{I}_H$ is $|M_I\rangle$, then

$$\hat{\mathscr{H}} = g\beta_e \boldsymbol{H} M_S + M_S M_I(\boldsymbol{n}^{\text{T}} \cdot \boldsymbol{A} \cdot \boldsymbol{n}) - g_n\beta_n \boldsymbol{H} M_I \tag{5.16}$$

Let $\boldsymbol{H}$ be in z-direction, then

$$\boldsymbol{n}^{\text{T}}\boldsymbol{A} \cdot \boldsymbol{n} = A_{zz} = A_0 + T_{zz}$$
$$T_{zz} = g\beta_e g_n \beta_n \left\langle \frac{3\cos^2\theta - 1}{r^3} \right\rangle \tag{5.17}$$

where θ is the angle between vectors $\boldsymbol{r}$ and $\boldsymbol{H}$, nucleus is located on the origin of the coordinates (x, y, z), vector $\boldsymbol{r}$ is the joining line between the nucleus and the unpaired electron, and $\boldsymbol{H}$ is an external magnetic field vector parallel to the $\boldsymbol{Z}$-axis; let α be the angle between vector $\boldsymbol{p}$ and vector $\boldsymbol{r}$ of the sp hybrid orbital where the unpaired electron is located; let Θ be the angle between vector $\boldsymbol{p}$ and the external magnetic field $\boldsymbol{H}$, that is, θ_P shown in Figure 5.3.

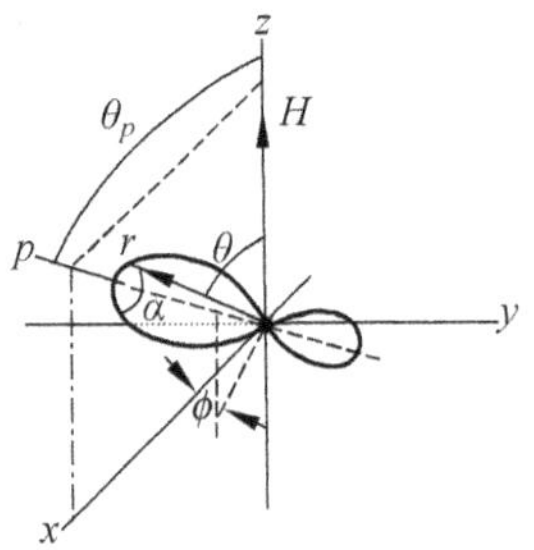

Figure 5.3: The angular relationship of the vector r, H, and p in coordinate axis.

It is thus proved that [6, 7]

$$\left\langle \frac{3\cos^2\theta - 1}{r^3} \right\rangle = \left\langle \frac{3\cos^2\alpha - 1}{2r^3} \right\rangle (3\cos^2\Theta - 1) \tag{5.18}$$

Let $B = g\beta_e g_n \beta_n \langle \frac{3\cos^2\alpha - 1}{2r^3} \rangle$, then Eq. (5.16) can be written as

$$\hat{\mathscr{H}} = g\beta_e H M_S + [A_o + B(3cos^2\Theta - 1)] M_S M_I - g_n \beta_n H M_I \tag{5.19}$$

which is the spin Hamiltonian expression for case 1($|\boldsymbol{H}| \gg |\boldsymbol{H}_{\text{hf}}|$).

Case 2:

($|\boldsymbol{H}| \ll |\boldsymbol{H}_{\text{hf}}|$). This is the most common one, as long as $|\boldsymbol{A}| > 1.0$ mT. Here $\hat{\boldsymbol{I}}$ and $\hat{\boldsymbol{S}}$ are quantization with different directions ($\hat{\boldsymbol{S}}$ along $\boldsymbol{H}$ direction and $\hat{\boldsymbol{I}}$ along $\boldsymbol{H}_{\text{hf}}$ direction). Thus, Eq. (5.5) becomes

$$\hat{\mathscr{H}}_{\text{dipolar}} = g\beta_e g_n \beta_n \left[\left\langle \frac{3z^2 - r^2}{r^5} \right\rangle S_Z I_Z + \left\langle \frac{3xz}{r^5} \right\rangle S_Z I_x + \left\langle \frac{3yz}{r^5} \right\rangle S_z I_y \right] \tag{5.20}$$

It proves that

$$g\beta_e g_n \beta_n \left\langle \frac{3xz}{r^5} \right\rangle = 3B \sin\Theta \cos\Theta \tag{5.21}$$

$$g\beta_e g_n \beta_n \left\langle \frac{3yz}{r^5} \right\rangle = 0 \tag{5.22}$$

Hence,

$$\begin{aligned} \hat{\mathscr{H}} &= g\beta_e H M_s + M_s\{[A_o + B(3cos^2\Theta - 1)]\hat{I}_z + 3B \sin\Theta cos\Theta \hat{I}_x\} \\ &= g\beta_e H M_s - g_n \beta_n \{H_{\|}\hat{I}_z + H_{\perp}\hat{I}_x\} \end{aligned} \tag{5.23}$$

where $H_{\|}$ and $H_{\perp}$ are given by

$$H_{||} = \frac{-M_s}{g_n\beta_n}[A_o + B3cos^2\Theta - 1)] \tag{5.24}$$

$$H_{\perp} = \frac{-M_s}{g_n\beta_n}[B3\sin\Theta\cos\Theta] \tag{5.25}$$

If the eigenfunction of $\hat{I}_z$ are $|\alpha_n\rangle$ and $|\beta_n\rangle$, take $|\alpha_n\rangle$ and $|\beta_n\rangle$ as basic function, the matrix representation of spin Hamiltonian is as follows:

$$\hat{\mathscr{H}} = \begin{array}{c} \langle\alpha_n| \\ \langle\beta_n| \end{array} \overset{\begin{array}{cc} |\alpha_n\rangle & |\beta_n\rangle \end{array}}{\begin{bmatrix} g\beta HM_S - \frac{1}{2}g_n\beta_n H_{||} & -g_n\beta_n H_{\perp} \\ -\frac{g_n\beta_n H_{\perp}}{2} & g\beta HM_S + \frac{g_n\beta_n H_{||}}{2} \end{bmatrix}} \tag{5.26}$$

Energy of the system:

$$\begin{aligned} E &= g\beta_e HM_s \mp \frac{g_n\beta_n}{2}[H_{||}{}^2 + H_{\perp}{}^2]^{\frac{1}{2}} = g\beta_e HM_s \mp \frac{g_n\beta_n}{2}|\boldsymbol{H}_{\mathrm{hf}}| \\ &= g\beta_e HM_s \pm \frac{M_s}{2}\{[A_o + B(3cos^2\Theta - 1]^2 + 9B^2sin^2\Theta cos^2\Theta\}^{1/2} \\ &= g\beta_e HM_s \pm \frac{M_s}{2}\{(A_o - B)^2 + 3B(2A_o + B)cos^2\Theta\}^{1/2} \end{aligned} \tag{5.27}$$

Let

$$A = \frac{1}{g\beta}\left\{(A_o - B)^2 + 3B(2A_o + B)cos^2\Theta\right\}^{1/2} \tag{5.28}$$

Resonant magnetic field for the EMR transition with the fixed microwave frequency is given by

$$H_r = \frac{hv}{g\beta} \pm \frac{1}{2}\boldsymbol{A} \quad = H_o \pm \frac{1}{2}\boldsymbol{A} \tag{5.29}$$

In practical operation, we often encounter two special situations:

(1) $|A_o| \approx 0$

$$\boldsymbol{A} = \frac{B}{g\beta}(1 + 3\cos^2\Theta)^{1/2} \tag{5.30}$$

(2) $|A_o| > B$

$$\begin{aligned} \boldsymbol{A} &= \frac{B}{g\beta}[A_o^2 - 2A_oB + 6A_oB\cos^2\Theta]^{\frac{1}{2}} \approx \frac{1}{g\beta}A_o\left[1 + \frac{2B}{A_o}(3cos^2\Theta - 1)\right]^{1/2} \\ &= \frac{1}{g\beta}[A_o + B\cos^2\Theta - 1] \end{aligned} \tag{5.31}$$

Case 3:

$|\boldsymbol{H}| \approx |\boldsymbol{H}_{hf}|$. This is the most complicated case and also the most general case [8, 9]. From Eq. (5.13),

$$\hat{\mathscr{H}} = g\beta_e \boldsymbol{H}\hat{S}_H - g_n\beta_n \boldsymbol{H}_{\text{eff}} \cdot \hat{I}$$

We can write that

$$\begin{aligned}\hat{\mathscr{H}} &= g\beta_e \boldsymbol{H}\hat{S}_H - g_n\beta_n \boldsymbol{H}_{\text{eff}}(M_S) \cdot \hat{I} \\ &= g\beta_e \boldsymbol{H}\hat{S}_H - g_n\beta_n |\boldsymbol{H}_{\text{eff}}(M_S)| \hat{I}_{H_{\text{eff}(M_S)}}\end{aligned} \tag{5.32}$$

Assume $|\alpha_e\alpha_n{}'\rangle$ and $|\alpha_e\beta_n{}'\rangle$ are the eigenfunctions of $\hat{\mathscr{H}}^\alpha$, $|\beta_e\alpha_n{}''\rangle$ and $|\beta_e\beta_n{}''\rangle$ are the eigenfunctions of $\hat{\mathscr{H}}^\beta$; here $|\alpha_n{}'\rangle$ and $|\beta_n{}'\rangle$ are the eigenfunctions of $\hat{\mathscr{I}}_{H^\alpha_{eff}}$, $|\alpha_n{}''\rangle$ and $|\beta_n{}''\rangle$ are the eigenfunctions of $\hat{\mathscr{I}}_{H^\beta_{eff}}$, that is,

$$\hat{\mathscr{I}}_{H^\alpha_{\text{eff}}} \begin{bmatrix} |\alpha_n{}'\rangle \\ |\beta_n{}'\rangle \end{bmatrix} = \begin{bmatrix} \frac{1}{2}|\alpha_n{}'\rangle \\ -\frac{1}{2}|\beta_n{}'\rangle \end{bmatrix} \tag{5.33}$$

$$\hat{\mathscr{I}}_{H^\beta_{\text{eff}}} \begin{bmatrix} |\alpha_n{}''\rangle \\ |\beta_n{}''\rangle \end{bmatrix} = \begin{bmatrix} \frac{1}{2}|\alpha_n{}''\rangle \\ -\frac{1}{2}|\beta_n{}''\rangle \end{bmatrix} \tag{5.34}$$

It should be pointed out that $|\alpha_n{}'\rangle|\beta_n{}'\rangle$ and $|\alpha_n{}''\rangle|\beta_n{}''\rangle$ are different. Let the angle between $\boldsymbol{H}^\alpha_{\text{eff}}$ and $\boldsymbol{H}^\beta_{\text{eff}}$ be ω; it could be proved that these two sets of nuclear spin wave functions has the following relationships:

$$\left|\alpha_n{}'\right\rangle = \cos\frac{\omega}{2}\left|\alpha_n{}''\right\rangle - \sin\frac{\omega}{2}\left|\beta_n{}''\right\rangle \tag{5.35}$$

$$\left|\beta_n{}'\right\rangle = \sin\frac{\omega}{2}\left|\alpha_n{}''\right\rangle + s\cos\frac{\omega}{2}\left|\beta_n{}''\right\rangle \tag{5.36}$$

Hence, after choosing $|\alpha_e\alpha_n{}'\rangle$, $|\alpha_e\beta_n{}'\rangle$, $|\beta_e\alpha_n{}''\rangle$, $|\beta_e\beta_n{}''\rangle$ as basic functions, energy of the system can be calculated:

$$E_{\alpha_e\alpha_n{}'} = \left\langle \alpha_e\alpha_n{}' \middle| \hat{\mathscr{H}}^\alpha \middle| \alpha_e\alpha_n{}' \right\rangle = \frac{1}{2}g\beta H - \frac{1}{2}g_n\beta_n|\boldsymbol{H}^\alpha_{\text{eff}}| \tag{5.37}$$

$$E_{\alpha_e\beta_n{}'} = \left\langle \alpha_e\beta_n{}' \middle| \hat{\mathscr{H}}^\alpha \middle| \alpha_e\beta_n{}' \right\rangle = \frac{1}{2}g\beta H + \frac{1}{2}g_n\beta_n|\boldsymbol{H}^\alpha_{\text{eff}}| \tag{5.38}$$

$$E_{\alpha_e\beta_n{}''} = \left\langle \beta_e\alpha_n{}'' \middle| \hat{\mathscr{H}}^\beta \middle| \beta_e\alpha_n{}'' \right\rangle = -\frac{1}{2}g\beta H - \frac{1}{2}g_n\beta_n|\boldsymbol{H}^\beta_{\text{eff}}| \tag{5.39}$$

$$E_{\beta_e\beta_n{}''} = \left\langle \beta_e\beta_n{}'' \middle| \hat{\mathscr{H}}^\beta \middle| \beta_e\beta_n{}'' \right\rangle = -\frac{1}{2}g\beta H + \frac{1}{2}g_n\beta_n|\boldsymbol{H}^\beta_{\text{eff}}| \tag{5.40}$$

According to the selection rule, $\Delta M_S = 1$, there are four allowed transitions:

$$\Delta E_1 = E_{\alpha_e\beta_n'} - E_{\beta_e\alpha_n''} = g\beta H + \frac{1}{2}g_n\beta_n\left\{|\boldsymbol{H}_{\text{eff}}^{\alpha}| + |\boldsymbol{H}_{\text{eff}}^{\beta}|\right\} \tag{5.41}$$

$$\Delta E_2 = E_{\alpha_e\beta_n'} - E_{\beta_e\beta_n''} = g\beta H + \frac{1}{2}g_n\beta_n\left\{|\boldsymbol{H}_{\text{eff}}^{\alpha}| - |\boldsymbol{H}_{\text{eff}}^{\beta}|\right\} \tag{5.42}$$

$$\Delta E_3 = E_{\alpha_e\alpha_n'} - E_{\beta_e\alpha_n''} = g\beta H - \frac{1}{2}g_n\beta_n\left\{|\boldsymbol{H}_{\text{eff}}^{\alpha}| - |\boldsymbol{H}_{\text{eff}}^{\beta}|\right\} \tag{5.43}$$

$$\Delta E_4 = E_{\alpha_e\alpha_n'} - E_{\beta_e\beta_n''} = g\beta H - \frac{1}{2}g_n\beta_n\left\{|\boldsymbol{H}_{\text{eff}}^{\alpha}| + |\boldsymbol{H}_{\text{eff}}^{\beta}|\right\} \tag{5.44}$$

Their relative intensity are given as

$$\begin{aligned}\mathscr{I}_1 &= |\langle\alpha_e\beta_n'|S^+|\beta_e\alpha''_n\rangle|^2 = |\langle\beta_n'|\alpha_n''\rangle|^2 \\ &= |\sin\tfrac{\omega}{2}|\langle\alpha_n''|\alpha_n''\rangle + \cos\tfrac{\omega}{2}\langle\beta_n''|\alpha_n''\rangle|^2 = \sin^2\tfrac{\omega}{2}\end{aligned} \tag{5.45}$$

$$\mathscr{I}_2 = |\langle\alpha_e\beta_n'|S^+|\beta_e\beta''_n\rangle|^2 = \cos^2\frac{\omega}{2} \tag{5.46}$$

Likewise, $\mathscr{I}_3 = \mathscr{I}_2, \mathscr{I}_4 = \mathscr{I}_1$. In this case, ω is between 30 ° and 70 °. Figure 5.4 shows (a) the energy levels of the system with $S = I = 1/2 (g_n > 0)$ and $\omega \approx 70$ and (b) the observed EMR spectrum.

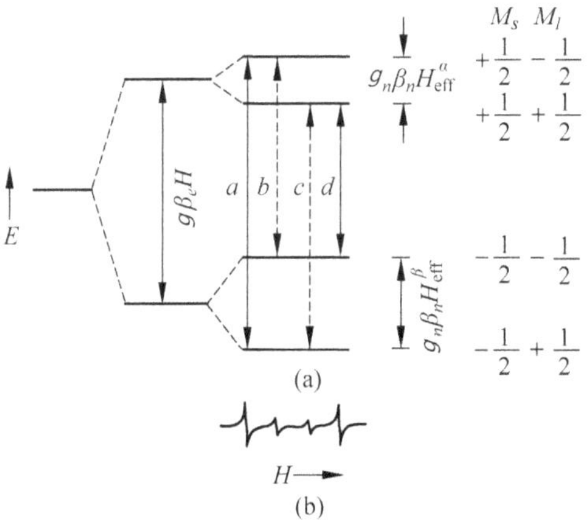

Figure 5.4: (a) Energy levels of the system with $S = I = 1/2$ and $\omega \approx 70$. (b) The observed EMR spectrum.

Now, let us switch back to see cases 1 and 2; actually they are two extreme cases of case 3:

(1) When $\omega \to 0$ (i.e., $|\boldsymbol{H}| > |\boldsymbol{H}_{\text{hf}}|$), this is case 1. Here $|\alpha_n'\rangle \approx |\alpha_n''\rangle$, then $\mathscr{I}_2 \approx \mathscr{I}_3 > \mathscr{I}_1 \approx \mathscr{I}_4$, which means 2, 3 are strong transitions and 1,4 are two lines that are very weak. $\Delta E_2 - \Delta E_3 = g_n\beta_n\left\{|\boldsymbol{H}_{\text{eff}}^{\alpha}| - |\boldsymbol{H}_{\text{eff}}^{\beta}|\right\} \approx 2g_n\beta_n|H_{\text{hf}}^*|$, where $|H_{\text{hf}}^*| = |\boldsymbol{n}^{\text{T}} \cdot \boldsymbol{A} \cdot \boldsymbol{n}/2g_n\beta_n|$.

(2) When $\omega \to 180°$ (i.e., $|\boldsymbol{H}| < |\boldsymbol{H}_{\mathrm{hf}}|$), this is case 2. Here $|\alpha_n'\rangle \approx |\beta_n''\rangle$, then $\mathcal{I}_1 \approx \mathcal{I}_4 > \mathcal{I}_2 \approx \mathcal{I}_3$, which means 1, 4 are strong transitions and 2, 3 are two lines that are very weak. $\Delta E_1 - \Delta E_4 = g_n\beta_n\left\{|\boldsymbol{H}^{\alpha}_{\mathrm{eff}}| + |\boldsymbol{H}^{\beta}_{\mathrm{eff}}|\right\} \approx 2g_n\beta_n|\boldsymbol{H}_{\mathrm{hf}}|$, where $|\boldsymbol{H}_{\mathrm{hf}}| = \left[\boldsymbol{n}^{\mathrm{T}} \cdot \boldsymbol{A} \cdot \boldsymbol{A}^{\mathrm{T}} \cdot \boldsymbol{n}/4g_n^2\beta_n^2\right]^{1/2}$.

(3) For case 3, four lines are commonly observed and their intensity ratio accords with Eqs. (5.45) and (5.46).

5.3 Example demonstration

Perfluoro succinic acid sodium salt that is monoclinic crystal to be irradiated by γ-ray can produce the radical $^{-}\mathrm{OOC} - \dot{\mathrm{C}}\mathrm{F} - \mathrm{CF}_2 - \mathrm{COO}^{-}$ as an actual example to analyze the hyperfine coupling tensor of its α fluorine atom.

Since it is monoclinic crystal, the crystal axes c and b are perpendicular to each other, and can be chosen as the two axes of the experimental orthogonal coordinate system. However, the angle between a- and c-axes is 106 °; therefore, it has to choose another one as an a^* axis perpendicular to the bc plane. The magnetic field $\boldsymbol{H}$ is rotated in the inner plane of the a^*b, bc, and a^*c respectively (magnetic field are fixed generally, and the crystals are glued on a rod with a dial to be rotated). The EMR spectra are recorded at various angles (Figure 5.5). The hyperfine splitting space are measured from the spectrum to seek the matrix element of $\boldsymbol{A}^2$ tensor(herein ignoring the nuclear Zeeman term). The data are collected in Table 5.1, and the angular dependence of hyperfine coupling tensor is shown in Figure 5.6.

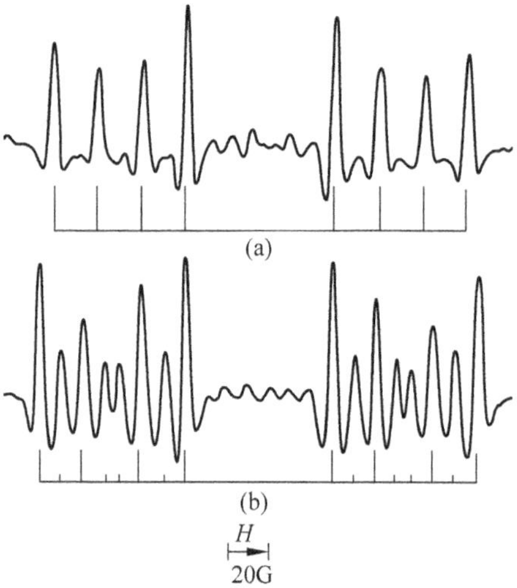

Figure 5.5: The measured second-derivative spectra of perfluoro succinic acid radical at 300 K for $H \parallel b$ at X-band (a) and Q-band (b).

The matrix form of $\boldsymbol{A}^2$ tensor in a^*bc coordinate system can be obtained from Table 5.1:

Table 5.1: Hyperfine coupling data obtained from spectra.

Plane	Angle	$(A/h)^2 \times 10^4$ MHz^2	Tensor element
	0 °	1.61	$(\boldsymbol{AA})_{a^*a^*}$
a^*b	90 °	16.24	$(\boldsymbol{AA})_{bb}$
	45 °	13.84	$(\boldsymbol{AA})_{a^*b} = (13.84 - 4.29)/2 = 4.78 \times 10^4\ (\text{MHz})^2$
	135 °	4.29	
	0	16.48	$(\boldsymbol{AA})_{bb}$
bc	90	2.72	$(\boldsymbol{AA})_{cc}$
	45	9.67	$(\boldsymbol{AA})_{bc} = (9.67 - 9.99)/2 = -0.16 \times 10^4\ (\text{MHz})^2$
	135	9.99	
	0	2.69	$(\boldsymbol{AA})_{cc}$
ca^*	90	1.59	$(\boldsymbol{AA})_{a^*a^*}$
	45	2.69	$(\boldsymbol{AA})_{a^*c} = (2.69 - 1.42)/2 = 0.64 \times 10^4\ (\text{MHz})^2$
	135	1.42	

*These data are collected from Ref. [10], and are obtained with *X*- band at 300 K ignoring nuclear Zeeman term.

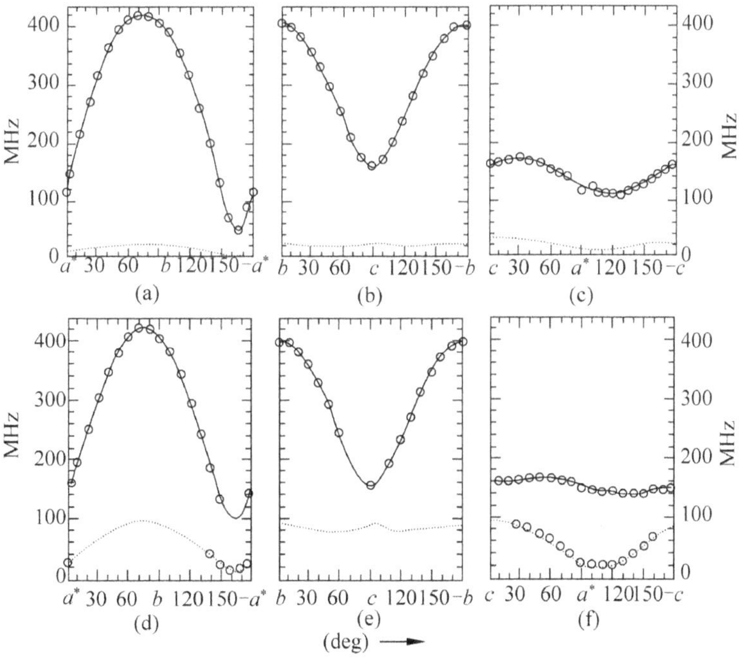

Figure 5.6: The angular dependence of hyperfine coupling tensor of perfluoro succinic acid radical.

$$\boldsymbol{AA}/h^2 \begin{bmatrix} 1.60 & \pm 4.78 & 0.64 \\ \pm 4.78 & 16.36 & \mp 0.16 \\ 0.64 & \mp 0.16 & 2.71 \end{bmatrix} \times 10^4 (\text{MHz})^2 \tag{5.4}$$

Since there are two possible radical sites that are symmetrical to each other in a single cell, the ± and ∓ in Eq. (5.47) are two radical tensor matrix of sites of mutual symmetry. The general form of the secular determinants of these two tensors is given by

$$\begin{bmatrix} a-\lambda & \pm d & e \\ \pm d & b-\lambda & \pm f \\ e & \pm f & c-\lambda \end{bmatrix} = 0 \tag{5.48}$$

On expanding, we get

$$\begin{aligned} &\lambda^3 - (a+b+c)\lambda^2 + (ab+bc+ca-d^2-e^2-f^2)\lambda \\ &- (abc+2def-e^2b-d^2c-f^2a) = 0 \end{aligned} \tag{5.49}$$

Substituting the specific data into Eq. 5.49 gives

$$\lambda^3 - (20.67\times 10^4)\lambda^2 + (51.56\times 10^8)\lambda - 1.30\times 10^{12} = 0 \tag{5.50}$$

Solving this algebra equation require three roots: 17.77, 2.87, and 0.025, the units are all $\times 10^4$ (MHz)^2:

$$d_{A^2} = \begin{bmatrix} 17.77 & 0 & 0 \\ 0 & 2.87 & 0 \\ 0 & 0 & 0.025 \end{bmatrix} \times 10^{-4} (MHz)^2 \tag{5.51}$$

$$d_A = \begin{bmatrix} 421.5 & 0 & 0 \\ 0 & 169.4 & 0 \\ 0 & 0 & 15.8 \end{bmatrix} (MHz)^2 \tag{5.52}$$

From Eq. 5.51, the square root yields Eq. (5.52), which can have positive or negative two possible signs. How to choose the sign is very important. We will discuss in detail next. The principle best agrees with the experimental spectrum.

The method of direction cosine seeks to solve the following equation set:

$$\left.\begin{aligned} (1.60-\lambda_i)\ell_{i1} \pm 4.78\ell_{i2} + 0.64\ell_{i3} = 0 \\ \pm 4.78\ell_{i1} + (16.36-\lambda_i)\ell_{i2} \mp 0.16\ell_{i3} = 0 \\ 0.64\ell_{i1} \mp 0.16\ell_{i2} + (2.71-\lambda_i)\ell_{i3} = 0 \\ \ell_{i1}^2 + \ell_{i2}^2 + \ell_{i3}^2 = 1 \end{aligned}\right\} \tag{5.53}$$

Only three of these four equations are independent. Substituting $\lambda_1, \lambda_2, \lambda_3$ yields

$$\mathscr{L} = \begin{bmatrix} \ell_{11} & \ell_{12} & \ell_{13} \\ \ell_{21} & \ell_{22} & \ell_{23} \\ \ell_{31} & \ell_{32} & \ell_{33} \end{bmatrix} = \begin{bmatrix} 0.282 & \pm 0.958 & 0.001 \\ 0.223 & \mp 0.069 & 0.972 \\ 0.933 & \mp 0.278 & -0.235 \end{bmatrix} \tag{5.54}$$

This $\mathscr{L}$ is the required direction cosine matrix. It is easy to verify whether it enables $\boldsymbol{A}^2$ matrix to be diagonalized, that is,

$$\mathscr{L}\, A^2 \mathscr{L}^\dagger = {}^{\mathrm{d}}\boldsymbol{A}^2 \tag{5.55}$$

A matrix that can has diagonalization of the $\boldsymbol{A}$ matrix should also be diagonalized making the $\boldsymbol{A}^2$ matrix, that is,

$$\mathscr{L}\, A^2 \mathscr{L}^\dagger = {}^{\mathrm{d}}\boldsymbol{A} \tag{5.56}$$

Hence,

$$\mathscr{L} A^2\, \mathscr{L}^\dagger = \mathscr{L} A\, \mathscr{L}^\dagger\, \mathscr{L}\boldsymbol{A}\, \mathscr{L}^\dagger = {}^{\mathrm{d}}\boldsymbol{A} \bullet {}^{\mathrm{d}}\boldsymbol{A} = {}^{\mathrm{d}}\boldsymbol{A}^2 \tag{5.57}$$

For the above calculation, the following should be pointed out:

(1) How do you know ± 4.87 corresponds to ∓ 0.16 but not to ± 0.16 from the experimental spectrum? Only from the experimental spectra of these three faces (a^*b, bc, ca^*), one still cannot determine the corresponding sign. The EMR spectrum of other concrete location should therefore also be analyzed.

(2) In the process of extraction of square root of ${}^d\boldsymbol{A}^2$ in Eq. (5.51) to yield ${}^d\boldsymbol{A}$ in Eq. (5.52), every principal value will have two possible signs of positive or negative. Hence, the signs of three principal values of ${}^d\boldsymbol{A}$ have four possible permutations as follows:

$$(+,\ +,\ +);\ (+,\ +,\ -);\ (+,\ -,\ +);\ (+,\ -,\ -)$$

Which is the right one? It is not easy to judge generally. If our instrument has two bands (X- and Q-bands), we can solve it from experiment: since the nuclear Zeeman energy would be increased, the hyperfine coupling term would be unchanged in the Q-band. In this way, we can use the obtained tensor data to calculate the splitting space of spectral lines; the calculating results are listed in Table 5.2.

Figure 5.6a–c shows hyperfine coupling parameter measured by X-band spectrometer at 300 K, depending on the relationship of the angle and the nuclear Zeeman term that can be ignored. While Figure 5.6d–f shows hyperfine coupling parameter measured by Q-band spectrometer under 300 K, depending on the relationship of the angle and the nuclear Zeeman item that cannot be ignored under this situation.

Table 5.2: Comparison of the experimental value of hyperfine coupling parameters of α-fluorine atom measured by Q-band spectrometer with the calculation value obtained by choice of different signs.

Direction cosine of magnetic field	Experimental value	Choice of different signs			
		(+++)	(++-)	(+-+)	(+--)
[1, 0, 0]	153	152(0.59)	154(0.58)	153(0.58)	154(0.58)
	29	31(0.41)	18(0.42)	21(0.42)	85(0.42)
[0, 1, 0]	407	407(1.00)	407(1.00)	407(1.00)	407(0.99)
	–	96(0.00)	96(0.00)	96(0.00)	96(0.01)
[0, 0, 1]	162	163(0.96)	164(0.95)	164(0.95)	163(0.96)
	–	97(0.04)	96(0.05)	96(0.05)	97(0.04)
[cos30 °, 0, cos60 °]	170	169(0.75)	171(0.73)	177(0.70)	176(0.70)
	65	61(0.25)	54(0.27)	25(0.30)	32(0.30)
[cos50 °, 0, cos40 °]	148	149(0.63)	152(0.62)	156(0.61)	154(0.61)
	48	54(0.37)	44(0.38)	28(0.39)	37(0.39)
[−cos20 °, cos70 °, 0]	–	110(0.18)	112(0.20)	112(0.20)	111(0.19)
	17	19(0.82)	1(0.80)	46(0.80)	14(0.81)
[0, cos60 °, cos30 °]	252	252(0.95)	253(0.94)	266(0.86)	266(0.86)
	–	84(0.05)	83(0.06)	22(0.14)	30(0.14)

Figure 5.5b clearly shows the spectral line of "forbidden transitions." Under the condition of Q-band and at 300 K, in crystal coordinate system, $\boldsymbol{A}$ matrix is as follows (these data are taken from Refs [10,11]):

$$\boldsymbol{A} = \begin{bmatrix} 46.9 & \pm 103.3 & 32.5 \\ \pm 103.3 & 392.7 & \mp 5.9 \\ 32.5 & \mp 5.9 & 157.7 \end{bmatrix} (MHz) \tag{5.58}$$

$$^{d}\boldsymbol{A} = \begin{bmatrix} 421 & 0 & 0 \\ 0 & 165 & 0 \\ 0 & 0 & 11 \end{bmatrix} (MHz) \tag{5.59}$$

$$\mathscr{L} = \begin{bmatrix} 0.267 & \pm 0.964 & 0.011 \\ 0.208 & \mp 0.068 & 0.976 \\ 0.941 & \mp 0.258 & -0.219 \end{bmatrix} (MHz) \tag{5.60}$$

For (1, 0, 0) direction, $Ms\frac{1}{2}$, thus

$$-\frac{1}{2}(1,0,0)\begin{bmatrix} 46.9 & \pm 103.3 & 32.5 \\ \pm 103.3 & 392.7 & \mp 5.9 \\ 32.5 & \mp 5.9 & 157.7 \end{bmatrix} = [-23.5 \ \mp 51.7 \ -16.2] \ (\text{MHz}) \tag{5.61}$$

Under Q-band situation,

$$\nu_n = \frac{g_n\beta_n}{h}|H| = 49.\ \text{MHz} \tag{5.62}$$

$$\frac{g_n\beta_n}{h}\left|H_{eff}^{\alpha}\right| = [(49.5 - 23.5), \quad \mp 51.7, \quad -16.2]\ \text{MHz} \tag{5.63}$$

$$\frac{g_n\beta_n}{h}\left|H_{eff}^{\beta}\right| = [(49.5 + 23.5), \quad \pm 51.7, \quad +16.2]\ \text{MHz} \tag{5.64}$$

$$\frac{g_n\beta_n}{h}\left|H_{eff}^{\alpha}\right| = \sqrt{(26)^2 + (\mp 51.7)^2 + (-16.2)^2} = 60.1\ \text{MHz} \tag{5.65}$$

$$\frac{g_n\beta_n}{h}\left|H_{eff}^{\beta}\right| = \sqrt{(73)^2 + (\pm 51.7)^2 + (+16.2)^2} = 90.9\ \text{MHz} \tag{5.66}$$

The angle Θ between $\left|H_{eff}^{\alpha}\right|$ and $\left|H_{eff}^{\beta}\right|$ is

$$\cos\Theta = \left(\frac{26}{60.1}, \quad \frac{\mp 51.7}{60.1}, \quad \frac{-16.2}{60.1}\right)\begin{pmatrix} \frac{73}{90.9} \\ \frac{\pm 51.7}{90.9} \\ \frac{16.2}{90.9} \end{pmatrix} = -0.1895 \tag{5.67}$$

$\Theta = 101°$, hence,

$$\mathscr{I}_1 = \mathscr{I}_4 = \sin^2\frac{\Theta}{2} = 0.595 \tag{5.68}$$

$$\mathscr{I}_2 = \mathscr{I}_3 = \cos^2\frac{\Theta}{2} = 0.405 \tag{5.69}$$

$$\frac{\Delta E_1 - \Delta E_4}{h} = 151\ \text{MHz} \tag{5.70}$$

$$\frac{\Delta E_2 - \Delta E_3}{h} = 30.8\ \text{MHz} \tag{5.71}$$

From Table 5.2 we can see that with the choice of (+, +, +), the sign of the three principal values of ${}^{d}\boldsymbol{A}$ would be more coincident with the experimental value.

Now we turn to the analysis of the anisotropic hyperfine EMR spectra of the perfluoro succinic acid ion radical. From room temperature 300 K to cooling down to liquid nitrogen temperature 77 K, measuring the EMR spectra of perfluoro succinic acid radical and its parameter matrix has brought great change [11,12]. Since at room temperature the molecules are in fast vibrational states, the data given by Eq. (5.58) can only represent the time average value of the dipole interaction in the case of distortion and vibration. Matrix **A** (Eq. (5.58)) cannot be interpreted as the orientation and angle of a static molecular bond. In fact, there are two unequivalent directional radicals in crystallography, radical (I) and (II), as shown in Figure 5.7. Using the lifetime τ, the transformation time between the two states can be expressed as follows:

$$\tau^{-1} = 9.9 \times 10^{12} \exp(-\Delta U^{*}/RT) \quad s^{-1}$$

where activate energy $\Delta U^{*} = 15.26$ kJ/mol (the data are taken from Ref. [11]).

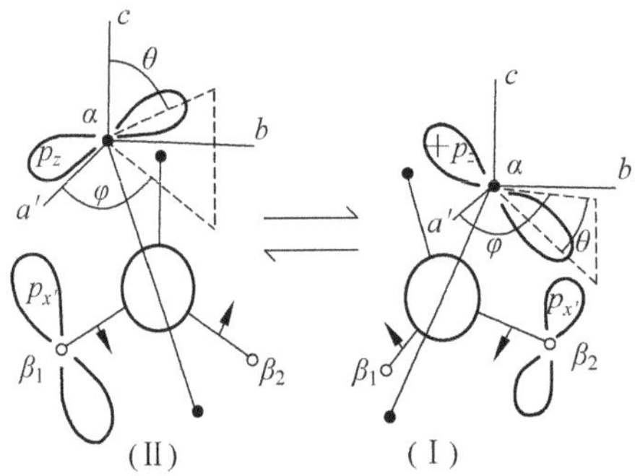

Figure 5.7: The projection of two configurations of perfluoro succinic acid radical along the C_α-C_β bond below 130 K [12].

The principal value and direction cosine measured at 77 K of ^{19}F nucleus hyperfine coupling tensor of configurations (I) and (II) of perfluoro succinic acid radical are listed in Table 5.3.

When the external magnetic field is parallel to the z-axis, the α-fluorine atom of these two configurations have the largest splitting value. This matrix form is the characterization of the electron spin density interaction with the nucleus on the p-orbital of the same atom.

5.4 Anisotropic hyperfine coupling tensor and structure of radical

5.4.1 Hyperfine coupling tensor of central atom

We call the unpaired electron locating atom as "central atom," such as the carbon atom in $\dot{C}H_3$, the nitrogen atom in $N(SO_3)_2^{2-}$, and so on. When the central atom is a magnetic nucleus, hyperfine interaction between the unpaired electron and the magnetic nucleus exist. The anisotropic part of the hyperfine tensor of the central

Table 5.3: The principal value and direction cosine of ^{19}F nucleus hyperfine coupling tensor of configurations (I) and (II) of perfluoro succinic acid radical at 77 K.

Radical configuration	Sign	Principal value (mT)	Spherical coordinate		Direction cosine ($a'bc$)		
			θ	ϕ			
(I)	A_α	21.7			−0.08059	± 0.87256	0.48090
		(+)0.7	61.26	±95.62	0.28079	± 0.44191	0.85198
		(+)0.2			0.95592	± 0.20822	−0.20704
	A_{β_1}	12.2			0.51634	± 0.59896	0.61207
		+4.1	52.3	± 49.2	0.52889	± 0.78516	0.32217
		+4.1			−0.67354	± 0.15737	0.72220
	A_{β_2}	0.9			0.42339	± 0.32094	0.84719
		−0.3	32.1	± 37.2	−0.67714	± 0.73336	0.06059
		−0.2			0.60185	± 0.59930	−0.52782
(II)	A_α	22.4			0.37758	± 0.73565	0.56236
		(+)0.5	55.78	± 62.38	−0.15214	± 0.54978	0.82134
		(+)0.4			0.91339	± 0.39568	−0.09566
	A_{β_1}	12.5			0.70222	± 0.13162	0.69969
		+4.4	45.6	± 10.6	−0.68619	± 0.13689	0.71442
		4.1			−0.18981	± 0.98180	0.00581
	A_{β_2}	1.4			0.34903	± 0.05819	0.93630
		−0.1	20.7	± 9.5	0.09252	± 0.99533	0.02739
		−0.2			0.93253	± 0.07697	−0.35278

atom depends on the electron density in its 2p-orbital, while the isotropic part depends on the electron density of the electrons in the s-orbital.

If the unpaired electron is located in $2\boldsymbol{p}_z$ orbital completely, the isotropic part of the hyperfine coupling tensor should be zero, that is $\boldsymbol{A} = \boldsymbol{T}$, where $\boldsymbol{T}$ is a traceless tensor. Since $2\boldsymbol{p}_z$ orbital is cylindrical symmetric, the principal value of ${}^d\boldsymbol{T}$ is as follows:

$$(T_{xx},\ T_{yy},\ T_{zz}) = (-B,\ -B,\ 2B) \tag{5.72}$$

where $B = \frac{2}{5} g\beta g_n\beta_n \left\langle \frac{1}{r^3} \right\rangle$
Let us prove it now:

Let the axis of p-orbital be z-axis; thus,

$$\psi_{2p_z} = \sqrt{\frac{3}{4\pi}} \cos\theta f(r) \tag{5.73}$$

$$
\begin{aligned}
T_{zz} &= \left\langle \psi_{2pz} \middle| g\beta g_n\beta_n \left(\frac{3\cos^2\theta - 1}{r^3} \right) \middle| \psi_{2pz} \right\rangle \\
&= \frac{3}{4\pi} g\beta g_n\beta_n \int_0^{\infty} \frac{f^2(r)}{r^3} r^2 \mathrm{d}r \int_0^{\pi} (3\cos^2\theta - 1)\cos^2\theta \sin\theta \mathrm{d}\theta \int_0^{2\pi} \mathrm{d}\phi \\
&= \frac{4}{5} g\beta g_n\beta_n \left\langle \frac{1}{r^3} \right\rangle = 2B
\end{aligned}
\tag{5.74}
$$

Then

$$
\begin{aligned}
T_{xx} &= T_{yy} = \left\langle \psi_{2pz} \middle| g\beta g_n\beta_n \left(\frac{3\sin^2\theta\cos^2\theta - 1}{r^3} \right) \middle| \psi_{2pz} \right\rangle \\
&= \frac{3}{4\pi} g\beta g_n\beta_n \int_0^{\infty} \frac{f^2(r)}{r^3} r^2 \mathrm{d}r \int_0^{\pi} \int_0^{2\pi} (3\sin^2\theta \cos^2\theta - 1)\cos^2 \sin\theta \, \mathrm{d}\theta \, \mathrm{d}\phi \\
&= -\frac{2}{5} g\beta g_n\beta_n \left\langle \frac{1}{r^3} \right\rangle = -B
\end{aligned}
\tag{5.75}
$$

Calculation results indicate that the sign of T_{zz} is positive, while T_{xx} and T_{yy} are negative. This is quite significant.

To calculate the value of T_{xx}, T_{yy}, T_{zz}, we need to know the specific form of $f(r)$. Using SCF atomic wave function, calculate the values of $\langle r^{-3} \rangle$ and B are listed in Table 5.4. Here are just a few of the common nuclei, for more information, please refer to constant table of appendix.

Table 5.4: Calculated values of the hyperfine coupling tensor of several common nuclei.

Nucleus	$\langle r^{-3} \rangle$(atomic unit)	B (MHz)
^{13}C	1.692	90.8
^{14}N	3.101	47.8
^{19}F	7.546	1515
^{17}O	4.974	144
^{31}P	3.318	287
^{35}Cl	6.795	137

Several examples are as follows:

(1) Malonic acid radical generated by γ – radiation

Using enriched isotropic ^{13}C synthesized single-crystal samples of malonic acid, $HOOC - \dot{C}H - COOH$ radical is generated by γ–radiation. There are two magnetic nuclei of this free radical: one is ^{13}C nucleus and another is α proton. Here ^{13}C is the central atom and the α proton will be discussed in the next section.

From the analysis of experimental spectra, we can obtain the principal value and relative sign of the hyperfine coupling tensor, but cannot get the absolute sign:

$$A_0^C = +92.6, \quad T_{xx} = -70, \quad T_{yy} = -50, \quad T_{zz} = +120(MHz)$$

or $(A_0^C, \quad T_{xx}, \quad T_{yy}, \quad T_{zz}) = (-, \quad +, \quad +, \quad -)$. However, from theoretical analysis, we have learned that the T_{zz} should be positive, so the former sign is correct. Thus, the sign of A_0^C should be positive, and the carbon atom ought to have positive spin density. Based on the spin polarization mechanism, we have pointed out previously that if an α spin unpaired electron is filled in $2p_z$ orbital, then, according to Hund's rules, on the sp^2 hybridized orbital of carbon atom, α spin unpaired electron should also predominate. That is, the carbon atom ought to have positive spin density, while the proton should have negative spin density. This experiment proved it again.

If this unpaired electron is entirely located on the $2p_z$ orbital, from Table 5.4, T_{zz} that is 120 MHz should be 181.6 MHz; therefore, the π-spin density of this unpaired electron on the carbon atom ought to be $\rho_p = 120/181.6 = 0.66$. Similarly, if the unpaired electron is completely on the s-orbital of this carbon atom, A_0^C that is only 92.6 MHz should be 3110 MHz; so, $\rho_S = 92.6/3110 = 0.0297$. The remaining ~30% of electron density is delocalized to the other atoms.

(2) γ – Radiation of sodium formate single crystal generated $O - \dot{C} - O^-$ radical

Sodium formate single crystal generated $O - \dot{C} - O^-$ radical by γ– radiation. The principal axis of its tensor is determined that place the radical on the *yz* plane and the bisector line of $\angle OCO^-$ on the *z*-axis. The obtained tensor principal values are $|A_{xx}| = 436\,\text{MHz}$, $|A_{yy}| = 422\,\text{MHz}$, $|A_{zz}| = 546\,\text{MHz}$, and they have the same sign. Hence,

$$A_0^C = \frac{1}{3}(A_{xx} + A_{yy} + A_{zz}) = 468\,MHz$$

Here the sign of A_0^C should be positive; hence,

$$T_{xx} = 436 - 468 = -32 \text{ MHz}$$
$$T_{yy} = 422 - 468 = -46 \text{ MHz}$$
$$T_{zz} = 546 - 468 = +78 \text{ MHz}$$

The unpaired electron density on the $2p_z$ orbital is 78/182 = 0.43. The unpaired electron density on the 2s orbital is 468/3110 = 0.15。 The total unpaired electron density on the central carbon atom is 0.43 + 0.15 = 0.58. The rest of the electron density is apparently extended to two oxygen atoms, and the average unpaired electron density on the each oxygen atom is

$$\frac{1}{2}(1 - 0.58) = 0.21$$

The unpaired electron density on the 1s orbital of carbon atom here has not been considered. The "hybrid ratio" can be calculated by the unpaired electron densities of 2s and 2p orbitals, that is,

$$\lambda^2 = \frac{\rho_p}{\rho_s} = \frac{0.43}{0.15} = 2.87$$

According to the hybrid orbital theory, the angle $\angle OCO^- = \phi$ can be calculated by

$$\phi = 2\cos^{-1}(\lambda^2 + 2)^{-(1/2)} = 2\cos^{-1}(2.87 + 2)^{-(1/2)} = 126°$$

It is known that CO_2^- is the same electron compound of NO_2; angle $\angle ONO$ of NO_2 is measured in the gas phase as 134 ° (very close to 126 °). It is seen that the EMR experiment of single-crystal samples can be obtained by the structural information of distribution of the unpaired electron in atomic orbitals, the hybrid ratio, the bond angle, and so on.

(3) $N(SO_3)_2^{2-}$ radical

Nitrogen atom is the central atom in this radical [13]. The principal value of ***T*** tensor is

$$(T_{xx},\quad T_{yy},\quad T_{zz}) = (-35,\ -35,\ +70)\text{MHz}$$

which satisfy the relationship of cylindrical symmetry very well. For the case of unpaired electron located in 2p orbital completely, it has

$$(T_{xx},\quad T_{yy},\quad T_{zz}) = (-47.8,\ -47.8,\ +95.6)MHz$$

Hence, $\rho_p = 0.73$.

(4) $\dot{C}HFCOONH_2$ radical

For this radical [13], the unpaired electron is mainly located on the carbon atom. However, as experiments have shown, the tensor principal values of *α* hydrogen atom and *α* fluorine atom with *z*-axis perpendicular to the radical plane are given respectively as follows:

$$(A_0^H,\quad T_{xx},\quad T_{yy},\quad T_{zz}) = (-63,\quad +32,\quad 0,\quad -33)MHz$$

$$(A_0^F,\quad T_{xx},\quad T_{yy},\quad T_{zz}) = (+158,\quad -203,\quad -169,\quad +372)MHz$$

There are two points worth noting here: (1) the negative sign of A_0^H accords with the spin polarization mechanism, since *α* proton should have negative spin density when the 2p orbital of carbon has α spin unpaired electron spin density. However, the sign of A_0^F is positive, this illustrates that the α fluorine nucleus hyperfine interaction with the unpaired electron comes not from the *spin polarization* mechanism, rather it is coupled to the α fluorine nucleus directly via $\mathrm{p}-\pi$ interaction. (2) The value of $\boldsymbol{T}$ tensor matrix element of α fluorine atom is in coincidence with the relation of $(-B,\ \ -B,\ \ 2B)$ on the whole, which indicates again that partial unpaired electron is located on the fluorine atom directly. That is, for this part of unpaired electron, α fluorine atom is the central atom. According to the principal value of $\boldsymbol{T}$ tensor of fluorine atom,

$$(T_{xx},\ \ T_{yy},\ \ T_{zz}) = (-1515,\ \ -1515,\ \ +3030)$$

it can obtain

$$\rho_F = \frac{372}{3030} = 0.1227$$

This is to say that it has 12.27% electron density coupling directly to fluorine atom via $\mathrm{p}-\pi$ interaction.

5.4.2 Hyperfine coupling tensor of α-hydrogen atom

γ-Irradiated malonic acid single crystal is the most appropriate model for study of the coupling tensor of α hydrogen atom, since the generated radical $\dot{\mathrm{C}}\mathrm{H(COOH)_2}$ has only one α hydrogen atom that interacts with the unpaired electron and causes two spectral lines with variable splitting space, depending on the orientation.

Experimental results indicate the principal axis coordinate system of hyperfine coupling tensor of malonic acid radical, as shown in Figure 5.8: $|A_{xx}| = 29$ MHz; $|A_{yy}| = 91$ MHz; $|A_{zz}| = 61$ MHz with the same relative sign. Hence,

$$|A_o| = \frac{1}{3}(29 + 91 + 61) = 60.3\,\mathrm{MHz}$$

The spin polarization mechanism can determine their absolute signs to be negative. Hence,

$$A_o = -60.3\,\mathrm{MHz},\quad A_{xx} = -29\,\mathrm{MHz},\quad A_{yy} = -91\,\mathrm{MHz},\quad A_{zz} = -61\,\mathrm{MHz}$$

For the proof of minus sign, please refer to literatures [14,15].

The results of malonic acid radical show an important fact that if $I = 1/2$ nucleus has the principal value of (–30, –60, –90) MHz, it should be α proton. And the principal axis of –30 MHz is the C–H direction, the principal axis of –60 MHz

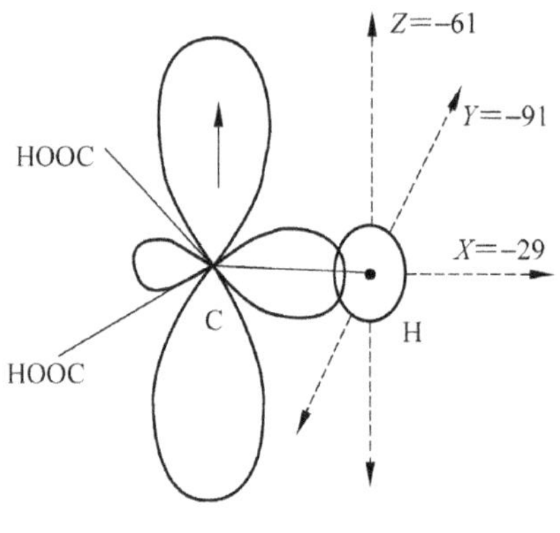

Figure 5.8: Principal axis coordinate system of α proton tensor of $\dot{C}H(COOH)_2$ [16].

is the direction of the symmetry axis of $2p_z$ orbital, the principal axis of −90 MHz is perpendicular to the plane composed of C–H bond and $2p_z$ orbital. Therefore, the diagonalization of $\boldsymbol{A}^2$ matrix can be obtained by the orientation of (X, Y, Z) coordinate system in (x, y, z) coordinate system; the spatial orientation of the radical should also be determined.

Malonic acid with γ-irradiation is also possible to produce another radical $\dot{C}H_2 - COOH$, where it has two α hydrogen atoms attached to the carbon atom. Unpaired electron coupled with these two protons causes four hyperfine spectral lines. Analyzing its EMR spectra gives the relationship of dependence on angles and thus two $\boldsymbol{A}^2$ tensor matrix is obtained: one belongs to H_a and the another pertains to H_b. After diagonalization, principal values and direct cosine are obtained:

(1) Their principal values are as follows:

$$A^a_{xx} = -30, \quad A^a_{yy} = -91, \quad A^a_{zz} = -55, \quad A^a_o = -62\,\text{MHz}$$
$$A^b_{xx} = -37, \quad A^b_{yy} = -92, \quad A^b_{zz} = -59, \quad A^b_o = -63\,\text{MHz}$$

Their principal axis is shown in Figure 5.9, the angle between Z_1 and Z_2 axes is 4.4 °, and the angle between X_1 and X_2 axes is 116 ± 5°. Based on general knowledge, the radical $\dot{C}H_2 - COOH$ should be a planar structure basically, the unpaired electron located at central carbon atom is sp^2 hybridized, the Z_1 and Z_2 axes should be parallel, and $\angle H\dot{C}H$ should be 120 ° with deviation less than 5 °. The experimental results prove that this judgment is correct. It is proven once again that EMR hyperfine spectrum of single-crystal radical can be used to study the molecular structure of radical and its orientation in the crystal.

The α proton hyperfine coupling tensor of $R - \dot{C}H - R'$ radical produced by γ-irradiation of saturated straight chain paraffin is basically similar to the result of malonic acid. However, if the unpaired electron is delocalized, the principal value will be more different. Take $HOOC - CH = \dot{C}H - COOH$ radical produced by γ-irradiation of glutaconic acid as an example [13]. Its three CH can be regarded as an allyl, the hyperfine coupling tensors of these three protons have the same principal coordinate

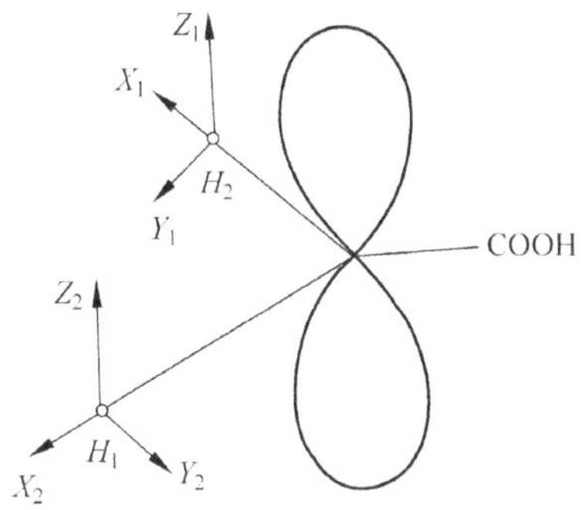

Figure 5.9: Direction of principal axis of hyperfine coupling tensor of two protons in $\dot{C}H_2 - COOH$ radical [16].

X, *Y*, *Z*, and the tensor principal values of two protons are identity that H_1 and H_3 are equivalent protons. The principal values of tensor are given as follows:

	A_{xx}	A_{yy}	A_{zz}	A_o
H_1, H_3	−18	−53	−36	−36
H_2	+12	+17	+7	+12

The absolute values of the above data are obtained from experiment, while the signs are determined on the basis of the following reasons: Negative signs of $\boldsymbol{H}_1$ and $\boldsymbol{H}_3$ are taken as the same with that of malonic acid. $A_0^\alpha = -63\,\text{MHz}$ of $\dot{C}H_2 - COOH$, so

$$\rho_1 = \rho_3 = \frac{-36}{-63} = +0.57$$

$$\rho_2 = \frac{+12}{-63} = -0.19$$

Here ρ_1 and ρ_3 have positive signs as we know the spin density of 1, 3 positions is positive. Likewise, ρ_2 has negative sign because (1) according to the normalizing condition, $\rho_1 + \rho_2 + \rho_3 = 1$, so it gives $\rho_2 = 1 - 2 \times 0.57 = -0.14$, explaining that the spin density of No. 2 carbon atom should have negative sign. (2) Since it is an allyl radical actually, ρ_2 of allyl radical in solution is negative. (3) Heller and others [16] have made quantitative calculation for $\mathbf{H}_1$ and $\mathbf{H}_3$ since

$$0.60(-30, -60, -90) = (-18, -36, -54)$$

is in coincidence with experimental value. The anisotropic hyperfine coupling tensor of H_1 basically results from ρ_1; while this is not case with ρ_2 and ρ_3 as the distance from ρ_3 to H_1 is too far and the value of ρ_2 is too small. However, ρ_1 and ρ_3 also contribute to the anisotropic hyperfine coupling tensor of H_2. If the calculation of Heller and others [16] gives ρ_2 as +0.19, then

$$(A_{xx},\quad A_{yy},\quad A_{zz}) = (+0.9,\quad -19.2,\quad -18.4)$$

which is not in coincidence with experimental results. If ρ_2 is −0.19, then

$$(A_{xx},\quad A_{yy},\quad A_{zz}) = (+12.7,\quad +16.8,\quad +5.6)$$

which accords well with experimental results. It proves once again that ρ_2 is negative and that it is actually an allyl radical.

5.4.3 Hyperfine coupling tensor of β-hydrogen atom

Most radicals generated by γ-irradiation of organic binary acids and amino acids also contain β protons, such as hydrogen in CH_2 of succinic acid radical $HOOC - CH_2 - \dot{C}H - COOH$ and hydrogen in CH_3 of $CH_3 - \dot{C}H - COOH$ are β protons. The β proton can also cause splitting by hyperfine coupling; however, the anisotropic level of their tensor is relatively small, since the anisotropic tensor $\boldsymbol{T}$ comes from the dipole–dipole interaction between each spin.

$$T_{zz} = g\beta g_n\beta_n\left\langle\frac{3\cos^2\theta - 1}{r^3}\right\rangle$$

When r is relatively large, this interaction reduces quickly. On the other hand, its isotropic hyperfine coupling increases. The reason is that the three protons of CH_3 that lie on the β position are precomposed of three "group orbitals" as follows:

$$\psi_1\frac{1}{\sqrt{3}}(a+b+c),\quad \psi_2\frac{1}{\sqrt{2}}(b-c),\quad \psi_3\frac{1}{\sqrt{6}}(2a-b-c)$$

Among them, the symmetry of $2p_z$ agrees well with ψ_3 generated by the hyperconjugation effect. Obviously, this effect has relationship with the location of the proton. If $C - H$ bond of β-position in CH_3 is parallel to $2p_z$ (perpendicular to the radical plane), this effect is strongest. When it is perpendicular to $2p_z$(in the radical plane), this effect is smallest. Suppose the angle between $C - H$ bond and $2p_z$ is θ, as shown in Figure 5.10, it gives

$$A_\beta = A_1 + A_2\cos^2\theta \tag{5.78}$$

We can obtain the experimental value of A_β and determine the angle θ according to Eq. (5.78). Knocking down amino group from α-alanine by γ-irradiation generates $CH_3 - \dot{C}H - COOH$. At room temperature, when CH_3 around the axis of $C_\alpha - C_\beta$ bond rotates freely, the three protons are equivalent, and its hyperfine coupling value is given by

$$\langle A_\beta\rangle = A_1 + A_2\cos^2\theta = A_1 + \frac{1}{2}A_2 \tag{5.79}$$

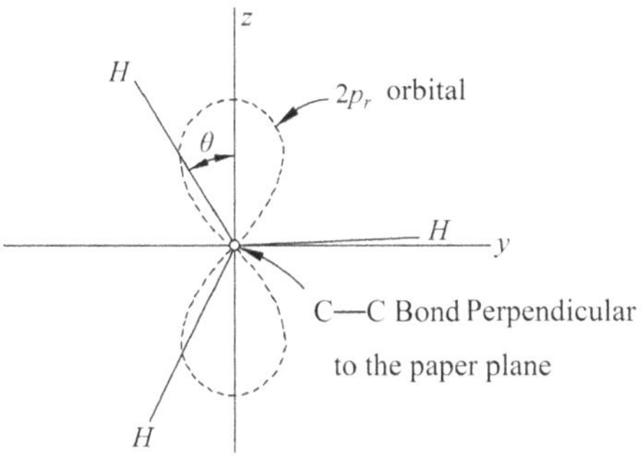

Figure 5.10: Orientation of methylic hydrogen atom of $R-\dot{C}H-CH_3$ radical.

and the value of $\langle A_\beta \rangle$ is equal to A_α by chance, that is, three β protons and one α proton are four protons equivalent mutually, and result in five hyperfine spectral lines with intensity ratio of 1:4:6:4:1, as shown in Figure 5.11a. However, when the

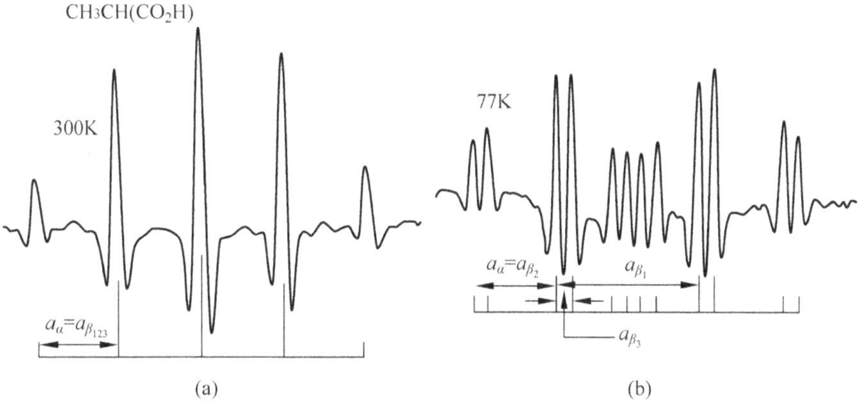

Figure 5.11: EMR spectra (second differential) of $CH_3-\dot{C}H-COOH$ radical generated by γ-irradiation of α–aminopropionic acid [17] at (a) 300 K and (b) 77 K.

temperature is lowered to 77 K, the rotation of the CH_3 around the axis of $C_\alpha - C_\beta$ bond causes obstacles and the three β protons are no longer equivalent, although β_2 and α protons are still equivalent. Therefore, its EMR spectrum appears to have 12 hyperfine spectral lines with intensity ratio [17] of 1:1:2:2:1:1:1:1:2:2:1:1, as shown in Figure 5.11b. The obtained average coupling tensors at 300 K are +67.0, +67.5, +76.5 MHz, and the isotropic value is +70.0 MHz. This tensor is cylindrically symmetric with $C_\alpha - C_\beta$ bond axis as its symmetry axis. The hyperfine coupling tensor of α proton are −25.0, −89.4, −49.8 MHz, which is consistent with the result of malonic acid basically. Furthermore, the angle between *X* principal axis of α proton (direction of $C-H^{(\alpha)}$ bond) and *Z*-axis of β-position CH_3 (direction of $C_\alpha - C_\beta$ bond) is 121° and is in coincidence with the preconceived value.

Anisotropic hyperfine tensors of the three protons on the β-position CH_3 at 77 K are listed as follows:

	A_{xx}	A_{yy}	A_{zz}	Average value
β_1	116	118	128	120
β_2	77	76	77	77
β_3	11	16	14	14

According to the isotropic values of these three tensors, it could be calculated as $\theta = 18°,\ 138°, 258°,\ A_1 = 9\text{MHz}, A_2 = 122\,\text{MHz}$, and

$$\frac{120 + 77 + 14}{3} \approx 70\,\text{MHz}$$

It is consistent with the result obtained at room temperature. This fact tells us that the measurement of EMR spectra at different temperatures can be used to study the interior rotational motion of radical molecule.

5.4.4 Hyperfine coupling tensor of σ-type organic radicals

We have already discussed the isotropic hyperfine interaction and structural problems of π-type organic radicals in detail. Since for π-type organic radicals the unpaired electron is mainly localized on $2p_z(\text{or}\pi)$ orbital of carbon atom, the observed values of isotropic hyperfine splitting are not too large generally. The nucleus causing hyperfine splitting is always localized on or nearby the nodal plane of $2p_z$ orbital. However, proton hyperfine splitting values of some radicals are very large, nearly 400 MHz. Indirect mechanism of spin polarization is how to explain such large split values. It is only caused by a direct mechanism, that is, the orbital that is to be occupied by the unpaired electron contain a certain amount of characteristics of s-orbital. It can also be said that the unpaired electron is localized on σ-orbital such as the $C-H$ bond, after loss of a hydrogen atom, the unpaired electron is localized on the σ-orbital. Most σ-orbitals have more s component; since the proton coupled directly with s component of -orbitals induced hyperfine splitting, the values are all relatively large, and the sign should be positive.

In formaldehyde radical $H-\dot{C}=O$, the proton hyperfine coupling constant is 384 MHz, and the spin density on the 1s orbital of the hydrogen atom is about 0.27. Except hydrogen atom, it is the largest value of the isotropic hyperfine splitting caused by the proton known [18]. For σ-radicals, the contribution of direct action to hyperfine splitting is always much more than indirect action.

The isotropic hyperfine splitting value of ^{13}C can be used to decide the content of s-orbital components. For example, $\underset{\bullet}{C}H_3$ (planar) radical, $a^C = 3.85\,\text{mT}$, and the $\underset{\bullet}{C}F_3$

radical [19], $a^C = 27.16\,\mathrm{mT}$. As $\dot{C}F_3$ radical is not strictly planar and has a great pyramid distortion, the orbital occupied by unpaired electron has a certain s-orbital characterization.

Comparison of hyperfine coupling constant of formaldehyde radical and ethylene radical establishes that the hyperfine coupling constants of the three protons of ethylene radical H_1, H_2, H_3 are 43.7, 95.2, 190 MHz, respectively. Even the largest value of 190 here is much smaller than the 384 MHz value of formaldehyde radical. From the hyperfine splitting value of the three protons of ethylene radical, it can be seen that they are relative to their bond angle. Since the hyperfine splitting value of H_1 is 43.7 MHz, the bond angle can be estimated to be 140 – 150°...

The hyperfine coupling of σ-radical appear in a relatively large anisotropy, such as the $F - \dot{C} = O$ radical; the principal value of hyperfine coupling tensor is 1437.5, 708.2, 662.0 MHz, respectively [18].

5.5 Anisotropy of the combination of *g*-tensor and *A*-tensor

The anisotropies of the g tensor and $\boldsymbol{A}$ tensor usually coexist in the system, and their principal axes are not coincident unless it is the case of a high degree of local symmetry [20]. The situation will be very complicated for the system with coexisted anisotropies of the g tensor and $\boldsymbol{A}$ tensor. In the energy expression, the combination matrix $g\cdot\boldsymbol{A}\cdot\boldsymbol{A}^{\mathrm{T}}\cdot g^{\mathrm{T}}$ should be used to express.

The best case is the system contains C, O, Mg, Si, and so on nuclei and the nuclear spin of the isotopes of the largest natural abundance are zero (nonmagnetic nuclei). Thus, we can use the method discussed in Section 3.2 of Chapter 3 to determine the g tensor, and then for the isotope of magnetic nucleus by means of enrichment to derive its $\boldsymbol{A}$ tensor.

The above system is not the only special technique to be used to solve the corresponding energy expressions with the aid of computer [21,22].

5.6 Anisotropy of *A* tensor in the irregular orientation system

The single-crystal samples are rarely encountered in our experiments. Most of the samples are irregular orientation system, such as polycrystalline, powder, or glass state. Therefore, discussing the problem of the anisotropic $\boldsymbol{A}$ tensor in the irregular orientation system has considerable practical significance. In order to facilitate the start, we choose the irregular orientation system with isotropic g tensor, $S = 1/2$, and $I = 1/2$.

From Eq. (5.28), we know

$$\cos\Theta = +\left[\frac{4g^2\beta^2(\Delta H)^2 - (A_o - B)^2}{3B(2A_o + B)}\right]^{\frac{1}{2}} \tag{5.80}$$

where $\Delta H = H_r - H_o$

$$\begin{aligned} P(H) \propto \frac{\sin\Theta}{d(\Delta H)/d\Theta} &= \pm\frac{4g^2\beta^2\Delta H}{3B(2A_o + B)\cos\Theta} \\ &= \pm\frac{2g\beta}{3B(2A_o + B)\cos\Theta}[(A_o - B)^2 + 3B(2A_o + B)cos^2\Theta]^{1/2} \end{aligned} \tag{5.81}$$

How to determine the sign? First, $P(H)$ must be positive. Thus, two separate envelope lines will appear, one corresponding to $M_I = +1/2$ and the another corresponding to $M_1 = -1/2$. Let $\xi = B/A_o$ then Eq. (5.81) becomes [23]

$$P(H) \propto \frac{[(1-\xi)^2 + 3\xi(2+\xi)cos^2\Theta]^{1/2}}{\xi(2+\xi)\cos\Theta} \tag{5.82}$$

EMR spectrum [24] of the irregular orientated $F - \dot{C} == O$ radical measured in the CO matrix at 4.2 K is shown in Figure 5.12. Although it is not strictly axial symmetric, we can treat it as axial symmetry approximately. Based on the spacing of the outside two lines $|A_o + 2B| \approx 51.4$ mT and that of the inside two lines $|A_o + 2B| \approx 24.6$ mT, we can obtain $A_o \approx \pm 940$ MHz; $B \approx \pm 250$ MHz or $A_o \approx \pm 20$ MHz; $B \approx \pm 710$ MHz. Further study indicates the former set data are correct [24].

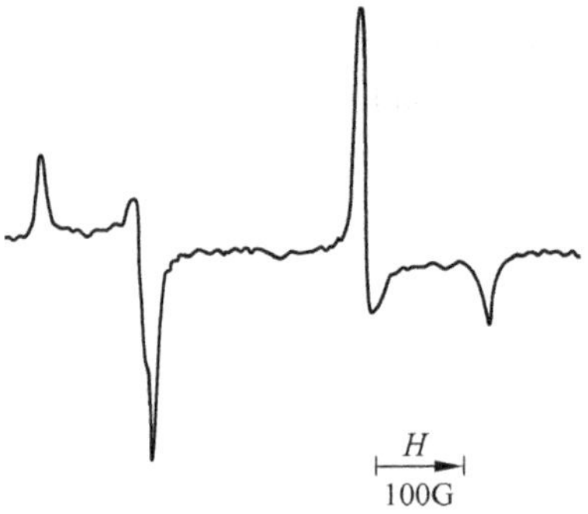

Figure 5.12: EMR spectrum of $F - \dot{C} = O$ radical measured in the CO matrix at 4.2 K

For the system with lower than axial symmetry, or more than one magnetic nucleus, or anisotropic *g* tensor, their theoretical treatment is very complicated. Only for a few relatively simple systems, the partial or total principal values of the *g* and ***A*** tensors are able to be determined. The differential EMR spectra of some ideal irregular orientated systems [25] are shown in Figure 5.13.

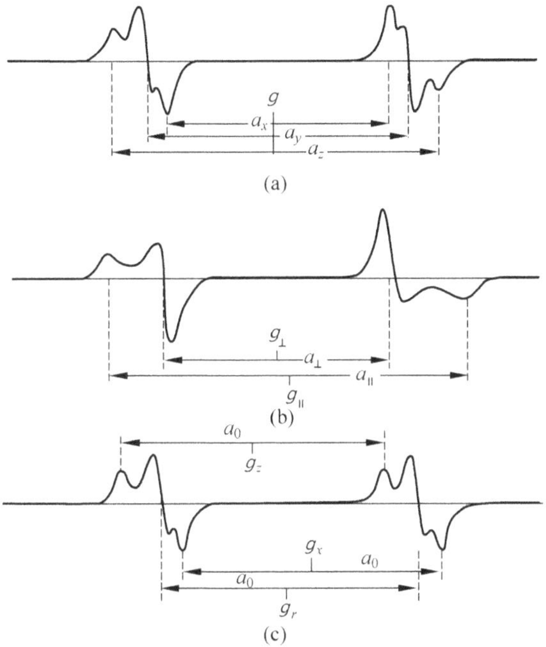

Figure 5.13: The first-order differential EMR spectra of several typical irregular orientation systems with a magnetic nucleus of $I = 1/2$ [25]. (a) g-Factor is isotropic and $A_{zz} > A_{yy} > A_{xx}$. (b) Axial symmetry with $g_\parallel < g_\perp; A_\parallel > A_\perp$. (c) A tensor is isotropic and $g_{xx} < g_{yy} < g_{zz}$.

References

[1] Kirklin P. W., Auzins P., Wertz J. E. *J. Phys. Chem. Solids*. 1965, **26**: 1067.
[2] Henderson B., Wertz J. E. *Adv. Phys*. 1968, **17**: 749.
[3] Cheston W. *Elementary Theory of Electric and Magnetic Fields*, John Wiley & Sons, Inc., New York, 1964, p. 151.
[4] Atkins P. W. *Molecular Quantum Mechanics*, 2nd edn, Oxford: Oxford Press, 1983, Section 14.7.
[5] Skinner R., Weil J. A. *Am. J. Phys*. 1989, **57**, 777.
[6] Zeldes H., Trammell G. T., Livingston R., Holmberg R. W. *J. Chem. Phys*. 1960, **32**: 618.
[7] Blinder S. M. *J. Chem. Phys*. 1960, **33**: 748.
[8] Weil J. A., Anderson J. H. *J. Chem. Phys*. 1961, **35**, 1410.
[9] Trammell G. T., Zeldes H., Livingston R. *Phys. Rev*. 1958, **110**, 630.
[10] Rogers M. T., Whiffen D. H. *J. Chem. Phys*. 1964, **40**: 2662.
[11] Kispert L. D., Rogers M. T. *J. Chem. Phys*. 1971, **54**: 3326.
[12] Bogan C. M., Kispert L. D. *J. Phys. Chem*. 1973, **77**: 1491.
[13] Qiu Z. W. Electron *Spin Resonance Spectroscopy*, Science Press, Beijing, 1980, Chapter 3.
[14] Ghosh K., Whiffen D. H. *Mol. Phys*. 1959, **2**: 285.
[15] McConnell M., Streethdee J. *Mol. Phys*. 1959, **2**: 129.
[16] Carrington A., McLachlan A. D. *Introduction to Magnetic Resonance*, Harper & Row, New York, 1967.
[17] Horsfield A., Morton J. R., Whiffen D. H. *Mol. Phys*. 1961, **4**: 425.

[18] Adrian F. J, Cochran E. L, Bowers V. A. *J. Chem. Phys.* 1965, **43**: 462.
[19] Fessenden R. W., Schuler R. H. *J. Chem. Phys.* 1965, **43**: 2705.
[20] Zeldes H., Livingston R. *J. Chem. Phys.* 1961, **35**: 563.
[21] Weil J. A. *J. Magn. Reson.* 1971, **4**: 394.
[22] Farach H. A., Poole C. P., Jr. *Adv. Magn. Reson.* 1971, **5**: 229.
[23] Blinder S. M. *J. Chem. Phys.* 1960, **33**: 748.
[24] Adrian F. J., Cochran E. L., Bowers V. A. *J. Chem. Phys.* 1965, **43**: 462.
[25] Atkins P. W., Symons M. C. R. *The Structure of Inorganic Radicals*, Elsevier, Amsterdam, 1967.

Further readings

[1] Poole C. P., Jr. Farach H. A. *Theory of Magnetic Resonance*, John Wiley & Sons, Inc., New York, 1987.
[2] Bleaney B. "Hyperfine structure and electron paramagnetic resonance", in Freeman A. J., Frankel R. B., Eds., *Hyperfine Interactions*, Academic Press, New York, 1967.

6 Fine structure

Abstract: This chapter discusses the systems with more than two unpaired electrons such as triplet, biradical, and so on. Due to the interaction of more than two unpaired electrons, the spin energy level in the zero magnetic fields could be split, and the EMR spectrum has a fine structure.

In the previous chapters, the systems of our discussion are the paramagnetic particles with only one unpaired electron $(S = s = 1/2)$ mainly, which could not produce zero-field splitting, and their spectrum have not fine structure too. If there are more than two unpaired electrons in the paramagnetic particle, $(S \geq 1)$, and suppose the number of electrons is odd, $\left(S = \frac{3}{2}, \frac{5}{2}, \frac{7}{2}, \cdots\right)$, the EMR signal can be observed in principle. However, assume the number of electrons is even, $(S = 1, 2, 3, \ldots)$, although this may not be the case always. In this chapter, we will discuss the system with more than two unpaired electrons.

We ignore all the nuclear spins for the time being, starting with the systems containing two unpaired electrons, which include (1) atoms or ions in the gas phase (as O atom); (2) small molecules in the gas phase (as O_2); (3) organic molecules containing more than two unpaired electrons (as to be excited metastable triplet naphthalene molecules); (4) inorganic molecules (as CCO and CNN in the rare gas); (5) point defects in crystals containing two or more unpaired electrons (as the F_t center in MgO); (6) biradicals in the fluid solution and solid; and (7) some transition ions (as V^{3+} and Ni^{2+}) and rare earth ions. The systems (1) and (2) will be discussed in Chapter 10.

If the energy level of the highest occupied orbital is nondegenerative, and the orbital is occupied with two paired electrons, then the ground state should be a spin singlet state (shown in Figure 6.1a). If one of the two electrons is excited to a nonoccupied orbital by absorbing the quantum of a certain energy, and still keeps spin paring, then the system is an excited singlet state (shown in Figure 6.1b), because the transition does not change the "multiplicity." However, the molecule may change the spin orientation by the nonradiation process, and become to a metastable triplet state (shown in Figure 6.1c), and their energy level is a bit lower than the singlet state. This nonradiation process may be caused by *spin–orbit coupling, molecular rotation*, and/or *hyperfine interaction in the presence of an external magnetic field*. There is also a four-electron system whose highest occupied orbitals are two degree degenerated; according to the Pauli principle, two electrons should be two unpaired electrons, occupying separately two orbits with same energy level and keeping spin parallel (shown in Figure 6.1d).

https://doi.org/10.1515/9783110568578-006

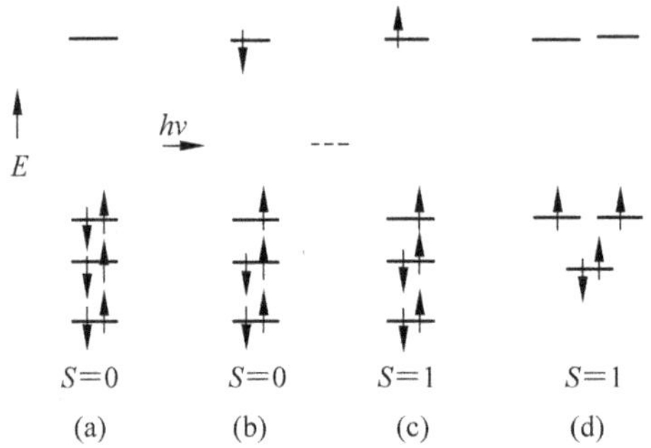

Figure 6.1: Schematic diagram of the electron energy level of six-electron system and four-electron system, and their spin structure.

6.1 Zero-field splitting

A system has two unpaired electrons, and each electron has α and β spin states; thus, there are four spin states of the following combination:

$$\alpha(1)\alpha(2) \quad \alpha(1)\beta(2) \quad \beta(1)\alpha(2) \quad \beta(1)\beta(2)$$

These four spin states, because of small interaction of the two unpaired electrons, can form two configurations: *symmetric* and *antisymmetric*. The combination functions are as follows:

Symmetric function	Antisymmetric function
$\alpha(1)\alpha(2)$	
$\frac{1}{\sqrt{2}}[\alpha(1)\beta(2)+\beta(1)\alpha(2)]$	$\frac{1}{\sqrt{2}}[\alpha(1)\beta(2)-\beta(1)\alpha(2)]$
$\beta(1)\beta(2)$	
Triplet state $S = 1$	Singlet state $S = 0$

The multiplicity of state is $2S+1$. If $S = 1$, then $2S+1=3$, hence the state is called *triplet state*; when $S = 0$, then $2S+1=1$, hence the state is called singlet state. If two electrons occupy a same space orbital, owing to the restraint of Pauli principle, only the *antisymmetric* or *singlet* state is possible. However, if every electron occupies each different orbital, then both the *triplet* and *singlet* states exist.

A system of two unpaired electrons, with the total spin angular momentum S, is the sum of the spin angular momentum of the two unpaired electrons $S=s_1+s_2$. Since $s_1=s_2=\frac{1}{2}$, $S=s_1+s_2=\frac{1}{2}+\frac{1}{2}=1$, thus $m_s = +1, 0, -1$. If there is no zero-field splitting, the relationship between the energy level of the system and the external magnetic field is shown as Figure 6.2a. Coupled with microwave, according to the selection rule of resonance transition $\Delta m_S = \pm 1$, there should be two transition lines: $-1 \to 0$ and

$0 \rightarrow +1$; however, they both occur at H_0. That is to say, although the two lines coincide with each other, only one line can be observed.

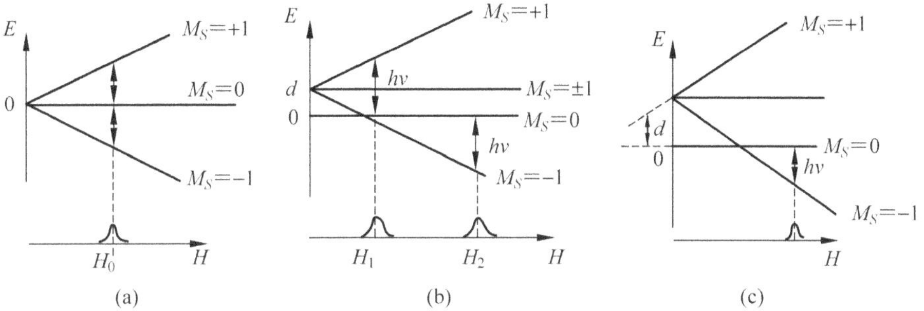

Figure 6.2: Schematic diagram of zero-field splitting and fine structure for the system $S = 1$.

If the $S = 1$ system is located in an axisymmetric crystal field (or ligand field), in the absence of an external magnetic field, the energy levels have been split into $m_S = 0$ and $m_S = \pm 1$, and the latter is double degenerate, which is called "zero-field splitting." Only when the external magnetic field is added, the Zeeman splitting will be produced; and when the microwave is added, two transition lines will be generated, which is "fine structure," as shown in Figure 6.2b. H_1 and H_2 in the figure are as follows:

$$H_1 = \frac{hv}{g\beta} - \frac{d}{g\beta} = H_0 - D \tag{6.1a}$$

$$H_2 = \frac{hv}{g\beta} + \frac{d}{g\beta} = H_0 + D \tag{6.1b}$$

$$|H_2 - H_1| = 2D \tag{6.2}$$

D is the splitting space of the fine structure corresponding to the splitting space of the hyperfine structure (A_o). Because the magnetic moment of the electron is 1,000 times greater than that of nuclear, the dipole–dipole interaction energy between electrons themselves can be up to three orders of magnitude higher than that between electrons and nuclei. Hence, D of the fine structure is much larger than A_o of the hyperfine structure. If the zero-field splitting $d > hv$, only one line can be observed (shown in Figure 6.2c), though the zero-field splitting exists. *Zero-field splitting* is mainly caused by the *dipole–dipole interaction* between electrons.

When the system has more than two unpaired electrons, there is a very important theorem called *Kramer theorem*, which states that "For a system with odd number of unpaired electrons (S is a half integer) in the zero field, every level should be kept at least two degrees of degeneracy"; this degeneracy is called *Kramer degeneracy*.

It is only under the action of the external magnetic field that the Kramer degeneracy can be relieved. That is to say, for the system with odd number of unpaired electrons, due to the existence of Kramer degeneracy, the EMR spectrum always could be observed. However, for the system containing even number of unpaired electrons, when it is in a field of very low symmetry, the degeneracy is relieved completely, its EMR signal could not be observed completely.

It can also be seen from Figure 6.2c that to observe the possible fine structure, the only way is to increase the microwave frequency, to reach $hv > d$. In other words, the other line of the fine structure, which originally is not observable in the X-band, may be observed in the Q-band.

It should also be pointed out that the zero-field splitting is dependent strongly on the orientation, which is anisotropic. Therefore, it is the best using the single crystal sample to study the fine structure caused by the zero-field splitting. The polycrystalline sample is measured with a broad envelope; if the signal amplitude decreases to a certain limit, the EMR signal may not be detected.

6.2 Spin hamiltonian of two-electron interaction

6.2.1 Exchange interaction of electron spin

Since electrons are all identical, when two electrons exchange positions between two different orbitals, the Hamiltonian operator of the system will not change.

$$\hat{\mathcal{H}}(1,2) = \hat{\mathcal{H}}(2,1) \tag{6.3}$$

Let $\hat{p}$ represent the permutation operator of the exchange between electrons 1 and 2, and act on both sides of Schrödinger equation:

$$ih\frac{\partial \Phi}{\partial t} = \hat{\mathcal{H}}\Phi \tag{6.4}$$

then we get

$$ih\frac{\partial}{\partial t}(\hat{p}\Phi) = \hat{p}\left(\hat{\mathcal{H}}\,\Phi\right) = \hat{\mathcal{H}}(\hat{p}\Phi) \tag{6.5}$$

That is to say, Φ is the solution of the Schrödinger equation, and $(\hat{p}\Phi)$ is also the solution of the Schrödinger equation. Owing to the identical principle of electrons, Φ and $(\hat{p}\Phi)$ descriptives are same state, and only the constant is different:

$$\hat{p}\Phi = \lambda\Phi \tag{6.6}$$

After $\hat{p}$ once more acts on both sides of Eq. (6.6), we get

$$\Phi = \lambda^2 \Phi \tag{6.7}$$

It can be seen that $\lambda^2 = 1$ and the eigenvalue of $\hat{p}$ is $\lambda = \pm 1$. That is to say, the state composed of all identical particles can be described by symmetric or antisymmetric wave functions. Experiments proved that the electron wave function should be composed of antisymmetric wave functions (i.e., $\lambda = -1$). In view of the changing sign property of the determinant value, using the determinant function to describe the antisymmetric wave function is more suitable.

Let $\phi_a = \psi_a(x, y, z,)\alpha$ denote the eigenfunctions of a single electron Hamiltonian operator, where the subscript a represents the a spin state of a set of quantum numbers (n, l, m); and let $\bar{\phi}_a = \psi_a(x, y, z)\beta$, which is different from ϕ_a, be the difference only in the spin state from α to β. Two electrons located on the different orbits can compose four determinant functions:

$$|\phi_a, \phi_b|, \cdots |\phi_a, \bar{\phi}_b|, \cdots |\bar{\phi}_a, \phi_b|, \cdots |\bar{\phi}_a, \bar{\phi}_b| \tag{6.8}$$

where

$$|\phi_a, \phi_b| \equiv \frac{1}{\sqrt{2}}[\phi_a(1)\, \phi_b(2) - \phi_a(2)\, \phi_b(1)] \tag{6.9a}$$

$$|\phi_a, \bar{\phi}_b| \equiv \frac{1}{\sqrt{2}}[\phi_a(1)\bar{\phi}_b(2) - \phi_a(2)\, \bar{\phi}_b(1)] \tag{6.9b}$$

$$|\bar{\phi}_a, \phi_b| \equiv \frac{1}{\sqrt{2}}[\bar{\phi}_a(1)\phi_b(2) - \bar{\phi}_a(2)\, \phi_b(1)] \tag{6.9c}$$

$$|\bar{\phi}_a, \bar{\phi}_b| \equiv \frac{1}{\sqrt{2}}[\bar{\phi}_a(1)\bar{\phi}_b(2) - \bar{\phi}_a(2)\, \bar{\phi}_b(1)] \tag{6.9d}$$

The operators $\hat{S}_z = \hat{s}_{1z} + \hat{s}_{2z}$ and $\hat{S}^2 = (\hat{s}_1 + \hat{s}_2)^2$. We note that the four functions of Eq.(6.8) are all the eigenfunctions of the operator $\hat{S}_z$, but not all of them are the common eigenfunction of $\hat{S}_z$ and $\hat{S}^2$. Only $|\phi_a, \phi_b|$ and $|\bar{\phi}_a, \bar{\phi}_b|$ are the common eigenfunctions of operators $\hat{S}_z$ and $\hat{S}^2$; however, $|\phi_a, \bar{\phi}_b|$ and $|\bar{\phi}_a, \phi_b|$ are not. And let's see

$$\hat{S}^2|\phi_a, \bar{\phi}_b| = |\phi_a, \bar{\phi}_b| + |\bar{\phi}_a, \phi_b|$$

$$\hat{S}^2|\bar{\phi}_a, \phi_b| = |\bar{\phi}_a, \phi_b| + |\phi_a, \bar{\phi}_b|$$

In order to ensure all are common eigenfunctions of $\hat{S}_z$ and $\hat{S}^2$, $|\phi_a, \bar{\phi}_b|$ and $|\bar{\phi}_a, \phi_b|$ should be linearly combined into a new function, thus,

$${}^3\Psi_1 = |\phi_a, \phi_b| \tag{6.10a}$$

$${}^3\Psi_0 = \frac{1}{\sqrt{2}}(|\phi_a, \bar{\phi}_b| + |\bar{\phi}_a, \phi_b|) \tag{6.10b}$$

$$^3\Psi_{-1} = |\bar{\phi}_a, \bar{\phi}_b| \tag{6.10c}$$

$$^1\Psi_0 = \frac{1}{\sqrt{2}}\left(|\phi_a, \bar{\phi}_b| - |\bar{\phi}_a, \phi_b|\right) \tag{6.10d}$$

If they are expressed as the product of the space function and spin function, the above formula can be written as follows:

$$^3\Psi_1 = \frac{1}{\sqrt{2}}[\psi_a(1)\psi_b(2) - \psi_a(2)\psi_b(1)]\alpha(1)\alpha(2) = \Psi_T(1,2)\alpha\alpha \tag{6.11a}$$

$$^3\Psi_0 = \frac{1}{\sqrt{2}}[\psi_a(1)\psi_b(2) - \psi_a(2)\psi_b(1)]\left[\frac{\alpha(1)\beta(2) + \beta(1)\alpha(2)}{\sqrt{2}}\right] = \Psi_T(1,2)\left[\frac{\alpha\beta + \beta\alpha}{\sqrt{2}}\right] \tag{6.11b}$$

$$^3\Psi_{-1} = \frac{1}{\sqrt{2}}[\psi_a(1)\psi_b(2) - \psi_a(2)\psi_b(1)]\beta(1)\beta(2) = \Psi_T(1,2)\beta\beta \tag{6.11c}$$

$$^1\Psi_0 = \frac{1}{\sqrt{2}}[\psi_a(1)\psi_b(2) + \psi_a(2)\psi_b(1)]\left[\frac{\alpha(1)\beta(2) - \beta(1)\alpha(2)}{\sqrt{2}}\right] = \Psi_T(1,2)\left[\frac{\alpha\beta - \beta\alpha}{\sqrt{2}}\right] \tag{6.11d}$$

From these four wave functions, it can be seen that if the space function is antisymmetric, the spin function should be symmetric. On the contrary, if the space function is symmetric, then the spin function should be antisymmetric, thereby guarantees the entire function is antisymmetric.

If the total Hamiltonian operator is composed of the sum of Hamiltonian operators of two single electrons, and does not include the spin term, then the four wave functions in Eq. (6.11) are all the eigenfunctions of the total Hamiltonian operator, and the energy is degenerated. That is,

$$E(^1\Psi_0) = E(^3\Psi_1) = E(^3\Psi_0) = E(^3\Psi_{-1}) = E_a + E_b$$

If the total Hamiltonian operator contains the electrostatic repulsion term $\frac{e^2}{r_{1,2}}$ of two unpaired electrons, then

$$\begin{aligned}
&\left\langle {}^1\Psi_0 \left| \frac{e^2}{r_{1,2}} \right| {}^1\Psi_0 \right\rangle \\
&= \frac{1}{2}\left\langle \psi_a(1)\psi_b(2) + \psi_a(2)\psi_b(1) \left| \frac{e^2}{r_{1,2}} \right| \psi_a(1)\psi_b(2) + \psi_a(2)\psi_b(1) \right\rangle \left\langle \frac{\alpha\beta - \beta\alpha}{\sqrt{2}} \middle| \frac{\alpha\beta - \beta\alpha}{\sqrt{2}} \right\rangle \\
&= \frac{1}{2}\left[2\left\langle \psi_a(1)\psi_b(2) \left| \frac{e^2}{r_{1,2}} \right| t\psi_a(1)\psi_b(2) \right\rangle + 2\left\langle \psi_a(1)\psi_b(2) \left| \frac{e^2}{r_{1,2}} \right| \psi_a(2)\psi_b(1) \right\rangle\right] \\
&= C + J
\end{aligned} \tag{6.12}$$

Here e is the electronic charge, and $r_{1,2}$ is the distance between electron 1 and electron 2.

$$C = \left\langle \psi_a(1)\psi_b(2)\left|\frac{e^2}{r_{1,2}}\right|\psi_a(1)\psi_b(2)\right\rangle, \text{ called coulomb integral}$$

$$J = \left\langle \psi_a(1)\psi_b(2)\left|\frac{e^2}{r_{1,2}}\right|\psi_a(2)\psi_b(1)\right\rangle, \text{ called exchange integral}$$

Similarly,

$$\begin{aligned}
&\left\langle {}^3\Psi_0\left|\frac{e^2}{r_{1,2}}\right|{}^3\Psi_0\right\rangle \\
&= \frac{1}{2}\left\langle \psi_a(1)\psi_b(2) - \psi_a(2)\psi_b(1)\left|\frac{e^2}{r_{1,2}}\right|\psi_a(1)\psi_b(2) - \psi_a(2)\psi_b(1)\right\rangle\left\langle\frac{\alpha\beta-\beta\alpha}{\sqrt{2}}\middle|\frac{\alpha\beta-\beta\alpha}{\sqrt{2}}\right\rangle \\
&= \frac{1}{2}\left[2\left\langle \psi_a(1)\psi_b(2)\left|\frac{e^2}{r_{1,2}}\right|\psi_a(1)\psi_b(2)\right\rangle - 2\left\langle \psi_a(1)\psi_b(2)\left|\frac{e^2}{r_{1,2}}\right|\psi_a(2)\psi_b(1)\right\rangle\right] \\
&= C - J
\end{aligned} \tag{6.13}$$

Furthermore,

$$\left\langle {}^3\Psi_1\left|\frac{e^2}{r_{1,2}}\right|{}^3\Psi_1\right\rangle = \left\langle {}^3\Psi_0\left|\frac{e^2}{r_{1,2}}\right|{}^3\Psi_0\right\rangle = \left\langle {}^3\Psi_{-1}\left|\frac{e^2}{r_{1,2}}\right|{}^3\Psi_{-1}\right\rangle = C - J \tag{6.14}$$

So far, we have seen that when the two unpaired electrons are close to each other, due to the electrostatic repulsion between electrons, the original four degenerate energy levels are divided into two groups: one is singlet ${}^1\Psi_0$, and the other is triplet ${}^3\Psi_{m_S}\,(m_S = 1, 0, -1)$. The energy gap between the singlet and triplet (Figure 6.3) is

$$E({}^1\Psi_0) - E({}^3\Psi_{m_S}) = (C + J) - (C - J) = 2J \tag{6.15}$$

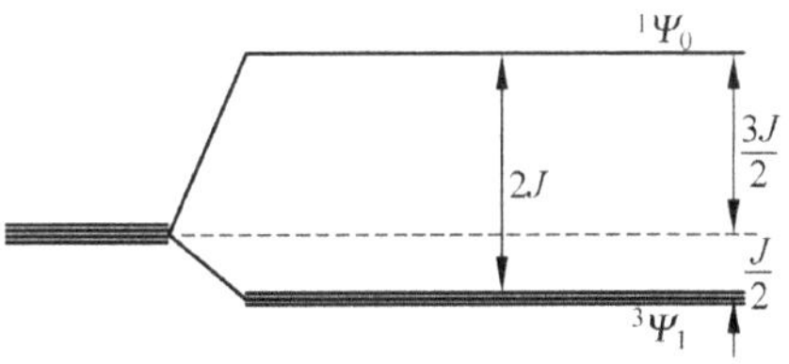

Figure 6.3: Illustrated diagram of singlet and triplet energy levels.

It can be concluded that the electrostatic repulsive interaction between electrons can relieve the degeneracy of the system energy level, and we can also use the spin Hamiltonian operator of the electron exchange interaction to act on these four wave functions; the same energy splitting results can be obtained. The spin Hamiltonian operator of the electron exchange interaction is as follows:

$$\hat{\mathscr{H}}_{\text{exch}} = \sum_{ij} J_{ij}\hat{S}_{1i}\hat{S}_{2j} \tag{6.16}$$

In the formula, $i, j = 1, 2, 3$ (or x, y, z) represent the spatial coordinates. For a detailed theory of exchange interaction of the system, please refer to Refs [1,2]. If the isotropic part is only considered, the spin Hamiltonian operator is

$$\left(\hat{\mathscr{H}}_{\text{exch}}\right)_{\text{iso}} = -2J_o\hat{S}_1 \cdot \hat{S}_2 \tag{6.17}$$

If we want to calculate the exchange interaction energy, then

$$\langle {}^1\Psi_0 | \hat{\mathscr{H}}_{\text{exch}} | {}^1\Psi_0 \rangle = \frac{3}{2}J_o \tag{6.18}$$

$$\langle {}^3\Psi_{m_S} | \hat{\mathscr{H}}_{\text{exch}} | {}^3\Psi_{m_S} \rangle = -\frac{1}{2}J_o \tag{6.19}$$

For the triplet molecule, J_o larger, such as naphthalene molecules in phosphorescence state, $2J_o = 14,600\text{cm}^{-1}$. For the biradical, due to the distance between two unpaired electrons rather far, the exchange energy is rather small, and J_o is relatively small. The sign of J_o depends on the ground state of the system. From Eqs. (6.18) and (6.19), we can see that if $J_o > 0$, the ground state is triplet; when $J_o < 0$, the ground state is singlet. If the value of $|J_o|$ is not very large, the number of molecules in triplet and singlet should obey the Boltzmann distribution.

6.2.2 Dipole interaction of electron–electron

In addition to the exchange interaction between the two electrons, which can divide the state of the system into single and triplet, there is yet another very important interaction, which is anisotropic magnetic *dipole–dipole* interaction. This interaction can eliminate the three-degree degeneracy of the triplet even in the zero magnetic field. Hence, it is also called "zero-field splitting."

6.2.2.1 The Hamiltonian operator of the dipole–dipole interaction of two electrons

The dipole–dipole interaction between the two unpaired electrons, which is similar to the dipole–dipole interaction between the unpaired electron and the nucleus, produces the anisotropic hyperfine splitting. The energy of dipole–dipole interaction between two unpaired electrons is similar to Eq. (6.3); the corresponding Hamiltonian operator is similar to Eq. (6.4):

Set $\boldsymbol{\mu}_{e_1} = g\boldsymbol{s}_1$; $\boldsymbol{\mu}_{e_2} = g\boldsymbol{s}_2$

$$E_{\text{dipolar}} = \frac{\boldsymbol{\mu}_{e_1} \cdot \boldsymbol{\mu}_{e_2}}{r_{1,2}^3} - \frac{3\left(\boldsymbol{\mu}_{e_1} \cdot \boldsymbol{r}_{1,2}\right)\left(\boldsymbol{\mu}_{e_2} \cdot \boldsymbol{r}_{1,2}\right)}{r_{1,2}^5} \tag{6.20}$$

$$\hat{\mathscr{H}}_{\text{ss}} = g^2\beta^2\left[\frac{\hat{\boldsymbol{s}}_1 \cdot \hat{\boldsymbol{s}}_2}{r^3} - \frac{3(\hat{\boldsymbol{s}}_1 \cdot \boldsymbol{r})(\hat{\boldsymbol{s}}_2 \cdot \boldsymbol{r})}{r^5}\right] \tag{6.21}$$

Note

$$\hat{\boldsymbol{s}}_1 \cdot \hat{\boldsymbol{s}}_2 = \left[\hat{\boldsymbol{s}}_{1x}\ \hat{\boldsymbol{s}}_{1y}\ \hat{\boldsymbol{s}}_{1z}\right]\begin{bmatrix}\hat{\boldsymbol{s}}_{2x}\\ \hat{\boldsymbol{s}}_{2y}\\ \hat{\boldsymbol{s}}_{2z}\end{bmatrix} = \hat{\boldsymbol{s}}_{1x}\ \hat{\boldsymbol{s}}_{2x} + \hat{\boldsymbol{s}}_{1y}\ \hat{\boldsymbol{s}}_{2y} + \hat{\boldsymbol{s}}_{1z}\ \hat{\boldsymbol{s}}_{2z} \tag{6.22}$$

$$\hat{\boldsymbol{s}}_1 \cdot \boldsymbol{r} = \left[\hat{\boldsymbol{s}}_{1x}\ \hat{\boldsymbol{s}}_{1y}\hat{\boldsymbol{s}}_{1z}\right]\begin{bmatrix}\boldsymbol{x}\\ \boldsymbol{y}\\ \boldsymbol{z}\end{bmatrix} = \hat{\boldsymbol{s}}_{1x}\boldsymbol{x} + \hat{\boldsymbol{s}}_{1y}\boldsymbol{y} + \hat{\boldsymbol{s}}_{1z}\boldsymbol{z} \tag{6.23}$$

$$\hat{\boldsymbol{s}}_2 \cdot \boldsymbol{r} = \left[\hat{\boldsymbol{s}}_{2x}\ \hat{\boldsymbol{s}}_{2y}\hat{\boldsymbol{s}}_{2z}\right]\begin{bmatrix}\boldsymbol{x}\\ \boldsymbol{y}\\ \boldsymbol{z}\end{bmatrix} = \hat{\boldsymbol{s}}_{2x}\boldsymbol{x} + \hat{\boldsymbol{s}}_{2y}\boldsymbol{y} + \hat{\boldsymbol{s}}_{2z}\boldsymbol{z} \tag{6.24}$$

Substituting Eqs. (6.22)–(6.24) into Eq. (6.21), we get

$$\hat{\mathscr{H}}_{SS} = \frac{g^2\beta^2}{r^5}\left[\left(r^2 - 3x^2\right)\hat{s}_{1x}\hat{s}_{2x} + \left(r^2 - 3y^2\right)\hat{s}_{1y}\hat{s}_{2y} + \left(r^2 - 3z\right)\hat{s}_{1z}\hat{s}_{2z}\right.$$
$$\left. - 3xy\left(\hat{s}_{1x}\hat{s}_{2y} + \hat{s}_{1y}\hat{s}_{2x}\right) - 3yz\left(\hat{s}_{1y}\hat{s}_{2z} + \hat{s}_{1z}\hat{s}_{2y}\right) - 3zx\left(\hat{s}_{1z}\hat{s}_{2x} + \hat{s}_{1x}\hat{s}_{2z}\right)\right] \tag{6.25}$$

Since

$$\hat{S} = \hat{s}_1 + \hat{s}_2 \tag{6.26}$$

$$\hat{S}_x^2 = \left(\hat{s}_{1x} + \hat{s}_{2x}\right)^2 = \hat{s}_{1x}^2 + \hat{s}_{2x}^2 + \hat{s}_{1x}\hat{s}_{2x} + \hat{s}_{2x}\hat{s}_{1x} \tag{6.27}$$

The eigenvalues of both $\hat{s}_{1i}^2$ and $\hat{s}_{2i}^2 (i = x, y, z)$ are $\frac{1}{4}$. (6.28)

Then,

$$\hat{s}_{1x}\hat{s}_{2x} = \frac{1}{2}\hat{S}_x^2 - \frac{1}{4} \tag{6.29a}$$

$$\hat{s}_{1y}\hat{s}_{2y} = \frac{1}{2}\hat{S}_y^2 - \frac{1}{4} \tag{6.29b}$$

$$\hat{s}_{1z}\hat{s}_{2z} = \frac{1}{2}\hat{S}_z^2 - \frac{1}{4} \tag{6.29c}$$

Let us see again

$$\hat{S}_x\hat{S}_y = \left(\hat{s}_{1x} + \hat{s}_{2x}\right)\left(\hat{s}_{1y} + \hat{s}_{2y}\right) = \hat{s}_{1x}\hat{s}_{2y} + \hat{s}_{2x}\hat{s}_{1y} + \hat{s}_{1x}\hat{s}_{1y} + \hat{s}_{2x}\hat{s}_{2y} \tag{6.30}$$

and focus on the commutation relation of angular momentum operators:

$$\hat{s}_{1x}\hat{s}_{1y} = \frac{i}{2}\hat{s}_{1z}, \quad \hat{s}_{2x}\hat{s}_{2y} = \frac{i}{2}\hat{s}_{2z} \tag{6.31}$$

Substituting into Eq. (6.30), we obtain

$$\hat{S}_x\hat{S}_y = \hat{s}_{1x}\hat{s}_{2y} + \hat{s}_{2x}\hat{s}_{1y} + \frac{i}{2}(\hat{s}_{1z} + \hat{s}_{2z}) \tag{6.32}$$

Similarly, we have

$$\hat{s}_{1y}\hat{s}_{1x} = -\frac{i}{2}\hat{s}_{1z}, \quad \hat{s}_{2y}\hat{s}_{2x} = -\frac{i}{2}\hat{s}_{2z} \tag{6.33}$$

Then,

$$\hat{S}_y\hat{S}_x = \hat{s}_{1x}\hat{s}_{2y} + \hat{s}_{2x}\hat{s}_{1y} - \frac{i}{2}(\hat{s}_{1z} + \hat{s}_{2z}) \tag{6.34}$$

Combining Eqs. (6.32) and (6.34), we obtain

$$\left(\hat{s}_{1x}\hat{s}_{2y} + \hat{s}_{2x}\hat{s}_{1y}\right) = \frac{1}{2}\left(\hat{S}_x\hat{S}_y + \hat{S}_y\hat{S}_x\right) \tag{6.35a}$$

$$\left(\hat{s}_{1y}\hat{s}_{2z} + \hat{s}_{2y}\hat{s}_{1z}\right) = \frac{1}{2}\left(\hat{S}_y\hat{S}_z + \hat{S}_z\hat{S}_y\right) \tag{6.35b}$$

$$\left(\hat{s}_{1z}\hat{s}_{2x} + \hat{s}_{2z}\hat{s}_{1x}\right) = \frac{1}{2}\left(\hat{S}_z\hat{S}_x + \hat{S}_x\hat{S}_z\right) \tag{6.35c}$$

Substituting Eqs. (6.29) and (6.35) into Eq. (6.25), we get

$$\begin{aligned}\hat{\mathscr{H}}_{SS} = \frac{1}{2}\frac{g^2\beta^2}{r^5}\Big[&(r^2-3x^2)\hat{S}_x^2 + (r^2-3y^2)\hat{S}_y^2 + (r^2-3z^2)\hat{S}_z^2 \\ &- 3xy\left(\hat{S}_x\hat{S}_y + \hat{S}_y\hat{S}_x\right) - 3yz\left(\hat{S}_y\hat{S}_z + \hat{S}_z\hat{S}_y\right) - 3zx\left(\hat{S}_z\hat{S}_x + \hat{S}_x\hat{S}_z\right)\Big]\end{aligned} \tag{6.36}$$

Note that

$$r^2 = x^2 + y^2 + z^2$$

Equation (6.36) can be represented by matrix:

$$\hat{\mathscr{H}}_{SS} = \frac{1}{2}g^2\beta^2\left[\hat{S}_x \quad \hat{S}_y \quad \hat{S}_z\right]\begin{bmatrix} \frac{r^2-3x^2}{r^5} & \frac{-3xy}{r^5} & \frac{-3xz}{r^5} \\ \frac{-3xy}{r^5} & \frac{r^2-3y}{r^5} & \frac{-3yz}{r^5} \\ \frac{-3xz}{r^5} & \frac{-3yz}{r^5} & \frac{r^2-3z}{r^5} \end{bmatrix}\begin{bmatrix}\hat{S}_x \\ \hat{S}_y \\ \hat{S}_z\end{bmatrix} \tag{6.37}$$

Let

$$D = \frac{1}{2}g^2\beta^2\begin{bmatrix} \frac{r^2-3x^2}{r^5} & \frac{-3xy}{r^5} & \frac{-3xz}{r^5} \\ \frac{-3xy}{r^5} & \frac{r^2-3y}{r^5} & \frac{-3yz}{r^5} \\ \frac{-3xz}{r^5} & \frac{-3yz}{r^5} & \frac{r^2-3z}{r^5} \end{bmatrix} \tag{6.38}$$

The element of matrix $\boldsymbol{D}$

$$D_{ij} = \tfrac{1}{2} g^2 \beta^2 \frac{r_{ij}{}^2 \delta_{ij} - 3ij}{r^5} \quad (ij = x, y, z) \tag{6.39}$$

Thus, Eq. (6.37) can be written as

$$\hat{\mathscr{H}}_{SS} = \hat{\mathbf{s}} \cdot D \cdot \hat{\mathbf{s}} \tag{6.40}$$

We can also obtain this result from the spin-orbit coupling interaction [3,4].

Note: The above $\boldsymbol{D}$ matrices are all in the *xyz* coordinate system. $\boldsymbol{D}$ is the second-grade traceless tensor and can be diagonalized as ${}^d\boldsymbol{D}$. Its principal axis coordinate system is *XYZ*.

$$^d\boldsymbol{D} = \begin{bmatrix} D_{XX} & 0 & 0 \\ 0 & D_{YY} & 0 \\ 0 & 0 & D_{ZZ} \end{bmatrix} \tag{6.41}$$

and

$$D_{XX} + D_{YY} + D_{ZZ} = \mathrm{T}_r\left({}^d\boldsymbol{D}\right) = 0 \tag{6.41a}$$

The Hamiltonian operator can be written as

$$\hat{\mathscr{H}}_{SS} = \begin{bmatrix} \hat{S}_X & \hat{S}_Y & \hat{S}_Z \end{bmatrix} \begin{bmatrix} D_{XX} & 0 & 0 \\ 0 & D_{YY} & 0 \\ 0 & 0 & D_{ZZ} \end{bmatrix} \begin{bmatrix} \hat{S}_X \\ \hat{S}_Y \\ \hat{S}_Z \end{bmatrix} = \hat{\mathbf{s}} \cdot {}^d\boldsymbol{D} \cdot \hat{\mathbf{s}} \tag{6.42}$$

Then,

$$\hat{\mathscr{H}}_{SS} = D_{XX} S_X^2 + D_{YY} S_Y^2 + D_{ZZ} S_Z^2 \tag{6.43}$$

Let

$$\mathscr{X} = -D_{XX}, \quad \mathscr{Y} = -D_{YY}, \quad \mathscr{Z} = -D_{ZZ}$$

Because of the $\boldsymbol{D}$ is the second-grade traceless tensor, hence

$$\mathscr{X} + \mathscr{Y} + \mathscr{Z} = 0 \tag{6.44}$$

Thus, Eq. (6.43) can be written as

$$\hat{\mathscr{H}}_{SS} = -\left(\mathscr{X} S_X^2 + \mathscr{Y} S_Y^2 + \mathscr{Z} S_Z^2\right) \tag{6.45}$$

6.2.2.2 State energy of $S = 1$ spin system (eigenvalue of the hamiltonian operator)

The matrix representation of spin operator in Eq. (6.45) is as follows:

$$\hat{S}_X^2 = \begin{pmatrix} \frac{1}{2} & 0 & \frac{1}{2} \\ 0 & 1 & 0 \\ \frac{1}{2} & 0 & \frac{1}{2} \end{pmatrix}, \quad \hat{S}_Y^2 = \begin{pmatrix} \frac{1}{2} & 0 & -\frac{1}{2} \\ 0 & 1 & 0 \\ -\frac{1}{2} & 0 & \frac{1}{2} \end{pmatrix}, \quad \hat{S}_Z^2 = \begin{pmatrix} 1 & 0 & 0 \\ 0 & 0 & 0 \\ 0 & 0 & 1 \end{pmatrix} \tag{6.46}$$

We define a set of spin wave functions from Eq. (6.11):

$$|+1\rangle = \alpha(1)\alpha(2), \quad |0\rangle = \frac{1}{\sqrt{2}}[\alpha(1)\beta(2) + \beta(1)\alpha(2)], \quad |-1\rangle = \beta(1)\beta(2) \tag{6.47}$$

With Eq. (6.47) as basis function, the matrix representation of $\hat{\mathscr{H}}_{SS}$ is as follows:

$$\begin{array}{c} \\ \langle +1| \\ \langle 0| \\ \langle -1| \end{array} \begin{array}{ccc} |+1\rangle & |0\rangle & |-1\rangle \\ \left[\begin{array}{c} -\frac{1}{2}(\mathscr{X}+\mathscr{Y})-\mathscr{Z} \\ 0 \\ \frac{1}{2}(\mathscr{Y}-\mathscr{X}) \end{array}\right. & \begin{array}{c} 0 \\ -(\mathscr{X}+\mathscr{Y}) \\ 0 \end{array} & \left.\begin{array}{c} \frac{1}{2}(\mathscr{Y}-\mathscr{X}) \\ 0 \\ -\frac{1}{2}(\mathscr{X}+\mathscr{Y})-\mathscr{Z} \end{array}\right] \end{array}$$

By using the traceless property of this matrix, that is, Eq. (6.44), their eigenvalues can be found. Assuming W is the solution of secular determinant, let us now solve the secular equation:

$$\begin{vmatrix} -\frac{1}{2}\mathscr{Z} - W & 0 & -\frac{1}{2}(\mathscr{X}-\mathscr{Y}) \\ 0 & \mathscr{Z} - W & 0 \\ -\frac{1}{2}(\mathscr{X}-\mathscr{Y}) & 0 & -\frac{1}{2}\mathscr{Z} - W \end{vmatrix} = 0 \tag{6.48}$$

Solution: $(-\frac{1}{2}\mathscr{Z} - W)^2(\mathscr{Z} - W) - (\mathscr{Z} - W)[-\frac{1}{2}(\mathscr{X}-\mathscr{Y})]^2$

$$= (\mathscr{Z} - W)\left\{\left(-\frac{1}{2}\mathscr{Z} - W\right)^2 - \left[-\frac{1}{2}(\mathscr{X}-\mathscr{Y})\right]^2\right\}$$

$$= (\mathscr{Z} - W)\left[-\frac{1}{2}\mathscr{Z} - W - \frac{1}{2}(\mathscr{X}-\mathscr{Y})\right]\left[-\frac{1}{2}\mathscr{Z} - W + \frac{1}{2}(\mathscr{X}-\mathscr{Y})\right] = 0$$

There are three roots of this equation:

When $(\mathscr{Z} - W) = 0$, we obtain $W_1 = W_Z = \mathscr{Z}$
when $[-\frac{1}{2}\mathscr{Z} - W + \frac{1}{2}(\mathscr{X}-\mathscr{Y})] = 0$, we obtain: $W_2 = W_Y = \mathscr{Y}$
when $[-\frac{1}{2}\mathscr{Z} - W - \frac{1}{2}(\mathscr{X}-\mathscr{Y})] = 0$, we obtain $W_3 = W_X = \mathscr{X}$
It is evident that the $\mathscr{X}$, $\mathscr{Y}$, and $\mathscr{Z}$ are eigenvalues of $\hat{\mathscr{H}}_{SS}$. Now assuming

$$D = \frac{1}{2}(\mathscr{X}+\mathscr{Y}) - \mathscr{Z} = -\frac{3}{2}\mathscr{Z} \tag{6.49}$$

$$E = -\frac{1}{2}(\mathscr{X}-\mathscr{Y}) \tag{6.50}$$

Substituting into Eq. (6.45), we obtain

$$\begin{aligned}\hat{\mathscr{H}}_{SS} &= D\left(\hat{S}_Z^2 - \frac{1}{3}\hat{S}^2\right) + E\left(\hat{S}_X^2 - \hat{S}_y^2\right)\\ &= D\left[\hat{S}_Z^2 - \frac{1}{3}S(S+1)\right] + E\left(\hat{S}_X^2 - \hat{S}_y^2\right)\\ &= D\left(\hat{S}_Z^2 - \frac{2}{3}\right) + E\left(\hat{S}_X^2 - \hat{S}_y^2\right) \end{aligned} \tag{6.51}$$

Thus, the eigenvalues of $\hat{\mathscr{H}}_{SS}$ can also be written as

$$W_X = \mathscr{X} = \frac{1}{3}D - E \tag{6.52a}$$

$$W_Y = \mathscr{Y} = \frac{1}{3}D + E \tag{6.52b}$$

$$W_Z = \mathscr{Z} = -\frac{2}{3}D \tag{6.52c}$$

Note: Do not confuse D of Eqs. (6.49), (6.51), and (6.52) with the $\boldsymbol{D}$ tensor in Eq. (6.38). Here D and E are *zero-field splitting constants*, and their values depend on the choice of the Z-axis (it is always able to choose the axis to meet $|E| \le |D/3|$) [5]. The sign of D and E can be positive but can also be negative; they can be same sign but can also be opposite, because the positions of the lines depend only on their relative signs [6–8]. Both D and E are usually given absolute values, and their units are generally used in magnetic field units (T or mT). Equation (6.51) is the ordinary expression of the electronic dipole–dipole interaction spin Hamiltonian operator in the literature. Its advantage is that it can reflect the symmetry directly, because $E = 0$, when it is in axial symmetry.

For the $S = 1$ system with the Z-axis parallel to the external magnetic field, the relation between energy level and magnetic field is shown in Figure 6.4. It can be seen that when $Z//H$, the width of the zero-field energy splitting d is equal to the zero-field splitting constant D. When the external magnetic field has not been applied, which is $H = 0$, the two energy levels of $|\pm 1\rangle$ are degenerated.

6.2.2.3 The eigenfunction of $S = 1$ spin system

Equation (6.47) is a set of wave functions chosen by us, but not the eigenfunction of the operator $\hat{\mathscr{H}}_{SS}$ in Eq. (6.40). Its eigenfunctions should be linear combinations of wave functions in Eq. (6.47). When $Z \parallel H$ and $H \to 0$, the eigenfunctions are as follows:

$$|T_X\rangle = \frac{e^{i\theta}}{\sqrt{2}}[|-1\rangle - |+1\rangle]; \quad |T_Y\rangle = \frac{e^{i\phi}}{\sqrt{2}}[|-1\rangle + |+1\rangle]; \quad |T_Z\rangle = |0\rangle \tag{6.53}$$

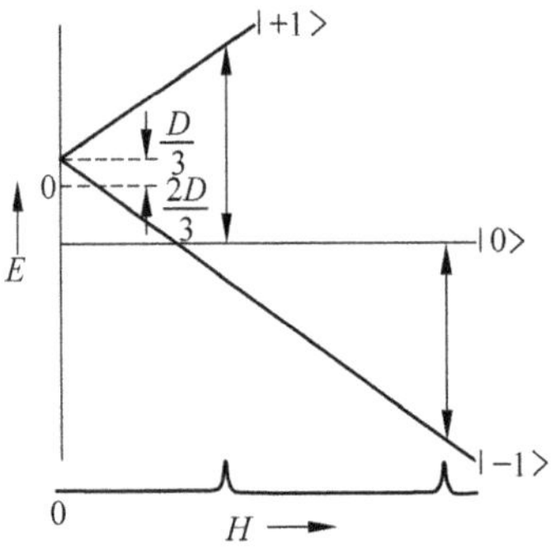

Figure 6.4: Illustrated diagram of the relationship between the energy level and the magnetic field; when *Z*-axis is parallel to the external magnetic field, $D > 0$, $E = 0$.

Here, $e^{i\theta}$ and $e^{i\phi}$ are phase factors, which can be chosen arbitrarily. The best choice is $\theta = \pi$ and $\phi = \frac{\pi}{2}$, then

$$|T_X\rangle = \frac{1}{\sqrt{2}}[|{-1}\rangle - |{+1}\rangle] = \frac{1}{\sqrt{2}}(\beta\beta - \alpha\alpha) \tag{6.54a}$$

$$|T_Y\rangle = \frac{\mathrm{i}}{\sqrt{2}}[|{-1}\rangle + |{+1}\rangle] = \frac{\mathrm{i}}{\sqrt{2}}(\beta\beta + \alpha\alpha) \tag{6.54b}$$

$$|T_Z\rangle = |0\rangle = \frac{1}{\sqrt{2}}(\alpha\beta - \beta\alpha) \tag{6.54c}$$

They are very like the wave functions $|p_x\rangle$, $|p_y\rangle$, and $|p_z\rangle$ of three orbital angular momentum of $\ell = 1$. So, they can also correspondingly satisfy the following relationships:

$$\begin{array}{lll}
\hat{S}_x|T_X\rangle = 0 & \hat{S}_y|T_X\rangle = -i|T_Z\rangle & \hat{S}_z|T_X\rangle = i|T_Y\rangle \\
\hat{S}_x|T_Y\rangle = i|T_Z\rangle & \hat{S}_y|T_Y\rangle = 0 & \hat{S}_z|T_Y\rangle = -i|T_X\rangle \\
\hat{S}_x|T_Z\rangle = -i|T_Y\rangle & \hat{S}_y|T_Z\rangle = i|T_X\rangle & \hat{S}_z|T_Z\rangle = 0 \\
\hat{S}_x^2|T_X\rangle = 0 & \hat{S}_y^2|T_X\rangle = |T_X\rangle & \hat{S}_z^2|T_X\rangle = |T_X\rangle \\
\hat{S}_x^2|T_Y\rangle = |T_Y\rangle & \hat{S}_y^2|T_Y\rangle = 0 & \hat{S}_z^2|T_Y\rangle = |T_Y\rangle \\
\hat{S}_x^2|T_Z\rangle = |T_Z\rangle & \hat{S}_y^2|T_Z\rangle = |T_Z\rangle & \hat{S}_z^2|T_Z\rangle = 0
\end{array}$$

$$\hat{\mathscr{H}}_{SS} = \begin{array}{c} \\ \langle T_X| \\ \langle T_Y| \\ \langle T_Z| \end{array} \begin{array}{c} |T_X\rangle|T_Y\rangle|T_Z\rangle \\ \begin{bmatrix} \mathscr{X} & 0 & 0 \\ 0 & \mathscr{Y} & 0 \\ 0 & 0 & \mathscr{Z} \end{bmatrix} \end{array} \tag{6.55}$$

With this set of wave function as the base function, the matrix representation of the operator $\hat{\mathscr{H}}_{SS}$ is shown in Eq. (6.55). At this stage, we obtain the Hamiltonian operator of the dipole–dipole interaction of two unpaired electrons, shown in Eq. (6.40). The eigenvalues of the operator $\hat{\mathscr{H}}_{SS}$ are $\mathscr{X}$, $\mathscr{Y}$, and $\mathscr{Z}$ as in Eq. (6.52), and the eigenfunctions of $\hat{\mathscr{H}}_{SS}$ are $|T_X\rangle$, $|T_Y\rangle$, and $|T_Z\rangle$ as in Eq. (6.54). Note that D and E are zero-field splitting constants, in the case of axisymmetry, when $Z \parallel H$, $E = 0$.

6.2.2.4 The transition of $\Delta M_S = \pm 2$

In the region of high field, the quantum numbers $M_S = +1, 0, -1$ as the eigenfunctions of spin Hamiltonian are very significant. The spin quantization in magnetic field direction M_S is a good quantum number; the energy state can be characterized by $|M_S\rangle$. When the microwave field is perpendicular to the direction of the external magnetic field, only the transition of $\Delta M_S = \pm 1$ can be produced. But the transition of "$\Delta M_S = \pm 2$" is forbidden. However, in the region of low field, the eigenfunctions become linear combination of the high-field state functions, as Eq. (6.55); the M_S is no longer a good quantum number. Thus, the usual selection rule of $\Delta M_S = \pm 1$ is not applicable at this condition. When the microwave field is parallel to the external magnetic field, the transition of " $\Delta M_S = \pm 2$ " will occur. Since the g value of this line is about 4, we usually call the transition line of " $\Delta M_S = \pm 2$ " as "half field line."

When $H \parallel Z$, and the microwave field is also parallel to Z, the transition of $\Delta M_S = \pm 2$ can occur; however, when the microwave field is perpendicular to Z direction, this transition cannot occur. The reason is

$$\left|\langle + |\hat{S}_Z| - \rangle\right|^2 = \left(\frac{\mathscr{X} - \mathscr{Y}}{\delta}\right)^2 \tag{6.56a}$$

$$\langle + |\hat{S}_X| - \rangle = \langle + |\hat{S}_Y| - \rangle = 0 \tag{6.56b}$$

From the above formulas, we can know that the intensity of the transition "$\Delta M_S = \pm 2$" is much weaker than the normal transition " $\Delta M_S = \pm 1$," because it is forbidden at normal conditions. But since the anisotropy is relatively small, on the contrary, it is very important in the disordered sample, because

$$\Delta E = E_+ - E_- = \left[\frac{1}{2}(\mathscr{X} + \mathscr{Y}) + \frac{1}{2}\delta\right] - \left[\frac{1}{2}(\mathscr{X} + \mathscr{Y}) - \frac{1}{2}\delta\right] = \delta \tag{6.57}$$

When the microwave frequency is fixed, its resonance magnetic field

$$H = \frac{1}{2g\beta}[(h\nu)^2 - (\mathscr{X} - \mathscr{Y})^2]^{1/2} = \frac{1}{2g\beta}[(h\nu)^2 - 4E^2]^{1/2} \tag{6.58}$$

Note: E of $4E^2$ in Eq. (6.58) is E of the zero-field splitting parameter (D and E).

When the system is axisymmetric and $E = 0$, the line will appear at $g = 4$, and

$$H(E_+ - E_-) = \frac{h\nu}{2g\beta} = \frac{1}{4}[H(E_0 - E_-) + (E_+ - E_0)] \tag{6.59}$$

This is the case when H//Z. For the conditions of $H//X$ *or* $H//Y$, and when the microwave field is parallel to the external magnetic field, the transition of $\Delta M_S = \pm 2$ also can be observed, while the resonance magnetic fields and relative intensities are listed in Table 6.1.

In the usual EMR spectrometer (the microwave field is perpendicular to the external magnetic field), the observable transition of $\Delta M_S = \pm 2$ is the single photon transition [9]. For irregular orientation of solid samples, the transition of $\Delta M_S = \pm 2$ appears in the low field, and the *X*, *Y*, and *Z* components cannot be distinguished, but there is a turning point $H_{\min}$ [9,10]. In the case of fixed microwave frequency ν, in order to make Eq.(6.58) to have a solution larger than zero, the system of isotropic *g* should maintain the minimum possible magnetic field ($H_{\min}$) [11] of the $\Delta M_S = \pm 2$ transition:

Table 6.1: The resonance magnetic fields and relative intensities of the transition $\Delta M_S = \pm 2$.

Orientation	Resonance magnetic field	Relative intensity
X	$\left(\frac{1}{2g\beta}\right)[(h\nu)^2 - (\mathscr{Y} - \mathscr{Z})^2]^{1/2}$	$[(\mathscr{Y} - \mathscr{Z})/\delta]^2$
Y	$\left(\frac{1}{2g\beta}\right)[(h\nu)^2 - (\mathscr{Z} - \mathscr{X})^2]^{1/2}$	$[(\mathscr{Z} - \mathscr{X})/\delta]^2$
Z	$\left(\frac{1}{2g\beta}\right)[(h\nu)^2 - (\mathscr{X} - \mathscr{Y})^2]^{1/2}$	$[(\mathscr{X} - \mathscr{Y})/\delta]^2$

$$H_{\min} = \frac{1}{g\beta}\left[\frac{(h\nu)^2}{4} - \frac{D^2 + 3E^2}{3}\right]^{1/2} \tag{6.60}$$

For the triplet system of irregular orientation, the low field side peak of the differential line can be used to estimate the value of $D^* = (D^2 + 3E^2)^{1/2}$, which is a method for measuring the *mean square root* values of zero-field splitting. In some cases, *D* and *E* can be determined approximately by line shape analysis of $\Delta M_S = \pm 2$ [12]. However, if the zero-field splitting parameter has larger energy than the microwave photon energy $h\nu$, the transition of $\Delta M_S = \pm 2$ cannot occur.

The triplet molecules in the liquid solution, owing to the rotary motion, especially the rapid rolling, lead to "spin-lattice" relaxation (τ_1 far less than that of the free radical with $S = 1/2$) [13]. If the zero-field splitting (*D*) is removed, it is very difficult to detect its EMR spectrum.

It should be pointed out that when microwave frequency is high enough (such as Q-band), the double quantum (double-photon) transition can be observed [14,15]. These occur between $|\pm 1\rangle$ states and near $g = 2$.

6.3 The triplet molecule ($S = 1$) system

We have already discussed all $S = 1$ systems in the absence of external magnetic field. Now we will talk about the variation of system and its EMR spectrum when the external magnetic field is added.

6.3.1 Energy levels and wave functions of triplet molecules under the action of external magnetic field

In order to simplify the problem, we focus the characteristic of variation caused by the dipole–dipole interaction produced by the spin of two unpaired electrons. Assume that g is isotropic, and the hyperfine interaction between the unpaired electrons and the magnetic nucleus is neglected, then the Hamiltonian operator of this system is

$$\hat{\mathscr{H}} = g\beta \boldsymbol{H} \cdot \boldsymbol{S} - \left(\mathscr{X}S_X^2 + \mathscr{Y}S_Y^2 + \mathscr{Z}S_Z^2\right) \tag{6.61}$$

In Eq. (6.61), the first term is the Zeeman interaction between the unpaired electron with external magnetic field, and the second term is the dipole–dipole interaction between the two unpaired electron spins, that is, $\hat{\mathscr{H}}_{SS}$. Then, Eq. (6.61) can be written as follows:

$$\hat{\mathscr{H}} = \hat{\mathscr{H}}_{\text{Zeeman}} + \hat{\mathscr{H}}_{SS}$$

Setting the direction cosines of the external magnetic field $\boldsymbol{H}$ in the principal axis coordinate system to be (l_X, l_Y, l_Z), and taking the $|T_X\rangle$, $|T_Y\rangle$, $|T_Z\rangle$ as base vector, the matrix representation of Eq. (6.61) is shown as follows:

$$\hat{\mathscr{H}} = \begin{array}{c} \\ \langle T_X| \\ \langle T_Y| \\ \langle T_Z| \end{array} \begin{array}{c} \begin{array}{ccc} |T_X\rangle & |T_Y\rangle & |T_Z\rangle \end{array} \\ \begin{bmatrix} \mathscr{X} & -\mathrm{i}g\beta H l_Z & \mathrm{i}g\beta H l_Y \\ \mathrm{i}g\beta H l_Z & \mathscr{Y} & -\mathrm{i}g\beta H l_X \\ -\mathrm{i}g\beta H l_Y & \mathrm{i}g\beta H l_X & \mathscr{Z} \end{bmatrix} \end{array} \tag{6.62}$$

The general solution of Eq. (6.62) is a function of (l_X, l_Y, l_Z). We consider several special cases:

At first, consider the external magnetic field is parallel to the direction of Z, namely, $(l_X, l_Y, l_Z) = (0, 0, 1)$, then the solution of the secular equation is as follows:

$$E_Z = E_0 = \mathscr{Z} \tag{6.63a}$$

$$E_+ = \frac{1}{2}(\mathscr{X} + \mathscr{Y}) + \left\{ \left[\frac{1}{2}(\mathscr{X} - \mathscr{Y})\right]^2 + (g\beta H)^2 \right\}^{1/2} \tag{6.63b}$$

$$E_- = \frac{1}{2}(\mathscr{X} + \mathscr{Y}) - \left\{ \left[\frac{1}{2}(\mathscr{X} - \mathscr{Y})\right]^2 + (g\beta H)^2 \right\}^{1/2} \tag{6.63c}$$

The corresponding eigenfunctions are as follows:

$$|0\rangle = |T_Z\rangle = \frac{1}{\sqrt{2}}(\beta\alpha + \alpha\beta) \tag{6.64a}$$

$$|+\rangle = \frac{1}{\sqrt{2}}\left\{ \left[1 + \left(\frac{\mathscr{X} - \mathscr{Y}}{\delta}\right)\right]^{1/2} |T_X\rangle + \mathrm{i}\left[1 - \left(\frac{\mathscr{X} - \mathscr{Y}}{\delta}\right)\right]^{1/2} |T_Y\rangle \right\} \tag{6.64b}$$

$$|-\rangle = \frac{1}{\sqrt{2}}\left\{ i\left[1 - \left(\frac{\mathscr{X} - \mathscr{Y}}{\delta}\right)\right]^{1/2} |T_X\rangle + \left[1 + \left(\frac{\mathscr{X} - \mathscr{Y}}{\delta}\right)\right]^{1/2} |T_Y\rangle \right\} \tag{6.64c}$$

When the external magnetic field $\boldsymbol{H}$ is very large, $\frac{\mathscr{X} - \mathscr{Y}}{\delta} \to 0$, then

$$|\pm\rangle \to |\pm 1\rangle \tag{6.65}$$

Its explanation is that the coupling of electron spins at high fields is relieved. Let us now see the transition of $\Delta M_S = 1$:

$$|-\rangle \to |0\rangle : E_0 - E_- = \mathscr{Z} - \frac{1}{2}(\mathscr{X} + \mathscr{Y}) + \frac{1}{2}\delta = \frac{3}{2}\mathscr{Z} + \frac{1}{2}\delta \tag{6.66a}$$

$$|0\rangle \to |+\rangle : E_+ - E_0 = \frac{1}{2}(\mathscr{X} + \mathscr{Y}) - \mathscr{Z} + \frac{1}{2}\delta = -\frac{3}{2}\mathscr{Z} + \frac{1}{2}\delta \tag{6.66b}$$

When the microwave frequency is fixed, the resonance magnetic field is

$$H(E_0 - E_-) = \frac{1}{g\beta}\left\{ \left[h\nu - \frac{3}{2}\mathscr{Z}\right]^2 - \left[\frac{1}{2}(\mathscr{X} - \mathscr{Y})\right]^2 \right\}^{1/2} \tag{6.67a}$$

$$H(E_+ - E_0) = \frac{1}{g\beta}\left\{ \left[h\nu + \frac{3}{2}\mathscr{Z}\right]^2 - \left[\frac{1}{2}(\mathscr{X} - \mathscr{Y})\right]^2 \right\}^{1/2} \tag{6.67b}$$

Taking into account the relation of $\mathscr{X}$, $\mathscr{Y}$, $\mathscr{Z}$ and D, E, that is, Eqs. (6.49) and (6.50) substituted into Eq. (6.67), it can be written as follows:

$$H(E_0 - E_-) = \frac{1}{g\beta}\left\{ [h\nu + D]^2 - E^2 \right\}^{1/2} \tag{6.68a}$$

$$H(E_+ - E_0) = \frac{1}{g\beta}\left\{[h\nu - D]^2 - E^2\right\}^{1/2} \tag{6.68b}$$

We take Eq. (6.63) to be written as Eq. (6.69) and Eq. (6.64) to be written as Eq. (6.70); when $H/\!/Z$, these can be summarized as follows:

$$E_0 = Z; \quad E_\pm = \tfrac{1}{2}(\mathscr{X} + Y) \pm \left\{\left[\tfrac{1}{2}(\mathscr{X} - Y)\right]^2 + (g\beta H)^2\right\}^{1/2} \tag{6.69}$$

$$|0\rangle = |T_Z\rangle \tag{6.64a}$$

$$|+\rangle = \frac{1}{\sqrt{2}}\left\{\left[1 + \left(\frac{\mathscr{X} - \mathscr{Y}}{\delta}\right)\right]^{1/2} |T_X\rangle + i\left[1 - \left(\frac{\mathscr{X} - \mathscr{Y}}{\delta}\right)\right]^{1/2} |T_Y\rangle\right\} \tag{6.64b}$$

$$|-\rangle = \frac{1}{\sqrt{2}}\left\{i\left[1 - \left(\frac{\mathscr{X} - \mathscr{Y}}{\delta}\right)\right]^{1/2} |T_X\rangle + \left[1 + \left(\frac{\mathscr{X} - \mathscr{Y}}{\delta}\right)\right]^{1/2} |T_Y\rangle\right\} \tag{6.64c}$$

$$H_\pm = \frac{1}{g\beta}\left\{\left(h\nu \pm \frac{3}{2}\mathscr{Z}\right)^2 - \left(\frac{\mathscr{X} - \mathscr{Y}}{2}\right)^2\right\}^{1/2} \tag{6.70}$$

For the case of $H/\!/Y$, the following solution can be obtained:

$$E_0 = \mathscr{Y}, \quad E_\pm = \tfrac{1}{2}(\mathscr{Z} + \mathscr{X}) \pm \left\{\left[\tfrac{1}{2}(\mathscr{X} - \mathscr{Y})\right]^2 + (g\beta H)^2\right\}^{1/2} \tag{6.71}$$

$$|0\rangle = |T_Y\rangle \tag{6.72a}$$

$$|+\rangle = \frac{1}{\sqrt{2}}\left\{\left[1 + \left(\frac{\mathscr{Z} - \mathscr{X}}{\delta}\right)\right]^{1/2} |T_Z\rangle + i\left[1 - \left(\frac{\mathscr{Z} - \mathscr{X}}{\delta}\right)\right]^{1/2} |T_X\rangle\right\} \tag{6.72b}$$

$$|-\rangle = \frac{1}{\sqrt{2}}\left\{i\left[1 - \left(\frac{\mathscr{Z} - \mathscr{X}}{\delta}\right)\right]^{1/2} |T_Z\rangle + \left[1 + \left(\frac{\mathscr{Z} - \mathscr{X}}{\delta}\right)\right]^{1/2} |T_X\rangle\right\} \tag{6.72c}$$

$$H_\pm = \frac{1}{g\beta}\left\{\left(h\nu \pm \frac{3}{2}\mathscr{Y}\right)^2 - \left(\frac{\mathscr{Z} - \mathscr{X}}{2}\right)^2\right\}^{1/2} \tag{6.73}$$

For the case of $H/\!/X$, the following solution can be obtained:

$$E_0 = X, \quad E_\pm = \tfrac{1}{2}(Y + Z) \pm \left\{\left[\tfrac{1}{2}(Y - Z)\right]^2 + (g\beta H)^2\right\}^{1/2} \tag{6.74}$$

$$|0\rangle = |T_X\rangle \tag{6.75a}$$

$$|+\rangle = \frac{1}{\sqrt{2}}\left\{\left[1 + \left(\frac{\mathscr{Y} - \mathscr{Z}}{\delta}\right)\right]^{1/2} |T_Y\rangle + i\left[1 - \left(\frac{\mathscr{Y} - \mathscr{Z}}{\delta}\right)\right]^{1/2} |T_Z\rangle\right\} \tag{6.75b}$$

$$|-\rangle = \frac{1}{\sqrt{2}}\left\{ i\left[1-\left(\frac{\mathscr{Y}-\mathscr{Z}}{\delta}\right)\right]^{1/2}|T_Y\rangle + \left[1+\left(\frac{\mathscr{Y}-\mathscr{Z}}{\delta}\right)\right]^{1/2}|T_Z\rangle \right\} \quad (6.75c)$$

$$H_\pm = \frac{1}{g\beta}\left\{\left(h\nu \pm \frac{3}{2}\mathscr{X}\right)^2 - \left(\frac{\mathscr{Y}-\mathscr{Z}}{2}\right)^2\right\}^{1/2} \quad (6.76)$$

6.3.2 Examples of triplet state excited by light

A large number of aromatic hydrocarbons (especially polycyclic aromatic hydrocarbons) can be excited by ultraviolet or visible light, for a long excited state. Their lifetime can be as long as a few minutes. It can be explained that this excited state is metastable state. Lewis et al. [16] assumed in 1941 that the long-lived metastable state is a "spin triplet state."Because of the direct excitation from singlet to triplet or vice versa, all spins are forbidden, and thus the life is quite long. The light excitation will excite to the singlet state at first, and then, owing to Hund's rule, the excited singlet state via nonradiative transition will make spin reversal to "excited triplet state" (as shown in Figure 6.1).

The hypothesis of Lewis et al. was confirmed soon by the magnetic susceptibility measurement. With UV irradiation, the paramagnetism is increased, irradiation stopped, and paramagnetism decreased. This is consistent with the case of phosphorescence decay. However, in early days, people studied it using EMR, but failed. The reason is that the dipole–dipole interaction between the two unpaired electrons is strongly anisotropic, but for research, the sample should be in the form of single crystals. If a single crystal of pure aromatic compound is produced, the energy of triplet state excitation could be transferred quickly between molecules. So that the triplet state is quenched, thus only the single crystal of the diluting solid solution can be studied successfully.

The first successful experiment [17] was that naphthalene was dissolved in 1,2,4,5-tetramethylbenzene to form single crystal. Owing to the similarity of their geometry structure, naphthalene can replace the molecule of tetramethylbenzene, and transfer of triplet excitation between naphthalene molecules cannot occur because of its low concentration. The functional relationship between the energy level of the lowest triplet state of naphthalene and the magnetic field are shown in Figure 6.5. When the external magnetic field is parallel to the three principal axes, we can obtain $D = 0.1003\ \mathrm{cm}^{-1}$; $E = -0.0137\ \mathrm{cm}^{-1}$; and the isotropic $g = 2.0030$ [6,17]. Under the prerequisite condition that the zero-field splitting is mainly caused by the contribution of $\hat{\mathscr{H}}_{SS}$, the values of D and E reflect the average distance between the two unpaired electrons.

When the external magnetic field is parallel to the principal axes X or Y of the naphthalene molecule, the five hyperfine lines with the intensity ratio of 1:4:6:4:1 can

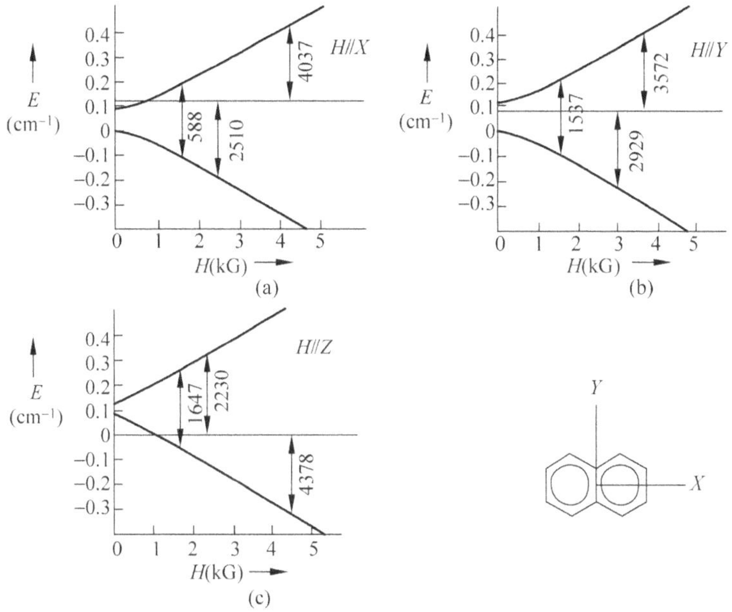

Figure 6.5: At 77 K, microwave frequency $v = 9.272$ GHz, the functional relationship of the lowest triplet state and the magnetic field [17] of naphthalene (the transition of lowest field can only be allowed in the case of a microwave field H_1 parallel to the external magnetic field H).

be identified at 77 K [18]. Deuterium-substituted naphthalene molecules detect that $a_\alpha = 0.561$ mT; $a_\beta = 0.229$ mT; and $a_\alpha / a_\beta = 2.45$, while for the solution of naphthalene anion, the $a_\alpha = 0.49$ mT; $a_\beta = 0.183$ mT; and $a_\alpha / a_\beta = 2.68$. The a_α / a_β ratio is very close, because in the triplet state of naphthalene, one of the electrons is located on the lowest antibonding orbit, and the other is located on the highest bonding orbit. According to the duality principle, the absolute value of the orbital coefficients should be same. Therefore, the ratio of the unpaired electron density at α and β positions of naphthalene also should be same as of the naphthalene anion.

In particular, it should be pointed out that in the system of $S = 1$, $M_S = \pm 1, 0$, where the energy level of $M_S = 0$ is nonsplitting. Hence, it is similar to the system of $S = \frac{1}{2}$; when $I = \frac{1}{2}$, the hyperfine structure of the system also has two lines, as shown in Figure 6.6.

6.3.3 Examples of thermal excitation triplet state

In the light of excitation triplet discussed above, it is assumed that the energy-level gap between triplet and singlet is large enough. Therefore, it does not need to consider the

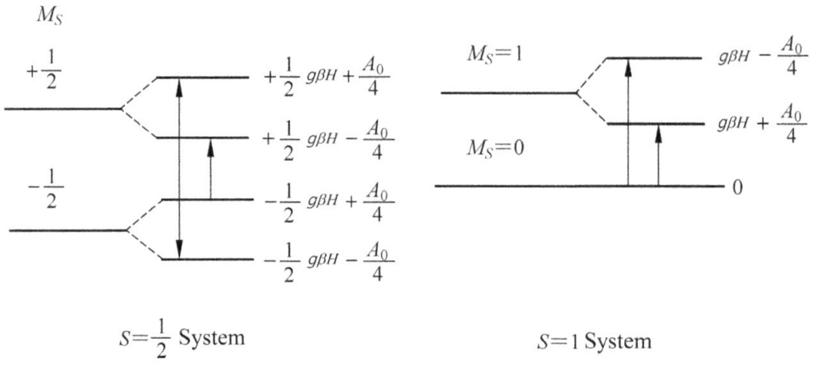

Figure 6.6: Schematic diagram of comparison between the hyperfine splittings in $S = 1$ and $S = 1/2$ systems.

problem of both states mixing. Now, we are interested in knowing in which case the energy gap between triplet and singlet states is less, and without state mixing. The difference between the singlet state and the triplet state correspond approximately to the exchange interaction energy between two electrons $|J_0|$ (refer to Section 6.2.1). When the triplet is slightly higher than singlet, the number of population between the two states should obey Boltzmann law. Generally, the intensity of the EMR absorption is proportional to the population number of the paramagnetic state, and is also proportional to the reciprocal of the absolute temperature (Curie's law). Therefore, the value of J_0 can be calculated from the area of the EMR absorption line.

The point defect in the MgO crystal (F center) provides a good example for the thermal excitation of triplet state [19]. This center is considered to be a neutral *trivacancy*, which is a defected linear fragment $(O-Mg-O)^{2-}$, and is filled by two electrons. At very low temperatures, the EMR spectrum cannot be observed, unless irradiated by UV light; however, when the temperature rises up to over 4 K, it will show the EMR spectrum. This is because the triplet energy level of these two electrons is over the singlet state, thus $J_0 > 0$. From the analysis of temperature dependence relation, we obtain $\bar{J}_0 = 56\,cm^{-1}$. This data analysis shows that the isotropic g =2.0030, D = 30.7 mT, and E = 0 of this system. The average distance between the molecules is 4.5Å, which is consistent with the corresponding distance between the O and O at 4.2 K [19].

Another interesting example is, in the Fremy salt $(K_4[(SO_3)_2NO]_2)$ powder sample, the EMR spectrum of the triplet state is observed [20]. Here, $D = \pm 0.076 cm^{-1}$; $E = \pm 0.0044 cm^{-1}$. It is found that the peak area of half field line is increased exponentially with a rise in temperature (250–350 K), and the energy level difference of the triplet with singlet is $J_0 = 2180 cm^{-1}$. Since the hyperfine splitting of ^{14}N has not been observed, the spectrum is identified as the contribution of the triplet exciton, which is fast flow excited state in the lattice.

6.3.4 Examples of other excited triplet

Due to the difficulty in culturing diluted solid solution single crystal samples, excited triplet state single crystals are used. The triplet EMR spectra of phenanthrene in phenylbenzene and quinoxaline in tetramethylbenzene are observed successfully. Two nitrogen atoms of quinoxaline even demonstrate hyperfine structures: $D = 0.1007\text{cm}^{-1}$ and $E = -0.0182\text{cm}^{-1}$, which are very similar to the D and E values of naphthalene.

Taking the phenanthrene and naphthalene dissolved in phenylbenzene and cultured as single crystals, the triplet EMR spectrum of naphthalene could be observed. From the triplet state of phenanthrene, via the triplet state of phenylbenzene, the naphthalene can be excited to triplet state. Because of the energy levels of the triplet states of phenanthrene, phenylbenzene and naphthalene are higher than their ground states, in order with the values of 21,410, 23,010, and 21, 110cm^{-1}, respectively. The energy levels of these three triplet states are also very close, thus it is favorable for energy transfer of the triplet exciton between molecules [21,22].

6.3.5 Examples of ground triplet state

The ground triplet state atoms such as C, O, Si, S, Ti, Ni, and so on, and the most famous diatomic molecules, O_2 and S_2, are all the examples of the ground triplet state.

6.3.5.1 Carbene

Carbene (CH_2), also called *methylene*, is one of the simplest molecular systems. Its spectrum has determined that it is ground triplet state. The EMR spectra of CH_2 and CD_2 have already been reported in Refs [23–25], the former: $D = 0.69\ \text{cm}^{-1}$, $E = 0.003\ \text{cm}^{-1}$; and the latter: $D = 0.75\ \text{cm}^{-1}$, $E = 0.011\ \text{cm}^{-1}$. Originally, both should be same, for the difference may come from the zero point fluctuation. Table 6.2 lists the zero-field splitting parameters of some carbene derivatives. Some molecules in the table have nonzero E values; thus, they have nonaxial symmetry and are even nonlinear. The maximum number of absorption peaks (six transitions of $\Delta M_S = \pm 1$) can be observed in the EMR spectra of this system in gas phase.

Fluorenylidene molecule [30] is also considered as a carbene derivative. It is formed by the irradiation on diazofluorene at 77 K to form the ground triplet molecule. If the zero-field splitting D is larger than the microwave quantum $h\nu$, according to selection rule, only some spectral lines of allowed transitions can be observed. When

Table 6.2: The zero-field splitting parameters of some carbene derivatives.

Molecule	$\lvert\bar{D}\rvert/cm^{-1}$	$\lvert\bar{E}\rvert/cm^{-1}$	References
H – C – H	0.69	0.003	[23–25]
D – C – D	0.75	0.011	[23–25]
H – C – C ≡ N	0.8629	0	[26]
H – C – CF_3	0.712	0.021	[27]
H – C – C_6H_5	0.5150	0.0251	[28]
H – C – C ≡ C – H	0.6256	0	[26]
H – C – C ≡ C – CH_3	0.6263	0	[26]
H – C – C ≡ C – C_6H_5	0.5413	0.0035	[26]
C_6H_5 – C – C_6H_5	0.4055	0.0194	[28]
N ≡ C – C – C ≡ N	1.002	< 0.002	[28]
N – C ≡ N	1.52	< 0.002	[29]

the external magnetic field H is parallel to X or Y axis, and under the case of microwave frequency $\nu = 9.7$ GHz, only one line can be observed. When merely the external magnetic field H is parallel to Z-axis, all the three transition lines can be observed. Its zero-field splitting parameter values $\lvert\bar{D}\rvert = 0.4078\ cm^{-1}$ and $\lvert\bar{E}\rvert = 0.0283\ cm^{-1}$ are larger than that of the triplet excited state of naphthalene.

Another, no matter what either experiments or theory are very worthy considerable ground state organic triplet molecule, is the trimethylenemethane radical. It is prepared by the γ -irradiation of methylenecyclopropane. EMR study reveals [31] that it is the ground triplet state rather than biradical. $\lvert J_0\rvert$ value is comparatively larger than its $\lvert D\rvert$ value. This radical is close to the uniaxial crystal (in a D_{3h} symmetrical plane); the zero-field splitting parameters at 77 K are $\lvert\bar{D}\rvert = 0.0248\ cm^{-1}$ and $\lvert\bar{E}\rvert \leq,\ 0.003\ cm^{-1}$ (Figure 6.7).

6.3.5.2 Dianion of symmetric aromatic hydrocarbon

A ground triplet state molecule may be a neutral molecule, a cation, or an anion, but a pair of degenerate orbitals has at least one unpaired electron. As shown in Figure 6.1d, on the two highest degenerate energy level of the lowest energy state (ground state), each occupy one electron, and these two electrons are spin parallel.

In the molecule with $n\,(n \geq 3)$ times symmetry axis, there exists degenerate orbital energy levels. It is not certain if these molecules have a ground triplet state, which depend on the sign of the electronic exchange integral. If J_0 is positive, then the ground state is the triplet state. Coronene bivalent anion is an example [32].

If bivalent anion is generated, an electron will result in every two degenerate orbits of the symmetrically substituted benzene. Many examples have proved that the symmetric molecule has triplet ground states. For example, 1,3,5-triphenyl benzene and triphenylene dianion. On the degenerate antibonding orbital of these ions, each

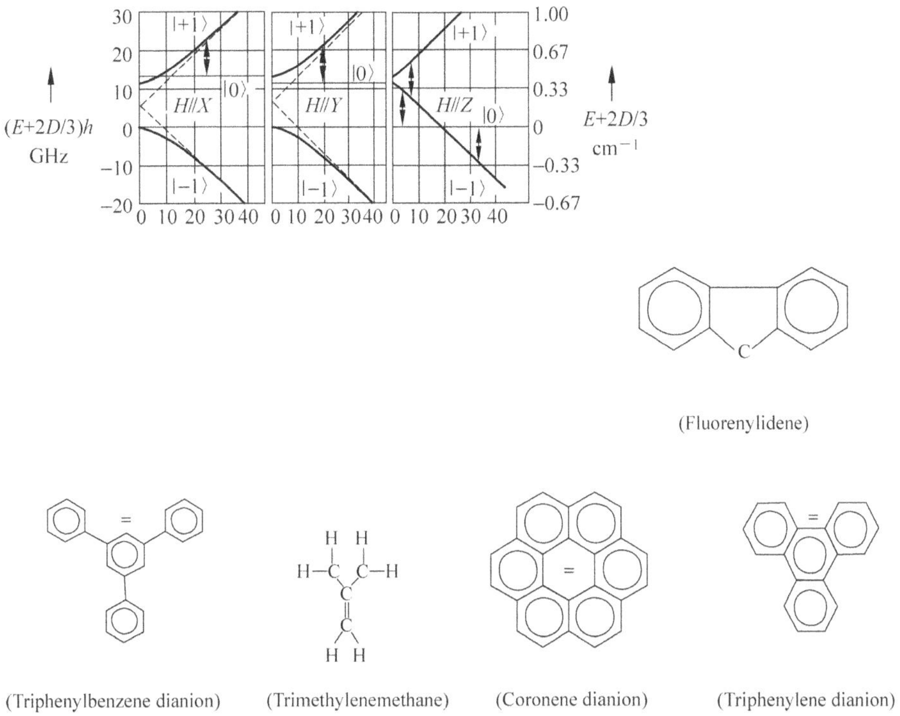

Figure 6.7: The relationship between the energy of the ground triplet state and the external magnetic field:**I** (fluorenylidene); **II** (triphenylbenzene dianion); **III** (trimethylenemathane); **IV** (coronene dianion); **V** (triphenylene dianion).

occupies an unpaired electron. $D(\bar{D} = 0.042\text{cm}^{-1})$ of triphenylmethyl dianion (ground triplet state) is less than $D(\bar{D} = 0.111\text{cm}^{-1})$ of the neutral excited state molecule [33], because the orbital electron filling of these two cases are quiet different, and the calculation shows that in the excited triplet molecules, the interaction between two electrons, each occupied on bonding and antibonding of two orbitals separately, is larger than between two electrons on antibonding degenerate orbitals of the ground triplet state dianion (which leads to larger D value).

6.3.5.3 Inorganic triplet state fragment

In addition to O_2, S_2, and some transition ions, the stable ground triplet state inorganic molecules is not common. Some unstable ground triplet state inorganic molecules (fragments) can be trapped in the frozen matrix at low temperature. It is a good example that the C atom react with CO or N_2 preparation of CCO or CNN molecules with the same number of electrons, and they are also captured in the frozen rare gas matrix at 4 K at the same time [34]. Both of these molecules have rather large values of D (CCO: $\bar{D} = 0.7392\text{cm}^{-1}$; CNN: $\bar{D} = 1.1590\text{cm}^{-1}$). In the X-band, their D values are larger than $h\nu$. Therefore, only one of their transition lines can be observed. Their hyperfine parameters of ^{13}C and ^{14}N have already been reported. The fact that $E = 0$ explains the linearity of these molecules.

The example of another quite different ground triplet state is the point defect of quasi-tetrahedral $[AlO_4]^+$ in the single crystal of quartz [35]. This center is regarded to be the two electron holes having the triplet spin system. Here, two unpaired electrons are placed separately on the $[AlO_4]^+$ tetrahedron of the two adjacent (and symmetric) oxygen atoms; the distance between the two oxygen atoms is about 2.6 $\AA$. At temperature ~35K, $D = -69.8$ mT ; $E = 6.3$ mT.

For the $S = 1$ system in transition metal ions, such as V^{3+} and Ni^{2+} in the 3d series, the latter ions are doped in K_2MgF_4 to replace the site of Mg^{2+}, which is surrounded by a slightly distorted octahedron composed of F^- ions. At temperature 1.6 K, $\bar{D} = -0.425\,\text{cm}^{-1}$ and $\bar{E} = -0.065\text{cm}^{-1}$ (isotropic $g = 2.275$) [36].

6.4 Triplet system of irregular orientation

All of the above discussed are regular orientation triplet systems, but most samples of experiments are of irregular orientation. Since the dipole–dipole interaction is strongly anisotropic, so, in the earlier research work, the EMR spectra of the irregularly oriented triplet systems are rarely reported. Later, in the magnetic field region of $g = 4$, the anisotropic EMR spectral line of rather weak $\Delta M_S = \pm 2$ transition was observed, which promoted the detection and study of the triplet states in many irregular systems [12]. In the less anisotropic system, the $\Delta M_S = \pm 2$ line with rather higher amplitude can be measured because the linewidth is narrower. Originally, the half field line can be observed only under the case of the microwave field parallel to the external magnetic field. But later it was found that the half field line can also be observed, even when perpendicular to the external magnetic field. It was verified soon [11] that at the "turning point" of the irregular system, the line of $\Delta M_S = \pm 1$ transition also can be detected.

Using the relationship of $\mathscr{X}$, $\mathscr{Y}$, $\mathscr{Z}$ and D, E, Eq. (6.63) can be rewritten as follows:

$$E_0 = -\frac{2}{3}D \tag{6.77a}$$

$$E_+ = \frac{D}{3} = [(g\beta H)^2 + E^2]^{1/2} \tag{6.77b}$$

$$E_- = \frac{D}{3} - [(g\beta H)^2 + E^2]^{1/2} \tag{6.77c}$$

There are two spectral lines of $\Delta M_S = \pm 1$ transition:

$$E_0 - E_- = [(g\beta H)^2 + E^2]^{1/2} - D \tag{6.78a}$$

$$E_+ - E_0 = [(g\beta H)^2 + E^2]^{1/2} + D \tag{6.78b}$$

There is only one spectral line of $\Delta M_S = \pm 2$ transition:

$$E_+ - E_- = 2[(g\beta H)^2 + E^2]^{1/2} \tag{6.79}$$

Observation by the spectrometer with fixed microwave frequency are as follows: $h\nu_0 = g\beta H_0$ for axisymmetric systems; zero-field splitting parameters: $E = 0$, $g_z = g_{\parallel}$; $g_x = g_y = g_{\perp}$. From Eq. (6.78), we obtain Eq. (6.80):

$$(H_1)_{\parallel} = H_0 + \frac{D}{g_{\parallel}\beta} \tag{6.80a}$$

$$(H_2)_{\parallel} = H_0 - \frac{D}{g_{\parallel}\beta} \tag{6.80b}$$

$$\Delta H_{\parallel} = (H_1)_{\parallel} - (H_2)_{\parallel} = 2D\left(g_{\parallel}\beta\right)^{-1} \tag{6.80c}$$

$$(H_1)_{\perp} = H_0\left[1 - \frac{D}{g_{\perp}\beta H_0}\right]^{1/2} \approx H_0 - \frac{D}{2g_{\perp}\beta} \tag{6.81a}$$

$$(H_2)_{\perp} = H_0\left[1 + \frac{D}{g_{\perp}\beta H_0}\right]^{1/2} \approx H_0 + \frac{D}{2g_{\perp}\beta} \tag{6.81b}$$

$$\Delta H_{\perp} \approx D(g_{\perp}\beta)^{-1} \tag{6.81c}$$

The unit of D in the above equation is Hz. If $D' = D(g\beta)^{-1}$, then the unit of D' is mT.

For the given zero-field splitting parameter D value and microwave frequency $\nu\,(g = g_e)$, the EMR theoretical absorption line of the triplet irregular system with an axial symmetry ($E = 0$) and $\Delta M_S = \pm 1$ is shown in Figure 6.8a; the computer simulation of the first-order differential spectrum is shown in Figure 6.8b. Obviously, if we can obtain an experimental spectrum as shown in Figure 6.8b, we can seek the value of the zero-field splitting parameter D.

The triphenylbenzene dianion in methyltetrahydrofuran (77 K frozen solid solution), the EMR spectrum of its triplet state is shown in Figure 6.9. R^- in the figure is the signal of monovalent anion radical. A set of $\Delta M_S = \pm 1$ spectral lines are very similar to Figure 6.8b. The amplitude of its half field line ($\Delta M_S = \pm 2$) is also strong [33]. It is a typical EMR spectrum of the irregularly oriented triplet state.

The EMR theoretical absorption spectrum $(g = g_e)$ and the computer simulating first-order differential spectrum of irregular oriented nonaxisymmetric $(E \neq 0)$ triplet system are shown in Figure 6.10a and b [37].

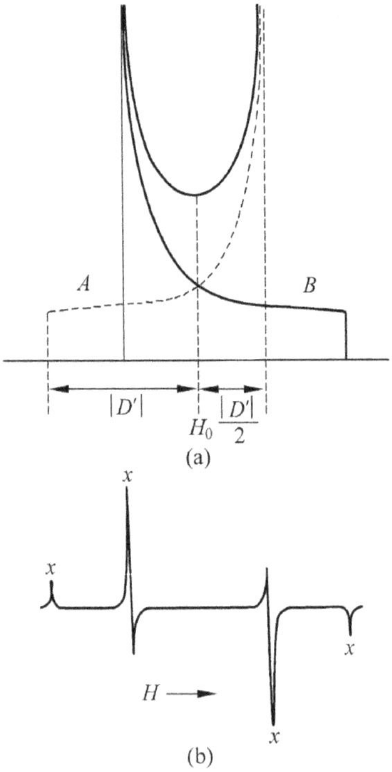

Figure 6.8: EMR theoretical absorption line of the triplet axisymmetric ($E = 0$) irregular systems (a) and computer simulation of the first-order differential (b) [37].

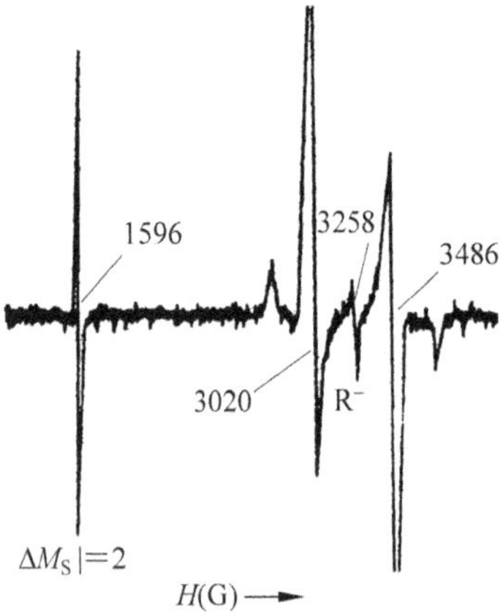

Figure 6.9: EMR spectrum of triphenylbenzene dianion in methyltetrahydrofuran at 77 K.

The EMR spectrum of full deuterated naphthalene ($C_{10}D_8$) organic solution at 77 K is shown in Figure 6.11 [12]. In order to eliminate the complex hyperfine structure, deuterium is used to substitute hydrogen. A line with $g = 2$ with very strong intensity

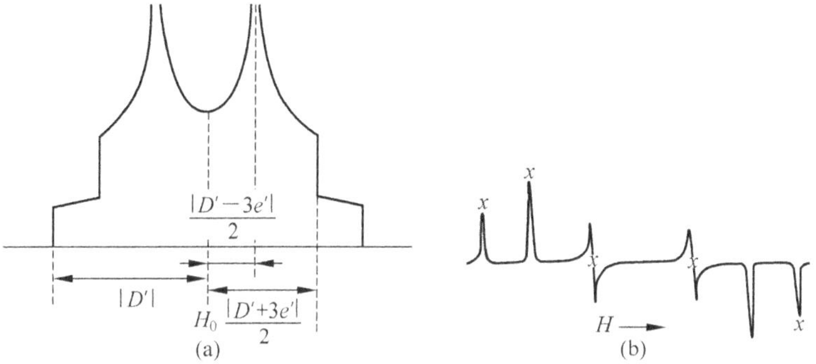

Figure 6.10: The EMR theoretical absorption spectrum (a) and the computer simulating first-order differential spectrum (b) for the irregular oriented nonaxisymmetric ($E \neq 0$) triplet system.

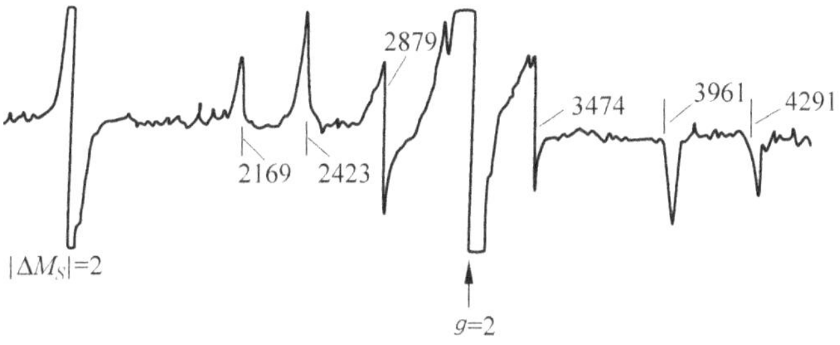

Figure 6.11: The EMR spectrum of the deuterated naphthalene ($C_{10}D_8$) solution at 77 K.

in the middle of the spectrum is from the free radical, and the line in the low field is a half field line ($\Delta M_S = \pm 2$). A set of the rest spectral lines are very similar to those of Figure 6.10b, which is a typical EMR spectrum of the irregularly oriented nonaxisymmetric ($E \neq 0$) triplet state system.

In order to prepare glass state, how to choose the structure of solvent is very clear. When the geometric configuration of the host and guest molecules is quite different, the EMR spectrum linewidth of the triplet state is many times broader than the host and guest molecules with very similar geometric configuration. If the diphenylmethylene radical $(C_6H_5 - \dot{C} - C_6H_5)$ was dissolved in diphenyldiazomethane $(C_6H_5 - CN_2 - C_6H_5)$, the linewidth was 1.7 mT; if it was dissolved in *n*-pentane, then the linewidth was 9.4 mT [38]. Hence, in the glass system with different degrees of diversity in molecular geometry configuration of the host and guest (within the allowable range), the different configurations showing the value of zero-field splitting parameters *D* and *E* would also be diverse.

6.5 Biradical

As discussed earlier, in a molecule containing two unpaired electrons, the difference in energy levels between the electrons is so large that the dipole–dipole interaction between the spins is very weak, and the tumbling motion of molecules in solution cannot provide strong relaxation mechanism; consequently, the gap becomes very wide and an unobservable EMR spectrum results. Here, we considered the biradicals to exist in a liquid solution system and to be isotropic. So the spin Hamiltonian [39,40] is

$$\hat{\mathcal{H}} = g\beta H \left(\hat{S}_{1z} + \hat{S}_{2z} \right) + a \left(\hat{S}_{1z}\hat{I}_{1z} + \hat{S}_{2z}\hat{I}_{2z} \right) - 2J\hat{S}_1 \cdot \hat{S}_2 \qquad (6.82)$$

There are only spin operators in the above spin Hamiltonian, so the basis functions only need to consider the spin function. In fact, the wave function can be written as a product of the spatial function and spin function. Here, it also contains the nuclear spin operator. The eigenfunction of nuclear spin operator should also be considered. The basis function for the symmetric biradical containing two magnetic nucleus in form is $|S, M_S, M_1, M_2\rangle$

$$|1, 1, M_1, M_2\rangle = \alpha\alpha|M_1M_2\rangle$$

$$|1, -1, M_1, M_2\rangle = \beta\beta|M_1M_2\rangle$$

$$|1, 0, M_1, M_2\rangle = \left(\frac{\alpha\beta + \beta\alpha}{\sqrt{2}} \right)|M_1M_2\rangle$$

$$|0, 0, M_1, M_2\rangle = \left(\frac{\alpha\beta - \beta\alpha}{\sqrt{2}} \right)|M_1M_2\rangle$$

In this set of basis functions, the matrix representation of $\hat{\mathcal{H}}$ is as follows:

	$\lvert 1, 1, M_1, M_2\rangle,$	$\lvert 1, 0, M_1 M_2\rangle,$	$\lvert 1, -1, M_1 M_2\rangle,$	$\lvert 0, 0, M_1 M_2\rangle$
$\langle 1, 1, M_1 M_2\lvert$	$g\beta H - \frac{1}{2}J + \frac{1}{2}a(M_1 + M_2)$	0	0	0
$\langle 1, 0, M_1 M_2\lvert$	0	$-\frac{1}{2}J$	0	$\frac{1}{2}a(M_1 - M_2)$
$\langle 1, -1, M_1 M_2\lvert$	0	0	$-g\beta H - \frac{1}{2}J - \frac{1}{2}a(M_1 + M_2)$	0
$\langle 0, 0, M_1 M_2\lvert$	0	$\frac{1}{2}a(M_1 - M_2)$	0	$\frac{3}{2}J$

$$(6.83)$$

When $J > 0$, the ground state is triplet, while for $J < 0$, the ground state is singlet. By solving the eigenequation $\lvert \hat{\mathcal{H}} - E\hat{I} \rvert = 0$, the energy levels and the corresponding eigenfunctions are obtained, as shown in Table 6.3.

Table 6.3: Energy levels and wave functions of the symmetric biradicals with two magnetic nuclei.

Energy level	Wave function
$E_1 = g\beta H - \frac{1}{2}J + \frac{1}{2}a(M_1 + M_2)$	$\psi_1 = \|1, 1, M_1M_2\rangle$
$E_2 = \frac{1}{2}J + \frac{1}{2}R$	$\psi_2 = \sqrt{\frac{1}{2R}}[\sqrt{R-2J}\|1, 0, M_1M_2\rangle + \sqrt{R+2J}\|0, 0, M_1M_2\rangle]$
$E_3 = -g\beta H - \frac{1}{2}J - \frac{1}{2}a(M_1 + M_2)$	$\psi_3 = \|1, -1, M_1M_2\rangle$
$E_4 = \frac{1}{2}J - \frac{1}{2}R$	$\psi_4 = \sqrt{\frac{1}{2R}}[\sqrt{R+2J}\|1, 0, M_1M_2\rangle - \sqrt{R-2J}\|0, 0, M_1M_2\rangle]$
$R = \sqrt{4J^2 + a^2(M_1 - M_2)^2}$	

Table 6.4: The transition energy and transition intensity of the symmetric biradicals with two magnetic nuclei.

Transition	Energy level	Relative intensity
4-3	$g\beta H + J - \frac{1}{2}R + \frac{1}{2}a(M_1 + M_2)$	$(R+2J)/4R$
2-3	$g\beta H + J + \frac{1}{2}R + \frac{1}{2}a(M_1 + M_2)$	$(R-2J)/4R$
4-1	$g\beta H - J + \frac{1}{2}R + \frac{1}{2}a(M_1 + M_2)$	$(R+2J)/4R$
2-1	$g\beta H - J - \frac{1}{2}R + \frac{1}{2}a(M_1 + M_2)$	$(R-2J)/4R$
	$R = \sqrt{4J^2 + a^2(M_1 - M_2)^2}$	

According to the energy level and wave function, we can obtain the transition energy and transition intensity, as shown in Table 6.4.
These transitions are divided into three cases for discussion:

1. $J \ll a$ (or $J = 0$: At this time the (4-3) and (2-1) transitions are reduced as $g\beta H + aM_2$, the (2-3) and (4-1) transitions are reduced as $g\beta H + aM_2$, and then Eq. (6.82) becomes

$$\hat{\mathscr{H}} \approx g\beta H\left(\hat{S}_{1x} + \hat{S}_{2x}\right) + a\left(\hat{S}_{1z}\hat{I}_{1z} + \hat{S}_{2z}\hat{I}_{2z}\right)$$
$$= \left(g\beta H\hat{S}_{1x} + a\hat{S}_{1z}\hat{I}_{1z}\right) + \left(g\beta H\,\hat{S}_{2x} + a\,\hat{S}_{2z}\hat{I}_{2z}\right) = \hat{\mathscr{H}}_1 + \hat{\mathscr{H}}_2 \tag{6.84}$$

It corresponds to both independent "single radicals" mutually. Only when both electrons in the molecule are so far apart, and there is interaction between both electrons, it can be obtained. The specific example is shown in Figure 6.12 [41].

Figure 6.12b shows three hyperfine structure lines of the EMR spectrum with the same intensity, which is produced by the unpaired electron interacting with an ^{14}N nucleus. Although the molecules of di(tetramethyl-2,2,6,6-piperidinyl-4-oxyl-1)terephthalate have two unpaired electrons and two ^{14}N nuclei, its EMR spectrum display is analogous to both independent "single radicals."

2. $J \gg a$: At this time the exchange interaction is very strong, $R \approx 2J$, thus the transitions (4-3) and (2-1) become

$$\Delta E = g\beta H + \frac{a}{2}(M_1 + M_2) \tag{6.85}$$

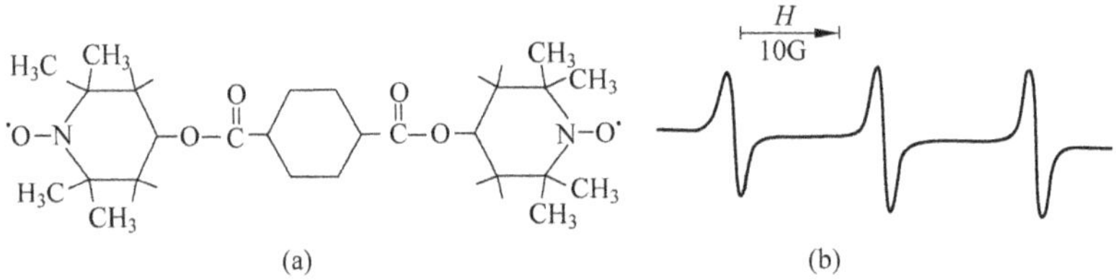

Figure 6.12: A specific example of $J \ll a$:(a) molecular structure of di(tetramethyl-2,2,6,6-piperidinyl-4-oxyl-1)terephthalate; (b) its EMR spectrum.

Because of the strong interaction between the two unpaired electrons, the time on the two nuclei is occupied one half by each electron, the relative intensity is 1, and the relative strength of transitions (2-3) and (4-1) is 0. Hence, if $I_1 = I_2 = 1$, it should have five lines, and the distance between each of the two lines is $a/2$. Figure 6.13a is the molecular structure of tetramethyl-2,2,5,5-pyrrolidoneazine-3 dioxyl- 1, 1′. Figure 6.13b is its EMR spectrum. This is a specific example of $J \gg a$ [42].

3. $J = 0{\sim}2a$ *(i.e., the intermediate situation:* Here, the difference between the above two cases is discussed. According to the results of Table 6.3 (see Figure 6.14), when $J = a/4$, $J = a/2$, and $J = a$, the spectrum is seen to be very complex [41,43,44]. However, this is rarely observed in the experiment.

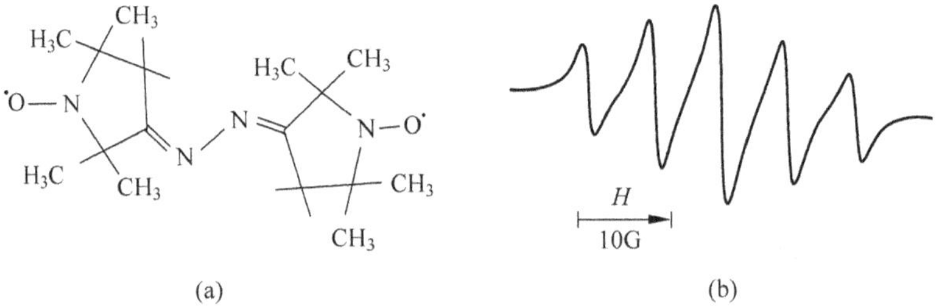

Figure 6.13: Molecular structure formula of tetramethyl-2,2,5,5-pyrrolidoneazine-3dioxyl-1′,1′ (a) and its EMR spectrum (b).

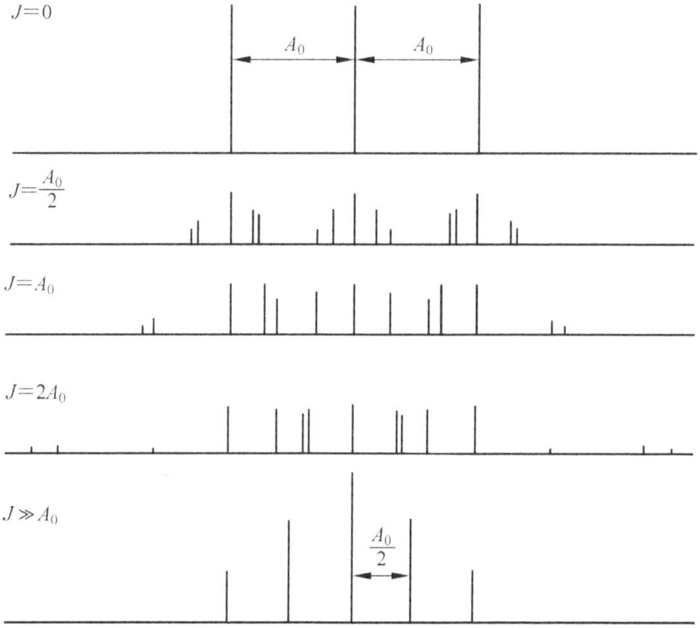

Figure 6.14: Theoretical rod spectrum of the biradical containing 2 nuclei of $I = 1$ When $J = 0$, the hyperfine structure is three lines with a strength ratio of 1:1:1, and the splitting space is a. When $J \gg a$, the hyperfine structure is five lines with a strength ratio of 1:2:3:2:1, and the splitting space is $a/2$.

References

[1] Bencini A., Gatteschi D. *EPR of Exchange Coupled Systems*, Springer, Berlin, 1990.

[2] Owen J., Harris E.A. "Pair Spectra and Exchange Interactions", in *EPR of Transition Ions*, Ed. S. Geschwind, Plenum Press, 1972.

[3] Abragam A., Bleaney B. *Electron Paramagnetic Resonance Ions*, Clarendon, Oxford, 1970, Section 9.3.

[4] Coffman J. E., Pezeshk A. *J. Magn. Reson.* 1986, **70**: 21.

[5] Hall P. L., Angel B. R., Jones J. P. E. *J. Magn. Reson.* 1974, **15**: 64.

[6] Hornig A. W., Hyde J. S. *Mol. Phys.* 1963, **6**: 33.

[7] Schuch H., Seiff F., Furrer R., Möbius K., Dinse K. P. *Z. Natureforsch.* 1974, **29a**: 1543.

[8] Yamaguchi Y., Sakamoto N. *J. Phys. Soc. Jpn.* 1969, **27**: 1444.

[9] de Groot M. S., van der Waals, J. H. *Mol. Phys.* 1960, **3**: 190.

[10] van der Waals, J. H., de Groot M. S. *Mol. Phys.* 1959, **2**: 333.

[11] Kottis P., Lefebvre R. *J. Chem. Phys.* 1963, **39**: 393; 1964, **41**: 379.

[12] Yager W. A., Wasserman E., Cramer R. M. R. *J. Chem. Phys.* 1962, **37**: 1148.

[13] Weissman S. I. *J. Chem. Phys.* 1958, **29**: 1189.

[14] de Groot M. S., van der Waals, J. H. *Physica* 1963, **29**: 1128.

[15] Grivet J.-Ph., Mispelter J. *Mol. Phys.* 1974, **27**: 15.

[16] Lewis G. G., Lipkin D., Magel T. T. *J. Am. Chem. Soc.* 1941, **63**: 3005.

[17] Hutchison C. A., Jr., Mangun B. W. *J. Chem. Phys.* 1961, **34**: 908.

[18] Hirota N., Hutchison C. A., Jr., Palmer P. *J. Chem. Phys.* 1964, **40**: 3717.
[19] Henderson B. *Br. J. Appl. Phys.* 1966, **17**: 851.
[20] Perlson B. D., Russell D. B. *J. Chem. Soc., Chem. Commun.* 1972, 69; *Inorg. Chem.* 1975, **14**: 2907.
[21] Hirota N., Hutchison C. A., Jr., *J. Chem. Phys.* 1965, **43**: 2869.
[22] Gutmann F., Keyzer H., Lyons L. E. *Organic Semiconductors*, Part B Krieger,, Malabar, 1983, Chapters 4, 5, 13.
[23] Bernheim R. A., Bernard H. W., Wang P. S., Wood L. S., Skell P. S. *J. Chem. Phys.* 1970, **53**: 1280; 1971, **54**: 3223.
[24] Bernheim R. A., Kempf R. J., Reichenbecher E. F. *J. Magn. Reson.* 1970, **3**: 5.
[25] Wasserman E., Kuck V. J., Hutton R. S., Anderson E. D., Yager W. A. *J. Chem. Phys.* 1971, **54**: 4120.
[26] Bernheim R. A., Kempf R. J., Gramas J. V., Skell P. S. *J. Chem. Phys.* 1965, **43**: 196.
[27] Wasserman E., Barash L., Yager W. A. *J. Am. Chem. Soc.* 1965, **87**: 4974.
[28] Wasserman E., Trozzolo A. M., Yager W. A., Murray R. W. *J. Chem. Phys.* 1964, **40**: 2408.
[29] Wasserman E., Barash L., Yager W. A. *J. Am. Chem. Soc.* 1965, **87**: 2075.
[30] Hutchison C. A., Jr., Pearson G A. *J. Chem. Phys.* 1967, **47**: 520.
[31] Classon O., Lund A., Gillbro T., Ichikawa T., Edlund O., Yoshida H. *J. Chem. Phys.* 1980, **72**: 1463.
[32] Glasbeek M., van Voorst J. D. W., Hoijtink G. J. *J. Chem. Phys.* 1966, **45**: 1852.
[33] Jesse R. E, Biloen P., Prins R., van Voorst J. D. W., Hoijtink G. J. *Mol. Phys.* 1963, **6**: 633.
[34] Smith G. R., Weltner W., Jr. *J. Chem. Phys.* 1975, **62**: 4592.
[35] Nuttall R. H. D., Weil J. A. *Can. J. Phys.* 1981, **59**: 1886.
[36] Yamaguchi Y. *J. Phys. Chem. Jpn.* 1970, **29**: 1163.
[37] Wasserman E., Snyder L. C., Yager W. A. *J. Chem. Phys.* 1964, **41**: 1763.
[38] Trozzolo A. M., Wasserman E., Yager W. A. *J. Chim. Phys.* 1964, **61**: 1663.
[39] Reitz D. C., Weissman S. I. *J. Chem. Phys.* 1960, **33**: 700.
[40] Luckhurst G. R. *Mol. Phys.* 1966, **10**: 543.
[41] Briere R, Dupeyre R. M., Lemaire H., Morat C., Rassat A., Rey P. *Bull. Soc. Chim. (France)* 1965, **11**: 3290.
[42] Dupeyre R. M., Lemaire H., Rassat A. *J. Am. Chem. Sco.* 1965, **87**: 3771.
[43] Glarum S. H., Marshall J. H. *J. Chem. Phys.* 1967, **47**: 1374.
[44] Nakajima A., Ohya-Nishigushi H., Deguchi Y. *Bull. Chem. Soc. Jpn.* 1972, **45**: 713.

Further reading

[1] McGlynn S. P., Azumi T., Kinoshita, M. *Molecular Spectroscopy of the Triplet State*, Prentice-Hall, Englewood, 1969.
[2] Weltner W., Jr. *Magnetic Atoms and Molecules*, van Nostrand Reinhold, New York, 1983, Chapters 3 and 5.
[3] Molin Yu N., Salikhov K. M., Zamaraev K. I. *Spin Exchange: Principles and Applications in Chemistry and Biology*, Springer, Berlin, 1980.
[4] Hutchison C. A., Jr. “Magnetic resonance spectra of organic molecules in triplet states in single crystals,” in *The Triplet State*, Ed. A B. Zahlan, Cambridge University Press, Cambridge, 1967, pp. 63–100.
[5] Weissbluth M. “Electron spin resonance in molecular triplet states,” in *Molecular Biophysics*, Eds. B. Pullman, M. Weissbluth, Academic Press, New York, 1965, pp. 205–238.
[6] Pratt D. W. “Magnetic properties of triplet states,” in *Excited States*, Ed. E C. Lim, Vol. 4, Academic Press, New York, 1979, pp. 137–236.
[7] Owen J., Harris E. A. “Pair spectra and exchange interactions,” *Electron Paramagnetic Resonance*, Ed. S. Geschwind, Plenum Press, New York, 1972, Chapter 6.

7 Relaxation and line shape and linewidth

Abstract: The line shape, linewidth, and intensity of the spectral lines reflect the intrinsic link of the EMR process, and the relaxation makes the EMR process to be continued. In this chapter, we will discuss various kinds of motions in the system, which reflect the variation of the line shape, linewidth, and intensity of the spectrum (including the change with time) and its mechanism.

In the previous chapters, only the implicit time dependencies (e.g., the Larmor frequency, the sine microwave excitation field H_1) are mentioned. In this chapter, we will discuss the time-dependent process in the sample, which is reflected by the nature of the EMR signal, when the amplitude of the continuous wave H_1 remains constant. For the case when the amplitude of the microwave excitation field H_1 varies with time, such as the pulse EMR, please refer to Appendix A1.3.

In this chapter, we will discuss the relaxation time problems that are characterized by the interaction between the electron spin and the surrounding environment, and the inter-electron spins interaction. In some cases, the lifetime of the individual spin orientation in the free radical or the lifetime of the free radical may be so short that it affects its linewidth. It is possible to obtain information about the dynamics from the line shape in this case. This information may be from the dynamic processes of electron exchange, intermolecular energy transfer, intramolecular motion, molecular limited tumbling in liquid or solid, and chemical reactions.

7.1 Model of spin relaxation

We start with a discussion of the performance of a two-level spin system (defined as the total spin angular momentum $J = \frac{1}{2}$). Essentially, the spin–spin interaction is ignored. It is assumed that under a uniform constant external magnetic field $\boldsymbol{H}$, the system can be divided into two spin levels. Only Zeeman term in the spin Hamiltonian is considered, thus, the difference of the two electronic energy levels $\Delta E = E_u - E_l = g\beta_e H$ (subscript u represents the upper energy level, and l represents the lower energy level).

7.1.1 Spin temperature and boltzmann distribution

We use the following relationship to define a thermodynamic parameter, called "spin temperature" （T_S):

$$\frac{N_u}{N_l} = \mathrm{e}^{-\frac{\Delta E}{k_B T_S}} \tag{7.1}$$

https://doi.org/10.1515/9783110568578-007

Here, N_u and N_l represent the population of spin particles in the upper and lower levels, respectively; ΔE is the energy difference between the upper and lower levels; k_B is the Boltzmann constant.

It is assumed that the spin system is tuned by a pulsed electromagnetic field $\boldsymbol{H}_1$ to make the photon energy match with ΔE. This causes the absorption of EMR energy by the spin system, which leads to the change of the population of spin particles in the upper and lower levels, that is, the change of N_u/N_l ratio. As the spin system absorbs the energy of the electromagnetic field $\boldsymbol{H}_l$, it can be considered that the spin system is "hotter" than the environment. Figure 7.1 shows a two-level spin system that contacts the environment at a temperature of T: (a) The spin system is in thermal equilibrium ($T = T_s = T_S$). (b) The temperature of the spin system is higher than the ambient temperature $T_S > T(T_S > T)$. (c) When the lower level spin particle absorbs energy to jump to the upper level, till $N_u/N_l > 1$, the spin temperature is negative ($T_S < 0$). (d) This figure represents the decay of part (b) or (c) excess energy with time. The total energy of the system is $E = N_l E_l + N_u E_u$. Here, E_u and E_l represent the energy of the upper and lower levels; τ_1 represents the corresponding relaxation time.

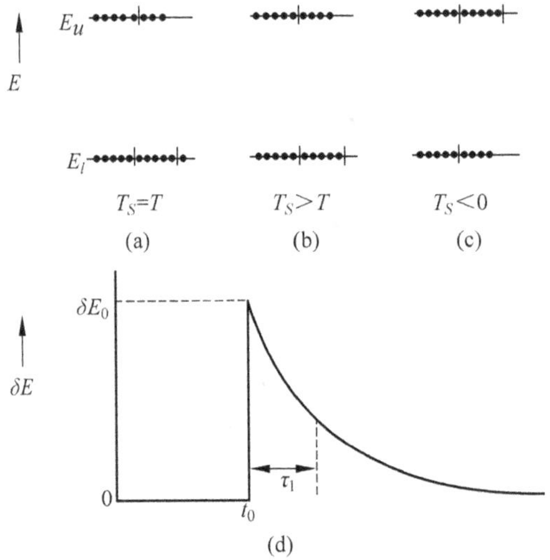

Figure 7.1: Schematic diagram of energy exchange between two-level spin system and environment.

In fact, through the interaction with the environment, the spin system"cools"the temperature of its own, and the spin temperature T_S eventually returns to the ambient temperature T. Any sufficiently simple thermodynamic system absorbs the external energy δE_0 at $t = t_0$, and this part of energy will be released to the environment in an exponential decay:

$$\delta E = \delta E_0 \exp[-(t - t_0)/\tau_1] \tag{7.2}$$

Here τ_1 is the characteristic time for energy flowing from the spin system to the environment (as shown in Figure 7.1d). The relaxation time τ_1 also reflects the degree of contact between the spin system and its environment. To let the spin system go back to its original state, that is, $T_S = T$, the time is $t \to \infty$. We note from Eq. (7.1) that the spin distribution is just the Maxwell–Boltzmann distribution at the temperature equal to T.

7.1.2 Spin particle transition dynamics

Now, from the population difference $\Delta N(H, T_S)$

$$\Delta N = N_l - N_u \tag{7.3}$$

we will study the spin transition dynamics; thus,

$$N_u = \frac{1}{2}(N - \Delta N) \tag{7.4a}$$

$$N_l = \frac{1}{2}(N + \Delta N) \tag{7.4b}$$

Here, $N = N_l + N_u$ is the total population of spin particles. When $|\Delta E/k_B T_S| \ll 1$,

$$\Delta N/N \approx \Delta E/2k_B T_S \tag{7.5}$$

To elaborate, in the presence of an excited microwave field, the probability of the transition from the lower level to the upper level in the unit time is $Z_\uparrow$; conversely, $Z_\downarrow$, as shown in Figure 7.2, ρ_ν is irradiation density of the microwave ν applied on the system. For the dynamic system of spin particles that are assumed isolated from each other, the differential rate equation is as follows:

$$\frac{d\Delta N}{dt} = -2N_l Z_\uparrow + 2N_u Z_\downarrow \tag{7.6}$$

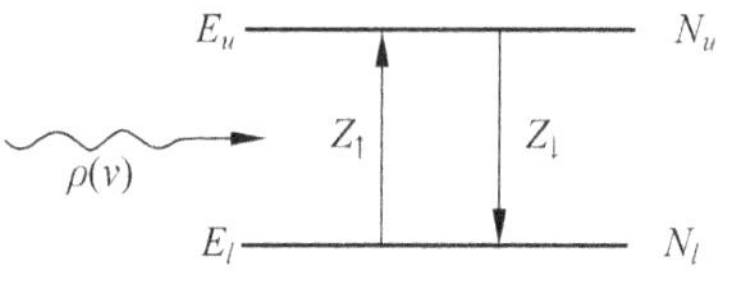

Figure 7.2: Schematic diagram of particle transition between energy levels in the two-level spin system.

The first term on the right-hand side is the rate of transition from the lower level to the upper level and the second term is the rate of transition from the upper level to the lower level; each term is multiplied by 2 because every transition of a particle can numerical change 2 of ΔN.Thus, Eq. (7.6) can be rewritten as follows:

$$\frac{d\Delta N}{dt} = N(Z_\downarrow - Z_\uparrow) - \Delta N(Z_\downarrow + Z_\uparrow) \tag{7.7a}$$

$$= \left(N\frac{Z_\downarrow - Z_\uparrow}{Z_\downarrow + Z_\uparrow} - \Delta N\right)(Z_\downarrow + Z_\uparrow) \tag{7.7b}$$

When the spin system tends to be stable, $d\Delta N/dt = 0$; so from Eq. (7.7b), one can obtain

$$\Delta N^{SS} = N_l^{SS} - N_u^{SS} = N\frac{Z_\downarrow - Z_\uparrow}{Z_\downarrow + Z_\uparrow} \tag{7.8}$$

The superscript SS represents the case of steady state. Then Eq. (7.7b) becomes

$$\frac{d\Delta N}{dt} = (\Delta N^{SS} - \Delta N)(Z_\downarrow + Z_\uparrow) \tag{7.9}$$

Here $(Z_\downarrow + Z_\uparrow)^{-1}$ is the dimension of time. We define it as τ_1. Then Eq. (7.9) becomes

$$\frac{d\Delta N}{dt} = \frac{\Delta N^{SS} - \Delta N}{\tau_1} \tag{7.10}$$

In the majority of cases, the ratio of $\Delta N^{SS}/N$ is very small, it can be approximated as $Z_\downarrow = Z_\uparrow \equiv Z$; then $\tau_1 \approx (2Z)^{-1}$. We note that τ_1 is a statistical parameter that is not directly related to the individual spin particles. Equation (7.10) is a first-order kinetic equation, and its solution is

$$\Delta N = (\Delta N)_0 + [\Delta N^{SS} - (\Delta N)_0]\{1 - \exp[-(t - t_0)/\tau_1]\} \tag{7.11}$$

Thus, ΔN is a variable that describes the change from $(\Delta N)_0$ to ΔN^{SS} in an exponential form with the rate constant $k_1 = \tau_1^{-1}$. Here, τ_1 is the evolution time of ΔN with $[\Delta N^{SS} - (\Delta N)_0][1 - e^{-1}]$ manner. Because the magnetization M in the Z-direction M_Z(parallel to the direction of the external magnetic field H) is proportional to ΔN (i.e., the unpaired electron without interaction in the volume V is proportional to $(g\beta_e/2V)\Delta N$), M_Z is also tending to the equilibrium value M_Z^o in an exponential form.

The definition of τ_1 in Eq. (7.10) is the reciprocal of the sum of the transition probabilities in the unit time, and it is related to the average lifetime of a given spin orientation state. The following argument will illustrate the effect of this lifetime limitation on the linewidth.

All quantum transitions have a *finite, nonzero* spectral width, the so-called *lifetime broadening*. Any excited state has a finite lifetime. Many monographs and papers attribute nonzero frequency shifts ($\Delta \nu \neq 0$) to the Heisenberg uncertainty principle, that is, $\tau \cdot \delta E \geq \hbar$. Here, δE is the uncertainty of the system energy due to the nonzero transition probability ($Z \neq 0$) and $\tau_1 = Z^{-1}$ is its average lifetime. $|\delta E| \approx \hbar/\tau$ is the energy level width. For example, it is assumed $\tau_1 = 10^{-9}$s, then $|\delta E| \approx 10^{-25}$ J or $\Delta \nu \approx 10^8$ s^{-1}, and the corresponding EMR linewidth is about 6.0 mT. Lifetime broadening is one of the contributions to the homogeneous linewidth, and is also the minimum linewidth of the given system.

7.1.3 Mechanism of the effect of relaxation time τ_1 on linewidth

Some mechanisms of the transition probabilities $Z_\downarrow$ and $Z_\uparrow$ can be described as follows:

$$Z_\uparrow = B_{lu}\rho_\nu + W_\uparrow \tag{7.12a}$$

$$Z_\downarrow = A_{ul} + B_{ul}\rho_\nu + W_\downarrow \tag{7.12b}$$

Here, ρ_ν is the time average of irradiation density on the spin system, and $W_\uparrow$ and $W_\downarrow$ are the probabilities of transitions from the lower level to the upper level, and from the upper level to the lower level, respectively, which were induced by the environment per unit time. A_{ul} is the Einstein coefficient of spontaneous emission; and B_{ul} and B_{lu} are the Einstein coefficients of stimulated emission and absorption [1], respectively. The definitions are as follows:

$$A_{ul} = \frac{64\pi^4\mu_0\nu_{ul}^3}{3hc^3}\left|\langle l|\hat{\mu}_{H_1}|u\rangle\right|^2 = \frac{8\pi h\nu_{ul}^3}{c^3}B_{ul} \tag{7.13}$$

$$B_{ul} = B_{lu} \tag{7.14}$$

Putting Eq. (8.12) into Eq. (8.10) and (8.11), we obtain

$$\Delta N^{SS} = N\frac{A_{ul} + W_\downarrow - W_\uparrow}{A_{ul} + 2B_{ul}\rho_\nu + W_\downarrow + W_\uparrow} \tag{7.15}$$

$$\tau_1 = (A_{ul} + 2B_{ul}\rho_\nu + W_\downarrow + W_\uparrow)^{-1} \tag{7.16}$$

where τ_1 is the sum of the relaxation times of each particle $\sum_i \tau_{1i}$. We consider now the following three cases:

(a) When the spin system is removed from its environment (including collision and radiation), then the B and W terms of Eqs. (7.15) and (7.16) are all zero. ΔN decays in constant rate $\tau_1 = A_{ul}^{-1}$. In this case, τ_1 is very long (for a set of spin particles unrelated to each other, when the external magnetic field H = 1 T, $\tau_1 \approx 10^4$ (a). But at the end ($t \to \infty$), the spin system decays to $\Delta N^{SS} \approx N$, which means that there are no spin particles in the upper order. In other words, the ambient temperature reaches 0K.

(b) The isolated spin system is exposed to a radiation source of T. At this time, the B term in Eqs. (7.15) and (7.16) is nonzero. When the radiation source is a blackbody, the value of the radiation density ρ_ν is given by the Planck blackbody law [2]. Then, the spin system is derived from the equilibrium with the radiation source with a temperature of T, and the ratio of the final spin population reaches to the value that is given by Eq. (7.1) at $T_S = T$. However, at 3 K, τ_1 is still up to 1000 a. When the radiation source provides a suitable excitation frequency, the effective temperature of the spin system obtained from the radiation source is T_S.

(c) Now, for the transition to the normal environment (electrons and nuclei), the transition probability term W in Eqs. (7.15) and (7.16) should exist and be dominant. The relaxation of the spin system is mainly spin–lattice relaxation (τ_1), which is induced by the spin tumblings of electrons, and the dynamic interaction with the environment (lattice). The "lattice" (i.e., liquid solvent, solid crystal, referred to as the "environment") is generally regarded as a large hot pool, and in the whole process of EMR experiment, its temperature can be regarded as constant. When the spin and the lattice reach the heat balance, τ_1 tends to zero. That is to say, the system transfers energy from ρ_ν to "lattice" "immediately."

If the density of the radiation (ρ_ν) is so dominant that $Z_\uparrow$ and $Z_\downarrow$ tend to be equal and $\Delta N \to \Delta N^{SS} \to 0$, then the spin system will no longer absorb energy from the radiation field, and the EMR signal will be lost. This is called "power saturation." Therefore, it is important to select the appropriate (moderate) radiation field energy (H_1) for the system with moderate spin–lattice relaxation mechanism.

From the viewpoint of experiment, the value of τ_1 is about 1 μs, so the W term has the magnitude of 10^6 s^{-1}. Because $A_{ul} \approx 3 \times 10^{-12}$ s^{-1}, it can be omitted when compared with W term, but for the term of $2B_{ul}\rho_\nu$, its value can be controlled as desired by adjusting the intensity of the radiation field. When the term is larger than W, the system tends to saturation.

Some spin–lattice interactions are carried out in the condensed phase, including the interaction between the spin system and the lattice vibration (phonon). The phonon density in the lattice obeys the Boltzmann distribution law. Therefore, the phonon density at the lower energy level is slightly higher than that of the upper energy level, which is the origin of the probability W inequality. Because $W_\uparrow \neq W_\downarrow$, this interaction results in the difference of absorption and emission by the photons (Einstein coefficient B). The inequality of the probability W fundamentally guarantees $\Delta N^{SS} \neq 0$.

The detailed description of the mechanism resulting in τ_1 is beyond the scope of this book. Hence, we outline only the most selective important mechanisms as follows:

1) *Direct process:* It is the nonradiative transition between spin states with the direct help of phonons. In the case of high-temperature approximation ($h\nu/k_BT \ll 1$), almost all of the experimental conditions are applicable. For the spin system $S = \frac{1}{2}$, τ_1 changes with $B^{-4}T^{-1}$; for the spin system $S > \frac{1}{2}$, τ_1 changes with $B^{-2}T^{-1}$[3]. This mechanism is only dominant at very low temperatures.
2) *Raman process:* As the Raman process in the electronic spectrum, this procedure includes the process of "real" excitation and the consequent process of "de-excitation" to be the phonon state. Its energy is much higher than that of spin.

τ_1 changes with the change of temperature from T^{-9} to T^{-5}. Therefore, the process becomes more important as the temperature increases.

3) *Orbach process:* If a low-lying spin level exists above the ground multiple state of energy Δ, the Raman process can control the spin–lattice relaxation. In this case, τ_1 changes with exp (Δ/k_BT), and thus Δ can also be obtained. The process was first proposed by Orbach and his collaborators [4,5].
4) *Other mechanisms:* Some of the mechanisms that have been proposed and verified by experiments are more complex than the above mechanisms [6]. In the gas phase, collision is an important relaxation mechanism.

Now we discuss spin relaxation from another point of view, using the famous Bloch equation [7] to describe the dependence of the total spin magnetization vector M on time, under the influence of the external static magnetic field and alternating magnetic field. Here, we only provide a brief overview of the processing method; for a comprehensive derivation, please refer to the relevant textbooks [8–12]. Although it mostly developed from the NMR, its basic theory can be applied to EMR.

In principle, the Bloch equation can be applied to any pair of energy levels, although there are some limitations. For example, it cannot be used to visualize individual spin particle quantum mechanics behavior, so the spin–spin coupling, such as ultrafine effect, is excluded; in addition, the effects of photon emission and radiation attenuation on the magnetization are ignored in Eq. [13]; simplifying to only two relaxation parameters is not strictly accurate, especially it is not correct for solid samples.

7.1.4 Magnetization in static magnetic field

When the material is magnetized in a magnetic field, the magnetization M is the magnetic moment in the unit volume. The relation of the magnetization M and the external magnetic field intensity is as follows:

$$M = \chi H \tag{7.17}$$

where χ is the "volume magnetic susceptibility" of the material. The spin set is S, which is in the thermal equilibrium state in the magnetic field H, the magnetic moment of each spin is $-g\beta S$. If there is no magnetic interaction between the spins, they should obey the Boltzmann distribution, and then the number of particles on the i level should be

$$p_i = \frac{\exp(-g\beta HM_i/kT)}{\sum\limits_{M_i=-S}^{S} \exp(-g\beta HM_i/kT)} \tag{7.18}$$

For N spin particles in the unit volume, the total magnetic moment per unit volume M is

$$M = N \sum_{M_i=-S}^{S} p_i(-g\beta M_i) = N \frac{\sum_{M_i=-S}^{S} (-g\beta M_i)\exp(-g\beta M_i/kT)}{\sum_{M_i=-S}^{S} \exp(-g\beta M_i/kT)} \quad (7.19)$$

When $S = 1/2$; $g = 2$; $H = 330$ mT, as long as $T > 22$ K, then $g\beta HM_i/kT < 10^{-2}$. We can put the above exponential function of the expansion in series, and only take the first two terms to get

$$M \approx N \frac{\sum_{M_i=-S}^{S} [-g\beta M_i + g^2 HM_i^2/kT]}{\sum_{M_i=-S}^{S} \left(1 - \frac{g\beta HM_i}{kT}\right)} \quad (7.20)$$

It is to be remembered that

$$\sum_{M_i=-S}^{S} (1) = 2S+1, \quad \sum_{M_i=-S}^{S} (M_S) = 0, \quad \sum_{M_i=-S}^{S} (M_i^2) = 2 \sum_{M_i=-S}^{S} (M_i)^2 = 2\frac{1}{6}S(S+1)(2S+1)$$

Then,

$$\chi = \frac{M}{H} = \frac{Ng^2\beta^2 S(S+1)}{3kT} \quad (7.21)$$

The equation is the famous Curie formula that is used to calculate the static susceptibility. The static magnetic susceptibility can also be measured experimentally. The experimental data of static magnetic susceptibility can be used to calculate the number of unpaired electrons of the transition metal ions and their complexes, where $g\sqrt{S(S+1)}$ is the "equivalent Bohr magneton number."

7.1.5 Bloch equation in the static magnetic field

In the absence of an external magnetic field, the magnetization M of the bulk phase sample is fixed in space, and the components in any Cartesian coordinate system are M_x, M_y, and M_z. When all the magnetic moments are exposed to a uniform static magnetic field H, the system is in a dynamic equilibrium without relaxation. However, here M is not fixed in space, but moves in accordance with the following equation:

$$\frac{d}{dt}M = \gamma_e H \times M \quad (7.22)$$

Here γ_e is the gyromagnetic ratio of the electron, and it equals to $g\beta_e/\hbar$.

Assuming H is parallel to the Z-axis direction, we get the following:

$$\frac{d}{dt}M_x = \gamma_e H M_y \tag{7.23a}$$

$$\frac{d}{dt}M_y = -\gamma_e H M_x \tag{7.23b}$$

$$\frac{d}{dt}M_z = 0 \tag{7.23c}$$

Their solutions are as follows:

$$M_x = M_\perp^o \cos\omega_H t \tag{7.24a}$$

$$M_y = M_\perp^o \sin\omega_H t \tag{7.24b}$$

$$M_z = M_z^o \tag{7.24c}$$

This set of equations shows that if $M_\perp^o \neq 0$, then the magnetization M is in precessional motion around the external magnetic field H with the angular frequency $\omega_H = -\gamma_e H$(Larmor frequency). The longitudinal component of magnetization M_z is a constant. Here we take the static magnetic field: if coupled with sine wave modulation, it will lead to more complex solution.

Now let us include the relaxation effect. If the system is subjected to a sudden change in the size or (and) direction of the external magnetic field $\boldsymbol{H}$, then M_x, M_y, and M_z (referring to the new direction of the magnetic field) relax to their new equilibria values. For example, when the magnetic field is suddenly added (when $t = t_o$, $\boldsymbol{H} = 0$), the initial state is $\Delta N = 0$; while the curve of M_z rises exponentially with time (as shown in Figure 7.3).

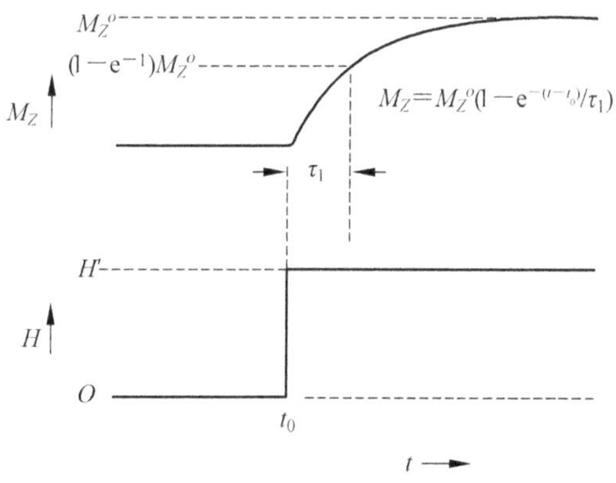

Figure 7.3: Schematic diagram of the change of magnetization M_Z, when the magnetic field H (Z-direction) is suddenly increased from 0 to H'.

If the transverse components M_x and M_y relax with the same rate constant (reciprocal of τ_2), then

$$\frac{\mathrm{d}}{\mathrm{d}t}M_x = \gamma_e HM_y - \frac{M_x}{\tau_2} \tag{7.25a}$$

$$\frac{\mathrm{d}}{\mathrm{d}t}M_y = -\gamma_e HM_x - \frac{M_y}{\tau_2} \tag{7.25b}$$

$$\frac{\mathrm{d}}{\mathrm{d}t}M_z = \frac{M_z^{\mathrm{o}} - M_z}{\tau_1} \tag{7.25c}$$

This is very important and famous Bloch equation. τ_2 is called "transverse relaxation time" or "spin–spin relaxation time"; correspondingly, τ_1 is called "longitudinal relaxation time" or "spin–lattice relaxation time." The mechanism of longitudinal relaxation and transverse relaxation is different, the changes of M_x and M_y does not change the total Zeeman energy of the spin system, but the change of M_z will lead to the energy exchange between the spin system and the environment (lattice). If there is no relaxation effect (i.e., both τ_1 and τ_2 tend to zero), then Eq. (7.25) is reduced to Eq. (7.23).

7.1.6 Bloch equation in the static magnetic field coupled with the oscillating magnetic field

In the magnetic resonance experiment, in addition to a static magnetic field in the z-direction, a linearly polarized magnetic field H_{1x} of a microwave or radio frequency should be added in the x-direction:

$$H_{1x} = 2H_1 \cos \omega t, H_{1y} = 0 \tag{7.26}$$

However, the linearly polarized magnetic field can be decomposed into two opposite circular polarized magnetic fields H_a and H_b:

$$H_a = iH_1 \cos \omega t + jH_1 \sin \omega t \tag{7.27a}$$

$$\begin{aligned} H_b &= iH_1 \cos(-\omega)t + jH_1 \sin(-\omega)t \\ &= iH_1 \cos \omega t - jH_1 \sin \omega t \end{aligned} \tag{7.27b}$$

Here, i, j, k are the unit vectors in x, y, z coordinate system. For H_a and H_b, only the component of the angular frequency $+\omega$ contributes to the induced EMR transition, while the contribution of the component $-\omega$ is small and can be neglected. That is, only H_a is the effective component, so the total magnetic field can be written as follows:

$$H = iH_1 \cos \omega t + jH_1 \sin \omega t + kH_o \tag{7.28}$$

Hence, $H \times M$ in Eq. (7.22) can be written as follows:

$$H \times M = \begin{vmatrix} i & j & k \\ H_1 \cos \omega t & H_1 \sin \omega t & H_o \\ M_x & M_y & M_z \end{vmatrix} = i(M_z H_1 \sin \omega t - M_y H_o)$$

$$+ j(M_x H_o - M_z H_1 \cos \omega t) + k(M_y H_1 \cos \omega t - M_x H_1 \sin \omega t) \tag{7.29}$$

Thus, the Bloch equation (7.25) becomes as follows:

$$\frac{\mathrm{d}}{\mathrm{d}t} M_x = \gamma_e(-M_y H_o + M_z H_1 \sin \omega t) - \frac{M_x}{\tau_2} \tag{7.30a}$$

$$\frac{\mathrm{d}}{\mathrm{d}t} M_y = \gamma_e(-M_z H_1 \cos \omega t + M_x H_o) - \frac{M_y}{\tau_2} \tag{7.30b}$$

$$\frac{\mathrm{d}}{\mathrm{d}t} M_z = \gamma_e(-M_x H_1 \sin \omega t + M_y H_1 \cos \omega t) + \frac{M_z^o - M_z}{\tau_1} \tag{7.30c}$$

In order to solve this set of Bloch equations, the most convenient way is to coordinate transformation. It is converted to a rotating coordinate system, which rotates around z-axis with the angular frequency ω (here ω is the angular frequency of the oscillating magnetic field H_1). Let the rotating coordinate system rotate an azimuth angle ϕ. The components in x, y directions of $\boldsymbol{M}$ in the original coordinate system are M_x, M_y, and in the new rotating coordinate system after transformation, the components in x_ϕ, y_ϕ directions are $M_{x\phi}$, $M_{y\phi}$. Then,

$$M_{x\phi} = M_x \cos \omega t + M_y \sin \omega t \tag{7.31a}$$

$$M_{y\phi} = M_x \sin \omega t - M_y \cos \omega t \tag{7.31b}$$

Here $M_{x\phi}$ and H_1 are the same phase, while $M_{y\phi}$ lag H_1 with phase of 90°. Putting Eq. (7.31) into Eq. (7.30), we get the following:

$$\frac{\mathrm{d}}{\mathrm{d}t} M_{x\phi} = -(\omega_H - \omega) M_{y\phi} - \frac{M_{x\phi}}{\tau_2} \tag{7.32a}$$

$$\frac{\mathrm{d}}{\mathrm{d}t} M_{y\phi} = (\omega_H - \omega) M_{x\phi} + \gamma_e H_1 M_z - \frac{M_{y\phi}}{\tau_2} \tag{7.32b}$$

$$\frac{\mathrm{d}}{\mathrm{d}t} M_z = -\gamma_e H_1 M_{y\phi} + \frac{M_z^o - M_z}{\tau_1} \tag{7.32c}$$

7.1.7 Stationary solutions of bloch equation

When the change of ω is very slow, we only need to find the stationary solutions for Eq. (7.31), that is,

$$\frac{\mathrm{d}}{\mathrm{d}t} M_{x\phi} = \frac{\mathrm{d}}{\mathrm{d}t} M_{y\phi} = \frac{\mathrm{d}}{\mathrm{d}t} M_z = 0$$

In this way, solution of a set of differential equations is transformed into solution of a set of algebraic equations:

$$M_{x\phi} = -M_z^o \frac{\gamma_e H_1(\omega_H - \omega)\tau_2^2}{1 + (\omega_H - \omega)^2\tau_2^2 + \gamma_e^2 H_1^2 \tau_1 \tau_2} \tag{7.33a}$$

$$M_{y\phi} = +M_z^o \frac{\gamma_e H_1 \tau_2}{1 + (\omega_H - \omega)^2\tau^2 + \gamma_e^2 H_1^2 \tau_1 \tau_2} \tag{7.33b}$$

$$M_z = +M_z^o \frac{1 + (\omega_H - \omega)^2\tau_2^2}{1 + (\omega_H - \omega)^2\tau_2^2 + \gamma_e^2 H_1^2 \tau_1 \tau_2} \tag{7.33c}$$

According to Eq. (7.32), we can obtain the stationary solutions of the Bloch equation in the *xyz* coordinate system as follows:

$$M_x = \frac{\gamma_e H_1 M^o}{\gamma_e H_1^2\left(\frac{\tau_1}{\tau_2}\right) + \left(\frac{1}{\tau_2}\right)^2 + (\omega_H - \omega)} \left\{(\omega_H - \omega)\cos\omega t + \frac{1}{\tau_2}\sin\omega t\right\} \tag{7.34a}$$

$$M_y = \frac{\gamma_e H_1 M^o}{\gamma_e H_1^2\left(\frac{\tau_1}{\tau_2}\right) + \left(\frac{1}{\tau_2}\right)^2 + (\omega_H - \omega)} \left\{(\omega_H - \omega)\sin\omega t - \frac{1}{\tau_2}\cos\omega t\right\} \tag{7.34b}$$

$$M_z = \frac{M^o\left\{\left(\frac{1}{\tau_2}\right)^2 + (\omega_H - \omega)^2\right\}}{\gamma^2 H_1^2\left(\frac{\tau_1}{\tau_2}\right) + \left(\frac{1}{\tau_2}\right)^2 + (\omega_H - \omega)^2} \tag{7.34c}$$

From Eq. (7.34), it is derived that the magnetization ***M*** in the direction of the static magnetic field M_z is not dependent on the time t; but in the xy plane, there are rotation components M_x and M_y that depend on the time t, and there is a maximum value when $\omega = \omega_H$.

7.2 Shape, width, and intensity of spectral line

7.2.1 Line shape function

In principle, the shape of EMR spectral lines should be of Lorentzian type, especially the free radical spectrum in dilute solution basically belongs to the Lorentzian shape function, the reason being the transverse component of the magnetization is attenuated in the exponential form. However, the results of the superposition of many Lorentzian lines tend to be a Gaussian function. The lines of the EMR spectrum

obtained in the experiment are mostly between the two types. Their general analytical forms are as follows:
Lorentzian function:

$$f(x) = \frac{a}{1 + bx^2} \tag{7.35}$$

Gaussian function:

$$f(x) = a\exp(-bx^2) \tag{7.36}$$

Both the line shape functions contain only two parameters a and b, which can be determined by two experiments. If the area is normalized, then the two parameters should meet a certain relationship, that is, there is only one independent parameter. The two line shape functions are shown in Figures 7.4 and 7.5, respectively.

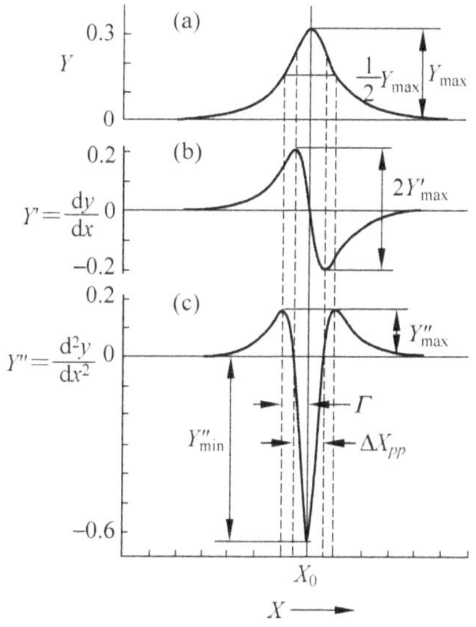

Figure 7.4: Lorentzian line shape.

Table 7.1 lists the line shape functions, peak heights, linewidths, and other mathematical expressions of the Lorentzian and Gaussian spectral lines.

Now we define a complex magnetic susceptibility:

$$\chi = \chi' - i\chi'' \tag{7.37}$$

In Eq. (7.37), χ' and χ'' are called Bloch susceptibility and H_{1x} in Eq. (7.26) is considered as the real part of the complex magnetic field H_c, that is,

$$H_c = 2H_1\exp(i\omega t) = 2H_1(\cos\omega t + i\sin\omega t) \tag{7.38}$$

In this way, we can also consider M_x as the real part of the complex magnetization M_c, then

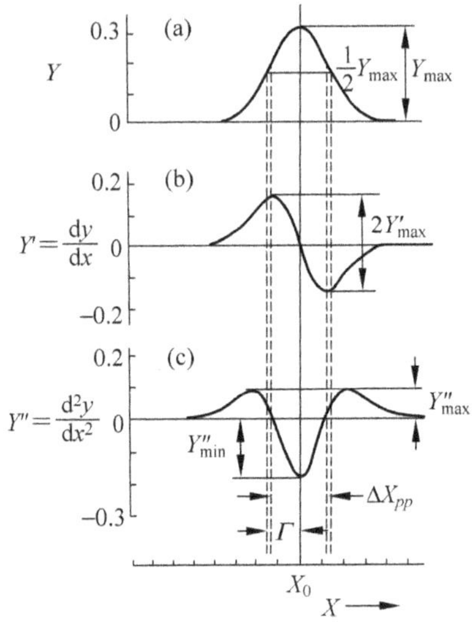

Figure 7.5: Gaussian line shape.

Table 7.1: Line shape functions, peak heights, and linewidths of the *Lorentzian* and *Gaussian* spectral lines.

	Properties of normalized Lorentzian line	**Properties of normalized Gaussian line**
Absorption line function	$Y = Y_{max}\frac{\Gamma^2}{\Gamma^2 + (x - x_r)^2}$	$Y = Y_{max}\exp\left[\frac{-(\ln 2)(x - x_r)^2}{\Gamma^2}\right]$
Peak height	$Y_{max} = Y\vert_{x=x_o} = \frac{1}{\pi\Gamma}$	$Y_{max} = Y\vert_{x=x_o} = \left(\frac{\ln 2}{\pi}\right)^{1/2}\frac{1}{\Gamma}$
Half peak width	$\Gamma = \frac{1}{2}\Delta x_{1/2}$	$\Gamma = \frac{1}{2}\Delta x_{1/2}$
First-order differential function	$Y' = -Y_{max}\frac{2\Gamma^2(x - x_r)}{[\Gamma^2 + (x - x_r)^2]^2}$	$Y' = -Y_{max}\frac{2(\ln 2)(x - x_r)}{\Gamma^2}\exp\left[\frac{-(\ln 2)(x - x_r)^2}{\Gamma^2}\right]$
Peak-peak hight	$2Y'_{max} = \frac{3\sqrt{3}}{4\pi}\frac{1}{\Gamma^2} = A_{pp}$	$2Y'_{max} = 2\left(\frac{2}{\pi e}\right)^{1/2}\frac{\ln 2}{\Gamma^2} = A_{pp}$
Peak–peak width	$\Delta x_{pp} = \frac{2}{\sqrt{3}}\Gamma$	$\Delta x_{pp} = \left(\frac{2}{\ln 2}\right)^{1/2}\Gamma$
Second-order differential function	$Y'' = -Y_{max}2\Gamma^2\frac{\Gamma^2 - 3(x - x_r)^2}{[\Gamma^2 + (x - x_r)^2]^3}$	$Y'' = -Y_{max}\frac{2\ln 2}{\Gamma^4}[\Gamma^2 - 2\ln 2(x - x_r)^2]\exp\left[\frac{-\ln 2(x - x)^2}{\Gamma^2}\right]$
Positive peak height	$Y''_{max} = Y_{max}\frac{1}{2\Gamma^2}$	$Y''_{max} = Y_{max}\frac{4e^{-3/2}\ln 2}{\Gamma^2}$
Negative peak height	$Y''_{min} = -Y_{max}\frac{2}{\Gamma^2}$	$Y''_{min} = -Y_{max}\frac{2\ln 2}{\Gamma^2}$

$$M_c = \chi H_c$$

Using Eqs. (7.37) and (7.38), we obtain

$$M_x = 2H_1\chi'\cos\omega t + 2H_1\chi''\sin\omega t$$

Comparing the above equation and Eq. (7.34a), we get

$$\chi' = \frac{1}{2}\frac{\gamma_e M^o(\omega_H - \omega)}{\gamma_e^2 H_1^2\left(\frac{\tau_1}{\tau_2}\right) + \left(\frac{1}{\tau_2}\right) + (\omega_H - \omega)^2} \tag{7.39a}$$

$$\chi'' = \frac{1}{2}\frac{\gamma_e M^o\left(\frac{1}{\tau_2}\right)}{\gamma_e^2 H_1^2\left(\frac{\tau_1}{\tau_2}\right) + \left(\frac{1}{\tau_2}\right)^2 + (\omega_H - \omega)^2} \tag{7.39b}$$

When the sample is subjected to magnetic resonance, the energy is transferred from the electromagnetic field to the sample. The question is: What is the relationship between the power consumed by the sample and the complex susceptibility?

The resonant cavity of the EMR spectrometer is equivalent to a coil composed of a *RL* series circuit. When the sample is not inserted into the resonator, the impedance of the resonant cavity is Z_o:

$$Z_o = R_o + i\omega\mathscr{L} \tag{7.40}$$

When the sample is inserted into the cavity, the impedance of the resonator is changed from Z_o to Z:

$$Z = R_o + i\omega L_o(1 + 4\pi\chi) \tag{7.41}$$

Substituting by Eq. (7.37), we get

$$Z = (R_o + 4\pi\omega L_o\chi'') + i\omega L_o(1 + 4\pi\chi') \tag{7.42}$$

When the sample is inserted in the resonant cavity, the resistance portion of the impedance of the resonant cavity is changed from R_o to $R_o + 4\pi\omega L_o\chi''$. It is shown that the additional power consumption caused by the insertion of the sample is proportional to χ'', that is, the resonance absorption is only related to the imaginary part of the complex susceptibility. Thus, the line shape function of EMR absorption should be proportional to χ''. The expression of line shape function $f(\omega)$ is as follows:

$$f(\omega) = \frac{\mathscr{C}\gamma_e M^o\left(\frac{1}{\tau_2}\right)}{\gamma_e^2 H_1^2\left(\frac{\tau_1}{\tau_2}\right) + \left(\frac{1}{\tau_2}\right)^2 + (\omega_H - \omega)^2} \tag{7.43}$$

$\mathscr{C}$ in the equation is a scale coefficient including instrumental factors.

It can be seen from Eq. (7.43) that when $\omega = \omega_H$, there is a maximum for $f(\omega)$. If both H_1 and τ_1 are large when $\omega = \omega_H, f(\omega)$ will become smaller. This is the so-called *saturation phenomenon*. In order to avoid saturation, the operation H_1 is always put to the minimum. At this time, compared with $(1/\tau_2)^2$, the term of $\gamma_e^2 H_1^2(\tau_1/\tau_2)$ can be ignored, so the line shape function can be written as

$$f(\omega) = \frac{\mathscr{C}\gamma_e M^o \left(\frac{1}{\tau_2}\right)}{\left(\frac{1}{\tau_2}\right)^2 + (\omega_H - \omega)^2} \tag{7.44}$$

$(1/\tau_2)^2$ divides the numerator and denominator at the right-hand side of Eq. (7.44). We thus obtain

$$f(\omega) = \frac{\mathscr{C}\gamma_e M^o \tau_2}{1 + \tau_2^2(\omega_H - \omega)^2} \tag{7.45}$$

Here $\mathscr{C}\gamma_e M^o \tau_2$ is exactly the a in Eq. (7.35), and τ_2^2 is the b in Eq. (7.35).

Equation (7.45) is a Lorentzian function, which is the most fundamental line shape function in EMR spectra.

7.2.2 Linewidth

7.2.2.1 Origin of linewidth

In theory, any spectral line should be an infinitely narrow $\delta-$ function, but the actual situation is that any spectral line has a certain width. For EMR spectra, different samples have different linewidth. Under different conditions, the linewidths of the same sample may be very different.

There are two reasons for the linewidth:

1) *Reason of the life time:* The residence time of an unpaired electron at an energy level is a finite value δt, that is, $\delta t \neq 0$. And the energy difference ΔE between the two levels cannot be a strict set value, that is, $\delta\Delta E \neq 0$. According to the Heisenberg principle in quantum mechanics,

$$\delta\Delta E \bullet \delta t \sim \hbar \tag{7.46}$$

where $\delta\Delta E = g\beta\delta H$, that is to say, $\delta H \bullet \delta t \sim \hbar/g\beta$.

The smaller the δt, the greater the δH, and thus the wider the linewidth. In the X-band, when $\delta t = 10^{-9}$s, δH is about 5.7 mT. The unpaired electron continues to transfer between the upper and lower levels, which is due to the energy coupling between the unpaired electron and the lattice (i.e. the environment), also known as the "spin–lattice" interaction. The stronger the interaction, the smaller the δt, and thus the wider the linewidth. Decreasing the ambient temperature can weaken the interaction and increase the lifetime of the unpaired electron at a certain level to narrow the line.

2) *Reason of secular:* A spin system is made up of a large number of spins. Each individual spin is like a small magnetic moment, and for a little moment, there are many small moments (including the magnetic moments of the electrons and nuclei) around it. These small magnetic moments constitute a local magnetic field H' for a specified small magnetic moment; the interaction between them is called "spin–spin" interaction. Since these small magnetic moments are constantly in motion, the additional local magnetic field H' also constantly changes. This change will lead to the change of energy level difference between two spin states. For a given microwave frequency, the resonance field H_r should be the sum of the external magnetic field H_o and the local magnetic field H', that is,

$$H_r = |H_o + H'| \tag{7.47}$$

At a given frequency, H_r should be fixed. However, because the value of H' is changing, the external magnetic field H_o is no longer a fixed value, and there is a distribution in a small scope around H_r. A spectral line with a certain width can be thus obtained by the superposition of many infinite narrow lines.

7.2.2.2 Definition of linewidth

In experiments, we often encounter the following three kinds of lines: (1) the normalized absorption line; (2) the first differential line; (3) the second differential line. The detailed descriptions are as follows:

1) Define the half width Γ at the half height $(Y_{max}/2)$ of the normalized absorption line as the linewidth. It is also called the "half height linewidth" of the normalized absorption line, as shown in Figure 7.4a. For Lorentzian line shape,

$$\Gamma = \frac{1}{|\gamma_e|\tau_2}(1 + \gamma_e^2 H_1^2 \tau_1 \tau_2)^{1/2} \tag{7.48}$$

2) Define the distance Δx_{pp} between the two extremes of the normalized first differential line (peak–peak) as the linewidth of the first derivative spectrum, as shown in Figure 7.4b.
3) Define half of the distance Γ between the two peaks of the normalized second differential spectrum as the linewidth of the second differential spectrum, as shown in Figure 7.4c.

The first differential line is the most common of the three kinds of spectral lines.

7.2.3 Line broadening

The variation of local magnetic field H' consists of two parts:

1) *Change with time*: It is also known as dynamic change. This change is due to the influence of thermal fluctuations, and each paramagnetic particle is subjected to a uniform-distributed and time-fluctuated local magnetic field, which leads to the broadening of the spectral line. Because the line broadening caused by this change is uniform, it is also known as *uniform broadening*.
2) *Change with the spatial position:* Because of the difference of the relative position between each paramagnetic particle and the adjacent other paramagnetic particles, the local magnetic field of each paramagnetic particle is different, which also causes the broadening of the spectral line. The interaction between the two paramagnetic particles is proportional to

$$\frac{1}{r^3}(1-3\cos^2\theta) \tag{7.49}$$

In the equation, r is the distance between paramagnetic particles and θ is the angle of connection between the two particles r and the magnetic field H. Because this part H' is changed along with the angle θ, its distribution in the space is not uniform, and as the broadening of the spectral line is uneven, it is also called *inhomogeneous broadening*. Because it rapidly decreases with the increasing r, increasing the distance r between particles can reduce the broadening effect. For crystal samples, using the isomorphous diamagnetic material to dilute paramagnetic molecules, such as a small amount of paramagnetic $CuSO_4$ doped into diamagnetic $ZnSO_4$ crystal, can weaken the line broadening effect of the spin–spin interaction between Cu^{2+} ions. For liquid samples, the diamagnetic solvents can be used to dilute.

7.2.4 Line intensity

The area of the spectral line is defined as the spectral line intensity, which is proportional to the total number of spin particles in the sample. There are different algorithms for different spectral lines. We used them to represent the relative strength of the spectral line:

1) The normalized absorption line: $\mathscr{I} = 2Y_{max}\Gamma = Y_{max}\Delta x_{1/2}$ (7.50)
2) The normalized first differential line: $\mathscr{I} = 2Y'_{max}\Delta x_{pp}$ (7.51)
3) The normalized second differential line: $\mathscr{I} = 2(Y''_{max} + Y''_{min})\Delta x_{1/2}$ (7.52)

These three equations are used to calculate the three different spectral line intensities, respectively.

7.3 Dynamic effects of line shape

The paramagnetic center and any dynamic process around it will affect the line shape. Molecules tumbling in thick liquid, interacting with other paramagnetic particles, and chemical reactions (e.g., acid–alkali balance and electron transfer reactions) hinder the free rotation of molecules and broaden the spectral line. This broadening also comes from the dynamic fluctuations of the local magnetic field around the unpaired electron. If the changes take place slowly enough, the observed lines can be assigned to the specified particles. However, with the increase of the fluctuation rate, the broadening of the EMR spectral line is finally merged into a (or a set of) spectral line(s), whose position is the weighted average position of the original spectral lines (as shown in Figure 7.6).

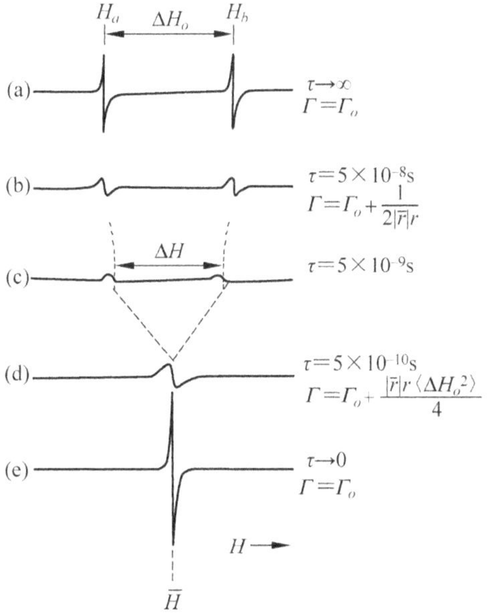

Figure 7.6: For paramagnetic tautomers of *a* and *b*, the effect of the tautomerism rate on the first differential line.

7.3.1 Generalized bloch equation

There are many theoretical models that can be used to simulate the influence of the fluctuation of dynamic magnetic field on the EMR spectrum. However, we choose the generalized Bloch equation model because it is easy-to-understand and easy-to-use computer.

Consider a radical with two unique forms of existence, *a* and *b* (i.e., each with its own unique EMR spectrum). For simplicity, assume two probability f_a and $f_b (f_a + f_b = 1)$ for two forms, respectively, and assume that the two forms of the EMR spectra are single Lorentzian lines. The magnetic field positions are H_a and H_b(as shown in Figure 7.6a), respectively. The two spectral line spacing $\Delta H_o = H_b - H_a$ depends on the magnetic field H. In other words, these two forms of particles have different g factors.

In the following, we will mention *exchange rate* (i.e. the local magnetic field fluctuation) "fast" and "slow," which refers to the comparison with characteristic parameters $|\gamma_e|\Delta H_o$.

Let us first define a complex transverse magnetization $M_{+\phi}$:

$$M_{+\phi} = M_{x\phi} + iM_{y\phi} \tag{7.53}$$

Equations (7.32a) and (7.32b) can be combined into

$$\frac{\mathrm{d}M_{+\phi}}{\mathrm{d}t} + \alpha M_{+\phi} = \mathrm{i}\gamma_e H_1 M_z \tag{7.54}$$

Here

$$\alpha = \tau_2^{-1} - i(\omega_H - \omega) \tag{7.55}$$

The present situation is that the microwave power is set low enough so that no power saturation occurs. The magnetization M_z can be substituted by M_z^o, and it is assumed that the linewidths of the line a and the line b in Figure 7.6 are same as τ_2^{-1}.

Let us simplify the notation $G = M_{+\phi}$, so that Eq. (7.54) can be written as follows:

$$\alpha_a = \tau_{2a}^{-1} - i(\omega_{Ha} - \omega) \tag{7.56a}$$

$$\alpha_b = \tau_{2b}^{-1} - i(\omega_{Hb} - \omega) \tag{7.56b}$$

The relaxation times τ_{2a} and τ_{2b} are the reciprocals of the a and b linewidths in the absence of dynamic process (and power saturation). They do not depend on the temperature. Note that $\gamma_a \neq \gamma_b$, which implies $g_a \neq g_b$.

The functions G_a and G_b can be considered to have the same meaning as the concentration in chemical kinetics. For a reaction

$$a \underset{k_b}{\overset{k_a}{\rightleftarrows}} b \tag{7.57}$$

we can add a first-order kinetic term into Eq. (7.53) by introducing the chemical or physical dynamics into the Bloch equation – a generalized Bloch equation is obtained as follows:

$$\frac{\mathrm{d}G_a}{\mathrm{d}t} + \alpha_a G_a = i\gamma_a H_1 M_{za} + k_b G_b - k_a G_a \tag{7.58a}$$

$$\frac{\mathrm{d}G_b}{\mathrm{d}t} + \alpha_b G_b = i\gamma_b H_1 M_{2b} + k_a G_a - k_b G_b \tag{7.58b}$$

Equations (7.58a) and (7.58b) are a pair of linear equations. Assuming $\frac{\mathrm{d}G_a}{\mathrm{d}t} = \frac{\mathrm{d}G_b}{\mathrm{d}t} = 0$, the stationary solution of G_a and G_b can be obtained. It is also assumed that in the case of thermal equilibrium between the spins, the relaxation times τ_{1a} and τ_{1b} are short enough; thus,

$$\frac{dM_{za}}{dt} = \frac{dM_{zb}}{dt} = 0 \tag{7.59}$$

Referring to Eqs. (7.32c) and (7.33c), we can use

$$M_{za} = f_a \gamma_a M_z^o / \bar{\gamma} \text{ and } M_{zb} = f_b \gamma_b M_z^o / \bar{\gamma} \tag{7.60}$$

Here,

$$\bar{\gamma} = f_a \gamma_a + f_b \gamma_b \tag{7.61}$$

The total complex transverse magnetization G is given:

$$G = G_a + G_b = iH_1 M_z^o \frac{f_a \gamma_a (\alpha_b + k_a + k_b) + f_b \gamma_b (\alpha_a + k_a + k_b)}{(\alpha_a + k_a)(\alpha_b + k_b) - k_a k_b} \tag{7.62}$$

Combining chemical equilibrium condition $f_a k_a = f_b k_b$, the population function obeys the relations of $f_a = \tau_a / (\tau_a + \tau_b)$ and $f_b = \tau_b / (\tau_a + \tau_b)$. Using $\tau_a^{-1} = k_a$ and $\tau_b^{-1} = k_b$ and defining the linewidth as the reciprocal of lifetime $\tau^{-1} = \tau_a^{-1} + \tau_b^{-1}$, Eq. (7.62) can be written:

$$G = iH_1 M_z^o \frac{\bar{\gamma} + \tau (f_a \gamma_a \alpha_b + f_b \gamma_b \alpha_a)}{\tau \alpha_a \alpha_b + f_a \alpha_a + f_b \alpha_b} \tag{7.63}$$

The intensity of the absorption line is proportional to the imaginary part of the magnetization G (Section 7.3.1), and the line shape is a function of the relaxation time τ, as shown in Figure 7.6. Before considering the generalized line shape function, we discuss the following two extreme cases:

Slow dynamics: Here the relaxation times τ_a and τ_b are longer than $|\gamma_e \Delta H_o|$. It is the spectrum with two separate lines as we expected, which is shown in Figure 7.6a. For example, when H is close to $H_a = -\omega/\gamma_a$, $G_b \approx 0$, the stationary solution of Eq. (7.58a) is as follows:

$$G_a = i f_a \gamma_a H_1 M_z^o \frac{1}{\alpha_a + k_a} \tag{7.64}$$

Taking the imaginary part, we can get

$$M_{y\phi a} = -f_a H_1 M_z^o \frac{\Gamma_{oa} + k_a / |\gamma_a|}{(\Gamma_{oa} + k_a / |\gamma_a|)^2 + (H_a - H)^2} \tag{7.65}$$

This is $\chi''(H)$ of the absorbed power, and it is Lorentzian line shape with half height width Γ_a:

$$\Gamma_a = \Gamma_{oa} + \left|\gamma_a \tau_a\right|^{-1} \tag{7.66}$$

Here, $\tau_a (= k_a^{-1})$ is the lifetime of configuration a. Configuration b has the same line shape as configuration a. Hence, each spectral line becomes wider (but not moving) after a gradually accelerated kinetic process, as shown in Figure 7.6b. The rate constant of the kinetic process can be determined by measuring the increase of linewidth.

Fast dynamics: When the tautomeric rate of the two configurations a and b is very fast, both τ_a and τ_b become very short, so τ in Eq. (7.63) can be ignored, then

$$G \approx i\bar{\gamma} H_1 M_z^o \frac{1}{f_a \alpha_a + f_b \alpha_b} \tag{7.67}$$

Taking the imaginary part, we get

$$M_{y\phi} = -H_1 M_z^o \frac{\bar{\Gamma}}{(\bar{\Gamma})^2 + (\bar{H} - H)^2} \tag{7.68}$$

Here, $\bar{\Gamma}$ and $\bar{H}$ are the weighted averages of Γ and H, respectively:

$$\bar{\Gamma} = f_a \Gamma_{oa} + f_b \Gamma_{ob} \tag{7.69}$$

$$\bar{H} = f_a H_a + f_b H_b \tag{7.70}$$

Equation (7.68) clearly shows the Lorentzian line with the average linewidth $\bar{\Gamma}$ and the convergence to the average magnetic field $\bar{H}$, as shown in Figure 7.6c.

More detailed analysis has shown that when the tautomeric rate of the two configurations is fast that tend to the limit, it becomes a Lorentzian line in the magnetic field $\bar{H}$, as shown in Figure 7.6d, and the linewidth is given by

$$\Gamma = \bar{\Gamma} + f_a f_b \tau |\bar{\gamma}| (\Delta H_0)^2 \tag{7.71}$$

Here, it is shown again that the kinetic rate constant can be obtained from the variation of the linewidth of the spectral line.

Intermediate dynamics: It is possible to derive a general expression of the line shape from the imaginary part of Eq. (7.62) [14,15]. As the system moves from the slow zone to the intermediate zone, it can be seen that the two lines not only become wide but also move inward (see Figure 7.5c). In a given magnetic field, from the minimum value in the denominator of the imaginary part of Eq. (7.62), the splitting distance of the two spectral lines can be deduced [14] as follows:

$$\Delta H = [(\Delta H_o)^2 - 2(\bar{\gamma}\tau)^{-2}]^{1/2} \tag{7.72}$$

When the right-hand side of the first term in square brackets is dominant, the equation is correct.

Finally, the two lines are combined into a wide line at $\bar{H}$, as shown in Figure 7.6d. The τ value of their convergent point is expressed in Eq. (7.73) as follows:

$$\tau = \frac{\sqrt{2}}{|\bar{\gamma}|\Delta H_o} \tag{7.73}$$

Note: The value (the dimension is s/rad, since the dimension of γH is the angular frequency) usually depends on the frequency used in the measurement of ΔH_0. The combination of the two spectral lines is the expression of lifetime broadening. If it is written as $\Delta t \Delta \omega \approx 1$, here $\Delta \omega$ is the splitting distance of the two spectral lines, of course, its dimension is the angular frequency; Δt is the minimum average time interval with which the configurations a and b can be identified. If the lifetime τ is less than Δt, then only a central line can be observed, because in this case it is unable to identify the configurations a and b.

If there are more than two configurations, or the stoichiometry of Eq. (7.57) is not 1:1, the line shape will be more complex. The form of the Bloch equation also need to be modified [16].

From the generalized Bloch equation, he paramagnetic particles movement has a great influence on the line shape and the linewidth of the spectral line, and paramagnetic particle movement is closely related to its configuration transformation, physical movement, the environment, and so on, such as the heat (temperature) effect sometimes becomes an important factor on EMR spectral line shape and linewidth. It should be pointed out that the peak–peak amplitude of the differential spectrum is proportional to the relative intensity of the transition. However, under certain conditions, the linewidth will change with temperature, and for a given spectrum, it may change from a spectral line to another line. As a result, the relation between the amplitude of the first differential line and the intensity of the spectral line is deviated, because the amplitude of the differential spectrum is inversely proportional to the *square* of the linewidth. Thus, a small change of the spectral linewidth will cause the big change of the relative amplitude of the line. Here we will look at several mechanisms that affect the *spectral line broadening*.

7.3.2 Chemical exchange broadening mechanism

7.3.2.1 Reaction between free radicals and diamagnetic molecules

Assume that there is a chemical reaction between a free radical $R^\bullet$ and a diamagnetic molecule L, occurring as follows:

$$R^\bullet + L \underset{k_-}{\overset{k_+}{\rightleftarrows}} R^\bullet L, K = \frac{[R^\bullet L]}{[R^\bullet][L]} = \frac{k_+}{k_-} \tag{7.74}$$

Here, k_+, k_- are the reaction rate constants, K is the equilibrium constant, $[R^\bullet]$, $[L]$, $[R^\bullet L]$ are the concentrations of $R^\bullet$, L, and $R^\bullet L$, respectively. When the reaction is carried out to the right, there is an average lifetime τ_R of $R^\bullet$, and when the reaction is carried out to the left, there is an average life $\tau_{R\cdot L}$ of $R^\bullet L$. If the exchange is done in an instant, then

$$\tau_R = \frac{1}{k_+[L]}, \tau_{R\cdot L} = \frac{1}{k_-} \tag{7.75}$$

Assume that there are two different g values $g_{R^\bullet}$ and $g_{R^\bullet L}$ for $R^\bullet$ and $R^\bullet L$, and none of them has a hyperfine structure. When the exchange is slow, there should be two g values $g_{R^\bullet}$ and $g_{R^\bullet L}$ for the EMR lines. Their linewidths are the natural linewidths $(\tau_2^o)_R^{-1}$ and $(\tau_2^o)_{R\cdot L}^{-1}$, and the integral intensities of the spectral lines are proportional to $[R^\bullet]$ and $[R^\bullet L]$. When the exchange rate is accelerated, the average lifetimes of $R^\bullet$ and $R^\bullet L$ become shorter. According to the principle of uncertainty relation, the spectral line should be broadened:

$$\Delta E \cdot \Delta t \sim \hbar, \text{that is}, \Delta\omega \cdot \Delta t \sim 1 \tag{7.76}$$

The spectral line broadening of $R^\bullet$ and $R^\bullet L$ should be

$$(\Delta\omega_{R^\cdot}) = \tau_{R^\cdot}^{-1} \text{ and } (\Delta\omega_{R^\cdot L}) = \tau_{R^\cdot L}^{-1} \tag{7.77}$$

Coupled with the natural width, they are the Lorentzian linewidths after the exchange broadening:

$$(1/\tau_2)_{R^\cdot} = (1/\tau_2^o)_{R^\cdot} + (1/\tau_{R^\cdot}), (1/\tau_2)_{R^\cdot L} = (1/\tau_2^o)_{R^\cdot L} + (1/\tau_{R^\cdot L}) \tag{7.78}$$

The above is a very slow exchange situation. When the exchange is carried out at a very fast rate, we see a sharp average line with a weighted average g of $g_{R^\bullet}$ and $g_{R^\bullet L}$, that is,

$$\langle g \rangle = x_{R^\cdot} g_{R^\bullet} + x_{R^\cdot L} g_{R^\bullet L} \tag{7.79}$$

In the equation,

$$x_{R^\cdot} = \frac{[R^\cdot]}{[R^\cdot] + [R^\cdot L]}, x_{R^\cdot L} = \frac{[R^\cdot L]}{[R^\cdot] + [R^\cdot L]} \tag{7.80}$$

The linewidth is

$$\langle (1/\tau_2^o) \rangle = x_{R^\cdot} (1/\tau_2^o)_{R^\cdot} + (1/\tau_2^o)_{\text{exch}} \tag{7.81}$$

Here, $(1/\tau_2^o)_{\text{exch}}$ represents the broadening caused by slow exchange.

It can be proved that

$$(1/\tau_2^o)_{\text{exch}} \sim (\Delta\omega)^2 \tau, \tag{7.82}$$

where

$$\Delta\omega \equiv \omega_{R^{\cdot}} - \omega_{R^{\cdot}L} = \frac{\beta H}{\hbar}(g_{R\bullet} - g_{R\bullet L}) \quad (7.83)$$

τ is the"restore lifetime," and is defined as

$$\tau = \frac{\tau_{R^{\cdot}}\tau_{R^{\cdot}L}}{\tau_{R^{\cdot}} + \tau_{R^{\cdot}L}} \quad (7.84)$$

Only when the exchange rate is very fast, $\tau \to 0$, and $(1/\tau_2^0)_{exch}$ is equal to zero. Please refer to Figure 7.6 for the specific lines.

7.3.2.2 Electron spin exchange between unpaired electrons on two radicals

The so-called"electron spin exchange" implies the unpaired electrons on two free radicals exchange their spin orientation. The first observed is the EMR spectrum of $(SO_3)_2NO^{2-}$ [17]. Here we show an example of EMR spectrum of a similar free radical: di-*t*-butyl nitroxide in ethanol solution at room temperature with different concentrations [18,19].

When the concentration is very low (10^{-4}M), three very narrow hyperfine splitting lines of (^{14}N) nucleus can be seen, as shown in Figure 7.7a. When the concentration increases to 10^{-2} M, the three spectral linewidths are obviously broadened, as shown in Figure 7.7b. The exchange of electronic spin state exists between the two radicals with the same nuclear spin state, but the resonance field is constant (g value). τ can be calculated from the linewidth by using the following equation:

$$\Gamma = \Gamma_o + |2\gamma_e\tau|^{-1} \quad (7.85)$$

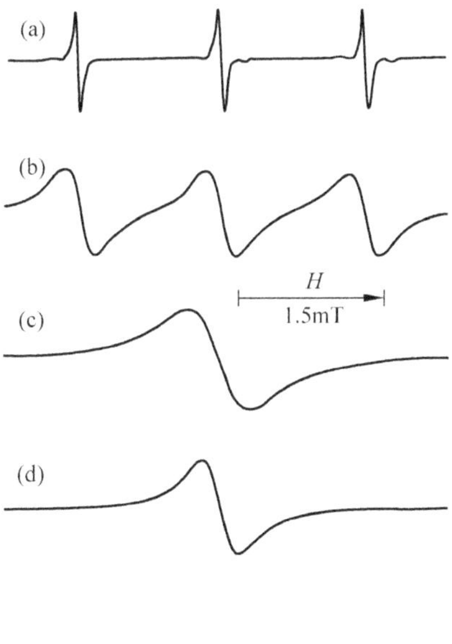

Figure 7.7: The EMR spectrum of di-*t*-butyl nitroxide in ethanol at different concentrations and room temperature:(a) 10^{-4} M; (b) 10^{-2} M; (c) 10^{-1} M; (d) pure liquid.

It must be pointed out here that the *electron spin exchange* rate of each molecule is $1/2\tau$ (this is important!). Therefore, τ is proportional to the concentration of free radicals $[R]^{-1}$. The two order rate constant is

$$k_{(2)} = \frac{1}{2\tau[R]} \tag{7.86}$$

For $(t-butyl)_2$ NO in DMF, $k_{(2)} = 7.5 \times 10^9 M^{-1} \cdot s^{-1}$[18]. Such a large value of $k_{(2)}$ suggests that the probability of a spin exchange must be very high, since the rate constant approaches a diffusion-controlled reaction.

When the concentration of $(t-butyl)_2$ NO continues to increase, the three hyperfine lines merge into one line, as shown in Figure 7.7c. If the concentration still increases, then this line will become narrow, as shown in Figure 7.7d. This is commonly known as the "exchange narrowing," because the electron spins exchange so fast that the time-averaged hyperfine field approaches zero. Usually, in order to obtain a good hyperfine structure, the exchange of electron spins is avoided.

For example, the dissolved molecular oxygen causes the spectral line broadening, which is linearly related with the temperature, but not related with the solvent concentration [20]. In recent years, this phenomenon has been used in biomedicine to determine the concentration of dissolved oxygen in blood [21].

The effect of electron spin exchange on the linewidth is different from that of intermolecular dipole–dipole interaction. Both of them affect on the broadening of the spectral line only when the collision occurs in the liquid. The electron spin exchange is a quantum mechanical effect, which produces a much larger broadening of the spectral line in the liquid than the dipole–dipole effect.

This can be seen from the following example: If the free radical concentration is 10^{-3} M, the rate constant of electron spin exchange $k_{(2)}$ is $10^{10} M^{-1} \cdot s^{-1}$. It can be calculated from Eq. (7.85) and Eq. (7.86) that $\Gamma - \Gamma_o = 0.06$ mT. However, at the same concentration, the contribution of the dipole broadening to the linewidth is less than 0.001 mT.

7.3.2.3 Broadening of the unpaired electron transfer from the free radical to the neutral molecule

There is a category of reactions wherein the reaction rate and the equilibrium constant can only be studied by EMR, and it is difficult to study with other methods. Electron transfer reactions belong to this category. Naphthalene anion can be obtained by reduction of naphthalene with alkali metal (Na or K), but in the solution there are excess of naphthalene molecules that have not been reduced. In this case, the electron transfer reaction occurs between naphthalene and naphthalene anion in the solution [22], as shown in Eq. (7.87):

$$\text{naph}(1)^- + \text{naph}(2) \rightleftarrows \text{naph}(1) + \text{naph}(2)^- \tag{7.87}$$

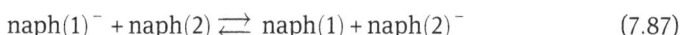

The role of cations (Na^+ or K^+) is completely ignored here. This kind of reactions can be generally expressed by the following expression:

$$A^- + A \rightleftarrows A + A^- \tag{7.88}$$

For the interaction between free radicals and diamagnetic particles, the effects of electron transfer and electron spin exchange on the spectrum are very similar, which may also broaden the spectral line. If there are hyperfine structures, the width of the different hyperfine lines will also be different. Usually, the spectral line at the outer side is more broadened, and the spectral line in the middle is less broadened. Take *para*-quinone as an example: It has four equivalent protons, the number of hyperfine lines is five, and the intensity ratio is 1:4:6:4:1. Their respective line broadenings are as follows:

$$\frac{15}{16}\tau,\ \frac{12}{16}\tau,\ \frac{10}{16}\tau,\ \frac{12}{16}\tau,\ \frac{15}{16}\tau \tag{7.89}$$

where

$$\tau = (k_e[A])^{-1} \tag{7.90}$$

It is assumed that the spin state of an electron remains constant before and after the spin exchange: The state that is α before the exchange is remains as α state even after the exchange; likewise, the state that is β before the exchange remains as β state even after the exchange. Nuclear spin states also remain unchanged. This assumption of spectral line broadening is called the secular broadening. Under the premise of this assumption,

$$\omega_\alpha = \frac{\beta H}{\hbar} g_\alpha + \frac{1}{2\hbar}\sum_i a_{i\alpha} M_{Ii} \tag{7.91}$$

For *para*-quinone,

$$\omega = \omega_o + Ma \tag{7.92}$$

The M in Eq. (7.92) is 2, 1, 0, −1, −2, and the corresponding eigenfunctions are as follows:

$M = 2$	$\alpha\,\alpha\,\alpha\,\alpha$	$M = 0$	$\alpha\,\alpha\,\beta\,\alpha$	$M = -1$	$\beta\,\beta\,\beta\,\alpha$
			$\alpha\,\beta\,\alpha\,\beta$		$\beta\,\beta\,\alpha\,\beta$
$M = 1$	$\alpha\,\alpha\,\alpha\,\beta$		$\beta\,\alpha\,\alpha\,\beta$		$\beta\,\alpha\,\beta\,\beta$
	$\alpha\,\alpha\,\beta\,\alpha$		$\alpha\,\beta\,\beta\,\alpha$		$\alpha\,\beta\,\beta\,\beta$
	$\alpha\,\beta\,\alpha\,\alpha$		$\beta\,\alpha\,\beta\,\alpha$		
	$\beta\,\alpha\,\alpha\,\alpha$		$\beta\,\beta\,\alpha\,\alpha$	$M = -2$	$\beta\,\beta\,\beta\,\beta$

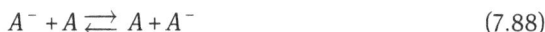

When the electron transfer occurs, the electron is originally associated with the state $\alpha\alpha\alpha\alpha$ of the nuclear $M = 2$, and the transferred electron can be associated with any of the above 16 states. Only when the system is connected with the state $\alpha\alpha\alpha\alpha$, the position of the resonance line does not change, which will not lead to spectral line broadening. The other 15 cases will lead to the broadening of the spectral line, and the relative broadening should be $\frac{15}{16}\tau$. Similarly, for the nuclear state of $M = 1$, the transferred electron may also be associated with any of the above 16 states. Only 4 states will not lead to the broadening of the spectral line, and the other 12 cases will lead to the broadening of the spectral line. The relative broadening should be $\frac{12}{16}\tau$. Owing to the same reason, for the nuclear state of $M = 0$, the relative broadening should be $\frac{10}{16}\tau$. Because the reaction of A^- and A is a two-order reaction, so $\tau = (k_e[A])^{-1}$.

If the broadening coefficient of the i hyperfine line is f_i, the intensity is I_i, and the total intensity of the hyperfine spectrum is I_Σ, then Eq. (7.89) can be written as follows:

$$f_i = \left(\frac{I_\Sigma - I_i}{I_\Sigma}\right)\tau \tag{7.93}$$

Equation (7.93) is used to calculate the spectral line broadening of the 4 lines among 25 hyperfine lines of the naphthalene anion radical, of which the relative intensities are 36, 24, 16, and 6. The results are compared with the experiments of Zandstra and Weissman [23], as listed in Table 7.2.

Table 7.2: The comparison of the calculated line broadening with the experimental data.

I_i	$f_i = \left(\frac{256-I_i}{256}\right)\tau$	Calculated $\frac{f_{36}}{f_i}$	Zandstra and Weissman's experimental value [23]
36	$\frac{220}{256}\tau$	1	1.000
24	$\frac{232}{256}\tau$	$\frac{220}{232} = 0.948$	0.947 ± 0.016
16	$\frac{240}{256}\tau$	$\frac{220}{240} = 0.917$	0.905 ± 0.026
6	$\frac{250}{256}\tau$	$\frac{220}{250} = 0.880$	0.883 ± 0.045

It can be seen from Table 7.2 that the calculated values are in good agreement with the experimental values. This fact shows that the above assumption – spectral line broadening is secular broadening – is reasonable.

In the fast exchange, because electrons can contact with all possible nuclear state in a very short period of time, the hyperfine splittings are totally averaged out and they become a line, but we can still determine the reaction rate from the spectral line broadening.

There are two noteworthy cases of electron transfer:
One kind is called *atom transfer*: If alkali metal (e.g., Na) is used to reduce aromatic hydrocarbon to obtain aromatic anion radical A^-, then Na^+ and A^- form the ion pair Na^+A^-. If an excess of A molecules are added in the solution, the exchange between Na^+A^- and A occurs:

$$A + Na^+A^- \rightleftarrows A^-Na^+ + A \tag{7.94}$$

The experiment shows the result of the fast exchange is not a single line, but four hyperfine spectral lines.^{23} Na is a magnetic core (I = 3/2). The reason is that average lifetime of the electron along with the Na^+ nucleus is longer, but that of the electron along with each A^- is very short, and it seems that the electron transfers with the cation, so it is called "atom transfer." There is also a better example: (Diphenyl ketone) $^-Na^+$ ion pair in the solvent of 1,2-dimethoxyethane. There are more than 80 hyperfine lines in the range of 2.8 mT, in which the hyperfine coupling constant of ^{23}Na is 0.1 mT. When an excess of diphenyl ketone is added, it becomes four lines of ^{23}Na, rather than a single line [24]. The study shows that the average lifetime of the electron along with ^{23}Na is longer than 3×10^{-7} s, but the average lifetime of the electron along with each ketone molecule is less than 10^{-8}s.

Another kind is called *cation exchange*. If there is an increase of the cation concentration in the ion pair solution, then the cation exchange phenomenon occurs:

$$M^+ + A^-M^+ \rightleftarrows M^+A^- + M^+ \tag{7.95}$$

In the fast exchange, the hyperfine lines of M^+ were averaged [25,26], while the hyperfine structure of A^- was preserved.

7.3.2.4 Proton transfer broadening

It has already been discussed that the effect of the magnetic environment change on the spectral line is caused by the electron spin state exchange, or the unpaired electron transfer from one molecule to another. However, the chemical exchange between one or more nuclei in the molecule and the nuclei in the solvent alters the magnetic environment – Does this effect change the spectral lines? Although this effect on the NMR spectrum is very significant, the effect on EMR spectral lines is not obvious because of the slow reaction rate. However, when the proton exchange exists, sometimes the reaction rate is fast enough to produce detectable effects on the spectral lines.

A good example of *proton transfer* is the proton exchange of free radicals $\dot{C}H_2OH$ in the aqueous solution of methanol. There are two equivalent protons in the radical; thus, there should be three hyperfine lines at pH = 1.03, as shown in Figure 7.8b. The proton of OH^- should be split into two lines, but only when the solution pH = 1.40, it

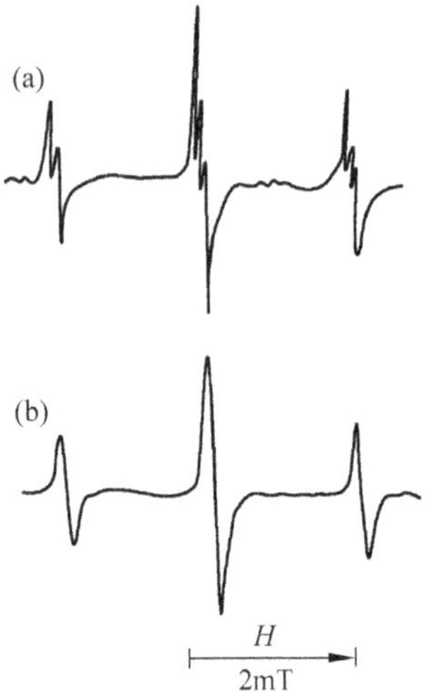

Figure 7.8: *X*-band EMR spectrum of the CH_2OH aqueous solution at room temperature.

can be detected, as the spectrum shown in Figure 7.8a. Its splitting distance $a_{OH} = 0.096$ mT and linewidth $\Gamma_o = 0.03$ mT. In this case, the protons of OH^{-1} exchange with the H^+ ions in the solvent very fast. The rate of proton exchange can be calculated from the spectrum by using Eq. (7.72). The two-order rate constant [27] is $k_{(2)} = 1.76 \times 10^8 M^{-1} \cdot s^{-1}$.

7.3.3 Mechanism of the spectral lines broadening caused by physical motion

7.3.3.1 Broadening caused by fluxional motion

The so-called *fluxional motion* refers to the intramolecular motion that can cause the molecular configuration change. Here, an inorganic crystal $Cu(H_2O)_6^{2+}$ $(S = 1/2)$ diluted in zinc fluorosilicate crystal is taken as an example. It is a crystal of octahedral coordination and due to the distortion caused by the Jahn–Teller effect, it forms the configuration of three equivalent quadrangular pyramids. The reorientation of these configurations causes the main line shape of $^{63,65}Cu$ to change from the normal hyperfine structure of four lines into one broad line (in fact, the overlap results in the offset). Figure 7.9 a shows the experimental angular dependence of EMR hyperfine spectrum of $Cu(H_2O)_6^{2+}$ in the zinc fluorosilicate crystal at 45 K, where the magnetic field is on the [1 1 0] crystal surface, rotating around the [0 0 1] axis with θ degree, and the measured microwave frequency is 9.5 GHz. Figure 7.9b shows the computer simulation of the theoretical spectrum. It is shown that the dynamic broadening depends strongly on M_I and is highly anisotropic. This effect has been simulated using the discrete jump density matrix [28].

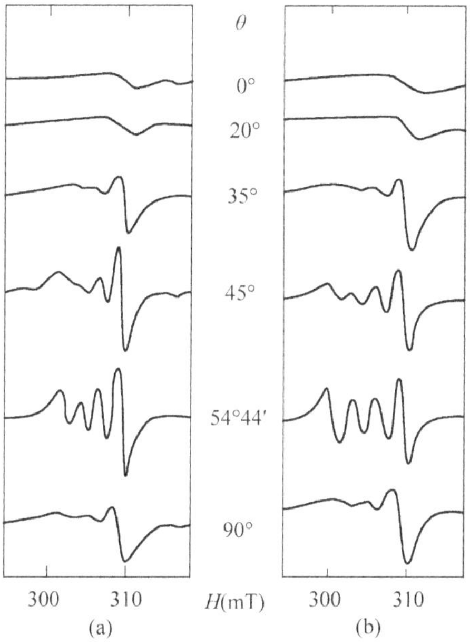

Figure 7.9: Angular dependence of the EMR spectra of $Cu(H_2O)_6^{2+}$ in zinc fluorosilicate crystal at 45 K.

7.3.3.2 Broadening caused by dynamic hyperfine splitting

Because of the chemical processor, the intramolecular rearrangement, such as the transformation of the *cis–trans* isomers, the ring inversion motion, the proton exchange, and so on, the EMR hyperfine splitting can be changed rapidly, which leads to the linewidth variation with M_I. Please refer to the literature [29] for a detailed discussion on this topic.

The case of single magnetic core: If free radicals have two configurations A and B at the same time, and the nuclear spin I = 3/2, then in the presence of configuration A, the hyperfine splitting is four lines and the splitting distance is a_A. In the presence of configuration B, the splitting distance is a_B. Set $a_A = 1\text{mT}$, $a_B = 0.1\text{mT}$, and let the configuration transformation be very slow; then the proportion of configuration A is $f_A = 0.75$ and the proportion of configuration B is $f_B = 0.25$. There are two sets of four spectral lines, the intensity ratio of which should be consistent with its proportion, that is, $I_A = 3I_B$; Figure 7.10a shows its theoretical rod spectrum. When the transformation rate is accelerated very fast, it becomes a set of four equal intensity lines, as shown in Figure 7.10b. The splitting distance $\bar{a} = f_A a_A + f_B a_B = 0.75 \times 1 + 0.25 \times 0.1 = 0.775\,\text{mT}$. Figure 7.10c shows the simulated spectrum. As can be seen from Figure 7.10a–c, because the small displacement of the spectral lines of $M_I = \pm 1/2$, the sharp spectral line can still be maintained after rapid transformation; however, the spectrum of $M_I = \pm 3/2$ has larger displacement, so the linewidth is broadened a lot after the rapid transformation. In the following we will prove the mathematical relation between the line broadening and M_I, where the

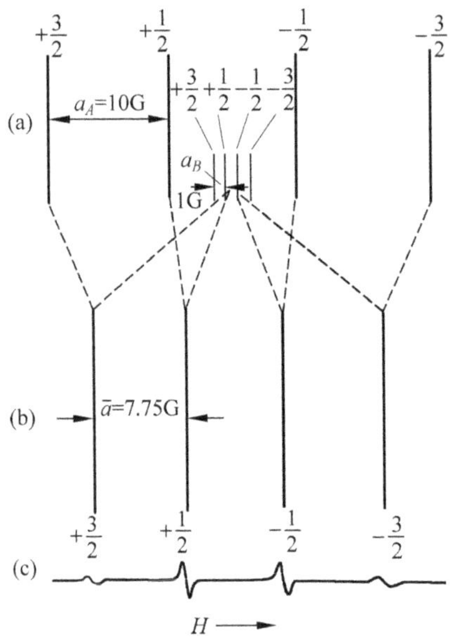

Figure 7.10: Diagram of the influence of configuration transformation on the hyperfine linewidth.

broadening is caused by configuration transformation. The line broadening caused by the configuration transformation should be as follows:

$$(1/\tau_2)_{\text{exchange}} = f_A f_B (\omega_A - \omega_B)^2 \tau \tag{7.96}$$

where

$$\tau = \frac{\tau_A \tau_B}{\tau_A + \tau_B}$$

because

$$\omega_A = \omega_o + Ma_A, \omega_B = \omega_o + Ma_B$$

Thus,

$$(1/\tau_2)_{\text{exchange}} = f_A f_B (a_A - a_B)^2 M^2 \tau \tag{7.97}$$

The line broadening of each hyperfine structure caused by the transformation is not equal, and it is proportional to M^2 of each line. Figure 7.11 is the EMR spectrum of naphthalene $^-Na^+$ ion pair in the mixed solvents of 25% tetrahydrofuran and 75% diethyl ether at low field. As seen from the spectrum, the transformation rate is very fast at −60°C, two sets of four hyperfine lines that represent the two ion pairs with different configurations caused by ^{23}Na nuclei are of equal strength. But at −75 °C, the transformation rate becomes slow; it is obvious as the broadening on the edge of the

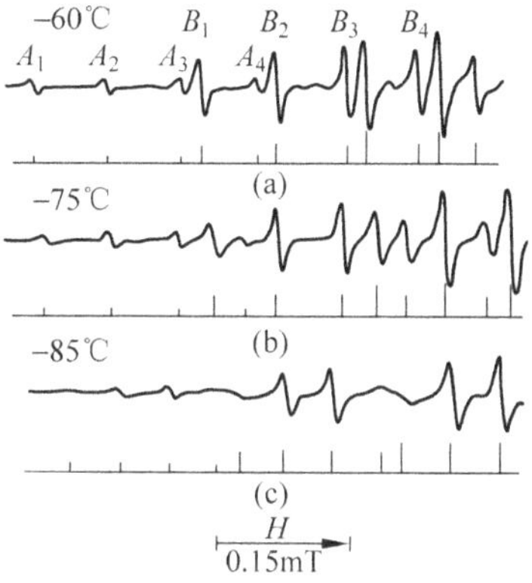

Figure 7.11: The low field part of EMR spectrum of naphthalene $^-Na^+$ ion pair.

spectral line ($I = 3/2$) is much larger than that on the middle line ($I = 1/2$). This effect is even more pronounced at −85°C [30].

The case of multiple magnetic cores: The ion pair of *cis*-1,2-dichloroethyleneanion radical and Na^+, for example, has two possible scenarios: One is shown in Figure 7.12a – because Na^+ is at the symmetry axis of dichloroethylene anion radical, the two protons always keep the equivalent coupling at any time; we call it "completely equivalent nuclei." The other scenario is shown in Figure 7.12b – When Na^+ is close to a certain side, the two protons are not equivalent (although they are equivalent due to the time-averaged effect, but in the transient case they are not equivalent). Hence, they are called "equivalent nuclei" and not "completely equivalent nuclei."

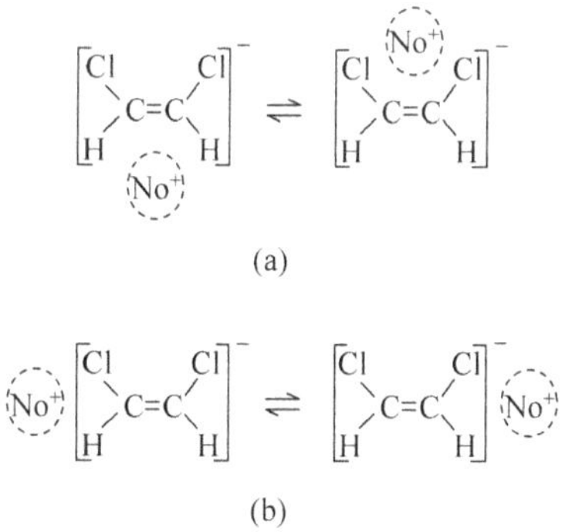

Figure 7.12: The structures of the ion pair of *cis*-1,2-dichloroethylene anion radical and Na^+.

For a set of "completely equivalent nuclei," they keep the equivalent coupling at any time, that is, $a_i(t) \equiv a$. Thus,

$$\omega_A = \omega_o + \sum_i a_i(t) M_i = \omega_o + a \sum_i M_i = \omega_o + aM$$

$$\omega_B = \omega_o + \sum_i a_i^{'}(t)M_i = \omega_o + a'M$$

$$(1/\tau_2)_{\text{exchange}} = f_A f_B (a - a')^2 M^2 \tau \tag{7.98}$$

Comparing Eq. (7.97) with Eq. (7.98), it is found that the spectral line broadening (depend on M^2) of the set of "completely equivalent nuclei" is the same as that of a single magnetic core.

However, in the case of non-"completely equivalent nuclei," the alternating linewidth effect is produced. For example, the ion pair consisting of 2,5-ditertbutyl-1,4-benzene quinone anion and cation M^+ has two configurations (as shown in Figure 7.13). In the configuration A, the coupling constant of proton H_a is a_1, that of H_b is a_2. In the configuration B, the coupling constant of proton H_a is a_2 and that of H_b is a_1. Thus,

$$\omega_A(M_a, M_b) = \omega_o + a_1 M_a + a_2 M_b \tag{7.99}$$

$$\omega_B(M_a, M_b) = \omega_o + a_2 M_a + a_1 M_b \tag{7.100}$$

$$\omega_A - \omega_B = (a_1 - a_2)(M_a - M_b) \tag{7.101}$$

$$(1/\tau_2)_{\text{exchange}} = f_A f_B (a_1 - a_2)^2 (M_a - M_b)^{2\tau} \tag{7.102}$$

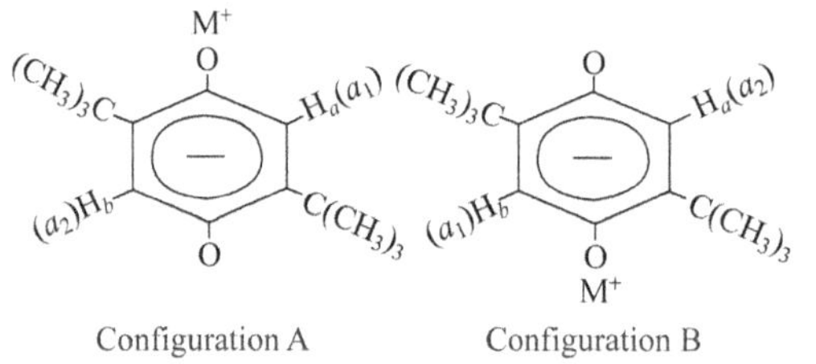

Figure 7.13: Two configurations of 2,5-ditertbutyl-1,4-benzene quinone anion.

From Eq. (7.102), when $M_a = M_b$, the spectral line will not become wider, so the two lines $|\alpha\alpha\rangle$ and $|\beta\beta\rangle$ will always remain sharp and unchanged, but the two lines $|\alpha\beta\rangle$ and $|\beta\alpha\rangle$ will be widened. Here the relative signs of a_1 and a_2 are very relevant – If a_1 and a_2 have the same sign (i.e., all are plus or all are minus), then

$$\omega_A\left(\frac{1}{2}, \frac{1}{2}\right) = \omega_B\left(\frac{1}{2}, \frac{1}{2}\right) = \omega_o + \frac{1}{2}(a_1 + a_2) \tag{7.103}$$

It is the two lines at the outer sides that are not broadened, as shown in Figure 7.14. However, if a_1 and a_2 have the opposite signs, the two lines at the inner sides are not broadened. So, from the line broadening of the two lines at the outside or inside, the relative sign of a_1 and a_2 can be determined.

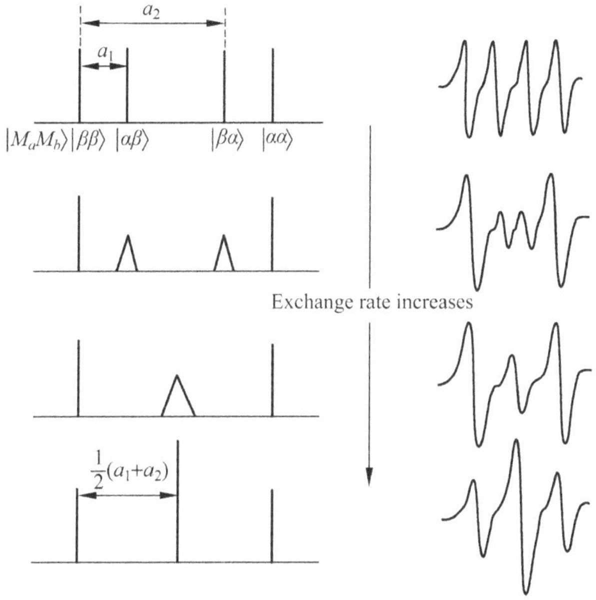

Figure 7.14: The EMR spectra of 2,5-ditertbutyl-1,4-benzene quinone anion and cation M^+ vary with the exchange rate.

Figure 7.15 is the EMR spectrum of the 1,4-dideuterated hydroxyl-2,3,5,6-tetramethyl benzene cation radical in the mixed solvent of $D_2SO_4 - CH_3NO_2$[31]. The main hyperfine structure is caused by 12 equivalent protons on the four methyl groups; there should be 13 set of lines, each set split into five lines 1:2:3:2:1 due to the two

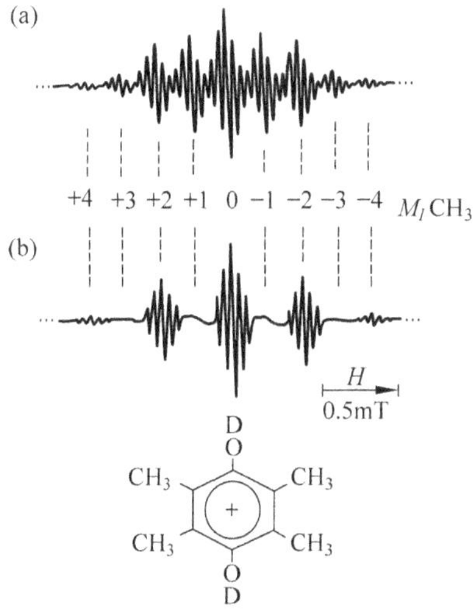

Figure 7.15: The variation of the EMR spectrum of the 1,4-dideuterated hydroxyl-2,3,5,6-tetramethyl benzene cation radical: (a) EMR spectrum at 60 °C; (b) EMR spectrum at −10 °C [31].

equivalent deuteron ($I = 1$). It has four *cis–trans* tautomers, when the transformation rate is fast (60 °C), as shown in Figure 7.15a; the methyl proton hyperfine splitting is 0.205 mT and deuteron splitting is 0.042 mT. Only nine sets can be seen, because two sets on each side are too weak to be observed. When the temperature drops to −10 °C, the transformation rate becomes slow, and there is an "alternating linewidth effect," as shown in Figure 7.15b. In every other set, the deuteron hyperfine splitting is blurred, and the strength of hydrogen nuclei splitting also drops to zero.

Another example is the EMR spectrum of 1,4-dinitro-2,3,5,6-tetramethyl benzene anion radical dissolved in DMF at room temperature, as shown in Figure 7.16 [32]. When two NO_2 in the free radical rotate around the C–N bond and if the plane of the rotation is same as that of the benzene ring, then the splitting a^N is very large (about 1.4 mT). However, when the plane of the NO_2 rotation is vertical to the benzene ring, the splitting a^N is very small (about 0.05 mT). The average result of the two motions is $a^N = 0.699$ mT. The splitting of protons on methyl is $a^H = 0.025$ mT. As can be seen from Table 7.3, in the case of two equivalent nitrogen nuclei, only when (M_a, M_b) is (1, 1), (0, 0), (−1, −1), the spectral line is not broadened. Thus, the further splitting can be seen in the figure. However, the spectral lines of $M = M_a + M_b$ being 1 or −1 are completely blurred due to the "alternating linewidth effect." The experimental result is fully consistent with the theoretical prediction.

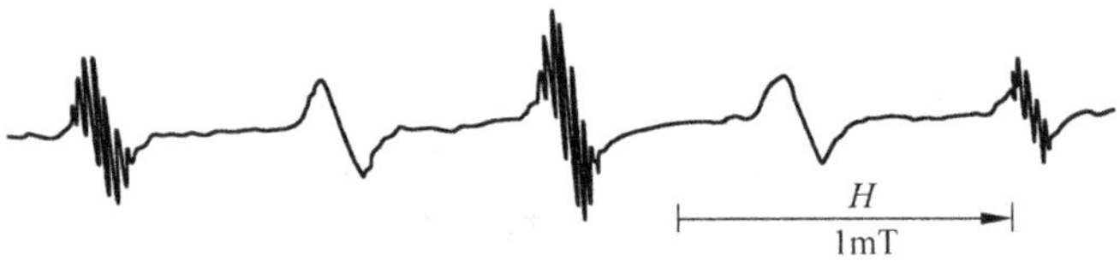

Figure 7.16: The EMR spectrum of 1,4-dinitro-2,3,5,6-tetramethyl benzene anion radical [32].

Table 7.3: The line broadening for equivalent nuclei.

Two equivalent nitrogen nuclei				Two sets of completely equivalent hydrogen nuclei			
M	***M_a***	***M_b***	$\|\Delta\omega\|$	***M***	***ab***	***cd***	$\|\Delta\omega\|$
2	1	1	0	2	$\alpha\alpha$	$\alpha\alpha$	0
					$\alpha\alpha$	$\alpha\beta$	Δa
1	1	0	Δa	1	$\alpha\alpha$	$\beta\alpha$	Δa
					$\alpha\beta$	$\alpha\alpha$	Δa
	0	1	Δa		$\beta\alpha$	$\alpha\alpha$	Δa
0	1	−1	$2\Delta a$	0	$\alpha\alpha$	$\beta\beta$	$2\Delta a$
					$\beta\beta$	$\alpha\alpha$	$2\Delta a$
	−1	1	$2\Delta a$		$\alpha\beta$	$\alpha\beta$	
					$\beta\alpha$	$\beta\alpha$	0
					$\alpha\beta$	$\beta\alpha$	
	0	0	0		$\beta\alpha$	$\alpha\beta$	0

7.3.3.3 Effect of molecular tumbling

Now let us go back to the system of one unpaired electron, and give priority to the effect caused by the anisotropic spin Hamiltonian parameter.

The rate of molecular tumbling depends on the shape and size of the molecule and the interaction with the environment (solvent), as well as on the temperature. The effect of molecular tumbling on EMR depends on the magnitude of the anisotropy of the spin Hamiltonian parameter. It is usually divided into two categories:

(1) *Isotropic tumbling:* The probabilities P_Ω of rotating around different axes at the same angle are equal.
(2) *Anisotropic tumbling:* The probability P_Ω depends on the axis of the molecular rotation.

The *theory of tumbling* can be dealt with by using the Brown motion model or discontinuous jump model. The Brown (spin diffusion) model assumes that each molecule rotates continuously and freely at an arbitrary angular velocity around a certain molecular axis. At any interval, the changes in the rotation axis and the rate are instantaneous and disordered, because each molecule collides with another molecule at any time and anywhere. For the discontinuous jump model, it is assumed that the arbitrary orientation of each molecule is fixed in a certain period of time, and the jump from one fixed orientation to another is instantaneous. Both the two models use the mean time interval as the perturbation, called the *mean residence time* or the *average lifetime of the orientation.*

7.3.3.4 Contributions of Zeeman and hyperfine interactions

In the previous chapters, we mentioned that the g factor of the solid sample and the hyperfine coupling parameter A strongly depend on the orientation. It should also be pointed out that the anisotropic interaction of free radicals in the low-viscosity solution is averaged to zero due to the rapid tumbling. But in the medium-viscous solvent, it will not be averaged to zero; thus, the magnetic resonance shifts ΔH_0.

The EMR spectra of the *para*-dinitrobenzene anion in the DMF solvent [33] are taken as the example of the solution spectrum change caused by decreasing molecular tumbling rate, as shown in Figure 7.17. At 12 °C, the "normal" EMR spectrum can be observed, and the amplitude of the first-order differential spectrum is proportional to the degeneracy of the corresponding transition energy level. However, when the temperature is reduced to −55 °C, the spectrum changes greatly; although the spectral line position does not change, the amplitude and width of the spectral line change. Note, these changes do not change symmetrically around the center line.

In order to understand the origin of these effects, the ditertbutyl nitroxide radical is taken as the example again. Its g matrix and the hyperfine tensor of the ^{14}N nucleus

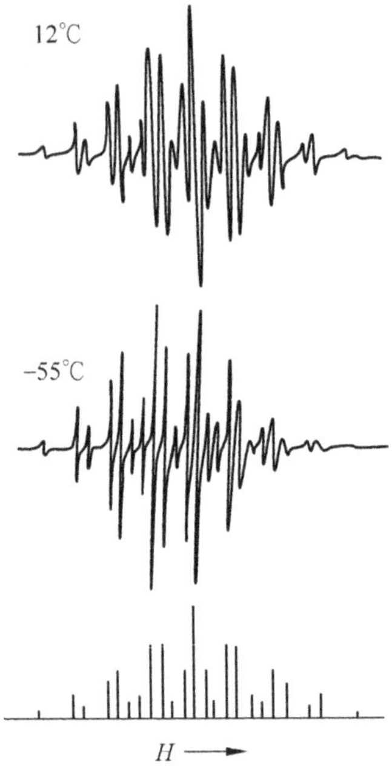

Figure 7.17: The EMR spectra of the *para*-dinitrobenzene anion in DMF change with the temperature.

have the same principal axis coordinate, and each matrix tends to a single axis structure [34].

Figure 7.18a is the EMR spectrum of this radical measured in the solid phase (77 K) with the random orientation. The linewidth of each hyperfine line has the following relation [35]:

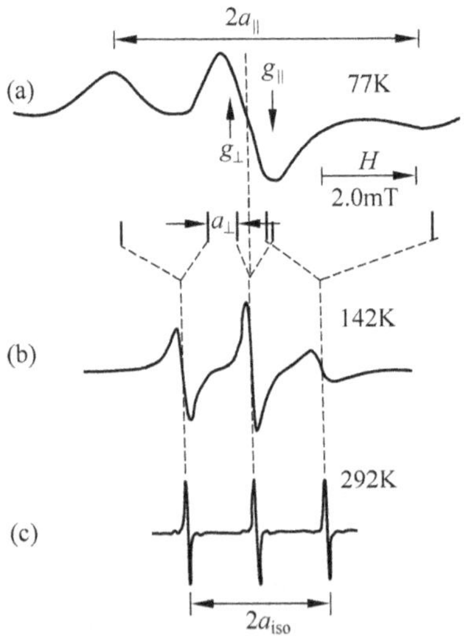

Figure 7.18: The EMR spectra of ditertbutyl nitroxide radical change with the temperature.

$$\Gamma = \alpha + \beta M_I + \gamma M_I^2 \tag{7.104}$$

The coefficients on the right-hand side of Eq. (7.104) depend on the anisotropy of g, the hyperfine splitting A, and the average tumbling rate (i.e., the viscosity of the solvent). With the increase of tumbling rate, the two parameters β and γ will tend to zero.

In fact, people have already found that the hyperfine structure of the solution EMR spectra of VO^{2+} and Cu^{2+} complexes at room temperature (the former has eight lines, while the latter has four lines) has asymmetric linewidth broadening, but the latter also has different contributions from ^{63}Cu, ^{65}Cu, two kinds of nuclei, to the splitting distance and linewidth. They also satisfy Eq. (7.104).

Figure 7.18b is the EMR spectrum of the medium thick dilute solution measured at 142 K. Figure 7.18c is the EMR spectrum measured at 292 K, in the situation of the fast enough tumbling rate, which is a completely average EMR spectrum, but the spectral line position has not changed, except that the linewidth has become narrow. It can be seen from Eq. (7.104) that the linewidth is related to the value and symbol of M_I. This is why the linewidths of the three lines are not same. Since $\beta < 0$, the line at the highest field ($M_I = -1$) is wider than the line at the lowest field ($M_I = +1$). The spectral parameters of the ditertbutyl nitroxide radical are measured: for ^{14}N nucleus, $a_{\parallel} = +3.18$mT, $a_{\perp} = +0.68$mT, $a_o = +1.51$mT; $g_{\perp} = 2.007$, $g_{\parallel} = 2.003$. From these parameters, it can be theoretically [36] obtained that $\beta < 0$. If both a_o and $a_{\parallel}/a_{\perp}$ are negative, the high field line is narrower than the low field line. When $g_{\parallel} > g_{\perp}$, all these conclusions have to be reversed. This phenomenon of linewidth is the basis of a method for measuring the sign of isotropic hyperfine splitting [37].

Since g and A are anisotropic, the position of the spectral lines will be changed when the paramagnetic particles are tumbling in the solution, which is ultimately reflected in the inhomogeneous broadening of the hyperfine lines. When paramagnetic particles are randomly tumbling in solution, $\Gamma \sim (\Delta\omega)^2 \tau_c$. Here $\Delta\omega$ is the variation of the spectral line frequency caused by the anisotropy and τ_c is the correlation time of tumbling. There are two contributions to $\Delta\omega$:

$$\Delta\omega = (\Delta\omega)_g + (\Delta\omega)_A$$

Then,

$$\Gamma \sim [(\Delta\omega)_g^2 + (\Delta\omega)_A^2 + (\Delta\omega)_g(\Delta\omega)_A]\tau_c \tag{7.105}$$

The variation of g only changes the central position of the whole spectrum; however, the shift of each hyperfine line is the same, and also the relative distance of the spectral line is the same, so it is homogeneous broadening. The variation of A only changes the relative distance of hyperfine lines without changing the central position of the whole spectrum, so it only produces symmetric broadening and the central spectral line remains sharp. When both g and A are changed, the asymmetric broadening will be caused. Therefore, $(\Delta\omega)_g^2$ in Eq. (7.105) is independent of M_I, $(\Delta\omega)_A$ is proportional to M_I, and $(\Delta\omega)_A^2$ is proportional to M_I^2.

Comparing with Eq. (7.104),

$$\alpha = (\Delta\omega)_g^2 \tau_c \tag{7.106a}$$

$$\beta M_I = (\Delta\omega)_g (\Delta\omega)_A \tau_c \tag{7.106b}$$

$$\gamma M_I^2 = (\Delta\omega)_A^2 \tau_c \tag{7.106c}$$

The above qualitative discussion gives a clear explanation of the physical basis of producing asymmetric broadening, but this is not complete unless the contribution of the secular term is considered. Please refer to the literature [38–41] for further theoretical analysis.

7.3.3.5 Contribution of spin–rotational interaction

In the study on EMR spectra of vanadyl acetylacetonate, Wilson and Kivelson [42] mentioned that α in Eq. (7.104) should also include a nonnuclear spin-dependent contribution to the linewidth, which is contribution of "spin–rotational interaction."

The molecular rotation causes the magnetic moment because the electrons do not accurately track the motion of the nucleus in a rotating molecule, which is always partially delayed and produces a magnetic moment. The magnetic moment is coupled with the spin magnetic moment, and the Hamiltonian equation is as follows:

$$\hat{\mathscr{H}} = \hat{S} \cdot C \cdot \hat{J} \tag{7.107}$$

$\boldsymbol{C}$ in Eq. (7.107) is *spin-rotational* coupling tensor. If the interaction is modulated, it contains a relaxation mechanism. There are two ways of modulation in the solution: One is that the rotation angular momentum is modulated when the molecule is passing through the neighboring molecules; the other is that the *spin-rotational* coupling tensor is modulated when the molecules change orientation, but their correlation times are different. If the fluctuation of $\boldsymbol{C}$ is expressed by the rotational correlation time τ_c, and the fluctuation of J is expressed by the correlation time τ_J, then $\tau_J \ll \tau_c$. That is to say, the main contribution of the spin relaxation is from the angular momentum modulation.

In the gas phase, the molecular rotation is completely free and the rotatory motion is quantized; if the molecule has inherent dipole moment, then the transition between rotational energy levels can be measured by microwave spectroscopy. However, in the gas phase, the dipole–dipole interaction between the electron spin magnetic moment and the rotational magnetic moment is not averaged to zero, because the rotational angular momentum and the magnetic moment vector are fixed on the same line in the space. Therefore, due to the *spin-rotational* interaction, the EMR spectrum of the gas phase is very complex.

For the low-viscosity liquid molecules or solutions, there are several rotations before the event of collision; hence, the rotational magnetic moments can be coupled with the electron spin magnetic moment [42–44]. This effect is equal for all spectral line broadening [42,43]. In the Brownian diffusion theory, there is Langevin equation:

$$m\frac{\mathrm{d}u(t)}{\mathrm{d}t} = -\zeta u(t) + mA(t) \tag{7.108}$$

In Eq. (7.108), m is the mass of the particle; u is the linear velocity of the particle diffusion; ζ is the coefficient of friction resistance; and $A(t)$ is the random force. Hubbard used a classical method to propose a similar equation for the rotating spherical particles:

$$I\frac{\mathrm{d}u(t)}{\mathrm{d}t} = -\zeta\omega(t) + IA(t) \tag{7.109}$$

In Eq. (7.109), I is the inertia moment of the particle and ω is the angular velocity of the particle rotation. The solution of Hubbard is Eq. (7.109):

$$\langle\omega_i(t)\omega_j(t+\tau)\rangle = \delta_{ij}(kT/I)\exp(-\tau/\tau_J) \tag{7.110}$$

Here τ_J is the correlation time:

$$\tau_J = I/\zeta = I/8\pi r^3\eta \tag{7.111}$$

where η is the viscosity coefficient. Equation (7.110) is derived from the classical theory – for the angular momentum of quantum mechanics, just change $I\omega$ as $\hbar J$, that is,

$$I\omega = \hbar J \tag{7.112}$$

Thus, we get

$$\langle J_i(t)J_j(t+\tau)\rangle = \delta_{ij}(kTI/\hbar^2)\exp(-\tau/\tau_J) \tag{7.113}$$

With the correlation function, the linewidth can be calculated:

$$\Gamma = (kTIC^2/\hbar^2)\left[\tau_J + \frac{\tau_J}{1+\omega^2\tau_J^2}\right] \tag{7.114}$$

where ω is the frequency of EMR, if τ_J is very short, $\omega^2\tau^2 \ll 1$; we apply Eq. (7.111) to get

$$\Gamma = (kI^2C^2/4\pi\hbar^2r^3)(T/\eta) \tag{7.115}$$

that is to say, the linewidth is proportional to (T/η).

Since the spin-rotational coupling constant cannot be measured directly, in order to apply this theory in a quantitative way, it must be expressed as an experimental quantity. Kivelson et al. [43,44] proved that

$$\Gamma = [k(\Delta g)^2/4\pi r^3](T/\eta) \tag{7.116}$$

here $\Delta g = g - g_e$; using the Debye equation

$$\tau_C = \frac{4\pi\eta r^3}{3kT}$$

we obtain

$$\Gamma = (\Delta g)^2/3\tau_C \tag{7.117}$$

Kivelson et al. estimated the contribution to the linewidth of vanadyl acetylacetonate was $2.9\times10^{-6}(T/\eta)$mT, and the experimental result was about $3.2\times10^{-6}(T/\eta)$mT. The theoretical and experimental results of copper acetylacetonate were both $2.95\times10^{-6}(T/\eta)$mT.

Therefore, as long as the molecule is large enough, the Brownian description of the rotational diffusion is reasonable. As long as the deviation of g from g_e is large enough, the spin-rotational interaction has an important contribution to the linewidth.

7.3.3.6 A comprehensive example

Here is an example [45] that has the following three factors: (1) the alternating effect of linewidth; (2) the line shape anisotropic distortion; and (3) the change of the compound concentration. This is a $4d^9(S=1/2)$ neutral Rh(0) complex bis [1,2-bis-(diphenylphosphino)ethan]rhodium(0), denoted as $[\mathrm{Rh(dppe)_2}]^0$. The center of the complex is Rh (0) atom, and there are four phosphorus atoms that form a distorted coordination system. Its EMR spectral line at 270 K shows five symmetry lines are produced by the hyperfine interaction between the four equivalent ^{31}P nuclei and the unpaired electron (the natural abundance of ^{31}P is 100%; I = 1/2). When the temperature is decreased to 259 K, the *alternating linewidth effect* is obvious, which is due to the fact that the four equivalent ^{31}P atoms are divided into two sets of completely equivalent nuclei, as shown in Figure 7.19a. The variation of the rate constant with the temperature is shown in Figure 7.20a. The activation enthalpy of the two configurations transformation is $\Delta H^\dagger = 14.7\,\mathrm{kJ/mol}$; the activation entropy $\Delta S^\dagger = -19\mathrm{J/(mol\,K)}$. Moreover, the coefficient of Eq. (7.104) changes with the decrease of temperature, thus causing the anisotropic broadening of the spectral line. The measured spectral line area reveals that the free radical concentration decreases with the decrease of temperature in accordance with the equilibrium constant $K = [(1-\alpha)\alpha^2]/2c_t$, as shown in Figure 7.20b. In the equation, α is the molecular fraction of $\mathrm{Rh(dppe)_2}$, which is

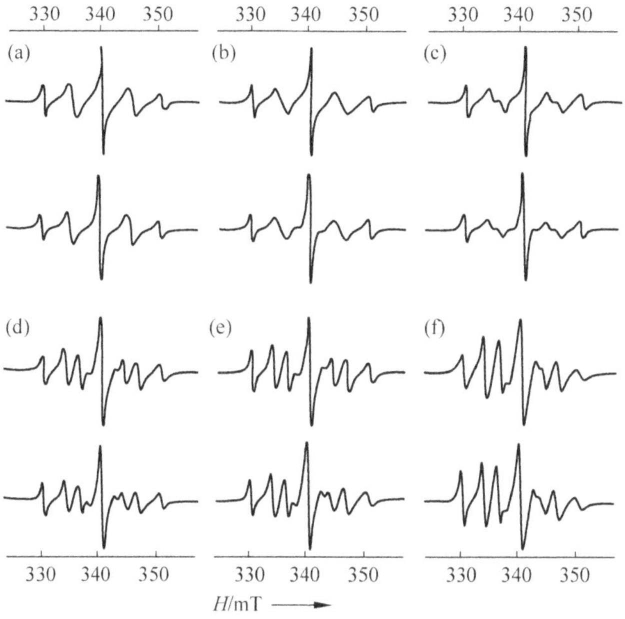

Figure 7.19: The EMR spectra of $[Rh(dppe)_2]^0$ in toluene solution. The upper layer is the calculated spectra and the lower layer is the experimental spectra:(a) 259 K; (b) 249 K; (c) 237 K; (d) 219 K; (e) 209 K; (f) 199 K.

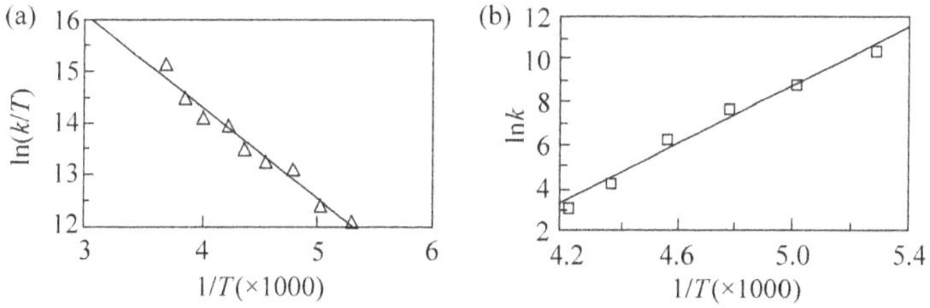

Figure 7.20: The configurations transformation rate, activation enthalpy, and activation entropy of $[Rh(dppe)_2]^0$.

the paramagnetic monomer in solution, and c_t is the total concentration of the paramagnetic monomer. The charge-transfer reaction in solution is presumed to be as follows:

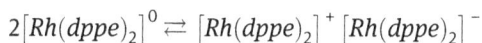

$$2[Rh(dppe)_2]^0 \rightleftarrows [Rh(dppe)_2]^+ [Rh(dppe)_2]^-$$

From the equilibrium constant (T), $\Delta H = -55.6\text{kJ/mol}$; $\Delta S = -207\text{J/(mol K)}$ can be obtained.

7.4 Saturation transfer of spectra

The so-called "saturation-transfer" refers to the diffusion of magnetization in the z-direction. Its efficiency depends sensitively on the dynamics of the unpaired electrons. By using sufficiently large microwave amplitudes H_1 as well as special techniques such as the observation of the first-order dispersion, special field modulation, or the use of two microwave sources H_1 without setting the frequency (ELDOR, see Appendix A1.2), we may obtain valuable information on a relatively slow molecular motion (correlation time between 10^{-11} and 10^{-3}s). As Hyde noted [46], this is particularly useful in biomedical systems.

Many literatures [47–49] have discussed the "saturation-transfer" EMR theory and the general picture of the experiments.

For example, in Ref. [47], it is found that there is a characteristic time constant $\tau_d = 20\,\mu$s between DPPH molecules dissolved in polystyrene, which may be due to the rebound of proton-spin on the neighboring solvent molecules.

7.5 Intensity of signal dependent on time

In some chemical processes (e.g., photochemistry), the free radical concentration changes with time, which is accompanied by *thermalization*, leading to the *non-equilibrium* population distribution of free radicals. This is called "chemically induced dynamic electron polarization" (CIDEP).

7.5.1 Free radical concentration changes with time

The intensity of the EMR spectrum (including line shape) is very useful for studying the kinetics of chemical reactions. We discuss the reaction rate in three ways:

1) *Usual reaction rate:* The diffusion-controlled reaction in a solid, for example, the whole reaction takes several weeks, and the intrinsic reaction time is measured in seconds. There are many reports of this kind of reaction; here only a few reviews are cited [50–52] for reference. The change of the free radical concentration can be studied by the intensity variation of the spectral lines measured by the CW-EMR spectrometer.
   Let us assume that the population on the electron-spin level is in a state of thermal equilibrium. If the linewidth is independent of the concentration, the amplitude of each EMR signal is proportional to the concentration of a kind of paramagnetic particles in the sample. However, it is not always true that the linewidth is independent of the concentration. The process to measure the signal amplitude varying with time is usually fixing the magnetic location of an extreme value of the first differential signal.

2) *Rapid reaction rate:* If the reaction product of paramagnetic material cannot be directly detected by using CW-EMR spectrometer, usually the spin trapping method is applied: Adding a suitable compound to react with a free radical generats a more stable free radical that can be detected by EMR and can also show distinguishable hyperfine pattern. A good example is the hydroxyl radicals ($\dot{O}H$) in solution can be captured by the dimagnetic spin trapping agent (*N-t*-butyl- α-phenyl-nitrone) *N-tert*-butyl phenyl ketone.

$$(CH_3)_3-\overset{+}{N}=\overset{\overset{\displaystyle H}{|}}{C}-C_6H_5+OH \longrightarrow (CH_3)_3-N-\overset{\overset{\displaystyle H}{|}}{C}-C_6H_5$$
$$\underset{O^-}{|} \qquad\qquad\qquad \underset{O\ \ OH}{|}$$

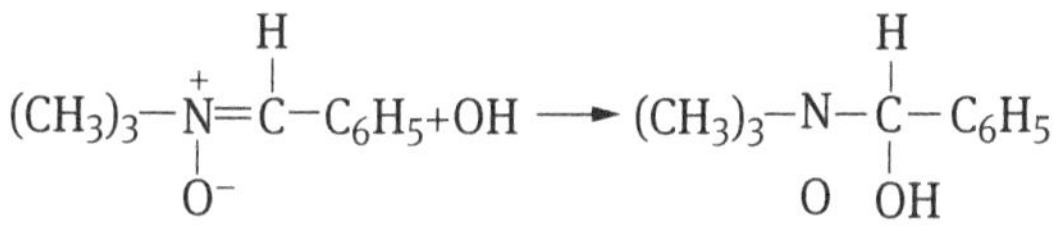

Because of the production of nitroxide, the three lines of ^{14}N nuclear splitting ($a = 1.53mT = 1.53\,mT$) at g = 2.0057 can be detected, and on the basis of the H atomic hyperfine interaction, each line is split into two lines ($a = 0.275mT$), but the hyperfine splitting of the proton on $\dot{O}H$ radical could not be seen [53]. Till now, many spin trapping agents have already been synthesized, they can react with different kinds of free radicals (e.g., OH^-, O_2^-, CH_3) to generate nitroxide free radicals, which have been listed in tables [54].

3) *Too rapid reaction rate:* It cannot be measured with the usual spectrometer and technique. Usually in the spectrometer, since the magnetic field scanning coil self-induced coupling, amplifier time constant and bandwidth reach the limit. In some systems, free radicals can be produced by using flash or pulse electron beam at a very fast rate; after each pulse, a fixed magnetic field H is selected to be sampled in a period (interval) and then the EMR measurement is done. In recent years, there have been a number of comprehensive reviews on this time-resolved technique [55–57]. They extend the available kinetic information to the nano-second range.

7.5.2 Chemical-induced dynamic electron polarization (CIDEP)

The paramagnetic particles produced in some cases reside in the electron spin level in the nonthermal equilibrium state. This may be due to pure photon emission that can lead to the excess population of the upper state, or is due to the increase of photon absorption that can lead to the excess population of the lower state; so the excess population tends toward the thermal equilibrium population in some characteristics rate. Because the amplitude of the EMR is directly proportional to the difference of the population number on the spin energy levels, the tendency toward the thermal equi-librium population is time dependent for the amplitude of the EMR signal.

Fessenden and Schuler [58] first observed the CIDEP effect. They decomposed liquid methane to form H atoms by using the pulsed electron emission with 2.8 MeV at a temperature below 100 K. They found that the phases of the first-order differential spectrum of the two hyperfine lines in the H nucleus were opposite: one was the emission (E) line, and the other was the absorption (A) line. The two populations of the M_I states are opposite, which is due to the polarization effect of the generated H and $\mathbf{CH_3}$ "free radical pairs."

Recently, the results of CIDEP research on $^1\mathbf{H}$ and $^2\mathbf{H}$ atoms in the ice provided information on the formation of the pair between the atomic hydrogen and hydroxyl radical [59].

The $\mathbf{HC}_{60}$ radical was in situ produced by the photolysis of fullerene $\mathbf{C}_{60}$ in the liquid benzene, and the CW-EMR spectra showed hyperfine structure of the *E–A* polarization of the hydrogen nuclei with a splitting of 3.3 mT [60].

A number of other free radical CIDEP phenomena have also been found, and have been used as a function of time lapse after pulse generation. Taking the free radical $(CH_3)\dot{C}OH$ as an example [55], the EMR spectrum and the time integral 2D spectrum of the CIDEP effect are shown in Figures 7.21 and 7.22.

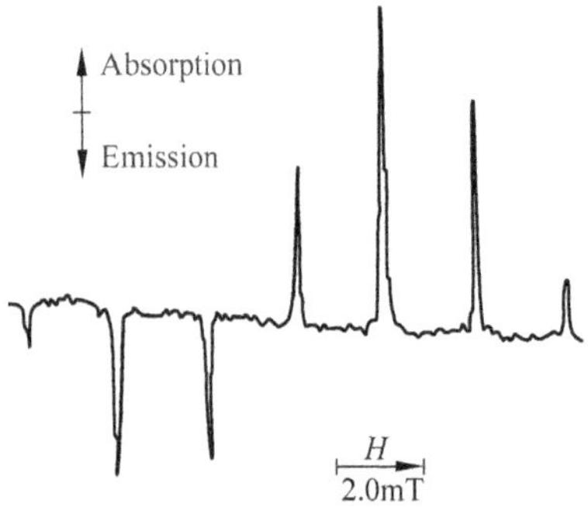

Figure 7.21: The EMR spectrum of CIDEP effect of the isopropanol radical.

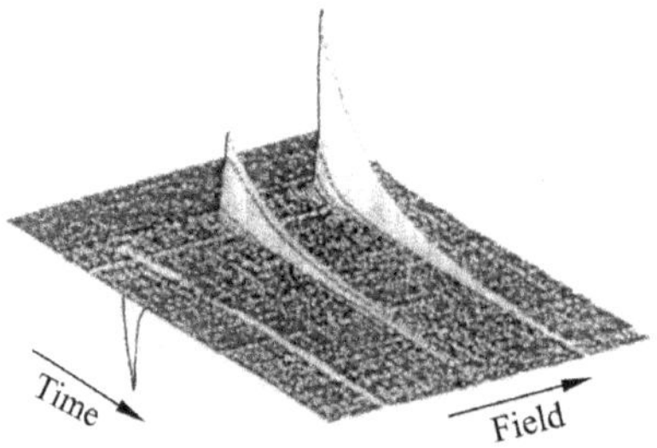

Figure 7.22: The 2D spectrum of CIDEP effect of the isopropanol radical.

A detailed study of the kinetic mechanism of the CIDEP effect can be seen in several reviews [61–64]. There are two mechanisms still being debated: one is the pairing mechanism of radical anion and cation; the other is the mechanism of the triplet state [65–67].

References

[1] Atkins P. W. *Molecular Quantum Mechanics*, 2nd edn, Oxford University Press, Oxford, 1983, pp. 199 and 443.
[2] Morse P. M. *Thermal Physics*, 2nd edn, Benjamin, New York, 1969, Chapter 25.
[3] Kronig R. de L. *Physica* 1939, **6**: 33.
[4] Finn C. B. P., Orbach R., Wolf W P. *Proc. Phys. Soc. (Lond.)* 1961, **77**: 261.
[5] Orbach R. *Proc. R. Soc. Lond.* 1961, **A264**: 458.
[6] Pake G. E., Estle T. L. *The Physical Principles of Electron Paramagnetic Resonance*, 2nd edn, Benjamin, New York, 1973, Chapter 8.
[7] Pople J. A., Schneider W. G., Bernstein H. J. *High-Resolution Nuclear Magnetic Resonance*, McGraw Hill, New York, 1959, Section 3-5.
[8] Bloch F. *Phys. Rev.* 1946, **70**: 460.
[9] Carrington A., McLachlan A. D. *Introduction to Magnetic Resonance*, Harper & Row, New York, 1967, Chapter 11.
[10] Pake G. E., Estle T. L. *The Physical Principles of Electron Paramagnetic Resonance*, 2nd edn, Benjamin, New York, 1973, Chapter 2.
[11] Bovey F. A., Jelinski L., Mirau P. A. *Nuclear Magnetic Resonance Spectroscopy*, Academic Press, San Diego, 1988, Section 1.7.
[12] Slichter C. P. *Principles of Magnetic Resonance*, 3rd edn, Springer, New York, 1990, Chapter 2.
[13] Barbara T. M. *J. Magn. Reson.* 1992, **98**: 608.
[14] Gutowsky H. S., Holm C. H. *J. Chem. Phys.* 1956, **25**: 1228.
[15] Rogers M. T., Woodbrey J. C. *J. Phys. Chem.* 1962, **66**, 540.
[16] Johnson C. S. Jr., *Adv. Magn. Reson.* 1965, **1**: 33.
[17] Lloyd J. P., Pake G. E. *Phys. Rev.* 1954, **94**: 579.
[18] Plachy W., Kivelson D. *J. Chem. Phys.* 1967, **47**: 3312.
[19] Miller T. A., Adams R. N. *J. Am. Chem. Sci.* 1966, **88**: 5713.
[20] Povich M. J. *J. Phys. Chem.* 1975,79, 1106; also see *Anal. Chem.* 1975, **47**: 346.
[21] Hyde J. S., Subczynski W. K. "Spin-label oximetry", in Berliner L J, Reuben J. Eds., *Biological Magnetic Resonance, Vol. 8, Spin Labeling Theory and Applications*, Plenum Press, New York, 1989, pp. 399425.
[22] Ward R. L., Weissman .S I. *J. Am. Chem. Soc.* 1957, **79**: 2086.
[23] Zandstra P. J., Weissman S. I. *J. Chem. Phys.* 1961, **35**: 757.
[24] Adam F. C., Weissman S. I. *J. Am. Chem. Sci.*1958, **80**: 1518.
[25] Adams R. F., Atherton N. M. *Trans. Faraday Soc.* 1968, **64**, 7.
[26] Rutter A. W., Warhurst E. *Trans. Faraday Soc.* 1968, **64**: 2338.
[27] Fischer H. *Mol. Phys.* 1965, **9**: 149.
[28] Zimpel Z. *J. Magn. Reson.* 1989, **85**: 314.
[29] Fraenkel G. K. *J. Phys. Chem.* 1967, **71**: 139.
[30] Hirota N. *J. Phys. Chem.* 1967, **71**: 127.
[31] Sullivan P. D., Bolton J. R. *Adv. Magn. Reson.* 1970, **4**: 39.
[32] Freed J. H., Fraenkel G. K. *J. Chem. Phys.* 1962, **37**: 1156.
[33] Freed J. H., Fraenkel G. K. *J. Chem.Phys.* 1964, **40**: 1815.
[34] Libertini L. J., Griffith O. H. *J. Chem. Phys.* 1970, **53**: 1359.
[35] Hudson A., Luckhurst G. R. *Chem. Rev.* 1969, **69**: 191.
[36] Carrington A., Hudson A., Luckhurst G. R., *Proc. R. Soc. Lond.* 1965, **A284**: 582.
[37] de Boer E., Mackor E. L. *J. Chem. Phys.* 1963, **38**: 1450.
[38] Freed J. H., Fraenkel G. K. *J. Chem. Phys.* 1964, **41**: 3623.
[39] Freed J. H, Fraenkel G. K. *J. Chem. Phys.* 1963, **39**: 326.

[40] Hudson A., Luckhurst G. R. *Chem. Rev.* 1969, **69**: 191.
[41] Carrington A., Lonquet-Higgins. *Mol. Phys.* 1962, **5**: 447.
[42] Wilson R., Kivelson D. *J. Chem. Phys.* 1966, **44**: 154.
[43] Atkins P. W., Kivelson D. *J. Chem. Phys.* 1966, **44**, 169.
[44] Nyberg G. *Mol. Phys.* 1967, **12**: 69; 1969, **17**: 87.
[45] Mueller K. T., Kunin A. J., Greiner S., Henderson T., Kreilick R. W., Eisenberg R. *J. Am. Chem. Soc.* **1987**, 109: 6313.
[46] Hyde J. S. "Saturation transfer spectroscopy," *Methods in Enzymology*,Academic Press, New York, 1978, **49**, Part G, Chapter 19.
[47] Boscaino R., Gelardi F. M., Mantegna R. N. *J. Magn. Reson.* 1986, **70**: 251.
[48] Boscaino R., Gelardi F. M., Mantegna R. N. *J. Magn. Reson.* 1986, **70**: 262.
[49] Galloway N. B., Dalton L. R., *Chem. Phys.* 1978, **30**: 445; *ibid.* 1978, **32**, 189; *ibid.* 1979, **41**: 61.
[50] Norman R. O. C., *Chem. Soc. Rev.* 1979, **8**: 1; *Pure Appl. Chem.* 1979, **51**: 1009.
[51] Fischer H., Paul H. *Acc. Chem. Res.* 1987, **20**: 200.
[52] Lebedev Ya S. *Prog. React. Kinet.* 1992, **17**: 281.
[53] Harbour J R, Chow V., Bolton J. R. *Can. J. Chem.* 1974, **52**: 3549.
[54] Buettner G. R. *Free Radical Biol. Med.* 1987, **3**: 259.
[55] McLauchlan K. A., Stevens D. G. *Mol. Phys.* 1986, **57**: 223.
[56] McLauchlan K. A., Stevens D. G. *Acc. Chem. Res.* 1988, **21**: 54.
[57] Trifunac A. D., Lawler R. G., Batels D. M., Thurnauer M. C. *Prog. React. Kinet.* 1986, **14**: 43.
[58] Fessenden R. W., Schuler R. H. *J. Chem. Phys.* 1963, **39**: 2147.
[59] Bartels D. M., Han P., Percival P. W. *Chem. Phys.* 1992, **164**: 421.
[60] Morton J. R., Preston K. F., Krusic P J, Knight L. B., Jr. *Chem. Phys. Lett.*, 1993, **204**: 481.
[61] Wan J. K. S., Wong S. K., Hutchinson D. A. *Acc. Chem. Res.* 1974, **7**: 58.
[62] Adrian F. J. *Rev. Chem. Intern.* 1979, **3**: 3.
[63] Buckley C. D, McLauchlan K. A. *Mol. Phys.* 1985, **54**, 1.
[64] Depew M. C., Wan J K S. *Magn. Reson. Rev.* 1983, **8**: 85.
[65] Hore P. J., McLauchlan K. A. *Mol. Phys.* 1981, **42**: 533.
[66] Yakimchenko O. E., Lebedev Ya S. *Russ. Chem. Rev.* 1978, **47**: 531.
[67] Wang Z., Tang J., Norris J. R. *J. Magn. Reson.* 1992, **97**: 322.

Further readings

[1] Allen L., Eberly J H. *Optical Resonance & Two-Level Atoms*,Dover, New York, 1987.
[2] Abragam A. *The Principles of Nuclear Magnetism*, Oxford University Press, Oxford, 1961.
[3] Weissbluth M. *Photon-Atom Interactions*, Academic Press, Boston, 1989, Chapter 3.
[4] Haar D. "Simple derivation of the Bloch equation", *Am. J. Phys.* 1966, **34**: 1164.
[5] Standley K. J., Vaughan R. A. *Electron Spin Relaxation Phenomena in Solids*,Adam Hilger, London, 1969.
[6] Hirota N, Ohya-Nishigushi H., "Electron paramagnetic resonance", in Bernasconi C F., Ed., *Techniques of Chemistry*, 4th edn, Vol. 4., *Investigations of Rates and Mechanisms of Reactions*, John Wiley & Sons, Inc., New York, 1986, Chapter 11.
[7] Freed J. H. "Molecular rotational dynamics in isotropic & oriented fluids studied by ESR", in Dorfmuller Th, Pecora R., Eds., *Rotational Dynamics of Small & Macromolecules*, Springer, Berlin, 1987, pp. 89142.
[8] Orbach R., Stapleton H J. "Electron spin-lattice relaxation", in Geschwind S., Ed., *Electron Paramagnetic Resonance*, Plenum Press, New York, 1972, Chapter 2.

[9] Bloembergen N. "The concept of temperature in magnetism", *Am. J. Phys.* 1973, **41**: 325.
[10] Muus L. T., Atkins P. W., McLauchian K. A., Pederson J. B. *Chemically Induced Magnetic Polarization*. NATO ASI Series, Vol. C34, Reidel, Dordrecht, 1977.
[11] Rabi I. I., Ramsey N. F., Schwinger J., "Use of rotating coordinates in magnetic resonance problems", *Rev. Mod. Phys.* 1954, **26**: 167.

8 Quantitative determination

Abstract: The main factors that affect the quantitative determination of the EPR spectrum will be discussed in this chapter. The chapter discusses how it can be made more accurate and how much accuracy of the quantitative measurement can be achieved.

The So-called quantitative determination that we want to know is the number of radicals in the given sample. Currently, one radical molecule, ion, or atom contains one unpaired electron. It means that is needed to determine the number of unpaired electrons in a determined sample, which is also called *spin number*. The experimental results show that the enclosed area under the EMR spectral curve is proportional to the spin number of the determined sample. Therefore, the problem of quantitative determination of EMR is summed up to measure the area under EMR spectral curves.

This area is not the spin number. It is just proportional to the spin number. For a simplest case, the sample has only one singlet spectrum. Measure the spectrum and the area under spectral curve many times repeatedly, and then take their average value. Then, a known signal sample (known the spin number/g and the *g*-factor) should be chosen as a standard sample and tested repeatedly using the same conditions, and the average value is obtained. Comparing the area of the unknown sample with the standard sample, the spin number of the unknown sample can be obtained.

In Section 1.2 of Chapter 1, the magnetic field modulation with phase-sensitive detection had been used to increase the ratio of signal/noise (sensitivity). Consequently, the absorption signal of the 100 kHz phase-sensitive detector output becomes a differential curve [1].

For the EMR spectrum of polycrystalline samples with anisotropy, its absorption curve is a broad line (dashed line in Figure 8.1), while its differential spectrum is displayed by two sharp peaks (solid line in Figure 8.1). The polycrystal samples of many transition metal ions or their complexes and their anisotropic EMR spectra have similar spectral graph. Figure 8.1 shows a spectrum of high-spin Fe(III) [1]. There are two peaks: one at $g = 2$ (near 3300 G) and another at $g = 6$ (near 1100 G). Integrating it once gives the dashed line of Figure 8.1. Integrating it a second time gives the area under the EMR curve.

For a first differential spectrum of EMR, two times integration is needed to quantify the area under the EMR curve (Figure 8.2).

The complexity of this problem is that multiple measurement results of the same sample always cannot be exactly the same. Moreover, it is very very difficult to test an unknown and a standard sample in the operating conditions exactly in the same way and get an accurate result. The main reason for this is that there are too many factors influencing the test. If the spectrum contains more than one peak, such as for a

https://doi.org/10.1515/9783110568578-008

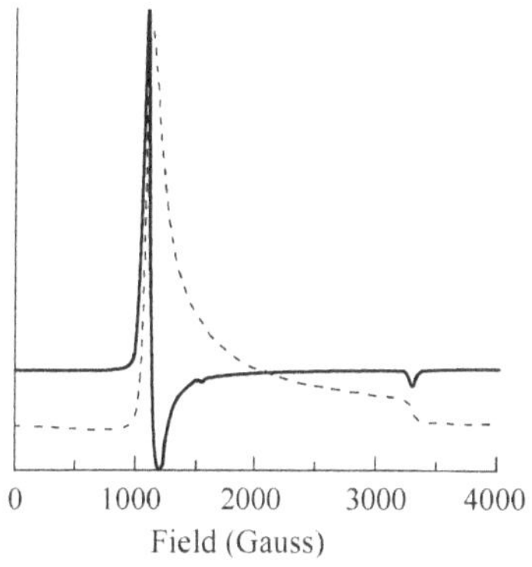

Figure 8.1: Diagram of energy exchange between a bilevel spin system and the environment.

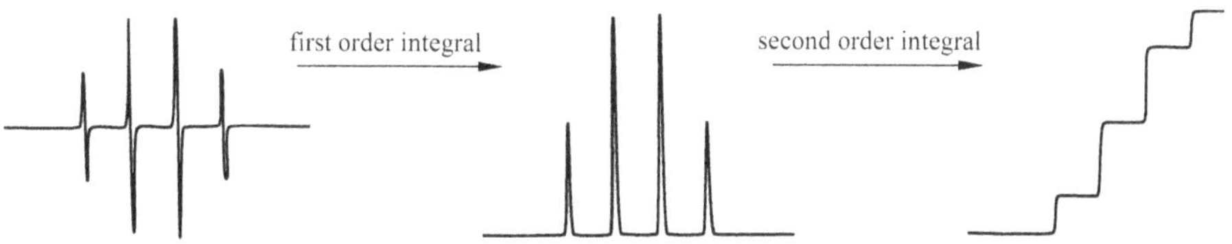

Figure 8.2: The double integration of an EMR spectrum to reveal the area.

hyperfine structure, the result will be much more complex. Therefore, if we desire to get as high as possible accuracy of the quantitative determination result, the main influencing factors of the quantitative determination result need to be looked into.

8.1 Main factors of influence for quantitative determination

There are several factors affecting quantitative determination that can be discussed roughly from the point of instrument, operation, and the selection of standard sample.

8.1.1 Factors of instrument

8.1.1.1 Effects of frequency band

The current commercial EMR spectrometer is usually configured *X*-band (the frequency is ca. 9.5 GHz). It is suitable for the samples of organic radicals. However, for transition element ions as well as their oxides and complexes, sometimes other frequency band should be chosen.

For the samples of S=1/2, all the hyperfine splitting lines of expectable sight can be observed clearly at *X*-band and above band (such as *Q*-band), and their relative intensities are also comparative. However, at lower microwave frequency band or radio frequency band, some low magnetic field hyperfine lines will be "lost." For example, Belford et al. [2] provided the EMR spectrum of ^{63}Cu in Pd(acac)$_2$ tested at different frequency bands (Figure 8.3); (a) *Q*-band (34.78 GHz): magnetic field scanning ranging from 10,500 to 12,300 G; (b) *X*-band (9.376 GHz): magnetic field scanning ranging from 2,500 to 3,500 G; (c) *S*-band (2.39 GHz): magnetic field scanning ranging from 200 to 1,400 G; (d) *L*-band (1.39 GHz): magnetic field scanning ranging from 0 to 1,000 G; (e) radio frequency band (560 MHz): magnetic field scanning ranging from 30 to 500 G. It is observed that the four parallel and four perpendicular peaks are obtained when measured by *Q*- and *X*-band. When the spectra is tested by *S*- and *L*-band, the four parallel and four perpendicular peaks are wholes or the parts overlap. It will introduce great errors when integrating seek area. The spectrum measured at radio frequency band that the signal is incomplete.

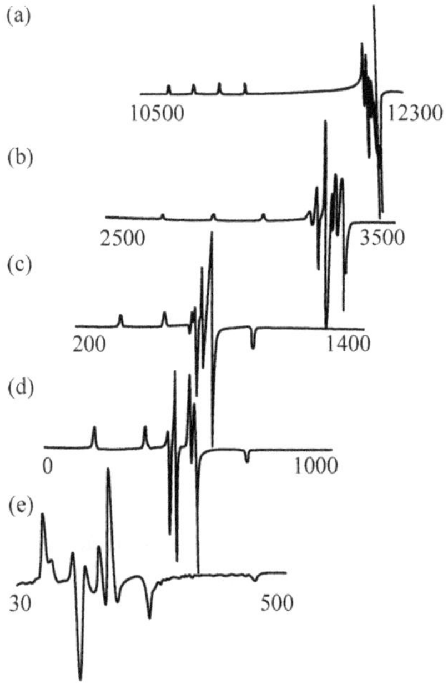

Figure 8.3: EMR spectra of ^{63}Cu in Pd(acac)$_2$ at different frequency bands.

Another example: The room temperature EMR spectra of VO(acac)$_2$ dissolved in toluene and chloroform (1:1) solution at different frequency bands [2] are shown in Figure 8.4. (a) *X*-band (9.76 GHz): magnetic field scanning ranging from 3000 to 4000 G; (b) radio frequency band (595MHz): magnetic field scanning ranging from 0 to 1200 G. When Figure 8.4b is compared with 8.4a, its hyperfine structure of EMR spectrum loses five lines. For doing quantitative determination, the spectrum must be complete. Otherwise, quantitative determination has some significance.

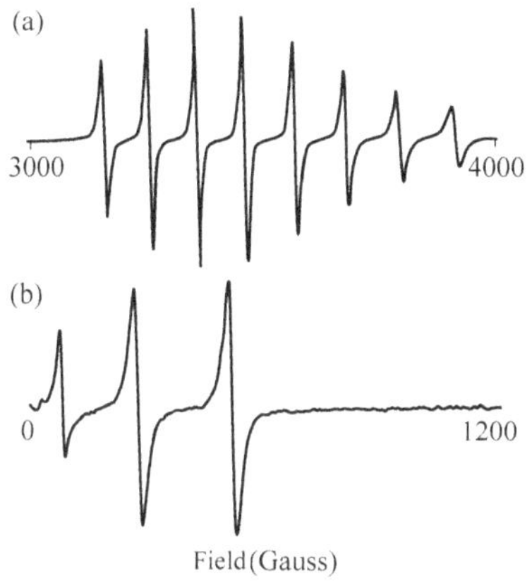

Figure 8.4: EMR spectra of $VO(acac)_2$ in toluene and chloroform (1:1) at different frequency bands: (a) X-band; (b) radio frequency band (595MHz).

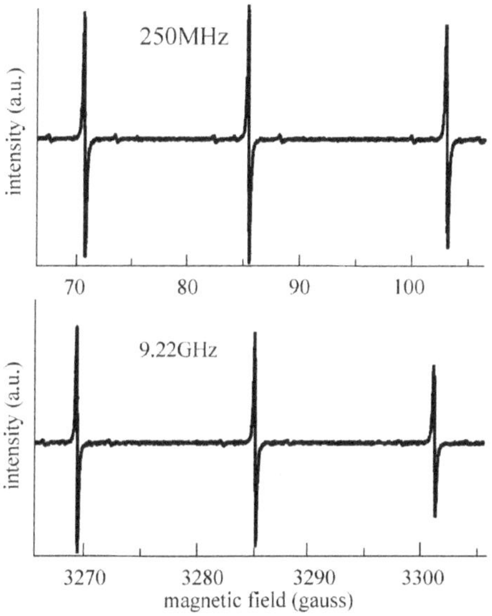

Figure 8.5: EMR spectrum of nitroxide radicals in water solution at (a) 250 MHz and (b) X-band [3].

In addition, the hyperfine coupling constant (the distance between hyperfine splitting line) is independent of the intensity of outer magnetic field. When the ratio of splitting distance to the intensity of outer magnetic field ($\mathscr{H}_0$) is quite small, the splitting distances are basically equal. They can also be measured quite easily from the experimental spectra. However, when this ratio increases to a certain level, it is found that the splitting distances are not equal. This situation occurs frequently in the spectral hyperfine splitting of metal ion at *X*-band as well as the organic radicals at *L*-band (1–2 GHz) or 250 MHz of radiation band. For the nitrogen–oxygen radicals in aqueous solution, the hyperfine coupling constant of the nitroxide radical is ~16.5 G. Their hyperfine splitting distances are equal basically at *X*-band.

At radiation band of 250 MHz, their splitting distances are not equal (Figure 8.5). This is the result because of Breit–Rabi effect [3].

Moreover, organic radicals, such as the perinaphthenyl anion radical, having the hyperfine structure of spectra in X-band do not display the forbidden transition. Some forbidden transitions are observed at L-band (Figure 8.6) [4]. This is very important for carrying out quantitative determination, since the area under the forbidden transition of spectral lines should not be counted as quantitative area. That is to say, the spectrum of quantitative measurement should not contain the forbidden transition line. Therefore, to do quantitative measurement of organic radicals, the *X*-band should be chosen rather than the *L*-band. The quantitative determination of transition metal ions (such as Mn^{2+}) in t *X*-band still display forbidden transition lines. So, to do the quantitative determination of transition metal ions, the *Q*- and above band should be chosen to avoid the appearance of forbidden transition line.

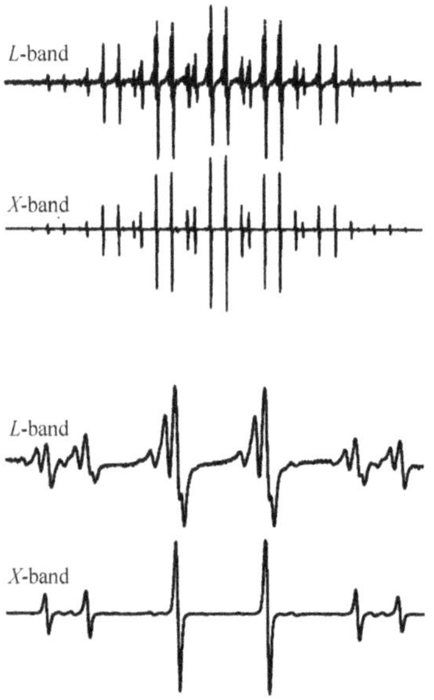

Figure 8.6: EMR spectrum of perinaphthenyl at L-band and X-band.

8.1.1.2 *Q*-value of resonator

The *Q*-value is used to measure the *quality factor* target of the resonator. Its relation with other factors depends on the design, material, manufacturing technology, and installation process of the resonator. Different types of resonators posses different *Q*-values. The *Q*-value of the H_{011} type cylindrical resonator is over 10,000, whereas the *Q*-value of the TE_{102} rectangle resonator ranges from 2000 to 4000.

The higher the *Q*-value of the resonator, the higher its sensitivity, as well as the stronger the signal of the sample (the larger the area under the EMR curve). When an EMR spectrum test is conducted, it is expected that the high the *Q*-value the better.

While doing a quantitative measurement, the factors that influence the variation of Q-value should also be considered.

After the manufacture and installation of a resonant cavity is complete, it can be said that its Q-value has been determined basically. This Q-value is called the unloaded Q-value (Q_U). When a sample tube (filling sample or not) is inserted into the resonator, its Q-value decreases greatly, to about half the Q_U value (expressed by Q_L).

There are many ways to express the Q-value of a resonator. Here, in terms of RLC, the Q-value of the resonator is expressed as follow:

$$Q_U = \frac{\omega_0 L}{R} \tag{8.1}$$

$$Q_L = \frac{\omega_0 L}{R + r_{\text{sample}}} \tag{8.2}$$

where ω_0 is the angular frequency of the resonator, L is the inductance of the resonator, R is the resistance of the resonator, and r_{sample} is the resistance of the sample.

In fact, the factors contributing to Q_L are very complicated. It could be expresses by following equation [5].

$$\frac{1}{Q_L} = \frac{1}{Q_U} + \frac{1}{Q_\varepsilon} + \frac{1}{Q_r} + \frac{1}{Q_\chi} + \frac{1}{Q_\mu} \tag{8.3}$$

Q_U is the unloaded Q-value; Q_ε is the effect of dielectric losses on the Q-value; Q_r is the effect of power lost through the cavity coupling hole (Q_r is also called the *radiation quality factor.*) on the Q-value; Q_χ is the effect of power absorbed by the sample at resonance on the Q-value; Q_μ is the effect on the Q-value that arises from surface currents in high-conductivity samples, it is proportional to the H_1^2.

Q_r effect on the Q-value of resonator is analogous to the sum of all other terms, so Eq. (8.3) can be rewritten approximately as follows:

$$\frac{0.5}{Q_L} = \frac{1}{Q_U} + \frac{1}{Q_\varepsilon} + \frac{1}{Q_\chi} + \frac{1}{Q_\mu} \tag{8.4}$$

Q_μ can be ignored for low conductivity samples, and for small samples the Q_χ also can be ignored, so that Eq. (8.4) can be rewritten as

$$\frac{0.5}{Q_L} = \frac{1}{Q_U} + \frac{1}{Q_\varepsilon} \tag{8.5}$$

When doing a quantitative measurement, in spite of using the same resonator for unknown and standard samples, however, due to different samples and a slight difference in operating conditions (such as the depth of the sample insertion and so on), different Q-values are obtained. Under a situation where large difference in Q-values is obtained, it is necessary to measure the Q-value of resonator in both cases. For method of measuring Q-value, please refer to Eaton *et al.* [4]

8.1.1.3 Filling factor

Filling factor (η) is another important element that affects the Q-value (that is, EMR signal) of the resonator. If the microview field H_1 is homogenous distribution in the resonator, the filling factor should be the ratio of the volume of sample to the volume of resonator. However, unfortunately, microview field H_1 distribution is not homogenous in a resonator; the calculation of filling factor becomes more complicated. Rigorous to say, the filling factor should be the ratio of square integration of microview field distribution on whole volume of the sample to square integration of microview field distribution on whole volume of the resonator. It can be expressed as follows:

$$\eta = \frac{\int_{\text{sample}} H_1^2 \mathrm{d}V}{\int_{\text{cavity}} H_1^2 \mathrm{d}V} \tag{8.6}$$

For a low dielectric constant sample filling in a 4 mm o.d. standard quartz tube, insert in an X-band rectangular resonator (TE_{102} cavity) the filling factor of about 0.01. For a sample with relatively greater dielectric constant, the filling factor is increased by the influence of the dielectric on the microwave distribution in the cavity [5]. For a small spherical sample in a rectangular cavity, the filling factor is given by Nagy [6] as

$$\boldsymbol{\eta}(\boldsymbol{\varepsilon}', R) = \frac{16\pi R^3}{3V_{\text{cavity}}}\left(1 - \frac{k^2R^2}{5}\boldsymbol{\varepsilon}'\right) \tag{8.7}$$

where R is the radius of the sample and $\boldsymbol{\varepsilon}'$ is the dielectric constant of the sample. Eq. (8.7) shows that the filling factor ($\boldsymbol{\eta}$) decreases with increase in either the radius or the dielectric constant($\boldsymbol{\varepsilon}'$). It is in contrast to the well-known fact that the microwave field (H_1) in TE_{102} cavity is increased in the cylindrical or flat sample. Nagy [6] gave Eq. (8.7) based on the experimental data of a spherical sample. Subsequently, Yordanov and Slavov [7] pointed out that the intensity of signal increases with increasing diameter of sample tube. The sample tubes used are all cylindrical quartz tubes.

To achieve a high filling factor, the resonator is designed such that it optimizes the volume ratio of sample tube with the cavity, and then the conversional efficiency is increased and the dielectric losses decreased.

To do the quantitative measurement, we need to compare the area under the spectral curves of unknown sample with standard sample. It is required to keep the

Q-value of the resonator both times the same. Actually, it is impossible to replicate the results of both samples, but a small difference is acceptable.

8.1.1.4 Size of sample tube

Under the common situation, for condensed state (solid powder or homogeneous liquid) samples, generally use quartz capillary of 1 mm inner diameter and filling sample height of ~10 mm. Then, place the quartz sample in a sealed capillary tube with 4 mm diameter. For quantitative measurement, the size of the sample tubes should as far as possible be the same for each measurement.

When the detected sample is a dilute aqueous solution, the nonresonant absorption of the sample in the resonator is quite large and cannot be measured. Using a flat-shaped quartz cell as sample tube can greatly improve the operation. The flat-shaped quartz cell should be used for both unknown and standard samples. This flat-shaped quartz cell is very sensitive for orientation. When the plane face of the flat cell is at right angle to the wave-nodal of microwave, the signal will be greatest [8, 9]. If there is a slight deviation, the signal is not the highest. Therefore, it is highly required for operation.

8.1.1.5 Double-cavity resonator (TE_{104})

A double-cavity resonator, TE_{104}, consists of two TE_{102} cavities joined together. The original intention of suggesting this design is a very beautiful idea. Because the frequency and power of microwave for both cavities are all equal, the magnetic field around double cavity becomes homogenous easily. That is to say, the intensity of magnetic fields of both are also equal. For quantitative measurement, one of these two cavities contains a reference standard sample, and the other contains the unknown sample. Thus, it can avoid introducing more errors owing to multi operations in the same cavity.

The experiment results, however, do not turn out as expected. The expectation is that when the two sample tubes filled with standard sample and unknown sample were inserted to these two cavities respectively for test, the microwave field would focus on the center of both sample tubes at the same time. Unfortunately, the focus point always deviates the sample tube center by 1–2 mm [10] due to the differences of situation and depth of the samples to be inserted (shown as Figure 8.7 [11].) So *the double-cavity resonator cannot be used for quantitative determination.*

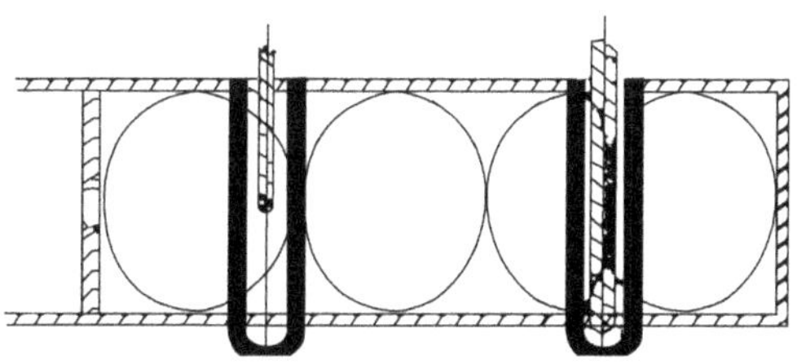

Figure 8.7: Double-cavity resonator (type TE_{104}).

8.1.2 Influence of operating factors

The influence of operating factors includes how to choose the appropriate operating parameters to optimize an EMR experiment when the spectrometer has been identified.

8.1.2.1 The choice of microwave power

If the experimental goal is to determine the quantitating spin concentration, in order to avoid power saturation, it must record the spectrum under as low microwave power as possible. Because of most EMR samples, almost all can be saturated under the case of the power levels provided by the commercial spectrometer. Thus, before the experiment is conducted, it is always important to check whether the microwave power has reached saturation or not.

In case saturation is not attained, the magnitude of EMR signal is increased linearly with increase in microwave field H_1. It is very important, the efficiency of microwave power convert to H_1 in the resonator. It of course is decided at first by design, material, and manufacturing technology of the resonator. However, the high Q-value and filling factor of the resonator in the operating case are also very important.

If the unsplitting hyperfine structural spectrum does not exist, the relationship between linewidth, H_1, and relaxation times is

$$(\Delta H_{pp})^2 = \frac{4}{3y^2T_2^2}(1 + y^2H_1^2T_1T_2) \tag{8.8}$$

where H_{pp} is peak-to-peak linewidth and γ is electronic magnetogyric ratio. If the H_1 is small enough that the $y^2H_1^2T_1T_2 \ll 1$, then the signal is unsaturated. The expression for saturation factor s is

$$s = \frac{1}{1 + y^2H_1^2T_1T_2} \tag{8.9}$$

The saturation factor $s < 1$; it means that the microwave power has not yet reached saturation. The s-values from 0.95 to 0.98 are all acceptable. In order to obtain stronger signals, larger microwave power should be used. However, when the microwave power is large enough to a certain degree, saturation can be reached. Before the saturation point, the signal amplitude increases proportional to the square root of the microwave power magnitude. After the saturation point, with increase in microwave power, the signal amplitude decreases (refer to Figure 8.8).

8.1.2.2 Choice of field modulation amplitude and microwave phase angular adjustment

The modulation amplitude has a great influence on the linewidth. The signal amplitude increases with increase in modulation amplitude, but over a certain limit, the

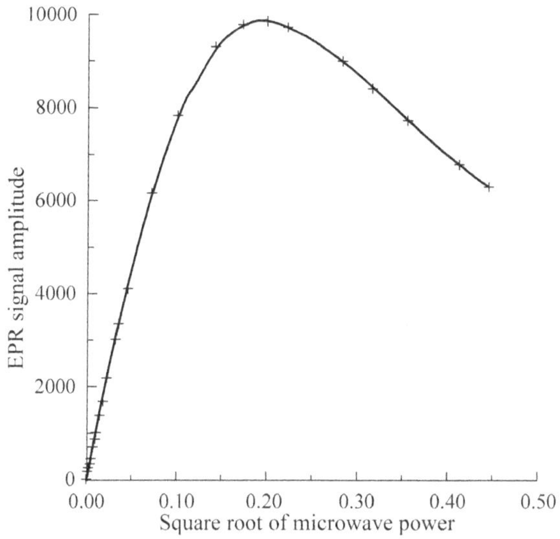

Figure 8.8: Plot of signal amplitude as a function of the square root of the microwave power.

spectral lines start broadening. When the modulation amplitude is about one-third of the linewidth, a Lorentzian spectral line is broadened by about 3%, and when the modulation is about two-thirds of the linewidth, the line is broadened by about 11% [12,13]. The spectral line shape would be distorted (Figure 8.9). Therefore, when making the quantitative measurements, the modulation amplitude should not be exceed one-fourth of the spectral linewidth.

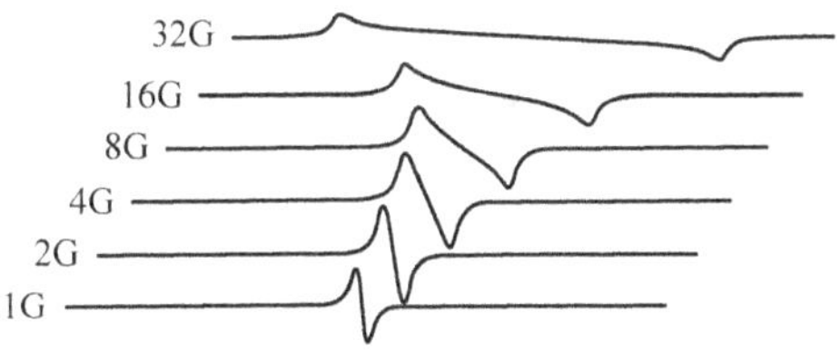

Figure 8.9: Signal shape of the DPPH EMR signal as a function of increasing modulation amplitude.

A dependent relationship exist between the phase of reference signal and the magnitude of modulation amplitude. So, when adjusting the magnitude of modulation amplitude simultaneously, one must pay attention to keep the phase angle constant (phase angle equal to 0∘) by timely turning off the knob of the phase angle. The relationship between signal intensity and phase angle is shown in Figure 8.10.

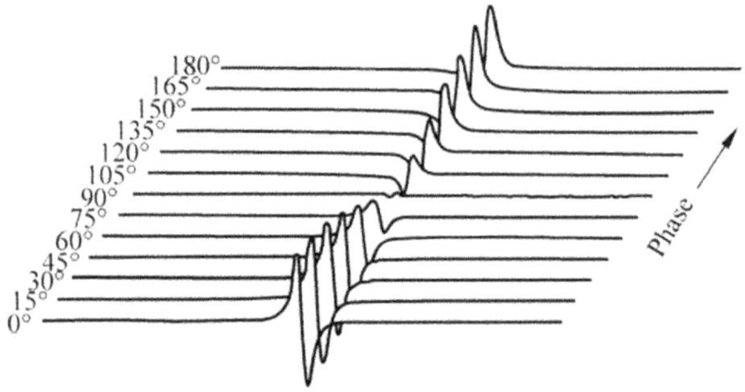

Figure 8.10: Relationship of the magnitude of signal amplitude with the variation of phase angle.

8.1.2.3 Choice of the field modulation frequency

Most *X*-band EMR spectrometers are using 100 kHz as the modulation frequency. However, for a spectrum with very narrow lines (<35 mG), such as the trityl radical, LiPc (lithium pthalocyanine), or N@C_{60}, the line shape can be distorted by the modulation sidebands. Figure 8.11 shows the spectra of N@C_{60} measurement with modulation frequencies of 100, 50, and 20 kHz. It can be seen that the measured spectrum with 100 kHz modulation frequency appear as modulation sideband. For spectrum with 50 kHz modulation frequency, although not appearing as modulation sideband, the spectral line was broadened. Only when 20 kHz modulation frequency is used does measured spectrum not distort. It is clear that for the sample of very narrow line, should be used more low modulation frequency such as 20 kHz.

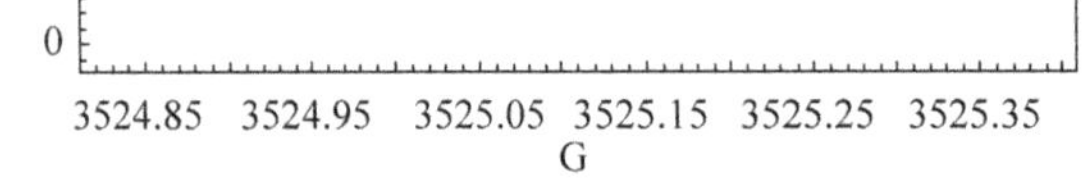

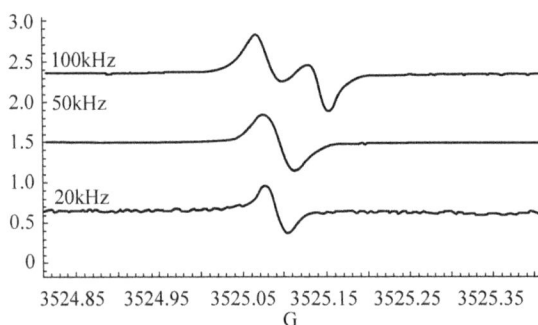

Figure 8.11: Signal distortion coursed by the increasing modulation frequency of magnetic field.

At the situation of high modulation frequency and high modulation amplitude, the rate of passage through the EMR line will be faster than the relaxation rate, which could arise as a result of so-called “passage effects” [14]. For example, a 1 G sinusoidal modulation at 100 kHz is a magnetic field scan of $\pi \times 10^5$ G/s at the point of maximum scan rate. It is easily demonstrated by the narrowline samples with long relaxation times. For slow-passage EMR, it is necessary to have the reciprocal of the modulation frequency much greater than T_1.

$$\nu_m^{-1} \gg T_1$$

This criterion is not met as often as it is assumed to be. The so-called *relaxation-dependent* phenomena [14] are easily accessible experimental situations, especially at cryogenic temperatures. Some samples at liquid nitrogen temperature, and many samples at liquid helium temperature, have long T_1 time. Due to the so-called *passage effects*, recognizable as distortions of line shapes, there is need to use a lower modulation frequency.

The previous section mentioned the selection of microwave power, modulation amplitude, and modulation frequency. The choice of these three will be mutual diversionary. A very typical example is the EMR spectrum of galvinoxyl radical (also called Coppinger’s radical) shown as Figure 8.12. This spectral figure provides two inspirations: First, for the resolution of spectrum, using 10 kHz modulation frequency is better than using 100 kHz. Second, an initial acquisition using 1.0 mW microwave power and 100 kHz modulation frequency would not reveal the additional hyperfine splitting due to the protons of *t*-butyl radical. In fact, the maximum spectral microwave power using 1.0 mW has been over the saturation point when the modulation frequency is 100 kHz. Thus, the hyperfine splitting of proton on the *t*-butyl radical can be removed by power saturation, while using microwave power of 0.1 mW; the modulation frequency of 10 kHz is the best condition choice. For the quantitative measurement of very narrow lines, operating conditions should be chosen with great caution.

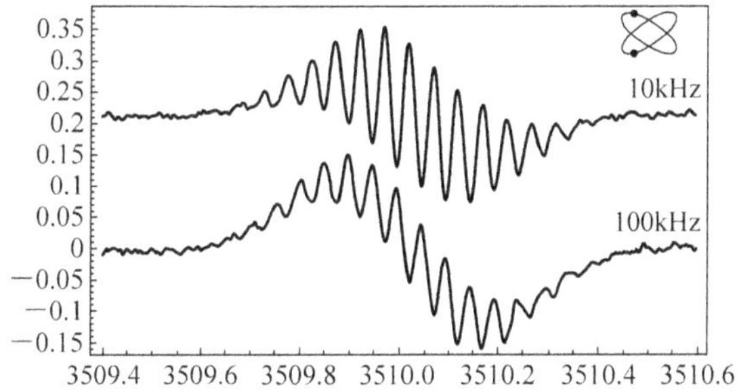

Figure 8.12: EMR spectrum of the Coppinger radical.

8.1.2.4 Automatic frequency control

The automatic frequency control (AFC) circuit is designed and manufactured in order to keep the frequency of microwave source (klystron or Gunn diode) locked to the resonant frequency of the resonator. The resonant frequency of the resonator can be changed by the temperature changes of resonator body, or changes of sample temperature or breed, and so on. The AFC is required to guarantee the frequency of resonator is kept constant. The relationship between the amplitude of the EMR signal and the frequency is that of cosinusoidal dependence, so it needs to be taken into account as a slight change in the AFC correction signal would cause a big change in EMR signal. However, the AFC correction signal has some dispersion signal that can merge with the EMR absorption signal. If the sample has been saturated, this merging should be noticed, because the dispersion signal could not be saturated again to the saturated absorption signal. For example, the irradiated fused silica standard sample [15] has long relaxation times (T_1 and T_2), and its CW absorption spectrum should be mixed with the dispersion spectrum unless the frequency is set and controlled very carefully. Hence, this sample is a good functional monitor of the AFC circuit of the spectrometer [16].

When the electron-spin relaxation time T_1 as well as the AFC correction signal is large enough, the dispersion mixing into absorption spectra can be perceived very sensitively, and the absorption signal is partially saturated by the dispersion signal. Figure 8.13 a and b are spectra of the same sample measured with two different spectrometers: spectra 1, 2, and 3 are the experimental results derived from microwave frequencies that differ by only a few megahertz [16].

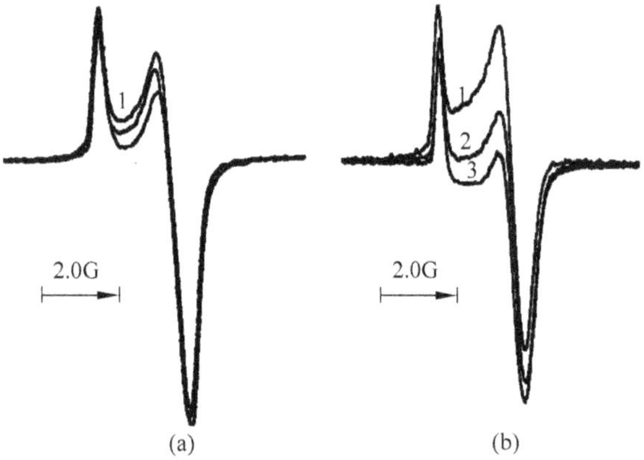

Figure 8.13: Absorption signal is partially saturated by the dispersion signal.

It is obvious that in the quantitative measurements, especially for the sample with longer relaxation time, the AFC needs to be very carefully set and controlled.

8.1.2.5 Detector current

The output current of the crystal detector in the microwave bridge depends upon the magnitude of the bias current in the crystal detector. Each spectrometer should be checked to determine the range of detector current values within which the signal amplitude is independent of detector current. If the detector current drifts, as can happen with lossy solvents, or when the temperature is changed, significant errors in signal amplitude can result; signal-to-noise (S/N) ratio is degraded, and quantitative measurements are prevented. The output of the detector crystal is dependent on temperature. Since the bridge warms up during the first hour or so after power is turned on, the accuracy of quantitative spectra may change during this period. Whether the accuracy of quantitative measurements can meet the requirements should be determined for long time testing using the spectrometer with bad radiation.

8.1.2.6 Optimum choice of the time constant and sweep time

Decisions concerning sweep time, time constant, and S/N ratio are correlated. The time constant and the sweep time are optimized to improve the S/N ratio. For a strong signal, it may be possible to achieve adequate S/N with a single scan, modest filtering, and a relatively fast scan. However, for weaker samples, decisions need to be made about trade-offs between filtering, scan rate, and signal averaging [3].

Signal-to-noise ratio can be improved by using a long time constant, slower scans, and fewer averages; or by using a small time constant, faster scans, and more signal averaging. With a perfectly stable sample and stable instrument, roughly equal time is involved in either method of S/N improvement. The problem is that perfect stability is not achieved, and the filtering discussion focuses on high-frequency noise. The long-term spectrometer drift due to air temperature changes, drafts, vibration, line voltage fluctuations, and so on limit the practical lengths of a scan. Drifts in the magnetic field magnitude are not averaged by filtering or slow scans, and always increase apparent linewidth. In addition, computer-collected spectra can subsequently be digitally filtered without changing the original data, whereas analog-filtered data are irreversibly modified.

If the sample decays with time, a separate set of problems emerges. Assume, for example, that the goal is to compare line shapes of two peaks in a noisy nitroxide EPR spectrum, and that the amplitude of the spectrum is changing with time due to chemical reaction (shifting equilibria, decay, oxygen consumption or diffusion, etc.). In this case it would be wise to scan the narrowest portion of the spectrum that will give the information of interest. The impact of the time dependence can be minimized more effectively by averaging rapid scans than by filtering a slow scan.

In older spectrometers, the sweep time was an independent parameter. In newer spectrometer, the sweep time is digitally controlled and it is determined by the dwell time at each field step. For Bruker spectrometers, an integrating digitizer is used and the integration time of the digitizer is the "conversion time," which is also the magnetic field dwell time. The integration time is the time during which the analog-to-digital converter (A/D) accumulates the signal and noise at each magnetic field step of the EPR experiment. The time for a magnetic field scan is the product of the conversion time (in seconds) and the number of data points acquired on the magnetic field axis. For example, a conversion time of 81.92 ms for 2048 magnetic field steps results in a scan time of 167.8 s.

The decrease in the standard deviation of stochastic noise is proportional to the square root of the time constant. Thus, if the filter time constant is increased by a factor of 4 from 20.48 to 81.92 ms, the noise will be reduced by a factor of 2. When the time constant is longer than the conversion time, the S/N improvement is proportional to the square root of the filter time constant.

If a time constant is selected that is excessively high relative to the sweep time, the signal may be filtered out! The time constant should be adjusted to be consistent with the A/D conversion time. The time constant should be sufficiently long to filter out undesirable noise, yet short enough that it does not distort the signal. Therefore, if a longer time constant is needed, the scan time (conversion time) should be increased. Figure 8.14 shows the effect of progressively increasing time constant (while maintaining the same sweep time).

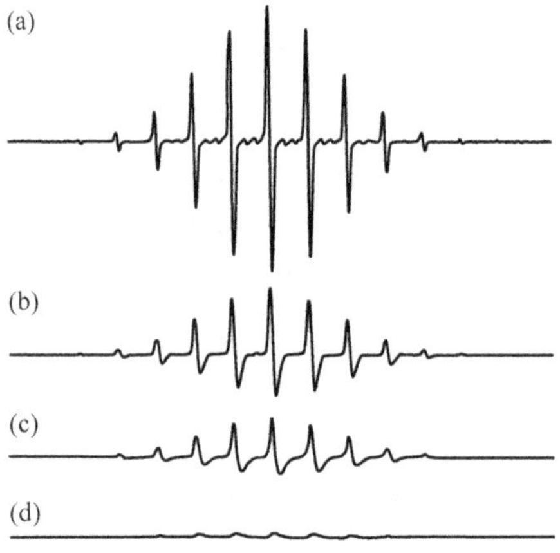

Figure 8.14: The effect of progressively increasing the time constant while maintaining the same sweep time.

Although the S/N increases when the time constant is increased, the spectrum is distorted if the time constant is too long (see Figure. 8.14c and d). As a "rule of thumb," the time constant should be chosen to be less than about 1/10 the time it takes to scan through the narrowest line in the spectrum. Consequently, scan rate and filter time constant are related to each other and to the CW linewidth in the following formula:

$$\frac{(\text{spectrum width in G})}{(\text{line width in G})} \times \frac{(\text{time constant in s})}{(\text{sweep time in s})} < 0.1$$

For example, to record a 20 G scan of a 0.1 G wide line in 84 s, the time constant should be less than 0.04 s: $(20 \times 0.04)/(0.1 \times 84) = 0.095 < 0.1$.

If a 40.96 ms conversion time is used, the scan will be 84 s if 2048 steps are used. Likewise, 1024 steps would result in a scan of 41 s, which would not be conservative with respect to line shape (and position) distortion.

This explains that in order to get a correct (undistorted) EMR spectrum, it is very important to optimize the scanning time, the time constant, the conversion time, and the relationship of these three. Especially, in the case of making quantitative measurements, the spectrum should be undistorted, the measurement conditions of the unknown as well as the standard sample should also be consistent, and both the integral data should be comparable.

Temperature has a great influence on the intensity of the signal. It is possible to test the unknown sample and the standard sample for the quantitative measurements at the strictly consistent temperature. It is self-evident.

8.2 Selection and preparation of standard samples

There is a paper in the literature that mentions a small crystal of $CuSO_4 \cdot 5H_2O$, properly prepared, was used as an intensity standard to measure the number of spins in an aqueous protein sample at room temperature, without paying any attention to the differences in resonator Q or H_1 and modulation amplitude distributions for the two samples with very different geometries and dielectric loss properties. Now it seems that the quantitative results are meaningless.

Absolute quantitation of EPR intensity is very difficult. In most cases, the spectral area of an unknown sample is compared with the spectral area of a known sample. Many samples proposed as standards for various purposes were discussed [17–20]. Note that weak pitch is a spectrometer performance standard, not an intensity standard. The derivative signal amplitude is calibrated under the stated spectrometer conditions for the "weak pitch S/N" test. Salesman of some spectrometer companies used to *flicker* users (acceptor) saying "this spectrometer has reached the required sensitivity on the contract." It is unscientific, since they are not at all known as yet how many spins in the unit weight of "weak coal," unless a known standard sample is used to calibrate the integrated area of "weak coal."

In the 1990s, the performing quantitative EPR measurements has been discussed extensively and deeply, including how to choose an appropriate reference standard sample [20–36].

As a prerequisite, the standard sample should be stable in general case and be measured accurately. From the above discussion, it follows that the choice of standard samples, the state (liquid, solid, glass, powder, crystal, etc.), and the properties (magnetic and electrical properties, especially dielectric properties) must be as far as possible close or similar to the unknown samples.

The *g*-value of DPPH is 2.0036 and this is being used as a *g*-value marker even today. Could it also be used as a quantitative standard sample? The problem is the purity of the solid or the stability of the solution of commercial DPPH is far less than the demand of standard sample [32,34,35] for quantitative measurement. It must be purified by recrystallization. Although the solid DPPH is stable enough, the stability of its solution is dependent on the solvent strongly [33,37]. As Slangen [38] reported, its acetonitrile solution is much more stable than in toluene solution. Yordanov [33,34] summarized prior literature on stability and spectroscopic molar absorptivity of DPPH, and reported $\varepsilon = 12350 \pm 650 M^{-1}cm^{-1}$ at 520 nm in ethanol should be used to determine its purity.

The high purity of nitroxide radicals can be sufficiently used as standards for quantitating organic radicals in fluid and frozen solutions. Nitroxide radicals can be selected for solubility in the solvent of interest (ranging from toluene to water) so that the dielectric properties of the known and unknown samples can be matched. Commercial Tempol is purer and easier to handle than tempone, since it is a higher melting crystalline solid, and is soluble in both water and alcohols.

K_3CrO_8 [39,40] has $g = 1.97$, which is a far different from the *g*-value of many organic radical spectra that can be measured at the same time, which is an aid to reducing uncertainties due to resonator *Q*-value, and so on.

For transition metal samples, such as $S = 1/2$ of Cu(II) and $S = 5/2$ of Mn(II) are commonly used as standard sample. $MnSO_4 \cdot H_2O$ is available with 99% purity. However, it is somewhat efflorescent, so uncertainty in the degree of hydration could cause uncertainties in spin quantitation. Yordanov [41] documented some of the problems with Mn(II) standards. It might be more practical for EMR labs to purchase analytical standard solutions such as Alfa Specpure standards.

Some labs try to use $CuSO_4 \cdot 5H_2O$ as a standard, but it is very difficult to prepare and store it in an atmosphere of relative humidity to ensure that it is really the pentahydrate. $CuCl_2 \cdot 2H_2O$ has also been cited as a primary standard, but its degree of hydration also depends strongly on the relative humidity of the air in which it is stored. For the most accurate work, it is better to dissolve a weighed amount of Cu metal to use as a standard. For nonaqueous samples, compounds of the highest available purity should be used, and for the most accurate results they should be analyzed for metal content.

$VOSO_4 \cdot nH_2O$ also has variable number of waters of hydration, so Dyrek et al. [23] titrated it with $KMnO_4$ in the usual volumetric analysis method to determine the V(IV) content.

The compound was used as the solid, and in some cases ground with NaCl or KCl. Dyrek et al. [24,25] found that the choice of diamagnetic diluent for a standard was not simple, and that the main problem is achieving a homogeneous distribution of paramagnetic component in the diamagnetic diluent. Grinding samples together to get a more homogeneous standard may result in chemical reactions that change the EPR spectra. The grinding process itself can produce radicals

[42–46]. For example, Dyrek et al. [25] observed that the $CuSO_4$ EMR spectrum changed when it was ground with NaCl, but it did not change when ground with SiO_2 that had been pulverized before the two were ground together. Traces of water in the materials changed the spectrum of a $VOSO_4/K_2SO_4$ sample that was sealed in a quartz tube for a few weeks. The line shape changed, but the integrated intensity did not change. Drying over P_2O_5 resulted in samples that were stable for 5 years. In spite of sample preparation difficulties, Dyrek et al. [24] concluded that preparation and measurement of the standard contributed only about 2% error to the overall quantitation of EMR spectra of metals.

For $S > 1/2$, it is important to take into account the differences in transition probabilities as demonstrated by Siebert et al. [47]. Chromium in FeS_2 and in $AlCl_3 \cdot 6H_2O$ was measured by EMR and by ICPMS and AAS, with good agreement between methods. The National Bureau of Standards (now NIST) produced an EPR standard sample using Cr^{3+} in Al_2O_3 in 1978 [48], which some labs still have (Standard Reference Material 2601). Nagy and coworkers [28–31] discussed choosing reference samples for EPR concentration measurements.

Cordischi et al. [49] compared a large series of compounds as possible primary standards for quantitative EPR of $S = 1/2$, 3/2, and 5/2 species. The $S = 1/2$ (listed in Table 8.1) species and $S = 5/2$ $MnSO_4 \cdot H_2O$ were reliable standards, but "none of the pure Cr^{3+} compounds proved useful as primary standards because of their large fine structure terms or high Néel temperature that invalidated the simple Curie law. Among the Cr compounds tested were $Cr(SO_4)_2 \cdot 12H_2O$; $(NH_4)_3[CrMo_6O_{24}H_6] \cdot 7H_2O$, and $Cr(acac)_3$。

As an internal standard for samples to be irradiated, Yordanov et al. [36] suggest Mn(II) doped into MgO. They concluded that pyrolyzed sucrose and Mn (II) in CaO were less suitable. Irradiation of CaO induced EMR signals that could interfere with measurements of organic radicals, such as in alanine. There was no observable change in the intensity of the EPR signal in pyrolyzed sucrose upon irradiation, but the EMR signal of this material would overlap organic radical signals.

8.3 Key parameters and its effect on the intensity of EMR signal

Many parameters of spectrometer and sample as well as the interactions between the sample and spectrometer have great influence on the quantitative accuracy. Some of these parameters are controlled by the operator, while others are not controlled, However, we should know how to perform quantitative measurements more accurately. For example:

The preparation of the sample (including the selection and preparation of sample tubes) can be controlled by the operator. *Selection of the resonator type* should reach minimum perturbation of the microwave fields in the resonator. For

Table 8.1: EMR intensity and *g*-value of standards.

Name of standard sample	Intensity	*g*-Value		References
DPPH		2.0037±0.0002		[53]
DPPH	No	2.0036±0.0001		[33]
DPPH	√			[49]
Bruker or Varian strong pitch	±10%	2.0028		
Wurster's blue perchlorate		2.00305±0.00002		[54,55]
Perylene cation in 98% H_2SO_4		2.00258±0.00002		[56]
Tetracence cation in 98% H_2SO_4		2.00260±0.00002		[56]
p-Benzosemiquinone in butanol-KOH at 290 K		2.004679±0.0000129		[52]
Naphthalene anion radical		2.002757±0.0000006		[52]
Fremy's salt $K_2NO(SO_3)_2$	√	2.0057	26.182G	[52,54,55]
Quinhydrone at pH 7.2	√			[57]
K_2CrO_8	√	1.97		[58,59]
$CuSO_4 \cdot 5H_2O$	√	$g_1 = 2.27$; $g_2 = 2.08$		[52]
$CuSO_4 \cdot 5H_2O$ solid, aqueous solution	√	2.09; 2.23; 2.27		[49]
$Cu(acac)_2$ solid or in toluene solution	√	2.13; 2.126	A=78G	[49]
Cu metal dissolved in acid	√			Eaton Lab
Cr^{3+} in Al_2O_3	√			[48]
Mn(II) doped into MgO	√			[36]
$MnSO_4 \cdot xH_2O$	√	2.0023		[52]
Tempo	√			[49]
Tempone	√			
Tempol	√			
3-Trimethylamino-methyl-2,2,5,5-tetramethyl-1-pyrrolidinyloxyl iodide	√			
3-Carbamoyl-2,2,5,5-tetramethylpyrrolinyloxyl	√			
3-Carbamoyl-2,2,5,5-tetramethylpyrrolidinyloxyl	√			[60]
$VOSO_4 \cdot nH_2O$	√			[23]
$VOSO_4 \cdot 5H_2O$	√	1.99		[49]
$VO(acac)_2$	√	2.00		[49]
VOTPP(tetraphenyl porphyrin)	√	1.958	*A*=183G	[49]
Vanadyl and copper sulfate	√			[24]
Fusinite	√			[21]
$MnSO_4$	√			[52]
Cu(diethyldithiocarbamate)$_2$	√			[52]
Pyrolyzed sucrose or dextrose	√	2.0028		[52]
Ultramarine blue	√	2.0294		[52]
F-centers in LiF	√			[52]
γ-Irradiated alanine	√			

accurate comparisons, the *Q-value of resonator must be measured.* While selecting the standard sample it should be considered that its spectral width, intensity, and dielectric loss are as far as possible similar to the unknown sample, and also an operator should be selected. However, the relaxation property of the sample is that the operator cannot control, but the sample is *not power-saturated* when making quantitative comparisons, or its power-saturated degree is as far as possible same or similar to the reference standard sample.

The following are the parameters (or operating conditions) that the operator must always keep in mind to select, tune, and control seriously, carefully, patiently:
- Modulation amplitude and frequency
- Scan time and time constant of detector
- Magnetic field scan width
- Gain of receiver
- Detector current
- AFC offset
- Tuning of microwave frequency and resonant cavity operating mode
- Microwave power and H_1 at the sample
- Sample position in the cavity
- Type of cavity
- Physical size of sample
- Dielectric properties of the solvent at the microwave frequency
- Temperature of the sample and its effect on sample concentration, species dissociation, paramagnetism, and dielectric loss

The early EMR experimental works did not pay much attention to quantitative accuracy of spin concentrations. Goldberg [50] documented the quantitative data based on older Varian spectrometer in 1978 that were not certain. Until 1994, better results were obtained by Yordanov and Ivanova [41]. They compared and contrast the methods and results of quantitative measurement presented by intra- and interlaboratory in the previous literatures report and focused on crucial problems substantially such as the resonator *Q*, sample positioning, reference standards, and then finally the accuracy of quantitative measurement could be improved.

8.4 Achievable accuracy of quantitative determination

Gancheva et al. [51] used current alanine dosimeters containing Mn^{2+}/MgO as an internal standard that were measured on 12 instruments of 6 different models by 3 manufacturers and run by 10 operators. With modern spectrometers, and careful attention to experimental conditions, interlaboratory comparisons have achieved 3% standard deviation on samples with internal standards. This 3% standard deviation was obtained only after a preliminary comparison showed the need to use a procedure that involved low power, low modulation amplitude, and a combination of sweep time and time constant that provide a distortion-free spectrum. Yordanov et al. [36] reported that the same operator on the same instrument can achieve about ±2% reproducibility on the same sample.

Here the so-called standard deviation is the upper limit of the achievable accuracy of the measurement, because there are many steps in the process of operation, and each step must be handled with care. A little mistake will lead to greater error. It is better if the final standard deviation of the quantitative measurement is 3%.

References

[1] Eaton S. S., Eaton G. R. "Electron paramagnetic resonance," in Cazes, J., Ed., *Analytical Instrumentation Handbook*, 3rd edn, Marcel Dekker, New York, 2005, pp. 349–398.

[2] Belford R. L., Clarkson R. B., Cornelius J. B., Rothenberger K. S., Nilges M. J., Timken M. D. *EPR over Three Decades of Frequency: Radiofrequency to Infrared*, in Weil, J A, Bowman, M. K, Preston, K. F., Eds., *Electron Magnetic Resonance of the Solid State*, Canadian Society for Chemistry, Ottawa, 1987, pp. 21–43.

[3] Eaton S. S., Eaton G. R., Barr D. P., Weber R. T. *Quantitative EPR*, Springer, New York, 2010, Chapters 7 and 10.

[4] Eaton S. S., Eaton G. R. Multi-frequency EPR Workshop, 24th International EPR Symposium, Denver, 2001, pp. 7–8.

[5] Dalal D. P., Eaton S. S., Eaton G. R. The effects of lossy solvents on quantitative EPR studies. *J. Magn. Reson.* 1981, **44**(3): 415–428.

[6] Nagy V. Quantitative EPR: Some of the most difficult problems. *Appl. Magn. Reson.* 1994, **6**: 259–285.

[7] Yordanov N. D., Slavov P. Influence of the diameter and wall thickness of a Quartz pipe inserted in the EPR cavity on the signal intensity. *Appl. Magn. Reson.* 1966, **10**: 351–356.

[8] Hyde J. S. A new principle for aqueous sample cells for EPR. *Rev. Sci. Instrum.* 1972, **43**: 629–631.

[9] Eaton S. S., Eaton G. R. Electron paramagnetic resonance cell for lossy samples. *Anal. Chem.* 1977, **49**: 1277–1278.

[10] Mazúr M., Valko M., Morri H. *Anal. Chim. Acta.* 2003, **482**: 229–248.

[11] Casteleijn G., Tenbosch J. J., Smidt J. *J. Appl. Phys.* 1968, **39**: 4375–4380.

[12] Poole C. P. *Electron Spin Resonance: A Comprehensive Treatise on Experimental Techniques*, Interscience, New York, 1967, pp. 398–413.

[13] Weil J. A., Bolton J. R, Wertz J. A. *Electron Paramagnetic Resonance: Elementary Theory & Practical Applications* John Wiley & Sons, Inc., New York, 1994, pp. 554–556.

[14] Weger M. Passage effects in paramagnetic resonance experiments. *Bull. Syst. Tech. J.* 1960, **39**: 1013–1112.

[15] Eaton S. S., Eaton G. R. *J. Magn. Reson.* 1993, **102**(3): 354–356.

[16] Ludowise P, Eaton S. S., Eaton G. R. *J. Magn. Reson.* 1991, **93**(2): 410–412.

[17] Poole C. P. *Electron Spin Resonance: A Comprehensive Treatise on Experimental Techniques*, Interscience, New York,. 1967, pp.589–595.

[18] Wertz J. A., Bolton J. R. *Electron Spin Resonance: Elementary Theory and Practical Applications*, McGraw-Hill, New York, 1972, pp. 462–466.

[19] Weil J. A., Bolton J. R., Wertz J. E. *Electron Paramagnetic Resonance: Elementary Theory & Practical Applications* John Wiley & Sons, Inc., New York, 1994, pp. 558–562.

[20] Eaton S. S., Eaton G. R. Signal area measurement in EPR. *Bull. Magn. Reson.* 1980, **1**(3): 130–138.
[21] Auteri F. P., Boyer S., Motsegood K. K., Smilnov A. I., Smilnova T., et al. *Appl. Magn. Reson.* 1994, **6**: 287–308.
[22] Czoch R. Quantitative EPR. *Appl. Magn. Reson.* 1996, **10**: 293–317.
[23] Dyrek K., Madej A., Mazur E., Rokosz A. *Colloids Surf.* 1990, **45**: 135–144.
[24] Dyrek K., Rokosz A., Madej A. *Appl. Magn. Reson.* 1994, **6**: 309–332.
[25] Dyrek K., Rokosz A., Madej A., Bidzinska E. *Appl. Magn. Reson.* 1996, **10**: 319–338.
[26] Nagy V. Y., Placek J. *J. Anal. Chem.* 1992, **343**: 863–872.
[27] Nagy V. Y. Quantitative EPR: some of the most difficult problems. *Appl. Magn. Reson.* 1994, **6**: 259–285.
[28] Nagy V. Y. *Anal. Chim. Acta* 1997, **339**: 1–11.
[29] Nagy V. Y., Komozin P. N. Desrosiers M. F. *Anal. Chim. Acta* 1997, **339**: 31–51.
[30] Nagy V. Y., Sokolov D. P, Desrosiers M. F. *Anal. Chim. Acta* 1997, **339**: 52–62.
[31] Nagy V. Y., Sokolov D. P. *Anal. Chim. Acta* 1997, **339**: 13–29.
[32] Yordanov N. D., Christova A. G. *Appl. Magn. Reson.* 1994, **6**: 351–345.
[33] Yordanov N. D.:*Appl. Magn. Reson.* 1996, **10**: 339–350.
[34] Yordanov N. D., Christova A G. *J. Anal. Chem.* 1997, **358**: 610–613.
[35] Yordanov N. D., Genova B. *Anal. Chim. Acta* 1997, **353**: 99–103.
[36] Yordanov N. D., Gancheva V., Pelova V. A. *J. Radioanal. Nucl. Chem.*, 1999, **240**: 619–622.
[37] Kolaczkowski S. V., Cardin J. T., Budil D. E. *Appl. Magn. Reson.* 1999, **16**: 293–298.
[38] Slangen H. J. M. Determination of the spin concentration by ESR. *J. Phys.* 1970, **E3**: 775–778.
[39] Dalal N. S., Suryan M. M., Seehra M. S. *Anal. Chem.* 1981, **53**: 938–940.
[40] Cage B., Cevc P., Blinc R., Brunel L. C., Dalal N. S. *J. Magn. Reson.* 1998, **135**: 178–184.
[41] Yordanov N. D., Ivanova M. *Appl. Magn. Reson.*, 1994, **6**: 347–357.
[42] Urbanski T. Formation of solid free radicals by mechanical action. *Nature* 1967, **216**: 577–578.
[43] More K. M., Eaton G. R., Eaton S. S. *J. Magn. Reson.* 1980, **37**(2): 217–222.
[44] Adem E., Munoz P. E., Gleason V. R., Murrieta S. H., Aguilar S. G., Uribe R. E. *Appl. Radiat. Isot.* 1993, **44**: 419–422.
[45] Tipikin D. S., Lazarev G. G., Lebedev Y. S. *Z. Fizich. Khim.* 1993, **67**: 176–179.
[46] Tipikin D. S., Lebedev Y. S., Rieker A. *Chem. Phys. Lett.* 1997, **272**: 399–404.
[47] Siebert D., Dahlem J., Nagy V. *Anal. Chem.* 1994, **66**: 2640–2646.
[48] Chang T. T., Foster D., Kahn A. H. *J. Res. Natl. Bur. Stand.* 1978, **83**: 133–164.
[49] Cordischi D., Occhiuzzi M., Dragone R. *Appl. Magn. Reson.* 1999, **16**: 427–445.
[50] Goldberg I. B. *J. Magn. Reson.* 1978, **32**: 233–242.
[51] Gancheva V., Yordanov N. D., Callens F., et al. *Radiat. Phys. Chem.* 2008, **77**: 357–364.
[52] Yordanov N. D. *Appl. Magn. Reson.* 1994, **10**: 339–350.
[53] Weil J. A., Anderson J. K. *J. Chem. Soc.* 1965: 5567–5570.
[54] Randolph M. L. Quantitative considerations in ESR studies of biological materials, in Swartz H. M., Bolton J. R., Borg D. C., Eds. *Biological Applications of ESR*, John Wiley & Sons, Inc., New York, 1972, pp. 119–155.
[55] Randolph M. L. Quantitative considerations in ESR studies of biological materials, in Swartz H. M., Bolton J. R., Borg D. C., Eds., *Biological Applications of ESR*, John Wiley & Sons, Inc., New York, 1972, Chapter 3.
[56] Segal B. G., Kaplan M., Frankel G. K. *J. Chem. Phys.* 1965, **43**: 4191–4200.
[57] Nami G., Mason H. S., Yamasaki I. *Anal. Chem.* 1966, **38**: 367–368.
[58] Dalal N. S., Suryan M. M., Seehra M. S. *Anal. Chem.* 1981, **53**: 938–940.
[59] Cage B., Cevc P., Brunel L. C., Dalal N. S. *J. Magn. Reson.* 1998, **135**: 178–184.
[60] Towell J., Kalyanaraman B. *Anal. Biochem.* 1991, **196**: 111–119.

9 Paramagnetic gases and inorganic radicals

Abstract: The EMR spectra of monoatomic and polyatomic gaseous paramagnetic particles, as well as the EMR spectra of inorganic radicals, semiconductors, and conductors have been discussed in this chapter.

Unlike the behavior observed in condensed phases, the translational motion of the atoms and molecules observed in the gas phase is almost completely free. The translational motion of this paramagnetic particle center of mass does not cause energy level splitting, nor is there a direct spectroscopical causal relationship between them. On the other hand, the rotational motion of free molecules produces quantized rotational energy splitting magnitude, which is often on the same order of magnitude as the Zeeman splitting of their spin states, continually causing *rotational–magnetic interaction*, which have a rather large effect on the EMR spectra of diatomic and polyatomic molecules.

Angular momentum is very important for understanding how the gas-phase system obtains EMR transition. There are four sources of angular momentum: orbital motion of (total) electron, spin motion of electron, rotational motion of nuclear skeleton, and spin motion of nuclei. Here, the total angular momentum vector set $\boldsymbol{F}$ of atoms or molecules can be considered as random orientation, but each of them is fixed before being subjected to a collision change. More precisely, the quantum number $\boldsymbol{M_F}$ is kept constant, but, in fact, before the action of the external magnetic field it is impossible to be measured out.

Collisions between atoms or molecules, or with the walls of the device, could have a perceptible effect on relaxation times and linewidths; however, no charged particles are observed by EMR in the gas phase. In general, this is because the concentration of ions in the gas-phase sample does not reach a level that is sufficiently observable by standard EMR experiments.

In the EMR spectrometer, detailed information on many of the target molecular parameters can be obtained from the multiplicity and position (g value) of the lines observed for paramagnetic gaseous samples, but also for real-time, qualitative and quantitative analyses, it is conducive to the development of gas reaction kinetics.

9.1 Spectra of paramagnetic gases

9.1.1 Monoatomic paramagnetic gases

Open-shell atoms are very easy to dimerize usually, and are not stable. Generally, corresponding molecules are obtained by means of discharge, pyrolysis, electron bombardment, or photolysis, and adjusting to the concentration of stable existence. Most of these species are prepared in advance and immediately introduced (by

https://doi.org/10.1515/9783110568578-009

diffusion or pumping) to the resonant cavity of EMR spectrometer. Obviously, the first and the best example of this kind of atom is the hydrogen atom; its spectral parameters (g, A, etc.) and its ground-state and lower excited-state parameters are already very much familiar to us.

The ground state of hydrogen atom is ^2S, and $g = g_e$. Due to the hydrogen nucleus (^{1}**H**) spin $I = \frac{1}{2}$, the hyperfine structure of hydrogen atom has two lines with the interval of 50.68 mT, and its hyperfine splitting is contributed by the Fermi contact term totally [1,2].

The nitrogen atom is an another example [3], whose ground state is ^4S, and due to nitrogen nucleus (^{14}**N**) $I = 1$, its hyperfine structure has three lines with the interval of 0.38 mT, $g = 2$, about 133 times smaller than the splitting distance of hydrogen atom, because the unpaired electron is in 2p orbital and the spin density caused by the spin polarization mechanism on ^{14}N nucleus is quite small.

The EMR spectrum of their isotopic ^{2}H and ^{15}N has also been early reported [1,2,4]. Their nuclear spins are 1 and 1/2, respectively, with three and two lines of their hyperfine structures, and the splitting distance coincides well with the nuclear gyromagnetic ratio.

In recent years, EMR studies of chemical reactions involving many hydrogen-like atoms (e.g., alkali metal atoms), including hydrogen atoms, have also been carried out. Most of these studies use atomic beam detection, which is better than standard EMR techniques.

There are also EMR studies of single atoms such as O, S, Sc, Te, P, As, Sb, F, Cl, Br, I, and Ar. The ground-state electronic configurations such as $^4S_{3/2}$, $^2D_{5/2}$, $^2D_{3/2}$ are easier to treat; however, owing to halogen atoms, whose unpaired electron is in the p-shell, its ground-state electronic configuration is $^2P_{3/2}$. Because of $J = \frac{3}{2}$, the transitions between M_J value of $-\frac{3}{2} \leftrightarrow -\frac{1}{2}$, $-\frac{1}{2} \leftrightarrow +\frac{1}{2}$, $+\frac{1}{2} \leftrightarrow +\frac{3}{2}$ occur in high field, and each absorption line could be split into $2I + 1$ lines.

Now we deal with a spherical symmetric system. In this system, the unpaired electron repartee is the central Coulomb field consisting of the nucleus and other electrons, so it is very important for the orbital angular momentum around the nucleus. The total angular momentum operator $\boldsymbol{J}$ of the electron should be the orbital angular momentum operator $\boldsymbol{L}$ plus the spin angular momentum operator $\boldsymbol{S}$. The role of nuclear motion has not yet been considered here. The natural abundance of atomic fluorine (produced by discharge [5]) ^{19}**F** is 100%, $I = \frac{1}{2}$, and the total angular momentum should be $\hat{\boldsymbol{F}} = \hat{\boldsymbol{J}} + \hat{\boldsymbol{I}}$ taking into account the spin motion of the nucleus. The quantum number J can take $2I + 1$ values, and the quantum number F should also be an integer, it takes the value from $|J - I|$ to $J + I$. The corresponding ground-state Zeeman energy levels and the allowable transitions of EMR experiment at the fixed frequency are shown as Figure 9.1. The first-order approximation of g factor for electron is given by the Landé formula:

$$g = 1 + [J(J+1) - L(L+1) + S(S+1)]/[2J(J+1)] \tag{9.1}$$

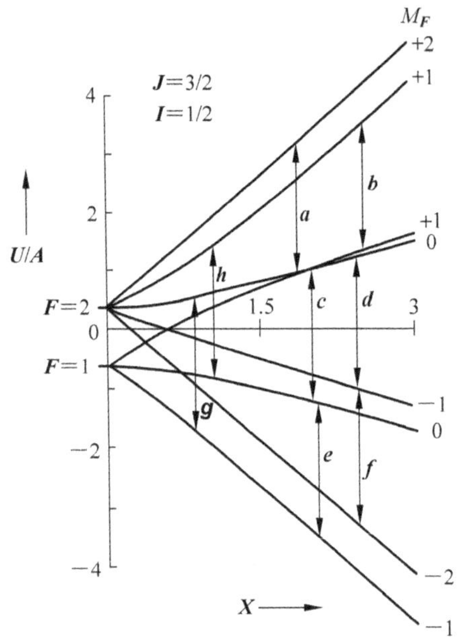

Figure 9.1: Zeeman-level splitting of ground-state $^2P_{3/2}$ fluorine atom, and possible EMR transitions at fixed frequency.

This formula is generally used in the case of nonrelativistic atoms [6]. For example, in the case of $J=1/2$, $g=0.66561(3)$. In this system [7], the EMR spectral lines of double-photon transition containing three spin levels could be observed. In the case of all ground-state $^2P_{3/2}$ halogen atoms, all transitions of selective rule as $\Delta M_J = \pm 1$, $\Delta M_I = \pm 1$, as well as $\Delta M_J = \pm 1$, $\Delta M_I = 0$ of the original EMR, could be observed [8].

The molecular iodine $\mathbf{I_2}$ passing through the following reaction can obtain atomic iodine of the ground state as $^2P_{3/2}$:

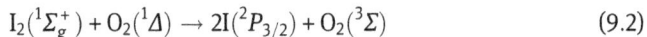

$$I_2(^1\Sigma_g^+) + O_2(^1\Delta) \rightarrow 2I(^2P_{3/2}) + O_2(^3\Sigma) \tag{9.2}$$

The atomic iodine of first excited state $I(^2P_{1/2})$ can be obtained via rapid equilibrium [9]:

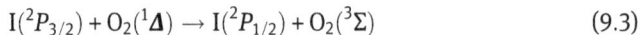

$$I(^2P_{3/2}) + O_2(^1\boldsymbol{\Delta}) \rightarrow I(^2P_{1/2}) + O_2(^3\Sigma) \tag{9.3}$$

The singlet dioxygen $O_2(^1\Delta)$ was prepared with chemical method. All except the diamagnetic species (I_2 at ground state) can be observable by EMR, and their relative concentrations can be obtainable by double-integration on their first-derivative lines. In Eq. (9.3), the equilibrium constant at 295 K was derived from EPR measurements to be $K=2.9$. The measured $g=0.6664$ for the excited iodine atom I ($^2P_{1/2}$) is within 0.2% of that derived by the Landé formula.

9.1.2 Diatomic paramagnetic gas

More angular momentum and more angular momentum coupling schemes need to be considered in diatomic linear molecules. There are five basic coupling schemes, called Hund coupling scheme, but only two kinds (a and b) appear in gas-phase EMR, while other three kinds disappear.

For the Hund coupling case (a), the electron spin angular momentum **S** and the orbital angular momentum **L** are strongly coupled on the internuclei axis, and their projections (components) on the internuclear axis are denoted by Σ and Λ and the sum by Ω.

$$\Omega = \Sigma + \Lambda \tag{9.4}$$

In addition to nuclear spin, the total angular momentum is composed of the angular momentum of the molecular rotation $\boldsymbol{N}$ and Ω, as shown in Figure 9.2a. N is perpendicular to the internuclear axis, and Ω is the projection of J on the internuclear axis. Therefore, the allowable value of J is $\Omega, \Omega+1, \Omega+2, \ldots$. If there are also nuclear spin in the system, then **J** and **I** would resynthesize to be the total angular momentum $\boldsymbol{F}$:

$$\boldsymbol{F} = \boldsymbol{J} + \boldsymbol{I} \tag{9.5}$$

The allowable value of F is $(J+I), \ldots, |(J-I)|$.

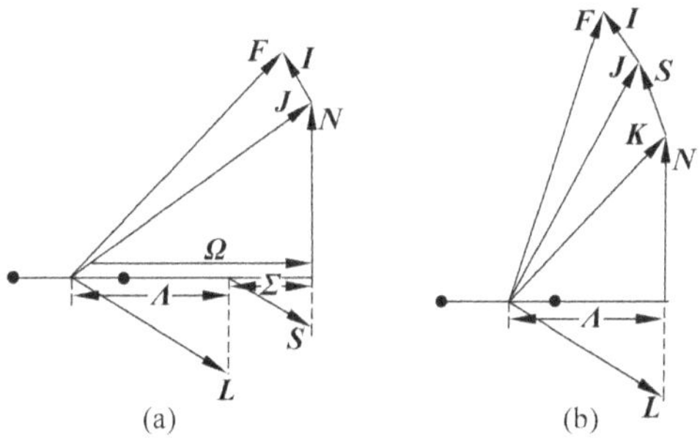

Figure 9.2: Hund coupling scheme of angular momentum.

For the Hund coupling case (b), as shown in Figure 9.2b, which is different from case (a), S is not coupled to the internuclear axis, so Ω is not defined. Now Λ is coupled with $\boldsymbol{N}$ into a vector $\boldsymbol{K}$, whose value can be taken as $\Lambda, \Lambda+1, \Lambda+2, \ldots$, and then $\boldsymbol{K}$ and **S** couple into the total angular momentum vector $\boldsymbol{J}$, whose value can be taken as $(K+S), \ldots, |(K-S)|$. If there is also nuclear spin $\boldsymbol{I}$ in the system, then $\boldsymbol{J}$ and $\boldsymbol{I}$ couple into the total angular momentum vector $\boldsymbol{F}$.

9.1.2.1 EMR spectrum of molecular oxygen O_2

The most important paramagnetic diatomic molecule may be the molecular oxygen O_2, because it plays an extremely important role in the biosphere. The electron ground state of O_2 is conventionally denoted by $^3\Sigma_g^-$. It is very important that the orbital angular momentum of the electron exists for name only (similar to the ground state of the H atom). According to the Hund rule, two unpaired electrons occupy two different atomic orbits separately to form a triplet state. Because there is a symmetric center, it does not have permanent electric dipole moment. EMR transitions are all contributed by the form of magnetic dipoles, so the intensity of the spectral line is much weaker than that of the electric dipole transition. There are three isotopes of oxygen: $^{16}O(99.75\%, I = 0$, $^{17}O(0.037\%, I = 5/2)$, $^{18}O(0.204\%, I = 0)$.

The oxygen molecules of gas phase are constantly dumbbell-shaped rotating tumbles with quantized rotational angular momentum, and are described by an appropriate spatial operator $\hat{\boldsymbol{N}}$. The total angular momentum operator is $\hat{\boldsymbol{J}} = \hat{\boldsymbol{N}} + \hat{\boldsymbol{S}}$ (here not including the spin angular momentum of the nucleus, since only ^{17}O is magnetic core and its natural abundance is very low). From the rule of quantum statistics derived, the rotating quantum number of the largest natural abundance $^{16}O^{16}O$ is always odd, because the energy is proportional to $N(N + 1)$, and thus N cannot be zero; so the rotating motion always exists.

From the triplet EMR spectra discussed previously, it is assumed that the spin action has no relation with rotation, although in actual molecules, the rotation of a portion charge produces a magnetic field and interacts with the spin moment. Therefore, the triplet state of spin energy level is entangled also with the numerous rotational energy levels. The result is that its EMR spectrum is actually composed of countless (but calculable) spectral lines [10,11]. Figure 9.3 shows a part (0–1.5 T) of the simulated EMR spectrum (X-band, 9.14456 GHz) of O_2 and Table 9.1 provides the positions of the spectral lines and the corresponding quantum numbers [12]. The relative intensities of the lines depend on the number of distribution in each state, that is, on the temperature of the gas [13].

Since energy of both spin and rotation are interdependent mutually, spin relaxation can be performed by molecular tumbling, which is very sensitive to the time of molecular collisions. Its result is the linewidth related to the pressure, but these lines are quite sharp in the case in which it can be obtained (e.g., 0.01 mT for pressures of ~0.2 torr) (Figure 9.3). Due to its stability, molecular O_2 is very useful as a standard for measuring the concentrations of other gas-phase radical [14].

The symmetry of the molecule O_2 produces a uniaxiality describing the EMR spectral parameter matrix. It is very important for understanding these parameters, as the molecules rolled cannot be averaged as in the situation of liquids. Thus, for example, $g_\perp$ and $g_\parallel$ both are all measurable. When the collisions of intermolecular and molecule with wall are neglected, the gaseous molecules do not tumble incoherently, and the orientation of the total angular momentum vector of each molecule is irregular and does not change with time. When a constant external magnetic field $\boldsymbol{H}$ is introduced, each $\boldsymbol{J}$ is quantized along a direction called the effective magnetic field. Since the *spin–orbit coupling* and the *spin–rotation coupling* produce the anisotropy

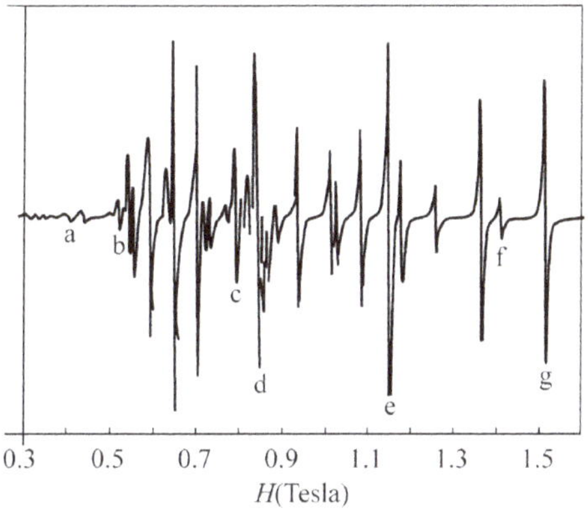

Figure 9.3: EMR spectrum of gas-phase O_2 at the temperature of 100 K (~ 0.1 Torr).

Table 9.1: Selected EPR lines of $^{16}O^{16}O$ at $v = 9.14459$ GHz (see Figure 9.3).

Line	Field (H/Tesla)	N	$J_1, M_{J_1} \leftrightarrow J_2, M_{J_2}$	
a	0.401842	5	6, −3	4, −2
b	0.541546	1	1, −1	1, 0
c	0.790642	5	4, 0	6, +1
d	0.863229	7	6, 0	6, +1
e	1.149089	9	4, +2	4, +3
f	1.408629	7	7, −7	7, −7
g	1.512532	5	6, +4	6, +5

of g, the direction of the effective magnetic field deviates from the direction of the external magnetic field $\boldsymbol{H}$ slightly.

The Zeeman part of the corresponding Hamiltonian can be written as follows:

$$H = \beta_e[g_{\perp}\hat{\boldsymbol{S}}^T \cdot \boldsymbol{H} + (g_z - g_{\perp})\hat{S}_Z H_Z + g_{\text{rot}}\hat{\boldsymbol{N}}^T \cdot \boldsymbol{H}] \tag{9.6}$$

where g_{rot} is the g parameter of rotational magnetic moment, while $g_{\perp}$ and $g_Z = g_{\parallel}$ are the g parameter of the electron spin. The direction of connective axis between appointive nuclear is z-direction. The g factor of electron is $g_z = g_e$ approximately, and

$$g_{\perp} = g_e - 2\lambda \sum_{n \neq G} \frac{\langle G|\hat{L}_x|n\rangle\langle n|\hat{L}_x|G\rangle}{E_n^{(0)} - E_G^{(0)}} \tag{9.7}$$

λ is the spin–orbit coupling parameter of molecule, n presents the electronic energy states, and G is the nondegenerate ground state.

The observational g_z in experiment is slightly less than g_e ($g_z - g_e \approx 10^{-4}$), owing to the relativistic effects. The spin–orbit coupling term mix into $g_\perp$ in relation with the distance between nuclei, because of the influence of mixing excited state (Π state) other than the ground state (Σ state). Since the magnitude of λ is not so big, its influence is quite small, about $g_\perp - g_e \approx 3\times10^{-3}$. The rotational g_{rot} factor is more smaller ($g_{rot} \approx 10^{-4}$); as for low rotational states, the magnetic moment associated with $\hat{\mathrm{N}}$ is almost negligible.

Through EMR measurement of nuclear quadrupole to study the enriching $^{17}O(I=5/2)$ molecule [15], we should understand in more detail the interaction between the magnetic moment of the molecule and the spin magnetic moment of the nucleus as well as the local electric field gradient. Due to mass differences between isotopes ^{16}O, ^{17}O, ^{18}O causing the difference in Hamiltonian parameters is very small. So we do not consider the contribution of molecular vibration.

As already pointed out, the molecular O_2 in the electronic excited state $^1\Delta_g$ can also be observed by EMR, since these particles are in the metastable state with respect to the ground state. Since they are spin forbidden, the transformation between singlet and triplet states is a slow process. The paramagnetism here has no spin components, and is entirely contributed by the orbital motion of electrons. Therefore, the total angular momentum operator is the sum of the projection of $\hat{\boldsymbol{L}}$ in the Z-direction and the rotation vector (perpendicular to the Z-direction). The EMR spectra of the states of $\boldsymbol{J}=2$ and 3, including the hyperfine effects of ^{17}O, are measured and analyzed [16,17].

The EMR investigation of $\mathbf{S}_2$, OS, OSe, and FN, which have the same valence electron configuration with the molecule $\mathbf{O}_2$, has also been reported. As for $\mathbf{S}_2$, the S atom is heavier than O atom, the spin–orbit coupling effect is more important. For the heteronuclear molecules, owing to the difference of rotation level ($\Delta\boldsymbol{N}\neq0$, $|\Delta\boldsymbol{M}_J|=1$) produced, the electric dipole transition is more dominant.

9.1.2.2 EMR spectrum of molecular NO

NO and O_2^- have the same valence electron configuration. The unpaired electrons are located in the π molecular orbit, so $\Lambda = \pm1$, $\Sigma = \pm\frac{1}{2}$, and then $\Omega = \frac{3}{2}, \frac{1}{2}$. There are two energy states – $^2\Pi_{3/2}$ and $^2\Pi_{1/2}$, of which $^2\Pi_{1/2}$ is ground state, while $^2\Pi_{3/2}$ is higher than $^2\Pi_{1/2}$: 124 cm^{-1}. Since $\Lambda = \pm1$, each state is double degenerate, called "Λ degeneracy." Added to a strong magnetic field, the degeneration will be relieved, followed by the energy level splitting, and its magnetic Hamiltonian operator is

$$\hat{\mathscr{H}} = \beta(\hat{\mathscr{L}} + g_e\hat{\mathbf{S}})\cdot\boldsymbol{H} \tag{9.8}$$

For the $^2\Pi_{1/2}$ state, spin and orbital angular momentum counteract with each other to become a "nonmagnetic" ground state, so its EMR spectrum is contributed by $^2\Pi_{3/2}$ state. Now, let us consider the lowest rotational energy level $N=0$, which is due to $\Omega=\frac{3}{2}$, $J=\frac{3}{2}$, $M_J=\frac{3}{2}, \frac{1}{2}, -\frac{1}{2}, -\frac{3}{2}$ of $^2\Pi_{3/2}$ state. The g-factor formula is

$$g = \frac{(\Lambda + 2\Sigma)(\Lambda + \Sigma)}{J(J+1)} \tag{9.9}$$

Therefore, for the rotational level $N = 0$ of the $^2\Pi_{3/2}$ state, $g = 4/5$. The $N = 1$ state, as well as other various rotational states of $N \neq 0$ except $N = 0$ should also be considered. Splitting case of these energy levels in magnetic field is shown in Figure 9.4. The splitting of the second-order Zeeman effect is presented in this figure. For $J = 3/2$, there are three transitions, while for $J = 5/2$ there are five, and the selection rule of transition is $\Delta M_J = 1$. Because of the second-order Zeeman effect, the resonant magnetic fields of these transitions are different. We can see from Eq. (9.9) that the value of g decreases when J increases (when $J = 5/2$, $g = 12/35$). So in the case of a fixed frequency, it could be observed only at a high magnetic field, or using a lower microwave frequency (S-band).

Figure 9.5 shows the EMR spectra[18] of $^{15}N^{16}O$ and $^{14}N^{16}O$ in gas phase, when $\nu = 2.8799$ GHz GHz (S-band), which display the electric dipole typical transition corresponding to $\Delta J = 0$, $\Delta M_J = \pm 1$, and $\Delta M_I = 0$, $\pm \leftrightarrow \pm$. Figure 9.5a is the rotation state of $J = 3/2$ and the resonant magnetic field is near 265 mT. For $^{15}N^{16}O$, owing to ^{15}N nuclei ($I = 1/2$) split into two lines and because of the second-order Zeeman effect, each line also splits into three, their intensity ratio being 3: 4: 3. For $\Lambda \neq 0$, there is also

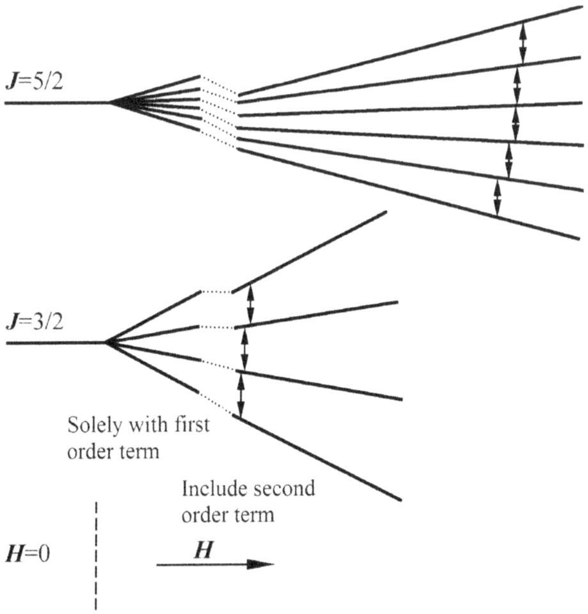

Figure 9.4: Splitting of selected energy levels ($J = 3/2$ and $J = 5/2$) of state $^2\Pi_{3/2}$ of molecular NO in magnetic field.

degree of degeneracy Λ, and under the interaction of the orbital angular momentum with the rotational angular momentum, the degeneracy is relieved causing "Λ double line splitting," denoted by "+" and "−". The Λ splitting is increased with the increasing J, so there are $2 \times 3 \times 2 = 12$ lines in total. For $^{15}N^{16}O$, due to ^{14}N ($I = 1$), each line splits into three and because of the second Zeeman effect, each line splits again into three, and thus, in addition to Λ double line splitting, there are $3 \times 3 \times 2 = 18$ lines in total.

Figure 9.5b is the spectrum of the rotation state $J = 5/2$ of $^2\Pi_{5/2}$, and its position is estimated roughly at

$$H_{5/2} = \frac{g_{3/2} H_{3/2}}{g_{5/2}} = \frac{(4/5)265}{12/35} \approx 620\ (\text{mT})$$

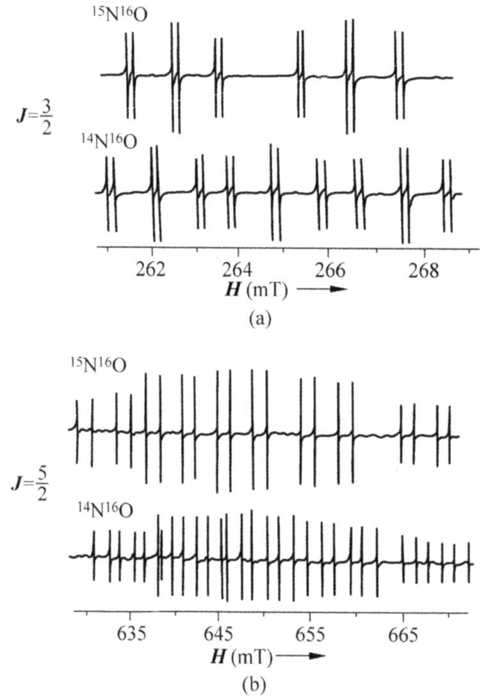

Figure 9.5: EMR spectra of $^{15}N^{16}O$ and $^{14}N^{16}O$ in the gas phase ($\nu = 2.8799$ GHz): (a) $J = 3/2$; (b) $J = 5/2$.

The experimental value is about 650 mT. For $^{15}N^{16}O$, due to ^{15}N nuclear ($I = 1/2$), each line splits into two, and then because of the second Zeeman effect, each line again splits into five, their intensity ratio is 5:8:9:8:5, and finally, Λ-double lines also split, so the total number of lines is $2 \times 5 \times 2 = 20$. There are $3 \times 5 \times 2 = 30$ lines for $^{14}N^{16}O$.

There are four types of transitions if the Λ degeneracy is relieved completely, of which two are electric dipole transition and the other two are magnetic dipole transition. Since the dipole transition is two to three orders of magnitude stronger than the magnetic dipole transition, we can only observe the doublet produced by the dipole transition. During experiment, the gas-phase molecules should be placed in the area having the largest electric field of resonant cavity. The resonant magnetic field positions of these lines are shown in Table 9.2.

9.1.2.3 EMR spectrum of ClO

The ClO diatomic molecules can be obtained by guiding the mixture gas of Cl^2 and O^2 into a microwave discharge chamber (microwave power 75 W, frequency 2000 MHz). Figure 9.6 shows the EMR spectrum of ClO[19,20] that is expressed as three sets of 12 strong lines, which are contributed by the ^{35}ClO radicals in the $\boldsymbol{J} = 3/2$ rotation level of $^2\Pi_{3/2}$, shown as the dotted line of stick spectrum. The weak spectrum is contributed

Table 9.2: Transitions of $^{15}N^{16}O$ and $^{14}N^{16}O$ between $\Delta J = 0(M_J,\ M_I,\ \pm) \leftrightarrow (M_J - 1,\ M_I,\ \mp)$ under S-band.

Resonance magnetic field (mT) $^{15}N^{16}O$ under f = 2.879926 GHz, J = 3/2			**Resonance magnetic field (mT) $^{14}N^{16}O$ under f = 2.879930 GHz, J = 3/2**		
$M_J\ M_I$	**$(M_J,\ M_I,+)$**	**$(M_J,\ M_I,-)$**	**$M_J M_I$**	**$(M_J,\ M_I,+)$**	**$(M_J,\ M_I,-)$**
$\frac{3}{2}, -\frac{1}{2}$	261.4728	261.6189	$\frac{3}{2}$, 1	260.9279	261.0930
$\frac{3}{2}, \frac{1}{2}$	265.2695	265.4067	$\frac{3}{2}$, 0	263.6375	263.7963
$\frac{1}{2}, -\frac{1}{2}$	262.4641	262.6137	$\frac{3}{2}$, −1	266.4786	266.6301
$\frac{1}{2}, \frac{1}{2}$	266.3115	266.4599	$\frac{1}{2}$, 1	261.9289	262.0852
$-\frac{1}{2}, -\frac{1}{2}$	263.4493	263.6027	$\frac{1}{2}$, 0	264.6444	264.8351
$-\frac{1}{2}, \frac{1}{2}$	267.3477	267.5097	$\frac{1}{2}$, −1	267.4165	267.5708
			$-\frac{1}{2}$, 1	262.9159	263.0850
			$-\frac{1}{2}$, 0	265.6668	26508430
			$-\frac{1}{2}$, −1	268.3410	268.5223
Resonance magnetic field (mT) $^{15}N^{16}O$ under f = 2.879903 GHz, J = 5/2			**Resonance magnetic field (mT) $^{14}N^{16}O$ under f = 2.879899 GHz, J = 5/2**		
$\frac{5}{2}, -\frac{1}{2}$	663.9699	665.3994	$\frac{5}{2}$, 1	664.3521	666.0022
$\frac{5}{2}, \frac{1}{2}$	667.8975	669.2521	$\frac{5}{2}$, 0	667.2897	668.7927
$\frac{3}{2}, -\frac{1}{2}$	653.5663	655.0013	$\frac{5}{2}$, −1	670.0826	671.6138
$\frac{3}{2}, \frac{1}{2}$	657.5004	658.8857	$\frac{3}{2}$, 1	654.3728	655.9601
$\frac{1}{2}, -\frac{1}{2}$	644.5492	645.9970	$\frac{3}{2}$, 0	657.1755	658.8124
$\frac{1}{2}, \frac{1}{2}$	648.4904	649.9233	$\frac{3}{2}$, −1	660.0493	661.5693
$-\frac{1}{2}, -\frac{1}{2}$	636.6356	638.1000	$\frac{1}{2}$, 1	645.6674	647.2373
$-\frac{1}{2}, \frac{1}{2}$	640.5777	642.0675	$\frac{1}{2}$, 0	648.3946	650.1077
$-\frac{3}{2}, -\frac{1}{2}$	629.6180	631.1083	$\frac{1}{2}$, −1	651.2917	652.8474
$-\frac{3}{2}, \frac{1}{2}$	633.5622	635.1201	$-\frac{1}{2}$, 1	637.8387	639.5776
			$-\frac{1}{2}$, 0	640.6918	642.4161
			$-\frac{1}{2}$, −1	643.5475	645.1890
			$-\frac{3}{2}$, 1	631.1208	632.7919
			$-\frac{3}{2}$, 0	633.8817	635.5415
			$-\frac{3}{2}$, −1	636.6297	638.4091

by ^{37}ClO, shown as the solid line of stick spectrum. Its relative intensity is also in coincidence with its natural abundance in the ratio of ^{35}Cl (75.53%) and ^{37}Cl (24.47%). The *g* value is 0.798 ± 0.01, consistent with the theoretical estimate value of $^2\Pi_{3/2}$. The weak line of the arrow is the spectral line of the O_2 molecule. Because the hyperfine interaction contains the contribution of the nuclear quadrupole moments, the four lines are unequal distances. The total Hamiltonian operator shown is as follows:

$$\hat{\mathscr{H}} = \hat{\mathscr{H}}_R + \hat{\mathscr{H}}_M + \hat{\mathscr{H}}_H \tag{9.10}$$

It corresponds to the Hund coupling (a), and the basic function is $|J, I, F, \Omega, \Sigma, M\rangle$, in which M is the projection of F in the field direction.

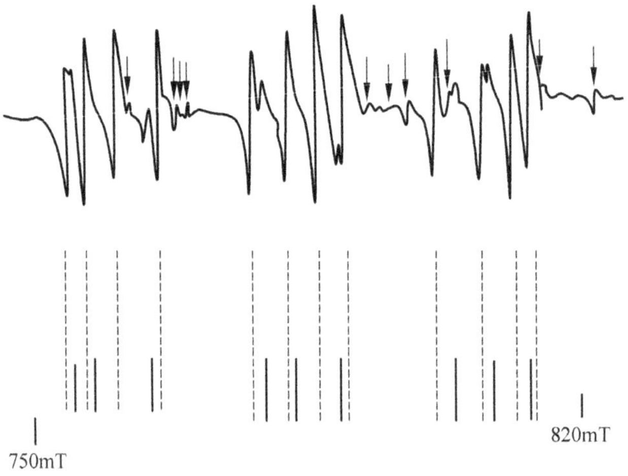

Figure 9.6: EMR spectrum[19,20] of ClO.

The operator $\hat{\mathscr{H}}_R$ containing the effect of *rotation* and *spin–orbit coupling* can be written as

$$\hat{\mathscr{H}}_R = B_o\,(\hat{\boldsymbol{J}} - \hat{\boldsymbol{L}} - \hat{\boldsymbol{S}})^2 + D\hat{\boldsymbol{L}} \cdot \hat{\boldsymbol{S}} \tag{9.11}$$

where B_o is the rotation constant and D is the constant of the fine structure. Putting it into a fixed coordinate system to expand and using Ω, Λ, Σ to replace $\hat{J}_z$, $\hat{L}_z$ and $\hat{S}_z$,

$$\begin{aligned}\hat{\mathscr{H}}_R = {} & B_o[\mathrm{J}(\mathrm{J}+1) + \mathrm{S}(\mathrm{S}+1) - 2\Omega\Sigma - \Lambda^2] - 2B_o[\hat{J}_x\hat{S}_x + \hat{J}_y\hat{S}_y] + D\Lambda\Sigma \\ & + (D + 2B_o)(\hat{L}_x\hat{S}_x + \hat{L}_y\hat{S}_y) - 2B_o(\hat{J}_x\hat{L}_x + \hat{J}_y\hat{L}_y) + B_o(\hat{L}_x^2 + \hat{L}_y^2)\end{aligned} \tag{9.12}$$

To analyze the spectrum of ClO, the last term $B_o(\hat{L}_x^2 + \hat{L}_y^2)$ in this Hamiltonian operator can be ignored, because its contribution to all levels is a constant displacement. In addition, the fourth and fifth terms can also be ignored.

The operator $\hat{\mathscr{H}}_M$ is the effective magnetic hyperfine Hamiltonian of the diatomic molecule. If the terms containing $\hat{L}_x$, $\hat{L}_y$, $\hat{S}_x$, $\hat{S}_y$ are ignored, the approximate Hamiltonian in the molecular coordinate system[21,22] will be

$$\hat{\mathscr{H}}_M = a\hat{I}_z\hat{L}_z + (b+c)\hat{I}_z\hat{S}_z \tag{9.13}$$

where a, b, and c are the coupling constants, and Eq. (9.13) can also be written as

$$\hat{\mathscr{H}}_M = [a\Lambda + (b+c)\Sigma]\hat{I}_z \tag{9.14}$$

The matrix element of this Hamiltonian operator depends on J. But from the observed value of the hyperfine splitting, we can only determine $\left[a + \frac{1}{2}(b+c)\right]$, so it is impossible to separate the dipole contribution from the Fermi contact term.

The effective magnetic hyperfine Hamiltonian operator should also contain the quadrupole moment term $\hat{\mathscr{H}}_Q$. After ignoring the several terms containing I_x, I_y,

$$\hat{\mathscr{H}}_Q = \frac{e^2Qq}{4I(2I-1)}[3\hat{I}_z^2 - I(I+1)] \tag{9.15}$$

In Eq. (9.10), $\hat{\mathscr{H}}_H$ is the Zeeman term, and its general form is

$$\hat{\mathscr{H}}_H = \beta\mathscr{H}\cdot(\hat{\boldsymbol{L}} + g_e\hat{\boldsymbol{S}}) + g_n\beta_n\mathscr{H}\cdot\hat{\boldsymbol{I}} + g_R\beta\mathscr{H}\cdot\boldsymbol{N} \tag{9.16}$$

The third term is the interaction between rotating magnetic moment and the magnetic field, which can be neglected in the spectrum of ClO because the value of g_R is quite small (usually on the order of 10^{-4}). If the terms containing $\hat{L}_x$ and $\hat{L}_y$ are also ignored, Zeeman term can be written as follows:

$$\hat{\mathscr{H}}_H = \beta\mathscr{H}\cdot[g_e\hat{\boldsymbol{S}}_x\boldsymbol{i} + g_e\hat{\boldsymbol{S}}_y j + (\Lambda + g_e\Sigma)\,\boldsymbol{k}] + g_n\beta_n\boldsymbol{H}\cdot\hat{\boldsymbol{I}} \tag{9.17}$$

With these Hamiltonian operators, the spectral position of the ClO spectrum can be calculated and its coincidence with experimental values is also comparatively well, as given in Table 9.3. For the rotational energy state of $J = 3/2$, its magnetic field position is 800–900 mT in X-band. As for the second rotational energy state $J = 5/2$, the spectrum can also be observed, and their spectral lines are in the range of 1.6–1.9 T.

9.1.2.4 EMR spectrum of SO molecule

The oxygen is passed through the microwave discharge chamber and the obtained products is mixed with COS to generate SO gaseous radicals. The gas-phase electron magnetic resonance spectra of most diatomic molecules belong to the Hund coupling case (a), while the spectrum of the state $^3\Sigma$ of the SO molecule belongs to the Hund coupling case (b). Its magnetic energy-level splitting is shown in Figure 9.7, and the five strong transition lines are the transition between levels of $K = 1$, $J = 1$ and $K = 2$,

Table 9.3: The magnetic field position of the EMR spectrum of gas-phase $^{35}Cl^{16}O$.

Zeeman modulation cavity v = 8.6664 GHz		Stark v = 9.6670 GHz	
Experimental (mT)	Calculated (mT)	Experimental.(mT)	Calculated (mT)
744.26	744.27	828.47	828.48
746.96	746.95	831.10	831.13
750.93	750.94	835.03	835.13
756.30	756.29	840.54	840.50
767.84	767.85	857.97	858.06
772.09	772.09	862.32	862.34
776.07	776.09	866.41	866.34
779.85	779.83	870.07	870.04
790.68	790.65	886.53	886.54
796.37	796.36	892.34	892.33
800.34	800.35	896.31	896.33
802.52	802.58	898.50	898.47
The following data are used for theoretical calculations: $B_0 = 0.622 \pm 0.001\,\text{cm}^{-1}$ $D = -282 \pm 9\,\text{cm}^{-1}$ $[a\Lambda + (b+c)\,\Sigma] = 111 \pm 2$ MHz $e^2Qq = -86 \pm 6$ MHz		The following data are used for theoretical calculations: $B_0 = 0.622 \pm 0.001\,\text{cm}^{-1}$ $D = -268 \pm 17\,\text{cm}^{-1}$- $[a\Lambda + (b+c)\,\Sigma] = 111 \pm 2$ MHz $e^2Qq = -90 \pm 6$ MHz	

$J = 1$. Figure 9.8 shows the stick spectrum[23,24], where the stick's height indicates the relative intensity. There are five lines marked "*", which is the spectrum of ground state ($^3\Sigma$) of molecular $^{32}S^{16}O$, and if using X-band microwave frequency, the spectrum will distribute in the range of 0.3–1.2 T. Its zero-field Hamiltonian operator is

$$\hat{\mathscr{H}} = B(\boldsymbol{K} - \boldsymbol{L})^2 + D\boldsymbol{L} \cdot \boldsymbol{S} + \gamma \boldsymbol{K} \cdot \boldsymbol{S} + \left(\frac{2\lambda}{3}\right)(3S_z^2 - \boldsymbol{S}^2) \tag{9.18}$$

The first term is the rotational energy, the second term is the spin–orbit coupling interaction energy, the third term is the spin–rotation interaction energy, and the fourth term is the spin–spin interaction energy. In the rotational energy term, using K rather than N, the calculation could be simplified. Their Zeeman interaction term $\hat{\mathscr{H}}_z$ include orbital angular momentum, spin angular momentum, and rotation angular momentum terms.

$$\hat{\mathscr{H}}_z = \beta \boldsymbol{H} \cdot (g_L \boldsymbol{L} + g_S \boldsymbol{S} + g_R \boldsymbol{K}) \tag{9.19}$$

For the Hund case (b), the basis function is $|J, K, \Lambda, S, M\rangle$, where M is the projection of J in the direction of the magnetic field.

The zero-field splitting constant can be obtained by analyzing the spectrum of SO, and the contribution of the primary term and secondary item can be

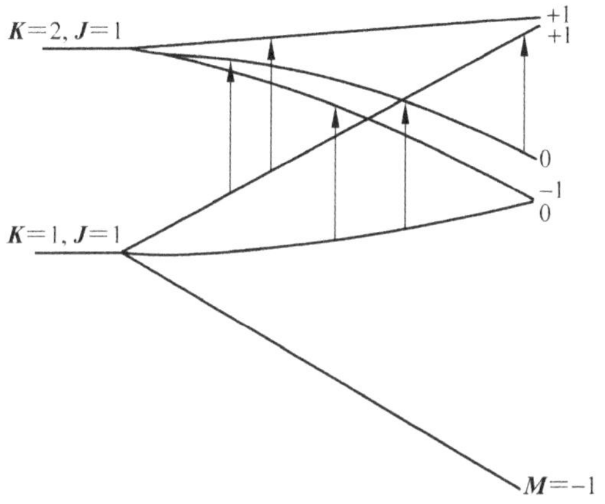

Figure 9.7: The magnetic energy level diagram of SO molecule, where the arrows indicate the 5 strong transition positions in X-band, and the highest field is larger than 1.2 Tesla.

distinguished. Then from research of the $^{33}S^{16}O$ satellite line, the hyperfine coupling constants b and c are separated. Since the natural abundance of the ^{33}S is quite small (0.76%), the line is weak and not shown in Figure 9.8. After obtaining b and c separately, the contribution of the dipole interaction term and the contact interaction term to the hyperfine interaction can be distinguished.

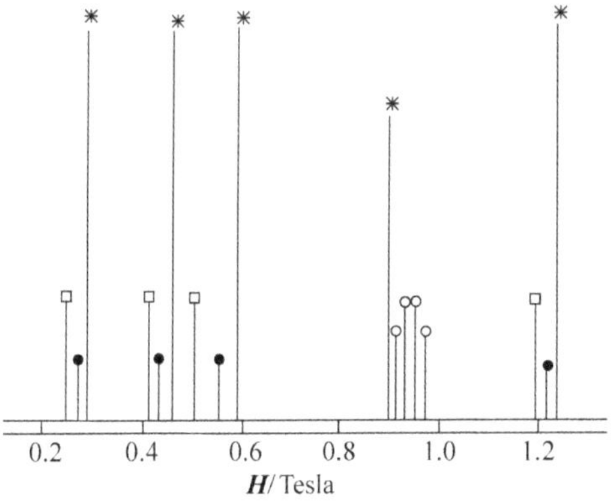

Figure 9.8: The stick EMR spectrum of gaseous SO radical.

Furthermore, Carrington et al.[25] studied the spectrum of gas-phase SO in the lowest state $^1\Delta$. This system does not have a total spin angular momentum, but has four lines corresponding to transition $\Delta M_J = 1$. Four lines marked "O" in Figure 9.8 belong to $^{32}S^{16}O$ in state $^1\Delta$, which are generated by the secondary Zeeman effect. In addition, the four lines marked "•" are wake lines of $^3\Sigma$ state of $^{32}S^{16}O$ in the first excitation vibration state. Lines marked with "□" is generated by $^{34}S^{16}O$ in state $^3\Sigma$.

The spectral lines of the state $^1\Delta$ belong to Hund case (a), and the Hamiltonian operator is

$$\hat{\mathscr{H}} = B_0(\hat{J}^2 - \Lambda^2) + \beta \boldsymbol{H} \cdot \hat{\boldsymbol{L}} \tag{9.20}$$

Some terms are ignored here, such as a constant displacement term that contributes to all levels and the terms contributing to relieve Λ degeneracy.

The heteronuclear diatomic molecules have electric dipole moment, and their electron magnetic resonance spectroscopy can display the Stark effect, and when the constant electric field is added, the line will undergo splitting. The splitting distance depends on the intensity of electric field and the magnitude of the electric dipole moment. So, the electric dipole moment can be calculated from the Stark effect. For example, the Stark splitting of the ClO molecule can be observed, and the electric dipole moment to be obtained is 1.26*D*. The magnetic resonance spectrum of a gas-phase molecule with nonzero dipole moment can be observed by using a resonant cavity of a modulated electric field. Some results of the Stark modulation cavity are listed in Table 9.3. For studying the weak lines of oxygen-containing molecular systems, it is very useful because many weak transitions to be observed are very difficult.

9.1.3 Gaseous molecules of triatom and polyatom

Only five triatomic molecules have been studied by EMR. NCO and NCS (ground-state $^2\Pi_{3/2}$) are two linear molecules; HCO, NF_2, and NO_2 are three nonlinear molecules. Using EMR, very interesting vibrational effects in both NCO and NCS linear molecules have been detected[26]. The spectrum of the ground state ($^2\Pi_{3/2}$) is very similar to that of the diatomic molecule. However, while meandering vibration makes the state alive, the linearity of molecule is destroyed. This nonlinear excited state mixed in the linear ground state would relieve the degeneracy of the ground state (called Renner effect).

Due to the Renner effect, the spectral analysis of the linear triatomic molecules is much more complex than that of the state $^2\Pi_{3/2}$ of diatomic molecular, although the basic spectral characteristics are the same. The lowest energy states of both NCO and NCS have three hyperfine lines, and each line splits into three lines owing to the second-order Zeeman effect. Besides the ground-state spectrum, the spectra of the excited states $^2\Delta_{5/2}$ of NCO and NCS and the state $^2\Phi_{7/2}$ of NCS are also observed.

Difluoramine (NF_2) radicals are dimerized into tetrafluoro hydrazine easily and there exists a chemical equilibrium between them; from the intensity of EMR[27] the enthalpy change of 81(4) kJ/mol can be obtained during the decomposition of N_2F_4 at 340–435 K.

The gaseous radicals with three or more atoms are almost never observed by EMR, with only one exception of $(CF_3)_2NO$ radical system diluted with different inert gases.

The relationship of its radical concentration and the total pressure P_t has been studied by EMR[28]. Since the *spin–rotational coupling* can be influenced by the increasing collision relaxation, the increase in total pressure P_t causes the spectral line to move and narrow, thereby improving the hyperfine structures of ^{14}N and ^{19}F resolution.

Small enough molecules of NH_2 and CH_3 can be captured in the inert gas freezing lattice. EMR observation indicates that they are actually free rotation in the lattice. At low temperatures, using nuclear spin statistical mechanics can elucidate the relative intensities of hyperfine lines[29].

9.2 Expanding of EMR technique for study on paramagnetic gas

Compared with the condensed state, the number of gaseous paramagnetic particles in the same volume of sample is less by several orders of magnitude. So it is much more difficult to detect them than the condensed sample. And the technique of detection must be developed.

9.2.1 Laser electronic magnetic resonance

Since the first laser electron magnetic resonance (LEPR) spectrum[30,31] appeared in 1968, much information on gaseous radicals has been provided. The principle of LEPR is that the radical is excited by suitable lasers, via far-infrared magnetic field; the resonance absorption lines of radical are detected by laser. The frequency of the laser must be very close to the radical zero field (rotation) frequency. The frequency in the range of 100–3000 GHz is only suitable for studies on the low-molecular-weight gas. The sensitivity of magnetic resonance increases rapidly with increasing frequency, and LEPR is 10^6 times more sensitive than the usual EMR. The absorption of the first LEPR was determined by using HCN lasers at a frequency of 891 GHz and at the magnetic field of 1.6418 T, and the transitions of ground-state O_2 between $N = 3, J = 4, M_J = -4; N = 5, J = 5, M_J = -4$ were also detected. Since then, many radicals (e.g., HC, HN, HS, and DS, HF^+, HCl^+, HBr^+, HSi, CH_2, HO_2, HSO, etc.) have been studied by LEPR; therefore, it is possible to obtain a large number of molecular parameters, including hyperfine coupling parameter.

9.2.2 Magnetic resonance induced by electron

We simply call magnetic resonance induced by electron as MRIE. The gas that entered into the cavity was bombarded by electron and excited to generate radicals. For example, ground-state $N_2(^1\Sigma_g)$ can be excited to be a paramagnetic metastable state

($^3\Sigma_u$), and then subjected to EMR studies. Similarly, $N_2(^1\Sigma_g)$ was first ionized to become N_2^+ ($^2\Sigma_u$), and then back to the ground state ($^2\Sigma_g$).

On the "electron-induced magnetic resonance" technology, a review has been published by Miller and Freund[32] in 1977. They classified MRIE into the following four kinds: (1) Microwave optical magnetic resonance induced by electron (MOMRIE); (2) anticrossing spectroscopy (ACS); (3) level crossing spectroscopy (LCS); (4) molecular beam magnetic resonance (MBMR). The information obtained by all of these four kinds are same as obtained with the usual EMR spectrum, but the detection sensitivity is improved greatly.

9.3 Inorganic radicals

The EMR spectrum analysis of inorganic radical is a very active research field. It is impossible to make a comprehensive introduction here; however, we give a few examples to sketch a contour.

Identification of Inorganic Radical Particles. As with the case of organic radicals, the principal value of hyperfine tensor can provide primary clues to identify the radical species after the inorganic material to be irradiated. For example, the single crystal of LiF irradiated by X-ray at 77 K generates radicals, the external magnetic field $\boldsymbol{H}$// [1 0 0] face measured by EMR spectrum shows three lines of hyperfine structure with intensity ratio of 1:2:1. This means that there are two equivalent nuclei of $I = 1/2$ with unpaired electron interaction. The principal values of their g tensors are 2.0234, 2.0227, 2.0031, indicating that it is similar to the axisymmetric structure. Its hyperfine splitting distances[33] are $a_\parallel = 88.7mT$ and $a_\perp = 5.9mT$. There is no doubt that this spectrum is contributed by F_2^- ion ($V_K$center). If in the experiment KCl is used, then the spectrum of molecular ion $(^{35}Cl^{35}Cl)^-$, $(^{35}Cl^{37}Cl)^-$, and $(^{37}Cl^{37}Cl)^-$ (Figure 9.9) can be obtained, which is clearly contributed by Cl_2^- ions ($V_K$center). How to analyze this spectrum[34] is left to readers as an exercise.

Sometimes the variety of radicals cannot be entirely determined only by EMR spectrum, such as KNO_3 single crystals irradiated by γ-ray, as shown in Figure 9.10. There are at least three different kinds of radicals, each containing a nitrogen atom, because they all exhibit a set of three lines hyperfine structure of equal intensity. However, in order to determine the species of radicals, further information is also necessary and possibly these are NO_2, NO_2^{2-}, NO_3, and NO_3^{2-}. The experimental values of the hyperfine coupling tensor of ^{14}N nucleus and g tensors are shown in Table 9.4.

Identifying these radicals requires the structure of each radical from theoretical and unpaired electrons to be orbital, as well as the properties of these radicals in other substances. For example, the anisotropic hyperfine coupling parameters of the NO_2 in various substrates are quite small and the isotropic coupling parameters are more

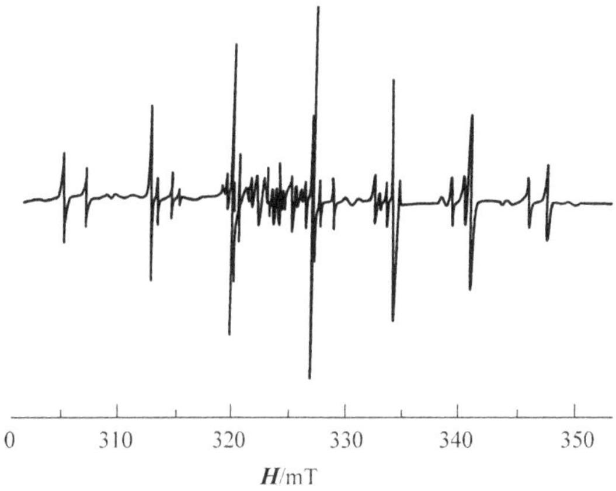

Figure 9.9: EMR spectrum of KCl single crystals irradiated by X-ray under 77 K.

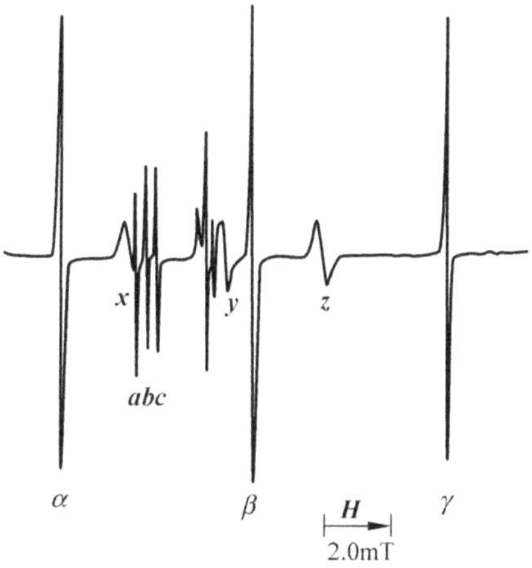

Figure 9.10: EMR spectrum of γ-ray irradiated KNO_3. *A*(α, β, γ) is NO_2; *B*(*a*, *b*, *c*,) is NO_3; *C*(*x*, *y*, *z*) is NO_3^{2-}.

large[35], about 150 MHz. The reason of small anisotropy is NO_2 rotates around the secondary axis frequently, although the "fixed" NO_2in solid state displays considerable anisotropy. The larger hyperfine coupling parameter results from the unpaired electron localized on the nonbond orbit of the nitrogen atom, the *g* tensor is very close to the isotropic value ($g_{iso} \approx 2.000$). Comparing the data in Table 9.4: $A_{\parallel} = 176\,\text{MHz}$ and $A_{\perp} = 139\,\text{MHz}$, it can be deduced that the species A may be NO_2 radical.

As in NO_3 (plane D_{3h} symmetry), the unpaired electron mainly lies on the orbit combined with nonbond p-orbits of oxygen atoms, which are perpendicular to molecular plane. These p-orbits of oxygen atoms overlap top to top mutually, while the probability of unpaired electron reaching the nitrogen nucleus would be quite small, so the hyperfine coupling parameter is also quite small. Analysis of the data of Table 9.4 gives that the species B may be NO_3 radical.

Table 9.4: g and A tensors of species generated from γ-ray irradiated KNO_3.

Species	*g* Tensor	Hyperfine coupling constants of ^{14}N(MHz)
A	$g_{\parallel} = 2.006$[36] $g_{\perp} = 1.996$	$A_{\parallel} = 176$ [36] $A_{\perp} = 139$
B	$g_{\parallel} = 2.0031$[37] $g_{\perp} = 2.0232$	$A_{\parallel} = 12.08$ [37] $A_{\perp} = 9.80$
C	$g_{\parallel} = 2.0015$ [36] $g_{\perp} = 2.0057$	$A_{\parallel} = 177.6$ [36] $A_{\perp} = 89.0$

The hyperfine structure of species C displays considerable interactions between isotropic and anisotropic; it is possible to be NO_3^{2-} because this ion has a less planar structure[38] and belongs to slightly distorted π-radical. Therefore, the differential value of $A_{\parallel}$ and $A_{\perp}$ is relatively large, and the distortion introduces in the unpaired electrons on the orbit some s-orbital characteristics, which can give rise to a larger isotropic hyperfine coupling (~120 MHz).

Structure information. After identifying radical species, g and $\boldsymbol{A}$ tensors can provide a certain information on the geometrical and electronic structures of radicals. The NO_2 radical observed in $NaNO_2$[39] is an excellent example. An electron on 2s orbit of the nitrogen produces isotropic hyperfine splitting constant a = 1540 MHz, while the experimental value of NO_2 is 151 MHz, so the spin density on the 2s orbit of nitrogen is $\rho_s = 151/1540 = 0.10$. In addition, from the maximum of anisotropic hyperfine coupling parameter, the spin density on the $2p_x$ orbit of nitrogen is $\rho_p = 12/48 = 0.25$, more than 2.5 times the 2s orbit. It is known from hybrid orbital theory that its bond angle should be at the intervals 130 ° and 140 °, which is in agreement with the result of gas-phase vibration analysis[40] as well as the result obtained by microwave spectroscopy experiments[41]. The sum of the electron spin densities on the 2s and 2p orbitals of nitrogen is not enough for 1.0, and the rest of the spin density is distributed to the 2p orbital of the oxygen atom.

If the isotropic hyperfine coupling parameter is quite small, the B radicals in Table 9.4 should be handled with care. The unpaired electron located on the orbit should not be mixed into s-component indeed, rather a percentage of s-component should be used. In addition, the indirect mechanism of the spin polarization effect can also contribute to the isotropic hyperfine coupling. In general, $|\rho_s| < 0.05$, although bond angle change cannot explain the phenomenon.

It is interesting to compare the EMR results of the same electron radicals. Some data of ClO_3, SO_3^-, PO_3^{2-}, NO_2, and CO_2^- radicals are listed in Table 9.5. It is clear that when the atomic order of the center atom decreases, the ρ_p/ρ_s ratio also decreases, the four-atoms radical tends more toward the triangular pyramid and the three-atoms radical becomes more curved.

Finally, we take the adsorbed oxygen radicals as an example [47]. The species O^-, O_2^-, and O_3^- of $S = 1/2$ adsorbed on the surface of various materials exhibit characteristic EMR spectrum, which is confirmed by the enrichment of ^{17}O and taking chemical tautomerism. It is very interesting and useful for catalytic research.

9.4 Point defects in solid states

This section is concerned with color center of g tensors of regular orientation sample discussed in Chapter 3. Here we will discuss the point defects in solid from the perspective of the inorganic radicals again.

Different from the "line defect" of solid dislocations, "point defects" are defects in the crystal, such as vacancy, impurity atoms or ions in the (substituted or gap) crystal sit, trapping electron centers capturing hole center, "broken bond," and so on. Most of the point defects are paramagnetic, generally the variety and structure of the defects can be identified from the EMR spectrum. Although the vacancy itself is not paramagnetic, its presence can form some paramagnetic centers.

The formation of the point defects. No matter how much high the purity of the crystal, some point defects still exist. It is particularly true for high-melting-point solids (metal oxides). It often has OH^- ion in alkali metal halide, which is produced by the water in the raw material during the hydrolysis reaction, and O_2^- is also often present in the halide crystals.

The antimagnetic material becomes paramagnetic after irradiation by γ-rays, X-rays, or UV, or it is paramagnetic originally, and it becomes another paramagnetic valence state after irradiation. Irradiation can produce many vacancies, and can produce much anion vacancies in the alkali metal halide. These vacancies can capture one or two free electrons in the radiation. On the other hand, irradiated solids can release electrons from certain crystal sites, which have lower electron affinity. The holes produced can be located on the same crystal sites, and can move in the crystal until it is trapped by impurity ions or anion vacancies. If it is irradiated by electron or rays, the EMR spectrum detects simultaneously the short-life free radical species. Many solids to be irradiated by γ-ray or X-ray cannot make the atoms in the crystal sites. This causes displacement for these substances; using high-energy proton beam or neutron beam irradiation can not only can produce various types of vacancies but also provide some electron.

Table 9.5: EMR Parameters of ClO_3, SO_3^-, PO_3^{2-}, NO_2 and CO_2^-.

Radicals	Bases	g Tensor			A Tensor			
		g_{xx}	g_{yy}	g_{zz}	T_{xx}	T_{yy}	T_{zz}	A_0
$^{35}ClO_3$	$KClO_4$	2.0132	2.0132	2.0066	−40.5	−40.5	81	342
$^{33}SO_3^-$	$K_2CH_2(SO_3)_2$	—	—	—	−37	−39	75	353
$^{31}PO_3^{2-}$	$Na_2HPO_3 \cdot 5H_2O$	2.001	2.001	1.999	−148	−148	297	1660
$^{14}NO_2$	$NaNO_2$	2.0057	2.0015	1.9901	−22.3	37.0	−14.8	153
$^{13}CO_2^-$	$NaHCO_2$	2.0032	2.0014	1.9975	−32.0	78.0	−46.0	468

Impurities. If the impurity is paramagnetic, their EMR spectra can always be observed. Even if it is not paramagnetic, as long as there is adjacent paramagnetic center and its nuclear spin is not zero, the EMR spectrum can be observed.

The simplest impurity defect is the atom that can be produced by irradiation with radiation, but its matrix must have sufficient hardness to prevent its rapid diffusion and make atoms coincide. Below 20 K, the hydrogen atom can be trapped in the acid (H_2SO_4, H_3PO_4, or $HClO_4$) [48]. However, when the temperature rises to a certain degree, the EMR spectrum would disappear. However, the hydrogen atom can be stored at room temperature for several years, if it is to be trapped in CaF_2[49] or $CaSO_4 \cdot 0.5H_2O$[50]. No matter what matrix is, the doublet splitting of proton could be always observed. Its hyperfine coupling parameters are close to the free hydrogen atom a_0 value 1420 MHz, while the experimental value is 1391–1460 MHz. When the hydrogen atom to be generated by the photolysis of HI is trapped in the matrix of the low-temperature rare gas[51], the wave function and the energy level of hydrogen atom are disturbed by the wave function of matrix, so the hydrogen atom occupying different crystal sites could be of different value of hyperfine coupling parameter. Moreover, in the Xe matrix, due to the 1s orbital of the H atom overlapping with the 5s orbital of the Xe atom, the hyperfine structures of ^{129}Xe and ^{131}Xe could be presented. When the hydrogen atom is very close to the particles of matrix, the electron cloud of the hydrogen atom overlaps with the electron cloud of the matrix particle and thus an exchange force is generated. This exchange force can lead to an increasing hyperfine coupling parameter, even larger than the parameter value of the free atom. The another is van der Waals force, which will lead to the decreasing of the hyperfine coupling parameters.

The hydrogen atom trapped in CaF_2 is a good example. In the presence of metallic aluminum, CaF_2 and H_2 are heated together, due to which hydrogen anion H^- is formed on the fluoride ion crystal. After irradiation by X-ray, H^- anion loses an electron and becomes hydrogen atom, it escapes the distance from vacancy to a gap crystal position. Figure 9.11 shows that the hydrogen atom is split into two lines at first, and then each of them splits into nine lines of the intensity ratio according to the

Table 9.5 (continued)

Spin density distribution				Reference
ρ_s	ρ_p	ρ_p/ρ_s	$\rho_s+\rho_p$	
0.076	0.34	4.5	0.42	[42]
0.13	0.49	3.8	0.62	[43]
0.16	0.53	3.3	0.69	[44]
0.099	0.44	4.4	0.54	[45]
0.14	0.66	4.7	0.80	[46]

distribution of binomial coefficient. It proves that there are eight equivalent ^{19}F nuclei around the hydrogen atom, and these eight equivalent ^{19}F nuclei are distributed at the eight vertexes of a normal cube, while the hydrogen atom is at the normal cubic center. When the external magnetic field is parallel to the [1 0 0] direction, as shown in Figure 9.11a, the weak line is the result of the "forbidden" transition. Since the hyperfine coupling is anisotropic, $A_{\parallel}^{F} = 173.8$ MHz and $A_{\perp}^{F} = 69.0$ MHz. When the external magnetic field is parallel to the [1 1 0] direction, the spectrum is as shown in Figure 9.11b.

Paramagnetic impurity defects are of different varieties. As an electron donor, phosphorus, arsenic, antimony, bismuth, and so on have an impurity that can be mixed into silicon, because each of their atoms has one electron more than the substrate. As an electron acceptor, boron, aluminum, gallium, indium, and so on have an impurity that can be mixed into the silicon. EMR and ENDOR technology can study the hyperfine splitting caused by the nucleus of donor, the ENDOR technology is concerned with the first main problem.

ENDOR technology can verify the location of adjacent nuclei, and it can also detect the influence of alkaline earth metal ions present in alkali metal halides. The ENDOR technology has been dealt with in detail in Appendix A1.2.

Many transition metals or rare earth ions are often present as substituting impurities in various matrices and are easily detected by EMR technology, but the analysis of these spectra requires using crystal field theory. We will discuss the problem of crystal field theory in Chapter 10.

Capturing electron center. Anion vacancy capturing an electron is commonly called *F*-center. Figure 9.12 is the *F*-center of the NaH [52], which can be considered as a "quasi-metal" halide. The unpaired electron hyperfine coupling with six nearest equivalent ^{23}Na ($I = 3/2$) nuclei gives 19 lines with an intensity ratio of 1: 6: 21: 56: 120: 216: 336: 456: 546: 580: 546: 456: 336: 216: 120: 56: 21: 6: 1. The hyperfine structure data of the nucleus near the *F*-center can provide a detailed situation of the spatial distribution of the captured electron wave function. Near the *F*-center, there are many progressive shells. Despite the fact that the

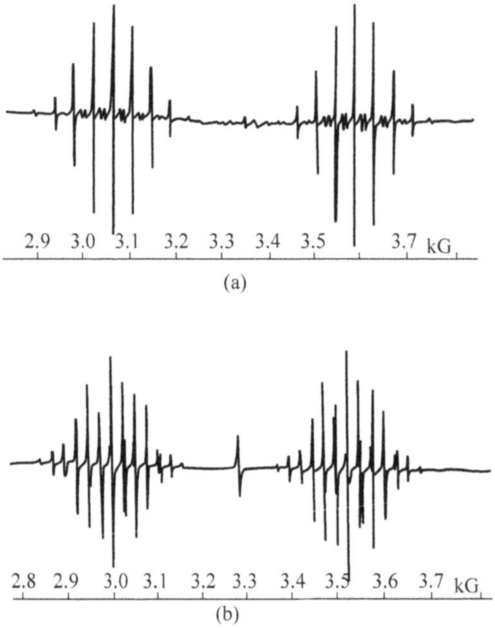

Figure 9.11: EMR spectrum of the hydrogen atom in CaF_2 after X-ray irradiation[49].

farther the distant, the weaker the hyperfine interaction, the EMR spectrum still becomes quite complicated. As most hyperfine lines overlap each other, only a wide envelope line could be observed. Of course, ENDOR technology can help resolve the complex hyperfine structure lines, such as LiF: the detailed information of the first to eighth shell can be obtained with ENDOR technology. Another example is NaF: there are only ^{23}Na and ^{19}F, because the natural abundance of both is 100%. In the case of KCl, there are ^{39}K, ^{41}K, ^{35}Cl, ^{37}Cl, which make the spectrum of the *F*-center all the more complicated.

Anionic vacancy in alkaline earth metal oxides such as MgO capturing an electron is also called *F*-center. Since these ions are divalent, compared to the alkali metal halides, the electrons to be trapped are located in the more deeper potential well and thus the electron wave function is more localized. The magnitude of the hyperfine splitting value can be estimated from the adjacent nucleus. In other alkaline earth metal oxides, sulfides, selenides, some salts or oxides, such as NaN_3[53], CaF_2, SrF_2, BaF_2[54], BeO, ZnO[55], and so on have also been observed in the EMR spectra of *F*-center.

Capturing hole center. The center without electrons form capturing hole center. An electron is removed from the anion to leave a net positive charge. This is the "positive hole," abbreviated as "hole." It can move in the crystal freely, and when encountered by an impurity atom of variable valence or cationic vacancy, it could be

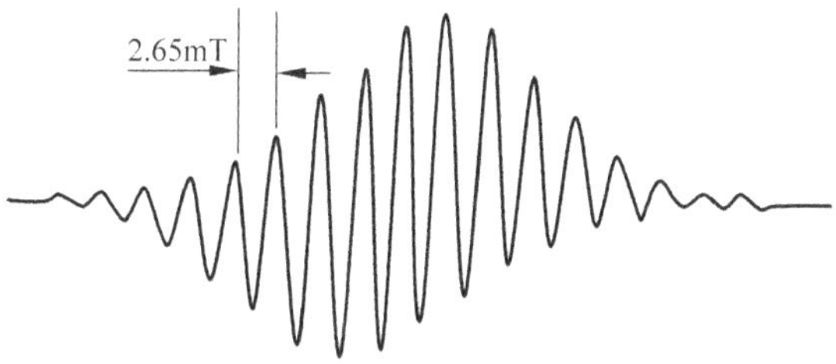

Figure 9.12: EMR spectrum of *F*-center of NaH at 77 K.

captured. It is called V_1 center, when the hole is captured by the cationic valence. When γ-ray irradiate the MgO or CaO crystals, the V_1 center is produced.

In the alkali metal halide, the halide ion (X^-) loses an electron and becomes an (X) atom, this atom of the center in combination with a neighboring (X^-) anion in the [1 1 0] direction is called V_K center. It is, in fact, X_2^-, such as $(^{35}Cl^{35}Cl)^-$, $(^{35}Cl^{37}Cl)^-$, $(^{37}Cl^{37}Cl)^-$ in KCl, the EMR spectrum of which is shown in Figure 9.9. In this case, the X_2^- molecule also occupy only one anion crystal position, which has a strong interaction with the nearest neighbor anion in [1 1 0] direction. In fact, it forms a linear X_4^{3-} ion, called the *H*-center. When the two atoms of both are outside the X_4^{3-} ion, they contribute to the secondary hyperfine splitting. The magnitude is about 1/10 times of the primary splitting of the original X_2^-. The spin density on these two atoms are 0.04–0.10.

9.5 Spectra of conductor and semiconductor

Metal. Metal can be seen as a matrix, like a highly delocalized "ocean" of electrons, which can fix the cations in it. Due to their high mobility, they will interact with each other. However, the EMR signal[56] at the layer close to the surface contributes because the ability of excited field H_1 (microwave) to penetrate the metal is quite weak (only ~1 μm).

Contrary to the normal unpaired electron causing paramagnetism in the magnetic field, the free electron in the metal with large delocalization under the action of the magnetic field makes circular motion. Therefore, the susceptibility of metal is diamagnetics, and the g factor of the observed EMR spectrum is very close to g_e. For example, in liquid- or solid-phase metallic sodium[57] $g - g_e = 9.7(3) \times 10^{-4}$. Due to the result of absorption and scattering mixed action, the line shape of the EMR spectrum is asymmetric (as shown in Figure 9.13)[58]. This effect, as explained by Dyson et al.[59,60], is due to the longer time of electron diffusion to the surface than the spin-relaxation time.

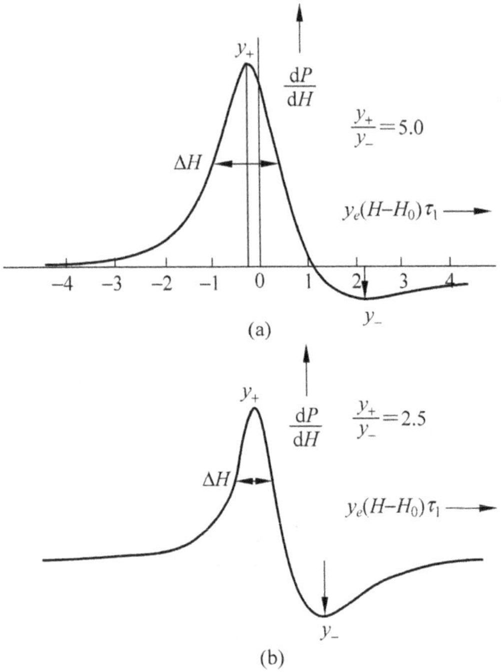

Figure 9.13: (a) First differential curve of the ideal Dyson absorption line. (b) Observed EMR spectrum of first differential curve of Na colloidal samples [58].

Metal in the ammonia and amine solution. When the alkali metal or alkaline earth metal (M) is dissolved in liquid ammonia or liquid amine, metal cations and ionized electrons is produced by ionization. These electrons appear in blue color when diluted in liquid ammonia, and display a very narrow EMR line of g = 2.0012(2) (linewidth only 0.002 mT!). It is neither related to the concentration of the solution (<1 M) nor to the concentration of the M^+ cation[61–63].

In the concentrated solution (bronze), their conductivity becomes more metallic than the electrolyte, and the EMR spectrum broaden and becomes Dysonian line shape [63]. In addition, they can separate solid-state cubic complexes $M(NH_3)_x[asLi(NH_3)_4]$ [64], whose EMR spectral line have Dysonian shape characterization[65] as normal metal behavior.

When the metal is in the dilute solution of amine, its EMR spectrum has a hyperfine splitting of ^{14}Nnucleus with high resolution, which gives the structure of the complex and the kinetic mechanism[66] of the electron interaction with peripheral molecules of solvent.

Semiconductor. The semiconductor actually has a continuous electron energy band, which is composed from all atomic orbitals in the crystal. Its highest occupied band (also called "valence band") is full of electrons, and if it is the second band or unoccupied band (also called as "conduction band"), it is separated by "energy gap," also called "band gap." In the band gap, there are only a few energy levels, or no energy level at all. In the insulator, this band gap is very large, even larger than 4 eV, so the electron from the valence band with thermal excitation to the conduction band is very difficult. As the band gap in the semiconductor is relatively small, only 1–3 eV, the electron transition between the valence band and the conduction band is relatively easy at adequate (gentle) temperature. Therefore, electron (or hole) has conductivity. It is possible to enhance conductivity greatly by mixing appropriate amount of "donor" impurities (n-type) or "acceptor" impurities (p-type), which produces paramagnetic particles.

EMR provides an important tool for semiconductor research, which could identify and clarify the structure of point defects and doping ions. The tetrahedral structure of the solid Si is bombarded by electron to generate defects (V^+, V^0, and V^- centers), where the electrons are captured by the next Si atom with a "rocking" bond"[67–69]. Neutral vacancies (V^0) have four interacting "rocking" bonds, which produce spin-paired; therefore, it is diamagnetics. V^+ and V^- have $S = 1/2$ and exhibit an EMR spectrum, often with a resoluble hyperfine splitting of ^{29}Si. The V^+ center exhibits an EMR spectrum characterized by three equal intensity peaks, there are two weak lines of ^{29}Si hyperfine splitting on both sides of each line [63,70].

The mixing type (Group III–V or II–VI elements) semiconductors have been extensively studied by using EMR / ENDOR. In the p-type semiconductor GaP[71,72], the "antisite" center of the anions, that is, an atom of Group V occupied the position of the Group III atoms, formed a "double donor"-type semiconductor. For example, the $P^{4+}(P^{3-})_4$ center exhibits an EMR spectrum of $g = 2.007(3)$ (see Figure 9.14), which is split into two isotropic hyperfine lines (splitting distance $a_o = 103$ mT) due to the unpaired electron interaction with the ^{31}P nucleus ($I = 1/2$) of the tetrahedral center

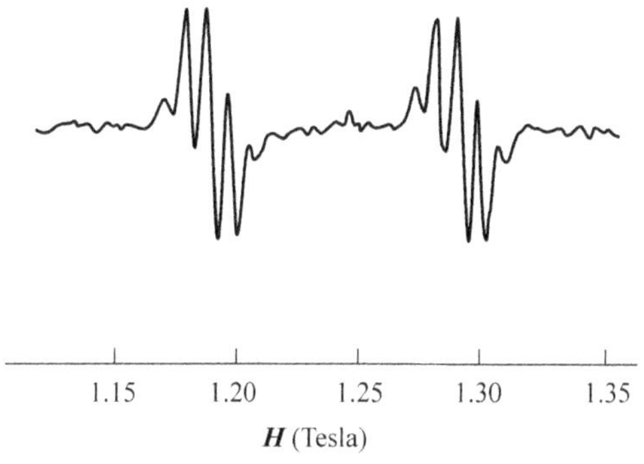

Figure 9.14: EMR spectrum of $^{31}P^{4+}$ ion at the antisite center $[P^{4+}(P^{3-})_4]$ of II–V-type semiconductor GaP (Q band, 20 K, $H \parallel [100]$) [72].

P^{4+} ion. Then, it further interacts with the four equivalent P^{3-} ions at the four vertices of the tetrahedron and split into five anisotropic hyperfine lines (splitting distance a_0~9 mT) with intensity ratio[71,72] of 1:4:6:4:1. It is preferable to study the point defects in semiconductors by photodetection using EMR and ENDOR techniques. When detecting the InP semiconductor doped with Zn by microwave modulation luminescence (0.8 eV), the hyperfine structure[73] of ^{31}P nucleus is the first shell around the antisite center of phosphorus, each of the four ^{31}P nuclei has splitting parameters $A_{\|} = 368.0(5)$ MHz and$A_{\perp} = 247.8(5)$ MHz.

9.6 Structure of a molecule structure of a molecule estimated from the data of EMR

Determining the structure of a molecule requires a lot of data and complicated technical methods; however, using experience and relatively simple methods to obtain information of the molecular structure is preferred. Even though these methods may be partial, relatively rough, but it is very useful. The following is a brief introduction about using electronic "spin–spin" interaction parameters D and J to obtain structural information method.

Newman overlap model. This empirical method[74–76] is used mostly in determining the symmetrical crystal structures (minerals) containing transition metal ions. Using the spin–spin coupling parameter D and taking into account the contribution of the nearest-neighbor ions to the axisymmetric crystal field give the coordination number, the ligand, and the local symmetry. This method is mostly used in oxides and halides of S-state ions (Mn^{2+}, Fe^{3+}, and Gd^{3+}). The Newman model for metal ions M and ligand X is as follows:

$$D = \frac{1}{2} D_o \sum_i (3\cos^2\theta_i - 1) \left(\frac{R_D}{R_i}\right)^{t_D} \tag{9.21}$$

The summation mark is for the nearest neighbor (coordinate atom), θ_i is the angle between the ligand X_i and the z-axis, and R_i is the distance between the center atoms M and X_i. Parameters D_o (related to M, X, R_D), R_D, and t_D are the data of empirical estimates: $t_D = 8 \pm 1$; $1.9\,\mathring{A} \le R_D \le 2.1\,\mathring{A}$. The equation for parameter E is different from Eq. (9.21), under the situation of axisymmetric crystal fields, $E = 0$, and these are expressed only in the case of appropriate angle factor.

The studies on MnX_6^{4-} (X is Cl, Br, or I) system explain that the parameter D_o increases monotonically with the increase of the covalent bond component of the Mn–X bond[77]. Another example of the successful application of the Newman model is the interpretation of the parameters S^2 and S^4 of the cationic position for Fe^{3+} in the Li_2O crystal, where the two adjacent positions are vacancies of Li^+ [78].

The quasi-cube method. This analysis method was developed by Michoulier and Gaite[79,80]. For example, in this way, it is possible to accurately identify that the position of Fe^{3+} ions in the $KTiOPO_4$ crystal is at Ti(1) rather than at Ti(2)[81].

Estimate the distance between two unpaired electrons from the D parameter. In the triplet system, the distance between the two unpaired electrons can be roughly estimated from the main value of the tensor ***D***. Refer to Eqs. (7.38) and (7.52), we can know that D and E are related to the average distance (i.e., on r^{-3}) between the two spin-parallel electrons:

$$D = \frac{3\mu_o}{16\pi}(g\beta_e)^2\left\langle\frac{r^2 - 3Z^2}{r^5}\right\rangle \tag{9.22}$$

$$E = \frac{3\mu_o}{16\pi}(g\beta_e)^2\left\langle\frac{Y^2 - X^2}{r^5}\right\rangle \tag{9.23}$$

where X, Y, Z are the components of the vectors between the two electrons expressed in the principal axis coordinate system. From the resulting experimental values, it is possible to provide information about the spatial deployment of the two electrons. This analysis is valid only when the dipole–dipole interaction is an absolute advantage and the spin–orbit coupling contribution to the D tensor is not significant.

Eaton's interspin distance formula. Eaton proposed a simple formula to estimate the average distance **r** of the interspin centers as follows:

$$\frac{\mathscr{A}(\Delta M_S = \pm 2)}{\mathscr{A}(\Delta M_S = \pm 1)} = k_r r^{-6} \tag{9.24}$$

The left-hand side of the above equation is the ratio of the absorption line area $\mathscr{A}$ of two transitions (see Section 6.3), which is proportional to the average distance r^{-6}, and the scale factor k_r can be obtained by appropriate method[82,83], usually $k_r = 19.5\,Å$. The above formula is effective over the anisotropic exchange interaction only in the case of the dipole–dipole interaction, with an absolute dominance and the average distance $r > 4$ Å. This method is successful in obtaining the distance between two electrons of the $[Cu^{2+}(3d^9) - NO]$ used as the spin label [82].

References

[1] Beringer R., Heald M A. *Phys. Rev.* 1954, **95**: 1474.
[2] Heald M. A., Beringer R. *Phys. Rev.* 1954, **96**: 645.
[3] Ultee C. J. *J. Phys. Chem.* 1960, **64**: 1873.
[4] Ultee C. J. *J. Chem. Phys.* 1965, **43**: 1080.
[5] Radford H. E., Hughes V. W., Beltrain-Lopez V. *Phys. Rev.* 1961, **123**: 153.
[6] Atkins P. W. *Molecular Quantum Mechanics*, 2nd edn, Oxford University Press, Oxford, 1983.

[7] Friedmann P., Schindler R J. *Z. Naturforsch.* 1971, **26a**: 1090.
[8] de Groot M. S., de Lange C. A., Monster A. A. *J. Magn. Reson.* 1973, **10**: 51.
[9] Lilenfeld H. V., Richardson R. J., Hovis F. E. *J. Chem. Phys.* 1981, **74**: 2129.
[10] Beringer R., Castle, J. G., Jr. *Phys. Rev.* 1945, **75**: 1963; 1951, **81**: 82.
[11] Tinkham M, Strandberg M W P. *Phys. Rev.* 1955, **97**: 937, 951.
[12] Weil J A, Bolton J R, Wertz J E. *Electron Paramagnetic Resonance: Elementary Theory and Applications*, 1st edn, John Wiley & Sons, Inc., New York, **1994**, p. 202.
[13] Goldberg I B, Laeger H O. *J. Phys. Chem.* 1980, **84**: 3040.
[14] Goldberg I B, Bard A J. "Electron spin resonance spectroscopy," in Elving P J., Ed., *Treatise on Analytical Chemistry*, 2nd edn, Vol. 10, John Wiley & Sons, Inc., New York, 1983, Chapter 3.
[15] Gerber P. *Helv. Phys. Acta.* 1972, **45**: 655.
[16] Arrington C. A., Jr., Falick A. M., Myers R. J. *J. Chem. Phys.* 1971, **55**: 909.
[17] Miller T. A. *J. Chem. Phys.* 1971, **54**: 330.
[18] Brown R. L., Radford H. E. 1966, *Phys. Rev.*, **147**: 6.
[19] Carrington A., Dyer P. N., Levy D. H. *J. Chem. Phys.*, 1967, **47**: 1756.
[20] Carrington A., Levy D. H. *J. Chem. Phys.* 1966, **44**: 1298.
[21] Frosch R. A., Foley H. M. *Phys. Rev.* 1952, **88**: 1337.
[22] Dousmanis G. C. *Phys. Rev.* 1955, **97**: 967.
[23] Carrington A., Levy D. H., Miller T. A. *Proc. R. Soc. Lond. A* 1967, **A298**: 340.
[24] Carrington A., Levy D. H., Miller T. A. *Mol. Phys.* 1967, **13**: 401.
[25] Carrington A., Levy D. H., Miller T. A. *Proc. R. Soc. Lond. A* 1966, **A293**: 108.
[26] Carrington A., Fabris A. R., Howard B. J., Lucas N. D. J. *Mol. Phys.* 1971, **36**: 20, 961.
[27] Piette L. H., Johnson F. A., Booman K. A., Colburn C. B. *J. Chem. Phys.*, 1961, **35**: 1481.
[28] Schaafsma T. J., Kivelson D. *J. Chem. Phys.* 1968, **49**: 5235.
[29] Weltner W., Jr. *Magnetic Atoms and Molecules*, Van Nostrand Reinhold, New York, 1983, pp. 120–121.
[30] Evenson K. M., Broida H. P., Wells J. S., Mahler R. J., Mizushima M. *Phys. Rev. Lett.* 1968, **21**:1038.
[31] Mizushima M., Evenson K. M., Mucha J. A., Jennings D. A., Brown J. M. *J. Mol. Spectrosc.* 1983, **100**: 303.
[32] Miller T. A., Freund R. S. *Adv. Magn. Reson.* 1977, **9**: 49.
[33] Castner T. G., Kanzig W. *J. Phys. Chem. Solids.* 1957, **3**: 178.
[34] Qiu Z. W. Electron Spin Resonance Spectroscopy. Science Press, Beijing. 1980: 178–180.
[35] Atkins P. W., Symons M. C. R. *J. Chem. Soc.* 1962, 4794.
[36] Zeldes H. "Paramagnetic species in irradiated," in Low W., Ed., *Paramagnetic Resonance*, Vol. 2, Academic Press, New York, 1963, p. 764.
[37] Livingston R., Zeldes H. *J. Chem. Phys.* 1964, **41**: 4011.
[38] Walsh A. D. *J. Chem. Soc.* 1953, 2296.
[39] Zeldes H., Livingston R. *J. Chem. Phys.* 1961, **35**: 563.
[40] Moore G. E. *J. Opt. Soc. Am.* 1953, **43**: 1045.
[41] Bird G. R. *J. Chem. Phys.* 1956, **25**: 1040.
[42] Atkins P. W., Brivati J. A, Keen N., Symons M. C. M., Trevalion P. A. *J. Chem. Soc.*, 1962, 4785.
[43] Chantry G. W., Horsfield A., Morton J. R., Rowlands J. R., Whiffen D. H. *Mol. Phys.* 1962, **5**: 233.
[44] Horsfield A., Morton J. R., Whiffen D .H. *Mol. Phys.* 1961, **4**: 475.
[45] Zeldes H., Livingston R. *J. Chem. Phys.* 1961, **35**: 563.
[46] Ovenall D. W., Whiffen D. H. *Mol. Phys.* 1961, **4**: 135.
[47] Lunsford J. H. *Catal. Rev.* 1973, **8**: 135.

[48] Livingston R., Zeldes H., Taylor E. H. *Dissc. Faraday Soc.* 1955, **19**: 166.
[49] Hall J. L., Schumacher R. T. *Phys. Rev.* 1962, **127**: 1892.
[50] Kon H. *J. Chem. Phys.* 1964, **41**: 573.
[51] Foner S. N., Cochran E. L., Bowers V. A., Jen C. K. *J. Chem. Phys.* 1960, **32**: 963; *Phys. Rev.* 1956, **104**: 846.
[52] Doyle W. T., Williams W. L. *Phys. Rev. Lett.* 1961, **6**: 537.
[53] King G. J., Carlson F. F., Miller B. S., McMillan R. C. *J. Chem. Phys.* 1961, **34**: 1499; 1961, **35**: 1441.
[54] Arends J. *Phys. Status Solidi* 1965, **7**: 805.
[55] Duvarney R. C., Garrison A. K., Thorland R. H. *Phys. Rev.* 1969, **188**: 657.
[56] Edmonds R. N., Harrison M. R., Edwards P. P. *Annu. Rep. Prog. Chem.* 1985, **C82**: 265.
[57] Devine R. A. B., Dupree R. *Philos. Mag.* 1970, **21**: 787.
[58] Vescial F, ven Vander N. S., Schumacher R. T.. *Phys. Rev.* 1964, **134A**: 1286.
[59] Dyson F. J. *Phys. Rev.* 1955, **98**: 349.
[60] Feher G., Kip A. F. *Phys. Rev.* 1967, **158**: 225.
[61] Hutchison C. A., Jr., Pastor R. C. *J. Chem. Phys.* 1953, **21**: 1959.
[62] Das T. P. *Adv. Chem. Phys.* 1962, **4**: 303.
[63] Alger R. S. *Electron Paramagnetic Resonance: Techniques & Applications*, John Wiley & Sons, Inc., New York, 1968, Section 6.3.
[64] Glaunsinger W. S., Sienko M J. *J. Chem. Phys.* 1975, **62**: 1873, 1883.
[65] Dye J. L. *Prog. Inorg. Chem.* 1984, **32**: 327.
[66] Edwards P. P. *J. Solution Chem.* 1985, **14**: 187; *J. Phys. Chem.* 1984, **88**: 3772.
[67] Stutzmann M. *Z. Phys. Chem. N. F.* (in English). 1989, **151**: 211.
[68] Henderson B. *Defects in Crystalline Solids*, E. Arnold, London, 1972, Section 5.3.
[69] Watkins G. D. "EPR studies of lattice defects in semiconductors," in Henderson B, Hughes A E., Eds., *Defects & Their Structure in Non-metallic Solids*, Plenum Press, New York, 1976, p. 203.
[70] Watkins G. D. "EPR & optical absorption studies in irradiated semiconductors," in Vook F L., Ed., *Radiation Damage in Semiconductors*, Plenum Press, New York, 1968, pp. 67–81.
[71] Kaufmann U., Schneider J., Rauber A. *Appl. Phys. Lett.* 1976, **29**: 312.
[72] Kaufmann U., Schneider J. *Festkorperprobleme* 1980, **20**: 87.
[73] Crookham H. C., Kennedy T. A., Treacy D. J. *Phys. Rev.* 1992, **B46**: 1377.
[74] Newman D. J. *Adv. Phys.* 1971, **20**: 197.
[75] Newman D. J., Urban W. *Adv. Phys.* 1971, **24**: 793.
[76] Moreno M. *J. Phys. Chem. Solids.* 1990, **51**: 835.
[77] Heming M, Lehmann G. *Chem. Phys. Lett.* 1981, **80**: 235; Heming M, Lehmann G. "Superposition model for the zero-field splittings of 3d-ion EPR: experimental tests, theoretical calculations and applications." in Weil J A, Ed., *Electron Magnetic Resonance of the Solid State*, Canadian Society for Chemistry, Ottawa, 1987, pp. 163–174.
[78] Baker J. M., Jenkins A. A., Ward R. C. R. *J. Phys. Condens. Matter* 1991, **3**: 8467.
[79] Michoulier J., Gaite J. M. *J. Chem. Phys.* 1972, **56**: 5205.
[80] Gaite J. M. "Study of the structural distortion around S-state ions in crystals, using the fourth-order spin Hamiltonian term of the EPR spectral analysis," in Weil J A., Ed., *Electron Magnetic Resonance of the Solid State*, Canadian Society for Chemistry, Ottawa, 1987, pp. 151–174.
[81] Nizamutdinov N. M., Khasanson N. M., Bulka G. R., Vinokurov V. M., Rez I. S., Garmash V. M., Ravlova N. I. *Sov. Phys. Crystallogr.* 1983, **32**: 408.
[82] Eaton S S, More K. M., Sawant B. M., Eaton G. R. *J. Am. Chem. Soc.* 1983, **105**: 6560.
[83] Coffman R. E., Pezeshk A. *J. Magn. Reson.* 1986, **70**: 21.

Further readings

Monatomic gaseous paramagnetic particles

[1] Carrington A. *Microwave Spectroscopy of Free Radicals*, Academic Press, London, 1974.
[2] Miller T. A. "The spectroscopy of simple free radicals," *Annu. Rev. Phys. Chem.* 1976, **27**: 127.
[3] Levy D. H. "Gas phase magnetic resonance of electronically excited molecules," *Adv. Magn. Reson.* 1973, **6**: 1.
[4] Westenberg A. "Use of ESR for the quantitative determination of gas phase atom & radical concentrations," *Prog. React. Kinet.* 1973, **7**: 23.
[5] Hills G. W. "Laser magnetic resonance spectroscopy," *Magn. Reson. Rev.* 1984, **9**(1–3): 15.
[6] Weltner W., Jr. *Magnetic Atoms and Molecules*, van Nostrand Reinhold, New York, 1983, Chapter 4.
[7] Hudson A., Root K. D. J.. "Halogen hyperfine interactions," *Adv. Magn. Reson.* 1971, **5**: 6–12.

Inorganic radicals

[1] Morton J. R. "Electron spin resonance spectra of oriented radicals," *Chem. Rev.* 1964, **64**: 453.
[2] Atkins P. W, Symons M. C. R.. *The Structure of Inorganic Radicals*, Elsevier, Amsterdam, 1967.

10 Ions of transition elements and their complexes

Abstract: This chapter discusses the EMR spectrum of transition elements (including rare earth) and their compounds, as well as their complexes. The ligand field (containing crystal fields) impact on the central ion energy-level splitting and relieving degeneracy finally reflect the variation of spectral parameters.

The transition elements generally refer to the elements whose 3d, 4d, and 5d shells are not full. From a broader perspective, lanthanum (4f) and actinium (5f) should also be included. Of the 107 known elements, 55 are transition elements. Principally, they are the main subjects for EMR research.

The EMR spectrum of transitional metal complexes and salts have the following features: (1) The spectrum generally has a broad linewidth; (2) They can usually be detected at low temperatures (liquid nitrogen or liquid helium); (3) The spectral explanation needs to consider not only the nature of the ion but also the environment of its surroundings, for example, the symmetry and intensity of the ligand (crystal) fields. Although the theoretical treatment is quite complicated, it nevertheless provides more useful information:

(1) Provides the number of unpaired electrons, valence of ions, and electron configurations.
(2) Identifies the symmetry and intensity of ligand (crystal) fields.
(3) Determines spin Hamiltonian parameters.

In the outer magnetic field, if the free ion is located in a spherical electrical field, then only the Zeeman effect is considered. However, practically (in solid or liquid phase), the symmetry of ligand field is usually less than spherical, such as cubic symmetry, monoclinic symmetry, rhombic symmetry, and so on. The orbital degenerates completely in the field of spherical symmetry. When the symmetry of field is decreasing, its degeneracy is relieved partly or totally, and the energy produced splits. The pattern of splitting depends on the symmetry and intensity of ligand fields. The "ligand field" referred to in this chapter, in a broad sense, considers the crystal field as one of the ligand fields. The research content of EMR spectrum of the transition element and its ion complexes and compounds are very fruitful. Many extensive monographs and reviews about EMR spectrum of transitional elements have already been discussed and published [1–11]. Only a brief introduction is provided in this book.

10.1 Electron ground state of transition element ion

Small alphabets s, p, d, f represent the orbital momentum of single electron $l = 0\hbar, 1\hbar, 2\hbar, 3\hbar$, and so on. Capital letters *S, P, D, F* represent the orbital momentum of whole

https://doi.org/10.1515/9783110568578-010

atom $L = 0\hbar, 1\hbar, 2\hbar, 3\hbar$, and so on. As in $3d^n$ family, the simplest ion is Ti^{3+} $(3d^1)$, its s $= 1/2$, $l = 2$, $M_{l(\max)} = 2$; $S = 1/2$, $L = 2$. ${}^{2S+1}L$ represents ground-state term; the ground-state term of $3d^1$ is ${}^2D$.

See the V^{3+} $(3d^2)$. It should adhere to two principles to determine ground-state term: (1) Pauli principle – there cannot be exactly same quantum number $\{S, M_s, l, M_l\}$ for two electrons; (2) Hund's rule – two electrons in ground state should have spin as parallel as possible. For the two electrons of $3d^2$, all of the M_s are 1/2 and the l are also same, equal to 2; the M_l should be different, for one electron, the $M_{l(\max)} = 2$ and for the other, $M_{l(\max)} = 1$, and $M_L = \Sigma M_l = 2 + 1 = 3$, the ground-state term of $3d^2$ should be ${}^3F$.

For light elements, the angular momentum depends on the spin–orbit coupling (L–S), that is, the total angular momentum vector $\boldsymbol{J}\hbar$ is the sum of angular momentum vector $\boldsymbol{S}\hbar$ and orbital angular momentum vector $\boldsymbol{L}\hbar$. The lengths of their vectors with the order are $\sqrt{S(S+1)}\,\hbar$, $\sqrt{L(L+1)}\,\hbar$, $\sqrt{J(J+1)}\,\hbar$, respectively. The possible values of $\boldsymbol{J}$ are

$$L + S,\ L + S - 1,\ L + S - 2,\ \ldots,\ |L - S|$$

For example, the ground-state term of Ti^{3+} ion is ${}^2D$, and the possible values of $\boldsymbol{J}$ are

$$2 + (1/2) = 5/2 \quad \text{and} \quad 2 - (1/2) = 3/2$$

${}^{2S+1}L_J$ represent the ground-state terms; for the Ti^{3+}, they have ${}^2D_{3/2}$ and ${}^2D_{5/2}$. Owing to existence of spin–orbit coupling, these two energy levels are different. When the d shell is less than half filled, the lowest energy state should be the smaller one of the $\boldsymbol{J}$ values, so that the ground-state term for Ti^{3+} ion is ${}^2D_{3/2}$. When the d shell is larger than half filled, the lowest energy state should be the maximum of the $\boldsymbol{J}$ values. For example, the ground-state term of both $3d^3$ and $3d^7$ ions is ${}^4F$. The possible values of $\boldsymbol{J}$ are

$$J_{\max} = 3 + (3/2) = 9/2,\ 7/2,\ 5/2; \quad J_{\min} = 3 - (3/2) = 3/2$$

Because the $3d^3$ ion is less than half filled, the ground-state term should be ${}^4F_{3/2}$, while for the $3d^7$ ion of more than half filled, the ground-state term should be ${}^4F_{9/2}$.

The ground-state terms of $3d^n$ ions are listed in Table 10.1. L has only three values of 0, 2, and 3, and are denoted the states S, D, and F, respectively.

The energy of free ion is W0 when external magnetic field $H = 0$, while it is not 0,

$$W = W_0 + g_J \beta M_J H$$

M_J is the component of J along H direction, g_J is Landé cofactor, when

$$J \neq 0 \quad \rightarrow \quad g_J = 1 + \frac{J(J+1) + S(S+1) + L(L+1)}{2J(J+1)} \tag{10.2}$$

when

Table 10.1: Ground-state terms of $3d^n$ ion.

d-Electron number	Ground state			Orbital degeneracy	Term	Examples
	S	L	J			
1	1/2	2	3/2	5	$^2D_{3/2}$	Sc^{2+}, Ti^{3+}, VO^{2+}, Cr^{5+}
2	1	3	2	7	$^3F_2$	Ti^{2+}, V^{3+}, Cr^{4+}
3	3/2	3	3/2	7	$^4F_{3/2}$	Ti^+, V^{2+}, Cr^{3+}, Mn^{4+}
4	2	2	0	5	$^5D_0$	Cr^{2+}, Mn^{3+}, V^+, Fe^{4+}
5	5/2	0	5/2	1	$^6S_{5/2}$	Cr^+, Mn^{2+}, Fe^{3+}, Co^{4+}
6	2	2	4	5	$^5D_4$	Mn^+, Fe^{2+}, Co^{3+}
7	3/2	3	9/2	7	$^4F_{9/2}$	Fe^+, Co^{2+}, Ni^{3+}
8	1	3	4	7	$^3F_4$	Fe^0, Co^+, Ni^{2+}, Cu^{3+}
9	1/2	2	5/2	5	$\beta^2D_{5/2}$	Ni^+, Cu^{2+}

$$J = 0, \quad \rightarrow \quad g_j = L + 2 \tag{10.3}$$

It should be noted that for different J values, the Landé cofactor g_J is different. It has been mentioned that the transition is permitted only when $\Delta M_J = \pm 1$. Free ions are not discussed in this chapter. In condensed phase, coordination field is always lower than spherical symmetry. The Landé cofactor equation is not appropriate and a new equation needs to be formulated.

10.2 Orbital degeneracy is rescinded in ligand field

Generally speaking, the orbit of free ion is degeneracy ($2L + 1$) (see Table10.1), while in electric field brought by ligands, it will be partly or totally removed. Spin degeneracy ($2S + 1$ for free ion) can be partly or totally avoided by spin–orbit coupling, and the whole degeneracy depends on the symmetry of the crystal field. The energy-level degeneracy of transitional $3d^n$ ion in different symmetrical crystal field is shown in Table 10.2.

Many types of interaction of different strength exist in the molecules; based on the interaction energy, an energy level diagram (the strongest interaction energy probably is the total energy 90%) is set up, and then rather weak interaction energy is introduced in proper order. The interaction energies of molecules contain following terms:

$$\hat{\mathscr{H}} = \hat{\mathscr{H}}_E + \hat{\mathscr{H}}_C + \lambda \boldsymbol{L} \cdot \boldsymbol{S} + \beta(\boldsymbol{L} + g_e\boldsymbol{S}) \cdot \boldsymbol{H} + \hat{\mathscr{H}}_{SS} + \hat{\mathscr{H}}_{SI} + \hat{\mathscr{H}}_{I\mathscr{H}} + \hat{\mathscr{H}}_Q \tag{10.4}$$

Here, the $\hat{\mathscr{H}}_E$ is the electronic kinetic energy and potential energy ($\sim 10^5 cm^{-1}$) in the atom; $\hat{\mathscr{H}}_C$ is the crystal field energy; $\lambda \boldsymbol{L} \cdot \boldsymbol{S}$ is the spin–orbit coupling energy of electron; the $\beta(\boldsymbol{L} + g_e\boldsymbol{S}) \cdot \boldsymbol{H}$ is the Zeeman interaction energy between magnetic moments of

Table 10.2: Energy-level degeneracy of transitional $3d^n$ ion in different crystal field symmetry [12][a].

Symmetry	d^1	d^2	d^3	d^4	d^5	d^6	d^7	d^8	d^9
	Orbital degeneracy in different symmetrical crystal fields[b]								
Free ion	5	7	7	5	1	5	7	7	5
Octahedron[c]	$2, 3^d$	$1{,}2 \cdot 3$	$1^d, 2 \cdot 3$	$2^d, 3$	1	$2, 3^d$	$1, 2 \cdot 3^d$	$1^d, 2 \cdot 3$	$2^d, 3$
Tetrahedron	$2^d, 3$	$1^d, 2 \cdot 3$	$1{,}2 \cdot 3$	$2, 3^d$	1	$2^d, 3$	$1^d, 2 \cdot 3$	$1{,}2 \cdot 3$	$2, 3^d$
Pyramid	$1{,}2 \cdot 2$	$3 \cdot 1{,}2 \cdot 2$	$3 \cdot 1{,}2 \cdot 2$	$1{,}2 \cdot 2$	1	$1{,}2 \cdot 2$	$3 \cdot 1{,}2 \cdot 2$	$3 \cdot 1{,}2 \cdot 2$	$1{,}2 \cdot 2$
Square pyramid	$3 \cdot 1{,}2$	$3 \cdot 1{,}2 \cdot 2$	$3 \cdot 1{,}2 \cdot 2$	$3 \cdot 1{,}2$	1	$3 \cdot 1{,}2$	$3 \cdot 1{,}2 \cdot 2$	$3 \cdot 1{,}2 \cdot 2$	$3 \cdot 1{,}2$
Rhombus	$5 \cdot 1$	$7 \cdot 1$	$7 \cdot 1$	$5 \cdot 1$	1	$5 \cdot 1$	$7 \cdot 1$	$7 \cdot 1$	$5 \cdot 1$
	Spin degeneracy in different symmetrical crystal fields[b]								
Free ion	2	3	4	5	6	5	4	3	2
Cubic	2	3	4	2,3	2,4	2,3	4	3	2
Pyramid	2	1,2	$2 \cdot 2$	$1{,}2 \cdot 2$	$3 \cdot 2$	$1{,}2 \cdot 2$	$2 \cdot 2$	1,2	2
Square pyramid	2	1,2	$2 \cdot 2$	$3 \cdot 1{,}2$	$3 \cdot 2$	$3 \cdot 1{,}2$	$2 \cdot 2$	1,2	2
Rhombus	2	$3 \cdot 1$	$2 \cdot 2$	$5 \cdot 1$	$3 \cdot 2$	$5 \cdot 1$	$2 \cdot 2$	$3 \cdot 1$	2

[a] Here, planar square symmetry is not included.
[b] The $\boldsymbol{m} \cdot \boldsymbol{n}$ means there are $\boldsymbol{m}$ state in $\boldsymbol{n}$ degeneracy.
[c] The order of these states are reverse of tetrahedron symmetrical field.
[d] Lower or the lowest state.

electronic spin and orbital with external magnetic field; $\hat{\mathcal{H}}_{SS}$ is the dipole–dipole interaction energy between electron spins (10^{-1}~$1\,cm^{-1}$); $\hat{\mathcal{H}}_{SI}$ is the hyperfine interaction energy between unpaired electron spin and magnetic nuclear ($\sim 10^{-1}\,cm^{-1}$); $\hat{\mathcal{H}}_{IH}$ is the Zeeman interaction energy between magnetic nuclear and external magnetic field $(10^{-4} - 10^{-2}\ cm^{-1})$; and $\hat{\mathcal{H}}_Q$ is the interaction energy between nuclear quadrupole moment with heterogeneous electrical field $(10^{-4} - 10^{-2}\ cm^{-1})$.

It is obvious that the first term $\hat{\mathcal{H}}_E$ is the strongest interaction energy in the energy diagram, Zeeman interaction energy is comparatively smaller, and $\hat{\mathcal{H}}_{SS}$, $\hat{\mathcal{H}}_{SI}$, $\hat{\mathcal{H}}_{IH}$, and $\hat{\mathcal{H}}_Q$ are smaller still. For the ion, which is located in the complex, the relative magnitude between the action energy of crystal electric field $\hat{\mathcal{H}}_C$ and the spin–orbit coupling action energy $\lambda L \cdot S$ is extremely important. Based on their relative magnitude, the crystal field can be classified into three types: (1) weak field; (2) middle field; and (3) strong field.

(1) *Weak field:* Defined the $\hat{\mathcal{H}}_c < \lambda \boldsymbol{L} \cdot \boldsymbol{S}$ as weak field. In the rare earth ions of lanthanide and actinium series, the 4f or 5f electrons are shielded by the outer shell electrons, where most crystals belong to the weak field action. So the crystal field effect is rather weak and the spin–orbit coupling effect is rather strong. Its $\boldsymbol{L}$ and $\boldsymbol{S}$ sum to the total angular momentum $\boldsymbol{J}$. For a given J, there are $2J+1$ values of M_J, they are $M_J = J, J-1, J-2, \ldots, -J+2, -J+1, -J$. Therefore, M_L and M_S are no longer significant quantum numbers for the ions of lanthanide and actinium series. In the case of weak field, spin–orbit coupling energy on the order of

$5\times10^3\ \mathrm{cm}^{-1}$, while in the crystal field, the splitting of M_J state is only about 100 cm^{-1}. Because of crystal field, these states split into doublet degenerate states as $\pm M_J$, which means $+M_J$ and $-M_J$ are on the same energy level. If J is an integer, there will also exist a nondegenerate state $M_J=0$. Because the crystal field is relatively more weak, the rare earth ions that are located in the crystal or solution have similar magnetic susceptibility to their free ions. The rare earth ions generally have trigonal symmetry rather than octahedron or tetrahedron.

(2) *Intermediate field:* Defined the $\hat{\mathscr{H}}_C > \lambda \boldsymbol{L}\cdot\boldsymbol{S}$ as middle field. Owing to the d electron in outer shell, there is a very strong interaction within crystal field, and since the *d* electrons are located in different orbitals, their strength of crystal field action is also different, leading to the energy-level splitting of d orbitals. For 3d, there are five degenerate orbitals: d_{z^2}, $d_{x^2-y^2}$, d_{xy}, d_{yz}, d_{zx} in the spherical field. When the symmetry of field changes to six coordination octahedron symmetrical field, the six ligands have negative ions; due to the electronic cloud of d_{z^2} and $d_{x^2-y^2}$ opposite to six negative ions directly, the repelling energy is rather high. So the energy level of d_{z^2} and $d_{x^2-y^2}$ is higher than that of d_{xy}, d_{yz}, d_{zx}. Then, the energy level is divided two: doublet degeneracy of d_{z^2} and $d_{x^2-y^2}$ on the upside, and triplet degeneracy of d_{xy}, d_{yz}, d_{zx} on the downside. Note that this energy level splitting is quite large, the crystal field splitting of $3d^n$ is about $10^4\ \mathrm{cm}^{-1}$ order, while spin–orbit coupling is about 50–850 cm^{-1} in this situation. The magnetic susceptibility of $3d^n$ in this crystal field should be sharply different from free ion.

(3) *Strong field:* The so-called strong field means that there exists a covalent bond between $3d^n$ ion and diamagnetic ligand. The $\hat{\mathscr{H}}_C$ is greater than not only the energy of spin–orbit coupling interaction but also the energy of electrostatic interaction $\hat{\mathscr{H}}_C > \hat{\mathscr{H}}_E \gg \lambda\boldsymbol{L}\cdot\boldsymbol{S}$. Thus, for the first-order approximate treatment, the ligand electrons should also be considered except the 3*d* electrons. It should be pointed out that the *J* is no longer a good quantum number; it is not significant to add a subscript *J* in spectral term.

The unpaired electron located in d orbital is the main subject of discussion in this chapter, which in most situations is found in the middle field. Strong field and weak field are very briefly discussed here.

Two important theorems introduced here are as follows:

Theorem I *For nondegenerate state, its state wave function must be real function. Whereas, if the wave function is complex function, the energy state is doublet degenerate at least.*

Proof: Assume $\psi=f+ig$ is the eigenfunction of operator $\hat{\mathscr{H}}$. Now we need to prove that its conjugate complex function $\psi^*=f-ig$ also is the eigenfunction of $\hat{\mathscr{H}}$. It also will be proved that the energy state is doublet degenerate at least.

For the $\psi=f+ig$ is the eigenfunction of operator $\hat{\mathscr{H}}$,

So $\hat{\mathcal{H}}(f + ig) = E(f + ig)$

$\hat{\mathcal{H}} f + Ef \;\; \hat{\mathcal{H}} g = Eg$

Then, $\hat{\mathcal{H}}(f - ig) = E(f - ig)$

That is, $\hat{\mathcal{H}}\psi^* = E\psi^*$

If the state is nondegenerate, its eigenfunction must be real function.

Theorem II *For nondegenerate state, its orbital angular momentum and orbital magnetic moment should be zero. That is, the orbital angular momentum must be quenched entirely.*

Proof: Assume $|n\rangle$ is the wave function of nondegenerate orbital. Now we want to prove that the angular momentum operator $\hat{L}_z$ applied to the $|n\rangle$ getting the eigenvalue M_l only can be zero. Please see the following equation:

$$\hat{L}_z = i\hbar\left(x\frac{\partial}{\partial y} - y\frac{\partial}{\partial x}\right)$$

It should be a pure imaginary operator, and it is also an Hermitian operator, so, its eigenvalue should be real number or zero. Since $|\mathrm{n}\rangle$ is an orbital nondegenerate state wave function, which must be real function, so, it makes the following equation to be tenable:

$$\hat{L}_z|n\rangle = M_l|n\rangle$$

and the $\mathrm{M_l}$ only can be pure imaginary number or zero, from the Hermitian property of operator $\hat{L}_z$ to know, the M_l cannot be imaginary number, so, it only can be zero.

When the ion has a spheric symmetry field, it generally always adopts the common eigenfunctions of $\hat{L}^2$ and $\hat{L}_Z$ as the wave functions of angular momentum. While in the crystal field, because energy level needs to be split, orbital degeneracy is partially or totally relieved, that means the orbital angular momentum partially or totally quenched. If it is entirely quenched, the eigenfunction should be a real function. Both real and complex forms of various orbital angular momentum wave functions are listed in Table 10.3.

10.3 Electric potential of ligand field

In crystal field theory, ligands are treated as negative point charges, and these negative point charges of regular arrangement produce a crystal field for central magnetic ion. Crystal fields of common sight are as follows:

Charge numbers	Polyhedron	Symmetry
4	Tetrahedron	Tetrahedron symmetry
6	Octahedron	Octahedron symmetry
8	Cube	Cube symmetry (octave symmetry)

All the forms are the base of cube, others as tetrahedron and octahedron all can be viewed as inscribed polyhedron of cube, so that it must use more definite names of symmetry. Octave symmetry is the base of dodecahedron, for example, $Mo(CN)_8^{3-}$.

The arbitrary point is R_j, whose radial distance from original point is r, where the negative point charge q_j gives rise to electric potential V_j:

$$V_j = \frac{q_j}{|R_j - r|} \tag{10.5}$$

All point charges of R_j give the total electric potential V:

$$V = \sum_j V_j = \sum_j \frac{q_j}{|R_j - r|} \tag{10.6}$$

Our purpose is to seek the interaction energy between this potential with the unpaired electron in a some specific orbit. If the unpaired electron at r_i is q_i, the potential energy of crystal field V_{crystal} is

$$V_{\text{crystal}} = \sum_i q_i V_i = \sum_i \sum_j \frac{q_i q_j}{|R_j - r_i|} \tag{10.7}$$

The electric potential as a result of these point charges is dependent on their arrangement in the crystal field and the distance R of central ion. The arrangement of six coordination octahedron symmetry is shown in Figure 10.1. There are six point charges at $x = \pm R$, $y = \pm R$, $z = \pm R$. Then,

$$V(x, y, z) = V_x + V_y + V_z$$

V_y is $y = \pm R$, the potential of two point charges at r_i:

$$V_y = q[(r^2 + R^2 - 2yR)^{-1/2} + (r^2 + R^2 + 2yR)^{-1/2}] \tag{10.8}$$

V_x and V_z can be obtained similarly. Adding all three equations and simplifying by using series expansion, we can get

$$V_{\text{Octah.}}(x, y, z) = \frac{6q}{R} + \frac{35q}{4R^5}\left[(x^4 + y^4 + z^4) - \frac{3}{5}r^4\right]$$
$$-\frac{21q}{2R^7}\left[(x^6 + y^6 + z^6) + \frac{15}{4}(x^2y^4 + x^2z^4 + y^2x^4 + y^2z^4 + z^2x^4 + z^2y^4) - \frac{15}{14}r^6\right] \tag{10.9}$$

Only the first two terms are needed for 3d electron, the third term should be included for 4f electron.

For the situation of distorted octahedron field, the six point charges at $x = \pm R$, $y = \pm R$, $z = \pm (R + \varepsilon)$, its crystal field potential V_{Tetrag} is

Table 10.3: Both real and complex forms of various orbital angular momentum wave functions.

Angular momentum	Complex form[a]
$l = 0$ s-orbit	
$l = 1$ p orbit *P* energy state	$-\sqrt{\frac{3}{8\pi}}\sin\theta e^{i\phi} = \|1, 1\rangle$
	$\sqrt{\frac{3}{4\pi}}\cos\theta = \|1, 0\rangle$
	$\sqrt{\frac{3}{8\pi}}\sin\theta e^{-i\phi} = \|1, -1\rangle$
$l = 2$ d orbit *D* energy state	$\sqrt{\frac{15}{32\pi}}\sin^2\theta e^{2i\phi} = \|2, 2\rangle$
	$-\sqrt{\frac{15}{8\pi}}\sin\theta\cos\theta e^{i\phi} = \|2, 1\rangle$
	$\sqrt{\frac{5}{16\pi}}(3\cos^2\theta - 1) = \|2, 0\rangle$
	$\sqrt{\frac{15}{8\pi}}\sin\theta\cos\theta e^{-i\phi} = \|2, -1\rangle$
	$\sqrt{\frac{15}{32\pi}}\sin^2\theta e^{-2i\phi} = \|2, -2\rangle$
$l = 3$f orbit *F* energy state	$-\sqrt{\frac{35}{64\pi}}\sin^3\theta e^{3i\phi} = \|3, 3\rangle$
	$\sqrt{\frac{105}{32\pi}}\sin^2\theta\cos\theta e^{2i\phi} = \|3, 2\rangle$
	$-\sqrt{\frac{21}{64\pi}}\sin\theta(5\cos^2\theta - 1)e^{i\phi} = \|3, 1\rangle$
	$\sqrt{\frac{7}{16\pi}}(5\cos^3\theta - 3\cos\theta) = \|3, 0\rangle$
	$\sqrt{\frac{21}{64\pi}}\sin\theta(5\cos^2\theta - 1)e^{-i\phi} = \|3, -1\rangle$
	$\sqrt{\frac{105}{32\pi}}\sin^2\theta\cos\theta e^{-2i\phi} = \|3, -2\rangle$
	$\sqrt{\frac{35}{64\pi}}\sin^3\theta e^{-3i\phi} = \|3, -3\rangle$

[a] Eigen wavefunctions of $\hat{L}_z$.
[b] Possessing symmetry as noted.
[c] *A, E, T* means nondegenerate, double degenerate, and triple degenerate, respectively; subscript means symmetry [13].

Table 10.3 (continued)

Real form[b]	Octahedron group[c]
$\frac{1}{\sqrt{4\pi}}$	A_1
$p_x = \sqrt{\frac{3}{4\pi}}\sin\theta\cos\phi = \frac{1}{\sqrt{2}}(\|1,-1\rangle - \|1,1\rangle)$	T_1
$p_y = \sqrt{\frac{3}{4\pi}}\sin\theta\sin\phi = \frac{i}{\sqrt{2}}(\|1,-1\rangle + \|1,1\rangle)$	
$p_z = \sqrt{\frac{3}{4\pi}}\cos\theta = \|1,0\rangle$	
$d_{xy} = \sqrt{\frac{15}{16\pi}}\sin^2\theta\sin^2\phi = \frac{-i}{\sqrt{2}}(\|2,2\rangle - \|2,-2\rangle)$	T_2
$d_{xz} = \sqrt{\frac{15}{16\pi}}\sin 2\theta\cos\phi = \frac{1}{\sqrt{2}}(\|2,-1\rangle - \|2,1\rangle)$	
$d_{yz} = \sqrt{\frac{15}{16\pi}}\sin 2\theta\sin\phi = \frac{i}{\sqrt{2}}(\|2,-1\rangle + \|2,1\rangle)$	
$d_{x^2-y^2} = \sqrt{\frac{15}{16\pi}}\sin^2\theta\cos 2\phi = \frac{1}{\sqrt{2}}(\|2,2\rangle + \|2,-2\rangle)$	E
$d_{z^2} = \sqrt{\frac{5}{16\pi}}(3\cos^2\theta - 1) = \|2,0\rangle$	
$f_{xyz} = \frac{-i}{\sqrt{2}}(\|3,2\rangle - \|3,-2\rangle)$	A_2
$f_{x(y^2-z^2)} = \frac{1}{4}[\sqrt{5}(\|3,-1\rangle - \|3,1\rangle) - \sqrt{3}(\|3,3\rangle - \|3,-3\rangle)]$	T_2
$f_{y(z^2-x^2)} = \frac{-i}{4}[\sqrt{5}(\|3,-1\rangle + \|3,1\rangle) - \sqrt{3}(\|3,3\rangle + \|3,-3\rangle)]$	
$f_{z(x^2-y^2)} = \frac{1}{\sqrt{2}}(\|3,2\rangle + \|3,-2\rangle)$	
$f_{x(2x^2-3y^2-3z^2)} = \frac{1}{4}[\sqrt{3}(\|3,1\rangle - \|3,-1\rangle) + \sqrt{5}(\|3,-3\rangle - \|3,3\rangle)]$	T_1
$f_{y(2y^2-3x^2-3z^2)} = \frac{-i}{4}[\sqrt{3}(\|3,1\rangle + \|3,-1\rangle) + \sqrt{5}(\|3,-3\rangle + \|3,3\rangle)]$	
$f_{z(2z^2-3x^2-3y^2)} = \|3,0\rangle$	

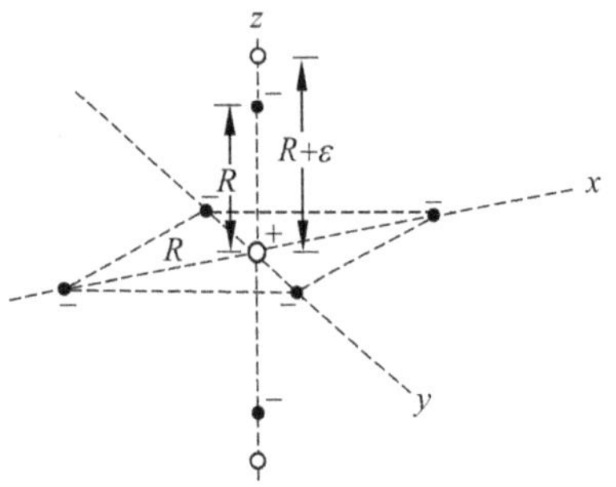

Figure 10.1: Octahedron symmetry arrangement of six charges (distance = R from central ion).

$$V_{\text{Tetrag.}} = A_t\left[(3z^2 - r^2) + \frac{1}{R^2}\left(\frac{35}{3}z^4 - 10r^2z^2 + r^4\right)\right] + B_c\left(x^4 + y^4 + z^4 - \frac{3}{5}r^4\right) \quad (10.10)$$

where $A_t = -\frac{3q\varepsilon}{R^4}$, $B_c = \frac{35q}{4R^5}$ (when the case of $\varepsilon \ll R$). It can be seen that the first term is tetragonal distortion, and the second term is octagonal potential. The potential of distorted octahedron field is the tetragonal distortion term add to the potential of octahedron field.

10.4 Energy-level splitting of transition metal ion in ligand field

Because of crystal field, the potential energy of d electron can be calculated using the matrix elements $\langle J, M_J|\hat{\mathscr{H}}_C|J, M_J\rangle$, where J and M_J are total angular momentum and its component of central magnetic ion, and $\hat{\mathscr{H}}_C$ consists of potential operator V_{Octah} or V_{Tetrag}. This method is very complex. Stevens[14] uses the permutation method for calculation, which is much easier (for details, please see Ref.[15]). Here, $\hat{\mathscr{H}}_{\text{Octah.}}$ and $\hat{\mathscr{H}}_{\text{Tetrag.}}$ can be calculated as follows:

$$\hat{\mathscr{H}}_{\text{Octah.}} = \frac{\beta_C}{20}[35\hat{L}_z^4 - 30L(L+1)\hat{L}_z^2 + 25L_z^2 - 6L(L+1) + 3L^2(L+1)^2] + \frac{\beta_C}{8}(\hat{L}_+^4 + \hat{L}_-^4) \quad (10.11)$$

$$\hat{\mathscr{H}}_{\text{Tetrag.}} = \hat{\mathscr{H}}_{\text{Octah.}} + \alpha_t[3\hat{L}_z^2 - L(L+1)] \quad (10.12)$$

For tetragonal field, its Hamiltonian operator $\hat{\mathscr{H}}_{Tetrag}$ is $\hat{\mathscr{H}}_{octah}$, which is added to the distortion term of tetragonal $\alpha_t[3\hat{L}_z^2 - L(L+1)]$; however, the sign and value of β_C are different from octagonal field: $(\beta_C)_{Tetrag} = -\frac{4}{9}(\beta_C)_{Octah}$, α_t is tetragonal distortion factor.

10.4.1 *P*-state ion in octahedron Field (L = 1)

Take the complex function $|1\rangle, |0\rangle, |-1\rangle$ of Table 10.3 as the base function, because for $L = 1$, $\hat{\boldsymbol{L}}_\pm^4$ is a zero matrix, so that $\hat{\boldsymbol{L}}_\pm^4$ does not contribute to $\hat{\mathscr{H}}_{Octah}$. The matrix representation of $\hat{\mathscr{H}}_{Octah}$ is

$$\hat{L}_z^4 = \hat{L}_z^2 = \begin{array}{c} \\ \langle 1| \\ \langle 0| \\ \langle -1| \end{array} \begin{array}{c} |1\rangle \ |0\rangle \ |-1\rangle \\ \begin{bmatrix} 1 & 0 & 0 \\ 0 & 0 & 0 \\ 0 & 0 & 1 \end{bmatrix} \end{array} \tag{10.13}$$

putting this in Eq. (10.11) we get

$$\hat{\mathscr{H}}_{Octah} = \begin{array}{c} \\ \langle 1| \\ \langle 0| \\ \langle -1| \end{array} \begin{array}{c} |1\rangle \ |0\rangle \ |-1\rangle \\ \begin{bmatrix} 0 & 0 & 0 \\ 0 & 0 & 0 \\ 0 & 0 & 0 \end{bmatrix} \end{array} \tag{10.14}$$

It is said that *the normal octahedron field cannot relieve the orbital degeneracy of P state ion.*

Let us see the distorted tetragonal field. Its Hamiltonian operator $\hat{\mathscr{H}}_{Octah}$ is added to the term $\alpha_t[3\hat{L}_z^2 - L(L+1)]$. Its matrix representation is

$$\hat{\mathscr{H}}_{Tetrag} = \begin{array}{c} \\ \langle 1| \\ \langle 0| \\ \langle -1| \end{array} \begin{array}{c} |1\rangle \quad |0\rangle \ |-1\rangle \\ \begin{bmatrix} \alpha_t & 0 & 0 \\ 0 & -2\alpha_t & 0 \\ 0 & 0 & \alpha_t \end{bmatrix} \end{array} \tag{10.15}$$

Thus, it can be seen that, because of octagonal field of tetragonal distortion, P state ion can be split. When $\alpha_t > 0$, the energy of state $|\pm 1\rangle$ will increase α_t, whereas the energy of state $|0\rangle$ will decrease $2\alpha_t$ (refer to Figure 10.2). $\alpha_t > 0$ means that the two point charges of up and down at z-axis is elongation distortion (as $\varepsilon > 0$ of Figure 10.1,), whereas $\alpha_t < 0$ means it is depression distortion.

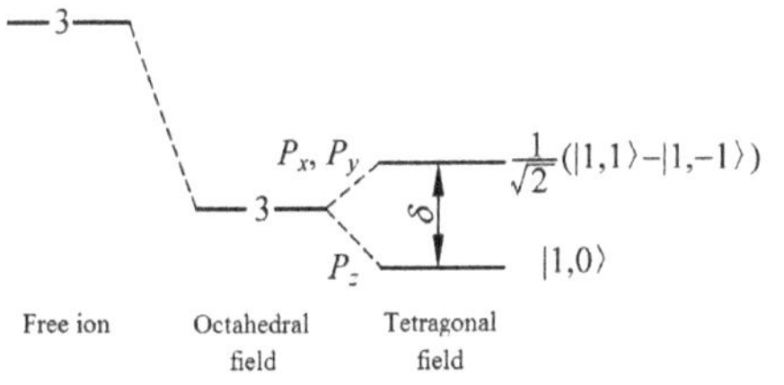

Figure 10.2: P-state energy splitting in octahedron field of tetragonal distortion.

10.4.2 *D*-state ion

In sphere field, *D*-state ion is quintet degeneracy. The degeneracy is relieved partially in octahedron or tetrahedron field. The relative negative point charges location of electron cloud distribution are shown in Figure 10.3 for five d orbits. Three (d_{xy}, d_{yz}, d_{zx}) of which are far from point charge relatively, the other two, $d_{x^2-y^2}$ and d_{z^2}, are partially facing the point charges directly. Then the five d orbits split into two groups: one to double degeneracy, and the other to triple degeneracy.

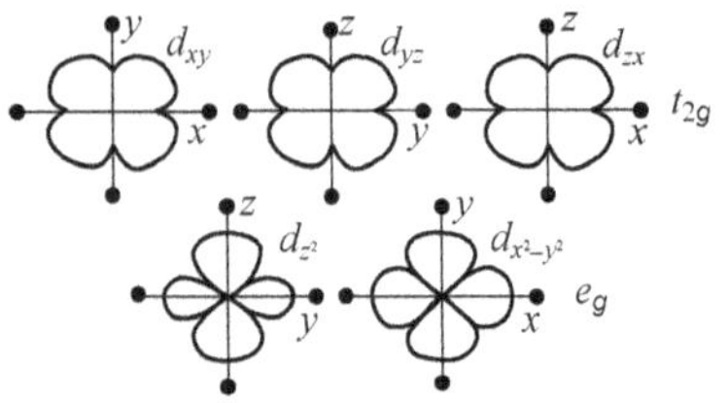

Figure 10.3: Relative location of 3d orbits and charges.

If desired to seek the matrix representation of D state ion $\hat{\mathscr{H}}_{\text{Octah}}$, you should have the matrix representation of $\hat{L}_z^4$, $\hat{L}_z^2$, $\hat{L}_+^4$, and $\hat{L}_-^4$ at first,

$$\hat{L}_+ = \begin{array}{c} \\ \langle 2| \\ \langle 1| \\ \langle 0| \\ \langle -1| \\ \langle -2| \end{array} \begin{array}{c} \begin{array}{ccccc} |2\rangle & |1\rangle & |0\rangle & |-1\rangle & |-2\rangle \end{array} \\ \begin{bmatrix} 0 & 2 & 0 & 0 & 0 \\ 0 & 0 & \sqrt{6} & 0 & 0 \\ 0 & 0 & 0 & \sqrt{6} & 0 \\ 0 & 0 & 0 & 0 & 2 \\ 0 & 0 & 0 & 0 & 0 \end{bmatrix} \end{array} \tag{10.16}$$

Then,

$$\hat{\mathscr{H}}_{\text{Octah}} = \begin{array}{c} \\ \langle 2| \\ \langle 1| \\ \langle 0| \\ \langle -1| \\ \langle -2| \end{array} \begin{array}{c} \begin{array}{ccccc} |2\rangle & |1\rangle & |0\rangle & \langle -1| & \langle -2| \end{array} \\ \begin{bmatrix} \frac{1}{10}\Delta & 0 & 0 & 0 & \frac{1}{2}\Delta \\ 0 & -\frac{2}{5}\Delta & 0 & 0 & 0 \\ 0 & 0 & \frac{3}{5}\Delta & 0 & 0 \\ 0 & 0 & 0 & -\frac{2}{5}\Delta & 0 \\ \frac{1}{2}\Delta & 0 & 0 & 0 & \frac{1}{10}\Delta \end{bmatrix} \end{array} \tag{10.17}$$

$\Delta = 6\beta_C$. Solve the secular determinant $\left|\hat{\mathscr{H}}_{Octah} - E\hat{1}\right| = 0$ that the eigenvalue and eigenfunction of $\hat{\mathscr{H}}_{octah}$ can be obtained.

Eigenvalue: $E(T_{2g}) = -\frac{2}{5}\Delta$ (triple degeneracy)

Eigenfunction: $|2,1\rangle$, $|2,-1\rangle$, $\frac{1}{\sqrt{2}}(|2,2\rangle - |2,-2\rangle)$

Eigenvalue: $E(E_g) = +\frac{3}{5}\Delta$ (double degeneracy)

Eigenfunction: $|2,0\rangle$, $\frac{1}{\sqrt{2}}(|2,2\rangle + |2,-2\rangle)$

The energy splitting diagram of D state ion for $3d^1$ and $3d^6$ ions in octahedron field is shown in Figure 10.4. For $3d^4$ and $3d^9$, ions can be viewed as half filled or full filled shell with a positive hole. It is only required to change the sign of the potential functions of

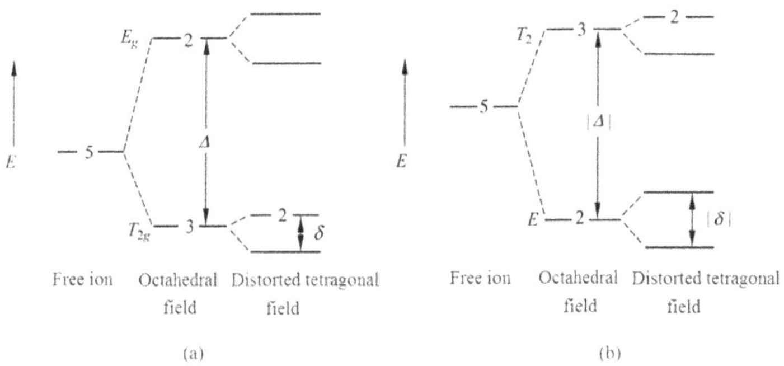

Figure 10.4: Energy splitting diagram in crystal field for *D* state ions.

crystal field and all of above derivation can be applicable entirely, so for the energy diagram of $3d^4$ and $3d^9$, only the order of level needs to be reversed in Figure 10.4.

After the tetragonal distortion term $\alpha_t[3\hat{\boldsymbol{L}}_z^2 - \boldsymbol{L}(\boldsymbol{L}+1)\hat{1}]$ is added, the orbit degeneracy will be further relieved, and the matrix representation of $\hat{\mathscr{H}}_{Tetrag}$ is given as follows:

$$\hat{\mathscr{H}}_{\text{Tetrag}} = \begin{array}{c} \\ \langle 2| \\ \langle 1| \\ \langle 0| \\ \langle -1| \\ \langle -2| \end{array} \begin{array}{c} \begin{array}{ccccc} |+2\rangle & |+1\rangle & |0\rangle & |-1\rangle & |-2\rangle \end{array} \\ \begin{bmatrix} \frac{1}{10}\Delta + 6\alpha_t & 0 & 0 & 0 & \frac{1}{2}\Delta \\ 0 & -\frac{2}{5}\Delta - 3\alpha_t & 0 & 0 & 0 \\ 0 & 0 & \frac{3}{5}\Delta - 6\alpha_t & 0 & 0 \\ 0 & 0 & 0 & -\frac{2}{5}\Delta - 3\alpha_t & 0 \\ \frac{1}{2}\Delta & 0 & 0 & 0 & \frac{1}{10}\Delta + 6\alpha_t \end{bmatrix} \end{array} \tag{10.18}$$

Solving its secular equations, the eigenvalue and eigenfunction can be obtained.

Eigenvalue	Eigenfunction
$E_1 = +\frac{3}{5}\Delta + \frac{2}{3}\delta$	$\psi_1 = \frac{1}{\sqrt{2}}(\lvert 2,2\rangle + \lvert 2,-2\rangle)$
$E_2 = +\frac{3}{5}\Delta - \frac{2}{3}\delta$	$\psi_2 = \lvert 2,0\rangle$
$E_3 = -\frac{2}{5}\Delta + \frac{2}{3}\delta$	$\psi_3 = \frac{1}{\sqrt{2}}(\lvert 2,2\rangle - \lvert 2,-2\rangle)$
$E_4 = E_5 = -\frac{2}{5}\Delta - \frac{1}{3}\delta$	$\psi_4 = \lvert 2,1\rangle = \lvert 2,1\rangle$
	$\psi_5 = \lvert 2,-1\rangle$

$\delta = 9\alpha_t$ is the energy splitting of tetragonal field in Figure 10.4. If $\alpha_t < 0$, the distortion is along z-axis depression, and E_3 is the lowest energy level. In this situation, the lowest energy-level orbital is nondegenerate. After splitting, only E_4 retains double degenerate in four level.

Now let us look at the ion located in the case of tetrahedron field. Connecting the nonadjacent apexes of a cube gives a tetrahedron as shown in Figure 10.5, and each occupies a negative point charge distributed at every apex. It is obvious that $d_{x^2-y^2}$ and d_{z^2} should be located at the lowest energy level for $3d^1$ and $3d^6$ ions, while for $3d^4$ and $3d^9$ ions, the d_{xy}, d_{yz}, d_{zx} orbits ought to be at the lowest energy level. For the same $3d^n$ ion, the order of level in tetrahedron field just right reversed as in octahedron field. In tetrahedron field, the signs of β_C and Δ are reversed.

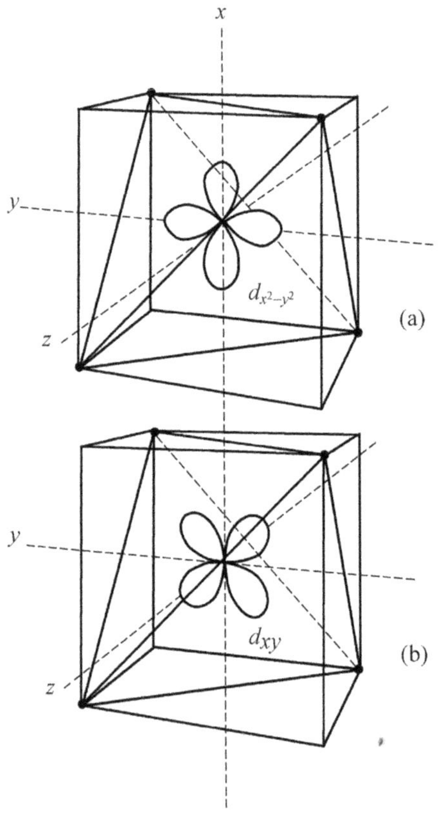

Figure 10.5: d Orbits and negative charges in tetrahedron field.

10.4.3 About *F*-state ion

The matrix representation of $\hat{\mathscr{H}}_{Octah}$ is

$$\hat{\mathscr{H}}_{\text{Octah.}} = \begin{array}{c} \\ \langle +3| \\ \langle +2| \\ \langle +1| \\ \langle 0| \\ \langle -1| \\ \langle -2| \\ \langle -3| \end{array} \begin{array}{c} \begin{array}{ccccccc} |3\rangle & |2\rangle & |1\rangle & |0\rangle & |-1\rangle & |-2\rangle & |-3\rangle \end{array} \\ \begin{bmatrix} \frac{3}{10}\Delta & 0 & 0 & 0 & \frac{\sqrt{15}}{10}\Delta & 0 & 0 \\ 0 & -\frac{7}{10}\Delta & 0 & 0 & 0 & \frac{1}{2}\Delta & 0 \\ 0 & 0 & \frac{1}{10}\Delta & 0 & 0 & 0 & \frac{\sqrt{15}}{10}\Delta \\ 0 & 0 & 0 & \frac{3}{5}\Delta & 0 & 0 & 0 \\ \frac{\sqrt{15}}{10}\Delta & 0 & 0 & 0 & \frac{1}{10}\Delta & 0 & 0 \\ 0 & \frac{1}{2}\Delta & 0 & 0 & 0 & -\frac{7}{10}\Delta & 0 \\ 0 & 0 & \frac{\sqrt{15}}{10}\Delta & 0 & 0 & 0 & \frac{3}{10}\Delta \end{bmatrix} \end{array} \tag{10.19}$$

$\Delta = 30\beta_C$. Solve the secular determinant of this matrix, and let the seats of basic function to be rearranged previously, to make the secular determinant with matrix form of broken block:

$$\begin{array}{c} \\ \langle 3| \\ \langle -1| \\ \langle 2| \\ \langle -2| \\ \langle 0| \\ \langle 1| \\ \langle -3| \end{array} \begin{array}{c} \begin{array}{ccccccc} |3\rangle & |-1\rangle & |2\rangle & |-2\rangle & |0\rangle & |1\rangle & |-3\rangle \end{array} \\ \begin{bmatrix} \frac{3\Delta}{10}-E & \frac{\sqrt{15}\Delta}{10} & & & & & \\ \frac{\sqrt{15}\Delta}{10} & \frac{\Delta}{10}-E & & & & & \\ & & -\frac{7\Delta}{10}-E & \frac{\Delta}{2} & & & \\ & & \frac{\Delta}{2} & -\frac{7\Delta}{10}-E & & & \\ & & & & \frac{3\Delta}{5}-E & & \\ & & & & & \frac{\Delta}{10}-E & \frac{\sqrt{15}\Delta}{10} \\ & & & & & \frac{\sqrt{15}\Delta}{10} & \frac{3\Delta}{10}-E \end{bmatrix} \end{array} = 0 \tag{10.20}$$

Here are three 2 × 2 subdeterminants that are easy to solve:

$$E(A_{2g}) = -\tfrac{6}{5}\Delta \qquad \psi_1 = \tfrac{1}{\sqrt{2}}(|3,2\rangle - |3,-2\rangle) = |\alpha\rangle$$

$$E(T_{2g}) = -\tfrac{1}{5}\Delta \text{ (triple degenerate)} \qquad \psi_2 = \tfrac{1}{\sqrt{2}}(|3,2\rangle + |3,-2\rangle) = |t''_0\rangle$$

$$\psi_3 = \sqrt{\frac{5}{8}}|3,1\rangle - \sqrt{\frac{3}{8}}|3,-3\rangle = |t''_{-1}\rangle$$

$$\psi_4 = \sqrt{\frac{5}{8}}|3,-1\rangle - \sqrt{\frac{3}{8}}|3,3\rangle = |t''_{+1}\rangle$$

$$E(T_{1g}) = \tfrac{3}{5}\Delta \text{ (triple degenerate)} \qquad \psi_5 = |3,0\rangle = |t'_0\rangle$$

$$\psi_6 = \sqrt{\frac{3}{8}}|3,1\rangle + \sqrt{\frac{5}{8}}|3,-3\rangle = |t'_{+1}\rangle$$

$$\psi_7 = \sqrt{\frac{5}{8}}|3,3\rangle + \sqrt{\frac{3}{8}}|3,-1\rangle = |t'_{-1}\rangle$$

The W (T_{1g}) is the lowest energy for $3d^2$ and $3d^7$, while W (A_{2g}) is the lowest energy level for $3d^3$ and $3d^8$. The ground state will be nondegenerate, such as Cr^{3+} and Ni^{2+}.

Same with D state ion, the order of energy should be reversed in tetrahedron field.

For S state ion (as Mn^{2+}), the ground state is originally orbital nondegenerate and does not relieve the orbital degeneracy; however, the crystal field probably arising spin degeneracy is to be relieved partially.

Table 10.4 lists out the orbital degeneracy and spin degeneracy in crystal field.

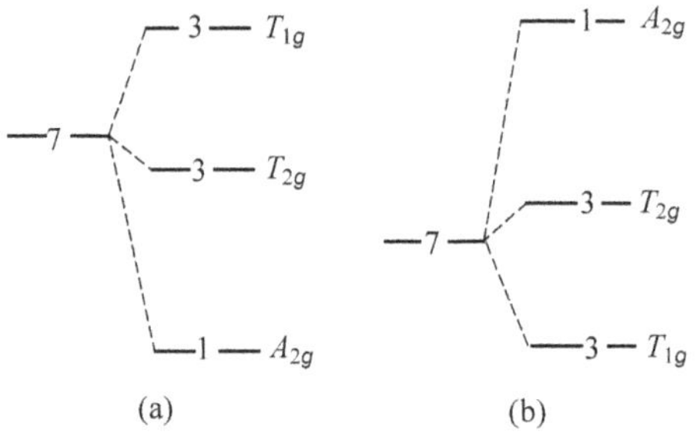

Figure 10.6: Energy splitting diagram of F state ion in crystal field.

10.5 Spin–orbit coupling and spin hamiltonian

For free atoms, g value should be calculated by Landé equation. However, for the free radical and ground-state ion of orbit nondegenerate, because of the orbital angular momentum already to be quenched entirely, their g value ought to be g_e. But their g value relative to g_e still has some deviation. The reason is that between the pure spin ground state and excited state exists configuration interaction, which leads to a few orbital angular momentum of excited state through spin–orbit coupling doped into ground state. Finally, make g value relative to g_e with a certain deviation. The magnitude of this configuration interaction is inversely proportional to the energy interval between ground-state and excited states. The larger energy interval leads to weaker configuration interaction; the g value is closer to g_e, but is stronger. Thus, the spin Hamiltonian of system, including the term of spin–orbit coupling interaction, should be considered:

$$\hat{\mathscr{H}}_S = \hat{\mathscr{H}}_{\text{Zeeman}} + \hat{\mathscr{H}}_{LS} = \beta \boldsymbol{H} \cdot (\hat{L} + g_e \hat{\mathbf{S}}) + \lambda \hat{\mathbf{L}} \cdot \hat{\mathbf{S}} \tag{10.21}$$

$\hat{\mathrm{H}}_{LS} = \lambda \mathbf{L} \cdot \mathbf{S}$ is the term of spin–orbit coupling interaction.

Presuming the ground state is orbit nondegenerate, the wave function is $|G, M_S\rangle$. If the first-order approximation is only considered, their energies are the matrix elements of the diagonal line.

$$E_G^{(1)} = \langle G, M_S | g_e \, \beta H \hat{S}_z | G, M_S \rangle + \langle G, M_S | (\beta H_z + \lambda \hat{S}_z) \hat{L}_z | G, M_S \rangle \tag{10.22}$$

The first term is pure spin Zeeman energy level, the second term can be decomposed into

$$\langle M_S | (\beta H_z + \lambda \hat{S}_z) | M_S \rangle \langle G | \hat{L}_z | G \rangle$$

Theorem II of Section 10.2 has proved that for the orbit nondegenerate state $\langle G | \hat{L}_z | G \rangle = 0$, so that using first-order approximate treat the orbit nondegenerate state, only the Zeeman energy term of pure spin electron is to be left. If the energy of configuration interaction is required, the second-order approximate should be considered. According to the perturbation theory,

$$(\hat{\mathscr{H}})_{M_S, M'_S} = -\sum_{n'} \frac{\left| \langle G, M_S | (\beta H + \lambda \hat{S}) \cdot \hat{L} + g_e \beta H \cdot \hat{S} | n, M'_S \rangle \right|^2}{E_n^{(0)} - E_G^{(0)}} \tag{10.23}$$

n' means the summation of all other terms except ground state. It is noted $\langle G | n \rangle = 0$, thus

$$\langle G, M_S | g_e \beta \boldsymbol{H} \cdot \hat{\mathbf{S}} | n, M'_S \rangle = g_e \beta M'_S \delta_{M_S, M'_S} \langle G | n \rangle = 0$$

$$(\hat{\mathscr{H}})_{M_S, M'_S} = -\sum_{n'} [\langle M_S | (\beta \boldsymbol{H} + \lambda \hat{\mathbf{S}} | M'_S \rangle \cdot \langle G | \hat{\boldsymbol{L}} | n \rangle]$$

$$\times [\langle n | \hat{\boldsymbol{L}} | G \rangle \cdot \langle M'_S | \beta \boldsymbol{H} + \lambda \hat{\mathbf{S}} | M_S \rangle] (E_n^{(0)} - E_G^{(0)})^{-1} \tag{10.24}$$

Table 10.4: Iron group ($3d^{m \cdot n}$) ion in crystal field, ground state, quantum number, and degeneracy[a].

Configuration	d^1 ^2D	d^2 ^3F	d^3 ^4F	d^4 ^5D	d^5 ^6S
	Sc^{2+}	Ti^{2+}	Ti^{+}	Cr^{2+}	Cr^{+}
	Ti^{3+}	V^{3+}	V^{2+}	Mn^{3+}	Mn^{2+}
	VO^{2+}	Cr^{4+}	Cr^{3+}		Fe^{3+}
	Cr^{5+}		Mn^{4+}		
S	1/2	1	3/2	2	5/2
L	2	3	3	2	0
J (free ion)	3/2	2	3/2	0	5/2
λ^b (cm^{-1})	145(Ti^{3+})	104(V^{3+})	56(V^{2+})	58(Cr^{2+})	
(free ion)	248(V^{4+})		91(Cr^{3+})	88(Mn^{3+})	
Orbital degeneracy in crystal fields					
Free ion	5	7	7	5	1
Octagonal[c]	2, 3[d]	1, 2·3[d]	1[d], 2·3	2[d], 3	1
Three square	1, 2·2	3·1, 2·2	3·1, 2·2	1, 2·2	1
Tetragonal	3·1, 2	3·1, 2·2	3·1, 2·2	3·1, 2	1
Rhombic	5·1	7·1	7·1	5·1	1
Single orbital spin degeneracy in crystal fields					
Free ion	2	3	4	5	6
Octagonal	2	3	4	2, 3	2, 4
Three square	2	1, 2	2·2	1, 2·2	3·2
Tetragonal	2	1, 2	2·2	3·1, 2	3·2
Rhombic	2	3·1	2·2	5·1	3·2

[a] $m \cdot n$ means that there are m sets, n degeneracy.
[b] Some books use Griffith spin–orbit coupling parameter ς, for one d electron, $\lambda = \varsigma$, but for more than two d electrons, $\lambda = \pm\varsigma/2S$, positive sign is for less than half filled, and negative sign is for larger than half filled.
[c] In the case of tetrahedron field, the order of level should be reversed.
[d] The lowest or lower energy state.

Let

$$-\sum_{n'} \frac{\langle G|\hat{L}|n\rangle\langle n|\hat{L}|G\rangle}{E_n^{(0)} - E_G^{(0)}} = \Lambda = \begin{bmatrix} \Lambda_{xx} & \Lambda_{xy} & \Lambda_{xz} \\ \Lambda_{yx} & \Lambda_{yy} & \Lambda_{yz} \\ \Lambda_{zx} & \Lambda_{zy} & \Lambda_{zz} \end{bmatrix} \tag{10.25}$$

Here, the two vector matrix elements multiply as exterior product, so the exterior product $\mathbf{\Lambda}$ is a second-order tensor, and its matrix element is

$$\Lambda_{ij} = -\sum_{n'} \frac{\langle G|L_i|n\rangle\langle n|L_j|G\rangle}{E_n^{(0)} - E_G^{(0)}} \tag{10.26}$$

Table 10.4 (continued)

d^6 ^5D	d^7 ^4F	d^8 ^3F	d^9 ^2D
Fe^{2+}	Fe^{+} Co^{2+} Ni^{3+}	Co^{+} Ni^{2+}	Ni^{+} Cu^{2+}
2	3/2	1	1/2
2	3	3	2
4	9/2	4	5/2
−103(Fe^{2+})	−178(Co^{2+})	−325(Ni^{2+})	−829(Cu^{2+})
5	7	7	5
2, 3[d]	1, 2·3[d]	1[d], 2·3	2[d], 3
1, 2·2	3·1, 2·2	3·1, 2·2	1, 2·2
3·1, 2	3·1, 2·2	3·1, 2·2	3·1, 2
5·1	7·1	7·1	5·1
5	4	3	2
2, 3	4	3	2
1, 2·2	2·2	1, 2	2
3·1, 2	2·2	1, 2	2
5·1	2·2	3·1	2

The subscripts i, j represent one of x, y, z. Substitution Eq. (10.25) in Eq. (10.24) and rearranging, we obtained

$$(\hat{\mathscr{H}})_{M_S, M'_S} = \langle M_S|\beta^2 \boldsymbol{H}\cdot\Lambda\cdot\boldsymbol{H} + 2\lambda\beta\boldsymbol{H}\cdot\Lambda\cdot\hat{\boldsymbol{S}} + \lambda^2\hat{\boldsymbol{S}}\cdot\Lambda\cdot\hat{\boldsymbol{S}}|M'_S\rangle \tag{10.27}$$

The first term is a paramagnetic term independent of temperature, which is a fixed constant and is not considered. And the second and third terms in the matrix element contain only spin operators, and combination with $g_e\beta\boldsymbol{H}\cdot\hat{\boldsymbol{S}}$ will give spin Hamiltonian operator $\hat{\mathscr{H}}_S$.

$$\begin{aligned}\hat{\mathscr{H}}_S &= \beta\boldsymbol{H}\cdot(g_e l + 2\lambda\Lambda)\cdot\hat{\boldsymbol{S}} + \lambda^2\hat{\boldsymbol{S}}\cdot\Lambda\cdot\hat{\boldsymbol{S}}\\ &= \beta\boldsymbol{H}\cdot g\cdot\hat{\boldsymbol{S}} + \hat{\boldsymbol{S}}\cdot D\cdot\hat{\boldsymbol{S}}\end{aligned} \tag{10.28}$$

where

$$g = g_e \boldsymbol{l} + 2\lambda\Lambda \tag{10.29}$$

$$D = \lambda^2 \Lambda \tag{10.30}$$

It should be pointed out that here the $\hat{\mathbf{S}}$ operator is represents an equivalent spin operator of ground state. If the system has only spin angular momentum, g should be isotropic, and its value equal to 2.002319. The g is a tensor and has some deviation from g_e value as arisen by the tensor $\boldsymbol{\Lambda}$, because that relates to the orbital angular momentum of the excited state. From Eq. (10.29), it can be known that the g value can be calculated if $\boldsymbol{\Lambda}$ can be sought.

Assuming that the P state ion is located in the quaternion symmetry field, its lowest energy state is $|1,0\rangle$, while its degenerate excited state can be expressed directly by the complex functions $|1,1\rangle$ and $|1,-1\rangle$. Since L of these three functions are all equal to 1, so, it can be abbreviated to $|0\rangle$, $|1\rangle$, $|-1\rangle$. Since in the quadratic symmetry axis the Z-axis should be the principal axis, the other two axes X and Y perpendicular to the Z axis are equivalent. In this principal axis coordinate system,

$$\Lambda_{ZZ} = -\frac{\langle 0|\hat{L}_z|1\rangle\langle 1|\hat{L}_z|0\rangle + \langle 0|\hat{L}_z|-1\rangle\langle -1|\hat{L}_z|0\rangle}{\delta} = 0$$

$$\Lambda_{XX} = -\frac{\langle 0|\hat{L}_x|1\rangle\langle 1|\hat{L}_x|0\rangle + \langle 0|\hat{L}_x|-1\rangle\langle -1|\hat{L}_x|0\rangle}{\delta}$$

$$= -\frac{\langle 0|\frac{1}{2}\hat{L}_-|1\rangle\langle 1|\frac{1}{2}\hat{L}_+|0\rangle + \langle 0|\frac{1}{2}\hat{L}_+|-1\rangle\langle -1|\frac{1}{2}\hat{L}_-|0\rangle}{\delta} = -\frac{1}{\delta} \tag{10.31}$$

Similarly, $\Lambda_{YY} = -\frac{1}{\delta}$.

Thus,

$$g_{\parallel} = g_{zz} = g_e + 2\lambda\Lambda_{ZZ} = g_e \tag{10.32}$$

$$g_{\perp} = g_{xx} = g_{yy} = g_e + 2\lambda\Lambda_{XX} = g_e - \frac{2\lambda}{\delta} \tag{10.33}$$

A concrete example is the V_1 center of point defect in the MgO crystal (O^- ion of around the cation vacancy). The experimentally observed value of $g_{\parallel} = 2{\cdot}00327$ is very close to g_e; and the value of $g_{\perp} = 2{\cdot}03859$. Since the λ value of the positive hole on oxygen is negative, so $g_{\perp} > g_{\parallel}$.

The excited state through the spin–orbit coupling interaction is doped into the ground-state wave function $|G, \frac{1}{2}\rangle$ and the ground-state wave function becomes another function denoted as $|+\rangle$. According to the perturbation theory,

$$|+\rangle = |G, \frac{1}{2}\rangle - \sum_{n'}\sum_{M'_s} \frac{\langle n, M'_s|\lambda\hat{L}\cdot\hat{S}|G, \frac{1}{2}\rangle}{E_n^{(0)} - E_G^{(0)}} |n, M'_s\rangle$$

$$= |G, \frac{1}{2}\rangle - \frac{\lambda}{2}\sum_{n'} \frac{\langle n|\hat{L}_z|G\rangle}{E_n^{(0)} - E_G^{(0)}}|n, \frac{1}{2}\rangle - \frac{\lambda}{2}\sum_{n'} \frac{\langle n|\hat{L}_+|G\rangle}{E_n^{(0)} - E_G^{(0)}}|n, -\frac{1}{2}\rangle \tag{10.34}$$

$$|-\rangle = |G, \frac{1}{2}\rangle + \frac{\lambda}{2}\sum_{n'} \frac{\langle n|\hat{L}_z|G\rangle}{E_n^{(0)} - E_G^{(0)}}|n, -\frac{1}{2}\rangle - \frac{\lambda}{2}\sum_{n'} \frac{\langle n|\hat{L}_-|G\rangle}{E_n^{(0)} - E_G^{(0)}}|n, \frac{1}{2}\rangle \tag{10.35}$$

Obviously, the function $|+\rangle$ and $|-\rangle$ are no longer an eigenfunction of real spin operator $\hat{S}_z$. For this, we define an equivalent spin operator $\hat{\mathbf{S}}$, assuming $|+\rangle$ and $|-\rangle$ on the right-hand side of the eigenfunction $\hat{S}_z$ and the operation rule are the same as the true spin operator is to $|\frac{1}{2}\rangle$ and $|-\frac{1}{2}\rangle$. Using the equivalent spin operator, spin Hamiltonian can be written as follows:

$$\hat{\mathscr{H}}_{\hat{S}} = \beta \boldsymbol{H} \cdot g \cdot \hat{\mathbf{S}} = \beta H\left(g_{zx}\hat{\mathbf{S}}_x + g_{zy}\hat{\mathbf{S}}_y + g_{zz}\hat{\mathbf{S}}_z\right)$$

The matrix representation of $|+\rangle$ and $|-\rangle$ as basic vectors are as follows:

$$\begin{array}{c} \\ \langle +| \\ \langle -| \end{array} \begin{array}{c} \quad |+\rangle \qquad\qquad\qquad |-\rangle \\ \begin{bmatrix} \frac{1}{2}\beta H g_{zz} & \frac{1}{2}\beta H(g_{zx} - ig_{zy}) \\ \frac{1}{2}\beta H(g_{zx} + ig_{zy}) & -\frac{1}{2}\beta H g_{zz} \end{bmatrix} \end{array} \tag{10.36}$$

The matrix representation of the Hamiltonian of real spin operator with $|+\rangle$ and $|-\rangle$ as basic vectors is

$$\beta H \begin{bmatrix} \langle +|L_z + g_e S_z|+\rangle & \langle +|L_z + g_e S_z|-\rangle \\ \langle -|L_z + g_e S_z|+\rangle & \langle -|L_z + g_e S_z|-\rangle \end{bmatrix} \tag{10.37}$$

Comparing these two matrix representations, we get

$$g_{zz} = 2\langle +|L_z + g_e S_z|+\rangle \tag{10.38a}$$

$$g_{zx} = \langle +|L_z + g_e S_z|-\rangle + \langle -|L_z + g_e S_z|+\rangle \tag{10.38b}$$

$$g_{zy} = \langle +|L_z + g_e S_z|-\rangle - \langle -|L_z + g_e S_z|+\rangle \tag{10.38c}$$

Expanding Eq. (10.38) to the first power of λ term

$$g_{zz} = g_e - 2\lambda\sum_{n'} \frac{\langle G|L_z|n\rangle\langle n|L_z|G\rangle}{E_n^{(0)} - E_G^{(0)}} = g_e + 2\lambda\Lambda_{zz}$$

This is the same as Eq. (10.29). The second term of Eq. (10.28) $\hat{\mathbf{S}} \cdot D \cdot \hat{\mathbf{S}}$ contributes only to the system $S \geq 1$, which is same entirely in form as Eq. (6.40), but they are different in mechanisms. In Chapter 6, the $\hat{\mathscr{H}}_{SS}$ is the contribution of the dipole–dipole interaction between electron spins, and here is the contribution of the spin–orbit coupling interaction. It is impossible experimentally to distinguish these two different contributions.

The term $\hat{\boldsymbol{S}} \cdot D \cdot \hat{\boldsymbol{S}}$ could also be written as

$$\hat{\boldsymbol{S}} \cdot D \cdot \hat{\boldsymbol{S}} = D_{XX}\hat{S}_x^2 + D_{YY}\hat{S}_y^2 + D_{ZZ}\hat{S}_z^2$$

$$= D\left[\hat{S}_z^2 - \frac{1}{3}S(S+1)\right] + E(\hat{S}_x^2 - \hat{S}_y^2) + \frac{1}{3}(D_{XX} + D_{YY} + D_{ZZ})S(S+1) \tag{10.39}$$

Here,

$$D = D_{ZZ} - \frac{D_{XX} + D_{YY}}{2} \tag{10.40a}$$

$$E = \frac{D_{XX} - D_{YY}}{2} \tag{10.40b}$$

The last term in Eq. (10.39), $\frac{1}{3}(D_{XX} + D_{YY} + D_{ZZ})S(S+1)$, is a constant term and is usually not included in spin Hamiltonian. Note that for Eq. (7.40a), that is $D_{xx} + D_{yy} + D_{zz} = T_r(^dD) = 0$, for the pure spin–spin dipole interaction, this term is zero.

For the system of containing magnetic nucleus, the Zeeman term of nuclei and the hyperfine interaction term should be added. Therefore,

$$\mathscr{H}_S = \beta \boldsymbol{H} \cdot g \cdot \hat{\boldsymbol{S}} + \hat{\boldsymbol{S}} \cdot \boldsymbol{D} \cdot \hat{\boldsymbol{S}} + \hat{\boldsymbol{S}} \cdot \boldsymbol{A} \cdot \hat{\boldsymbol{I}} - g_n\beta_n \boldsymbol{H} \cdot \hat{I} \tag{10.41}$$

Equation 10.41 is only applicable to the system of $S = 1$, while the system of $S > 1$ is more complicated [16,17]. For example, $S = 3/2$ of $3d^7$ ion should add the $\hat{\boldsymbol{S}}^3 \cdot \boldsymbol{H}$ item, if there is magnetic nucleus, the $\hat{\boldsymbol{S}}^3 \cdot \hat{\boldsymbol{I}}$ term else should be added. For the system of $S \geq 2$, the terms of $\hat{\boldsymbol{S}}^4$ should also be added.

10.6 Ground-state ion with orbital nondegeneracy

Generally speaking, for free ions of different states, their energy variation with the magnitude of crystal field is not a linear relationship. In the above discussion, the excited states of free ion are always neglected. In fact, in the crystal field, some excited states are possibly very close to the ground state, so they also can give a certain contribution to the g value or zero field splitting values D and E. For some d^n ions (d^4, d^5, d^6, and d^7 in particular), with the increasing crystal field, the energy of some excited states is decreasing faster than the ground state, hence a new ground state is generated with a value greater than the critical value. Figure 10.7a shows the variation of the energy level of the d^4 ion and the octahedron field intensity D_q (here $D_q = \Delta/10$). The left side of the dotted line (i.e., critical line) is the intermediate field, and the right side is the strong field.

Here, we are concerned that the variation of interval of the energy levels between the excited state and the ground state, energy level diagram is shown in Figure 10.7b. From Figure 10.7b, on the left side of the critical line, the ground state is ^5E_g. The interval between the excited state $^3T_{1g}$ and ground state ^5E_g decreases rapidly with

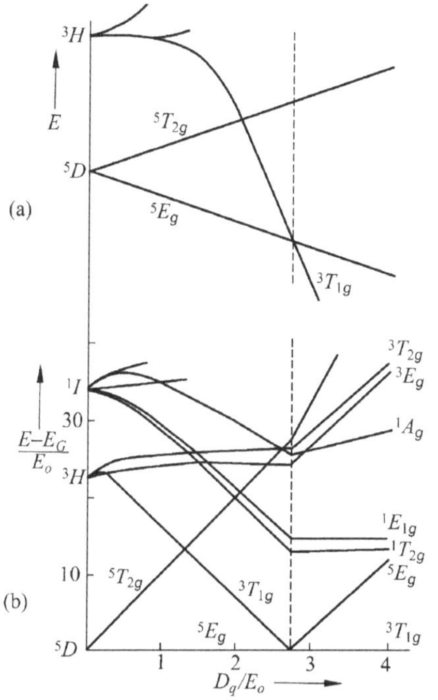

Figure 10.7: The $3d^4$ ion in the octahedron field related to the energy-level splitting with the field intensity. This is Tanabe-Sugano Diagram [18], where the dash line is the critical line between the intermediate field and the strong field.

the intensity of the crystal field increasing. At the critical line, the two energy levels $^3T_{1g}$ and ^5E_g overlap (equal), and then $^3T_{1g}$ becomes a new ground state, while the energy level of the original ground state ^5E_g increases and becomes the excited state. On the left of the critical line is the intermediate field, the d^4 ion is a high-spin state, and the strong field is a low-spin state.

For a certain special ion-matrix system, there is a specific reference energy (E_o). So, D_q/E_o is used as the abscissa and $(E-E_G)/E_o$ as the ordinate in Figure 10.7b. This figure is called Tanabe–Sugano diagram[18]; it can be applied to all the d^n ions.

10.6.1 *D*-state ions of ground-state orbital nondegenerate

10.6.1.1 $3d^1$ in the tetrahedral field + tetragonal distortion or cubic field + tetragonal distorted crystal field

The lowest energy state of the $3d^1$ ion in the tetrahedral field is E_g. If the tetragonal distortion is added, the ground-state orbital degeneracy is relieved completely (Figure 10.4). For $\alpha_t > 0$, the lowest state is $|0\rangle$ or d_{z^2}. Conversely, for the $\alpha_t < 0$, the lowest state is $(|2\rangle + |-2\rangle)/\sqrt{2}$ or $d_{x^2-y^2}$.

For the case of $\alpha_t > 0$: Let the tetragonal distortion axis be Z-axis, then

$$g_{zz} = g_{\parallel} = g_e + 2\lambda\Lambda_{zz} = g_e \tag{10.42}$$

When $|n\rangle \neq |0\rangle$, $\langle 0|\hat{L}_z|n\rangle = 0$. Therefore, $\Lambda_{zz} = 0$. For $H \parallel X$

$$g_{xx} = g_{\perp} = g_e + 2\lambda\,\Lambda_{xx}$$

$$= g_e - [\langle 0|\hat{L}_x|1\rangle\,\langle 1|\hat{L}_x|0\rangle + \langle 0|\hat{L}_x|-1\rangle\langle -1|\hat{L}_x|0\rangle] = g_e - \frac{6\lambda}{\Delta} \quad (10.43)$$

A concrete example of a $3d^1$ ion in the tetrahedral field + tetragonal distortion is the CrO_4^{3-} ion in Ca_2PO_4Cl single crystal. Here, the Cr^{5+} is $3d^1$ ion. The $g_{\parallel} = 1{\cdot}9936$, $g_{\perp} = 1{\cdot}9498$ are measured experimentally. Here, $g_{\parallel} > g_{\perp}$, explanation $\alpha_t > 0$ is real fact. The unpaired electron should be in the d_{z^2} orbit.

For the case of $\alpha_t < 0$:

$$g_{\parallel} = g_e - \frac{8\lambda}{\Delta} \quad (10.44a)$$

$$g_{\perp} = g_e - \frac{2\lambda}{\Delta} \quad (10.44b)$$

There is a compressed tetrahedral configuration of ion in $CaWO_4$. When W of WO_4^{2-} ion is replaced by V^{4+}, V^{4+} should be $3d^1$ ion (tetrahedral + tetragonal distortion); the g value 2.0245 of experimental measurement is isotropic, which does not agree with the above-mentioned theory [19]. The reason is that the above theory adoptive is a rough point charge model, while practically the W–O bond has the character of covalent bond. Therefore, when the W is replaced by V^{4+}, the V–O bond also has a certain character of covalent bond. A small amount of covalent bonds can reduce the contribution of excited state to the g value, while a large number of covalent bonds can give the g value a positive deviation for g_e; here, the g value is greater than g_e. In addition, the characteristic of covalent bond can also reduce the hyperfine coupling parameter. When measured experimentally, $A_{\parallel} = 0{\cdot}00179\,\text{cm}^{-1}$, and $A_{\perp} = 0{\cdot}00190\,\text{cm}^{-1}$, while the hyperfine coupling parameters of V^{4+} ions range from 0.0070 to 0.0100 cm^{-1}. All of these explain that the V–O bond has characterization of covalent bond and cannot adopt the point charge model.

The V^{4+} ions also can replace Ca^{2+} ions of $CaWO_4$; here, V^{4+} ion has eight $g_{\perp} = g_e$ oxygen atoms, and forms a stretched oxygen tetrahedron. This is not a simple $3d^1$ (tetrahedral + tetragonal distortion) ion, but its environment is ionicity rather strong and gives rise to a negative deviation for the g_e value, $A = 0{\cdot}0087\,\text{cm}^{-1}$.

10.6.1.2 $3d^7$ (Low spin) and $3d^9$ ions in the octahedral field + tetragonal distorted crystal field

The g value of $3d^7$ (low spin) and $3d^9$ ions in the octahedral fields with quadratic distortion are also the following:

$$g_{zz} = g_{\parallel} = g_e \quad (10.45a)$$

$$g_{\perp} = g_e - \frac{6\lambda}{\Delta} \tag{10.45b}$$

Since both $3d^7$ and $3d^9$ ions are more than half filled, so $\lambda < 0$, when $\alpha_t > 0$, $g_{\perp} > g_{\parallel}$.

Cobalt phthalocyanine is one of the factual examples of $3d^7$ ions (low spin) in an octahedral field with tetragonal distortion. Its unpaired electrons are located mainly in d_{z^2} orbit, and this orbital is just perpendicular to the plane of the complex molecules. In the pyridine solvent, the upper and lower sides of the plane can be coordinated with each one pyridine molecule, and the hyperfine splitting of ^{14}N nucleus on the pyridine molecule can be observed. Their $g_{\parallel} = 2{\cdot}016$, $g_{\perp} = 2{\cdot}268$. This basically agrees with the results of theoretical analysis[20].

The $Fe(CN)_5NOH^{2-}$ in the antimagnetic $Na_2Fe(CN)_5NO \cdot 2H_2O$ matrix is another example of the $3d^7$ ion (low spin) in the octahedral field with quadratic distortion. While their $g_{\parallel} = 2{\cdot}0069$ and $g_{\perp} = 2{\cdot}0374$ also agree with theoretical analysis, it explains that its unpaired electron is also located on the d_{z^2} orbital.

The Ag^{2+} ion in the HNO_3 or H_2SO_4 solution is an example of the $3d^9$ ion in the octahedral field with tetragonal distortion. The EMR spectrum measured at room temperature is a singlet line with $g = 2{\cdot}133$. When measured at 77 K, the EMR spectrum is shown in Figure 10.8 [21]. It is $g_{\parallel} = 2{\cdot}265$, $g_{\perp} = 2{\cdot}065$. Hyperfine splitting are contributed by both nuclei of ^{107}Ag and ^{109}Ag commonly, their nuclear spin I are all of 1/2, the natural abundance are also very close (the former is 51.35%, the latter is 48.65%), the magnetic moment are also near (0.1130 β_n and 0.1299 β_n). Because $g_{\parallel} > g_{\perp}$, so it is the case of $\alpha_t < 0$, that is, the unpaired electron is located on the $d_{x^2-y^2}$ orbit, and the g value should be calculated from Eq. (10.44).

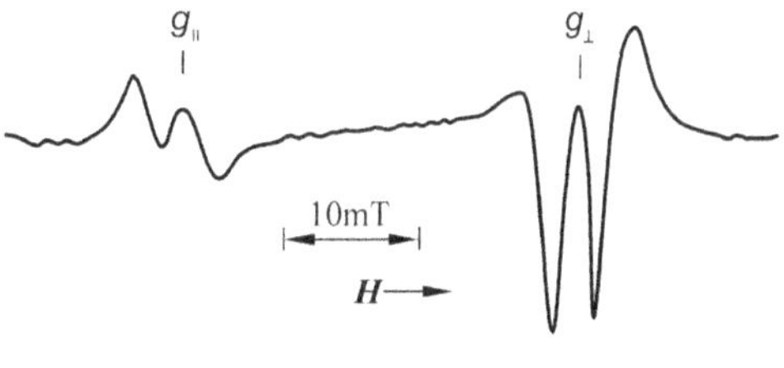

Figure 10.8: The EMR secondary differential spectrumof Ag^{2+} in HNO3 solution at 77 K.

When the λ value of free ion ($\lambda = -840 cm^{-1}$) and the $g_{\parallel}$ and $g_{\perp}$ value obtained from the experiment are substituted into Eq. (10.44), and the Δ value is calculated, we find that using $g_{\parallel}$ value, we get $\Delta = 53500\,cm^{-1}$, and using $g_{\perp}$ value, we get $\Delta = 48750\ cm^{-1}$. But from the absorption spectrum we proved that the Δ value should be $25560 cm^{-1}$, far different from the value obtained in the calculations. It can be concluded that the λ value of the transitional group ion in the complex is much smaller than the λ value of free ion. So, we cannot adopt the λ value of free ion to calculate the Δ value. In addition, it is also not possible that the model can assume a positive hole when located in the $d_{x^2-y^2}$ orbit of silver ion.

In the octahedral field of tetragonal distortion, the orbital energy-level splitting of the $3d^4$ ion is same as that of the $3d^1$ ion in the tetrahedral field of the tetragonal distortion.

10.6.2 *F*-state ions of ground-state orbital nondegenerate

10.6.2.1 $3d^8$ Ions in the octahedral field

The ^3F ground state of $3d^8$ free ions has degeneracy of seven degrees. In the octahedral field, it splits into two triple degenerate states and one nondegenerate state. The lowest energy state is $^3A_{2g}$ (Figure 10.6a) (for eigenfunction and eigenvalue, refer to Section 10.4.3).

In a zero-order approximation, the ground-state wave function influenced by the spin–orbit coupling is ignored. In the case of $\boldsymbol{H} \parallel Z$, the Zeeman interaction of the spin Hamiltonian is

$$\hat{\mathscr{H}}_{\text{Zeeman}} = \beta H_Z(\hat{L}_z + g_e\hat{S}_z) \tag{10.46}$$

Since the ground state is an orbit nondegenerate, the contribution of $\hat{L}_z$ is zero. Ground-state wave function $|G\rangle$ is

$$|G\rangle = \frac{1}{\sqrt{2}}(|2, M_s\rangle - |-2, M_s\rangle) \tag{10.47}$$

Note that at ^3F state $S = 1$, $Ms = \pm 1, 0$, in the magnetic field, its energy is as follows:

$$E_\pm = \pm\ 2\beta H_Z \tag{10.48a}$$

$$E_0 = 0 \tag{10.48b}$$

In the normal octahedral symmetric field, the zero order approximation of g is isotropic, that is, $g_{zz} = g_{yy} = g_{xx} = 2$.

Since $S = 1$, at a field lower than the octahedral symmetry field, the spin degeneracy is also relieved possibly, and we can calculate the contribution of the spin–orbit coupling by using tensor $\boldsymbol{\Lambda}$. Since it is an octahedral field, we need to calculate only Λ_{zz}, while in the excited state, only the $|t_0''\rangle = \frac{1}{\sqrt{2}}(|2\rangle + |-2\rangle)$ of T_{2g} can contribute to Λ_{zz}, so:

$$\Lambda_{zz} = -\frac{\left[\frac{1}{\sqrt{2}}(\langle 2| - \langle -2|)L_Z\frac{1}{\sqrt{2}}(|2\rangle + |-2\rangle)\right]^2}{\Delta} = -\frac{4}{\Delta} \tag{10.49}$$

The Δ here is the energy interval between T_{2g} and A_{2g} (see Figure 10.9)

$$g_{zz} = g_e + 2\lambda\Lambda_{zz} = g_e - \frac{8\lambda}{\Delta} = g_{\text{iso}} \tag{10.50}$$

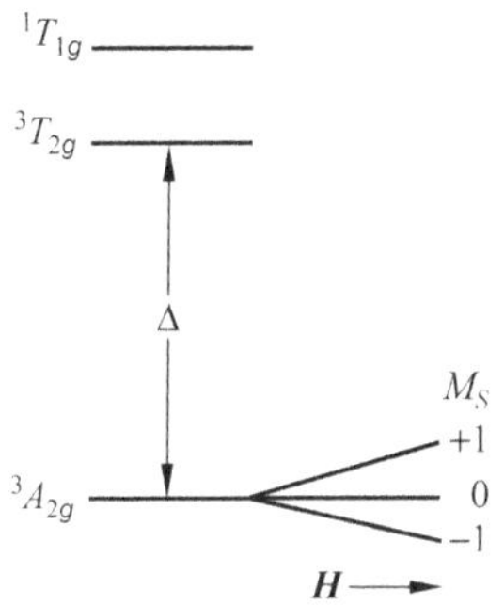

Figure 10.9: Energy splitting in the octahedral field of d^8 ions.

Since the excited state is far from the ground state and the relaxation action is rather weak, so that the EMR spectrum of the $3d^8$ ion can be observed at room temperature or 77 K usually.

The Ni^{2+} ion is the most important model example of $3d^8$ (octahedral) ions. The g value can be estimated from the spectral data by using Eq. (10.50). An example of the absorption band of $Ni(NH_3)_6^{2+}$ (assuming the transition of $^3T_2 \leftarrow {}^3A_2$) is 10700 cm^{-1}; using the λ value of free ion ($\lambda = -325 cm^{-1}$), g value is found to be 2.245. The experimental value [22] is 2.162. The difference in the values is due to existence of some covalent properties in the complex. In addition, the equivalent λ value can be calculated from the experimental g value and Δ value, which is 211 cm^{-1}. In most octahedral environments, the g values are in the range of 2.10–2.33. Cu^{3+} ions have similar property.

The EMR spectra of Ni^{2+} ions are rather wide. In the MgO crystal, the EMR spectrum of other substituted ions is usually about 0.05 mT; however, the EMR spectrum of Ni^{2+} ions is 4.0 mT. Since Ni^{2+} ions have even number of electrons, they are not restricted by Kramer's theorem. The remaining lattice tension making the relative interval of $|0\rangle$ and $|+1\rangle$ states with $|-1\rangle$ states have a deviation of different degrees; zero-field splitting parameters can be positive or negative, so, it has a heterogeneous broad line. For a high microwave power, a sharp double-quantum transition line over the wide line can be observed, which explains that it is heterogeneous broadening (as shown in Figure 10.10). Since the natural abundance of ^{61}Ni is only 1.25%, it is very difficult to observe its hyperfine structure. In the enriched ^{61}Ni

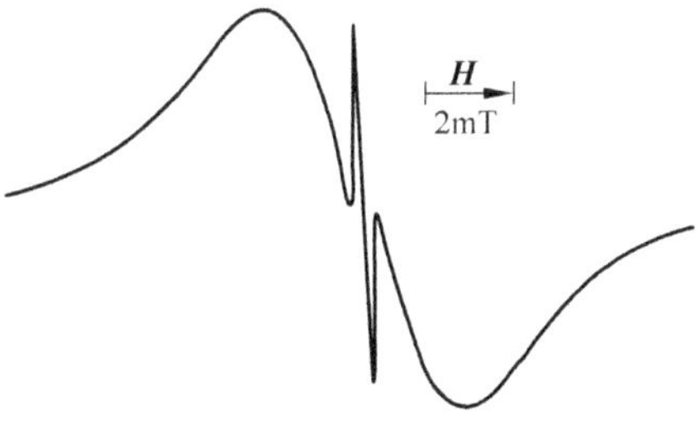

Figure 10.10: EMR spectra of Ni^{2+} ions in MgO crystal.

sample, its hyperfine structure of the quadruplet line can be observed, and the hyperfine coupling parameter [23] is 0·00083cm^{-1}.

10.6.2.2 The $3d^2$ ions in the tetrahedral field

Because of the tetrahedral field, the seven degree degeneracy of the ^3F state of the $3d^2$ ion is relieved. It is similar to the case of $3d^8$ ion in the octahedral field; its lowest level is orbital singlet. In the pure tetrahedral field, the g value is isotropic, that is,

$$g_{zz} = g_e - \frac{8\lambda}{\Delta} = g_{\text{iso}}$$

The EMR spectra of Ti^{2+} ions in the tetrahedral ZnS matrix are shown in Figure 10.11 [24]. Since Ti^{2+} is less than half filled, λ is larger than zero; therefore, $g < g_e$, and the experimental value is 1.9280. Its spin Hamiltonian wants to add hyperfine item, because ^{47}Ti and ^{49}Ti are magnetic nuclei. Their nuclear spin are 5/2 and 7/2, respectively, but their magnetogyric ratios are almost equal ($\gamma(^{47}\text{Ti}) = 1{\cdot}5079 \times 10^3$ rad – Gauss^{-1}, $\gamma(^{49}\text{Ti}) = 1{\cdot}5083 \times 10^3$ rad – Gauss^{-1}). So the six lines of ^{47}Ti and eight lines of ^{49}Ti almost completely overlap. The broad line in the middle diagram is the heterogeneous broadening line caused by the transition between $|-1\rangle \leftrightarrow |0\rangle$ and $|0\rangle \leftrightarrow |+1\rangle$. Because $3d^2$ ions are also even electron and are also not restricted by the Kramer theorem, the narrow line of the center is the double quantum transition of $|-1\rangle \leftrightarrow |+1\rangle$. Six hyperfine structures in the middle are contributed by the ^{67}Zn nucleus ($I = 5/2$, natural abundance 4.12%).

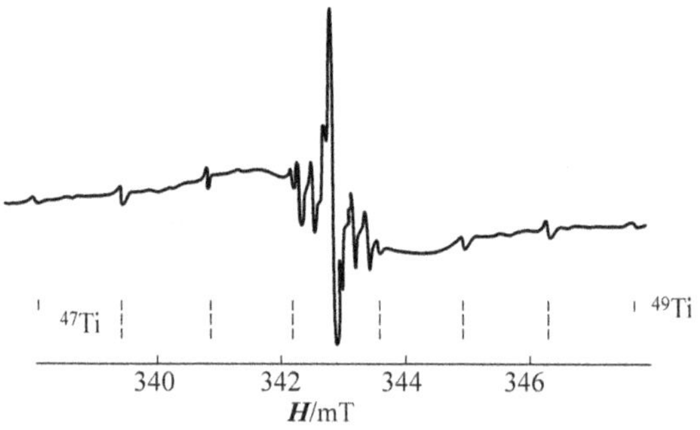

Figure 10.11: EMR spectra of Ti^{2+} ions in tetrahedral ZnS matrix.

10.6.2.3 $3d^8$ ions in the case of octahedral field + tetragonal distortion

For $3d^8$ ion located in the octahedral field, if the direction of the Z-axis is given a tetragonal distortion, in T_{1g} and T_{2g}, the energy level of doublet degeneracy will be further split; the interval of energy level after splitting is $\delta \ll \Delta_{\|}$ or $\Delta_{\perp}$, as shown in Figure 10.12.

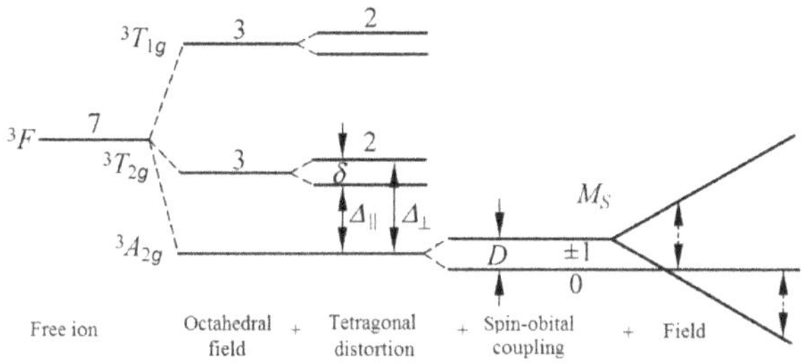

Figure 10.12: Energy splitting of $3d^8(^3F)$ ions in the crystal field.

$$g_{zz} = g_\parallel = g_e - \frac{8\lambda}{\Delta_\parallel} \tag{10.51}$$

The $\hat{L}_x$ can couple the $|t''_1\rangle$ and $|t''_{-1}\rangle$ of T_{2g}. so,

$$\Lambda_{xx} = -\frac{\left[\frac{1}{\sqrt{2}}(\langle 2| - \langle -2|)\hat{L}_x\left(\sqrt{\frac{5}{8}}|-1\rangle - \sqrt{\frac{3}{8}}|3\rangle\right)\right]^2}{\Delta_\perp} - \frac{\left[\frac{1}{\sqrt{2}}(\langle 2| - \langle -2|)\hat{L}_x\left(\sqrt{\frac{5}{8}}|1\rangle - \sqrt{\frac{3}{8}}|-3\rangle\right)\right]^2}{\Delta_\perp}$$

$$= -\frac{4}{\Delta_\perp} \tag{10.52}$$

Then,

$$g_{xx} = g_\perp = g_e - \frac{8\lambda}{\Delta_\perp} \tag{10.53}$$

If $\delta \ll \Delta_\parallel$ or $\Delta_\perp (\Delta_\parallel \approx \Delta_\perp)$, the g value also is almost isotropic. In the case of axisymmetry, according to Eq. (10.40a), the zero field splitting parameter is given as follows:

$$D = D_{zz} - D_{xx} = \lambda^2(\Lambda_{zz} - \Lambda_{xx}) = -4\lambda^2\left(\frac{1}{\Delta_\parallel} - \frac{1}{\Delta_\perp}\right)$$

$$= -4\lambda\left(\frac{\Delta_\perp - \Delta_\parallel}{\Delta_\perp \Delta_\parallel}\right) \approx -\frac{4\lambda^2\delta}{\Delta^2} \tag{10.54}$$

In the pure octahedral field, because $\delta = 0$, there is no zero-field splitting.

In the crystal field of the oblique or lower symmetry, we need to use the spin Hamiltonian of Eq. (10.39). In experimentally observed spectrum of $3d^8$ ions, most of the g values are almost isotropic. Spin Hamiltonian can be written approximately as

$$\hat{\mathscr{H}} = g\beta \boldsymbol{H} \cdot \hat{\boldsymbol{S}} + D\left(\hat{S}_z - \frac{1}{3}\hat{S}^2\right) + E\left(\hat{S}_x^2 - \hat{S}_y^2\right) \tag{10.55}$$

Ni^{2+} ion in the lower than octahedral symmetry crystal field shows a strong zero field splitting indeed. Because of the local variation of the D value, the spectral line is very wide. In the $Zn_3La_2(NO_3)_{12} \cdot 24H_2O$, $D = 0{\cdot}043\text{cm}^{-1}$, $E = 0$; in the TiO_2, $D = -8{\cdot}3\,\text{cm}^{-1}$, $E = 0{\cdot}137\,\text{cm}^{-1}$. If the ground states are not mixed with the states of different L and S values, the D and E values can be estimated from the g value as follows:

$$D = \frac{1}{2}\lambda\left[g_{zz} - \frac{1}{2}(g_{xx} + g_{yy})\right] \tag{10.56a}$$

$$E = \frac{1}{4}\lambda(g_{xx} - g_{yy}) \tag{10.56b}$$

10.6.2.4 $3d^2$ Ions in the tetrahedral field of tetragonal distortion

The g and D values can also be calculated using Eqs. (10.51), (10.53), and (10.54). The V^{3+} ion in CdS is a good example [25]. The experimentally measured values are $g_{\parallel} = 1{\cdot}934$, $g_{\perp} = 1{\cdot}932$, and $D = 0{\cdot}1130\,\text{cm}^{-1}$. If we adopt 70% of the free ion λ value, estimated from the measured g value, $\Delta_{\parallel} = 8600\,\text{cm}^{-1}$, while the V^{3+} ion in the Al_2O_3, and the $\Delta_{\parallel} = 18000\,\text{cm}^{-1}$, then the ratio of both is about 4:9. It could be known from the crystal field theory that the ratio of the splitting value of the tetrahedral field to the splitting value of the octahedral field should be 4:9. It can be seen that both match very well. The g value deviation from the g_e value should also be about twice the deviation from the value in the octahedral field.

For the Fe^{6+} ion [26] in the K_2CrO_4, $g = 2{\cdot}000$, $D = 0{\cdot}103\text{cm}^{-1}$, and $E = 0{\cdot}016\text{cm}^{-1}$; here, the $E \neq 0$ explains that it has a certain degree deviation to axis symmetry.

10.6.2.5 $3d^3$ Ions in octahedral field

The energy-level splitting in the octahedral field of ^4F ground-state $3d^3$ free ion is same with the $3d^8$ ion, and the lowest level still is the A_{2g} state of orbital non-degenerate (as Figure 10.6). The main difference between $3d^3$ and $3d^8$ is the spin multiple degree degeneracy. For the $S = 3/2$ of $3d^3$ ion, its zero-order approximation energy in the magnetic field is given as

$$E_{\pm 3/2} = \pm\frac{3}{2}g_e\beta H_z \tag{10.57a}$$

$$E_{\pm 1/2} = \pm\frac{1}{2}g_e\beta H_z \tag{10.57b}$$

Since the case of orbital wave function is exactly the same as the $3d^8$ ion, Eq. (10.50) is applicable to the case of $3d^3$ ion in the octahedral field.

Several reports on the EMR of $3d^3$ ions and their complexes in octahedral fields are already available. Their linewidths are generally narrower and can be easily observed at room temperature. The Cr^{3+} ion is the $3d^3$ ion studied by EMR in detail. Figure 10.13 shows the EMR spectrum of Cr^{3+} ion in MgO, which is the most typical example of $3d^3$ ion in the octahedral field, and shown as an isotropic spectrum at 290 K of $g = 1{\cdot}9796$. A sharp strong line at the center [50,52,54] is contributed by nonmagnetic Cr ($I = 0$) nucleus, both sides of which there are two weak quadruplet lines contributed by ^{53}Cr nucleus ($I = 3/2$, natural abundance of 9.54%); hyperfine coupling parameters $a = 0{\cdot}00163\text{cm}^{-1}$.

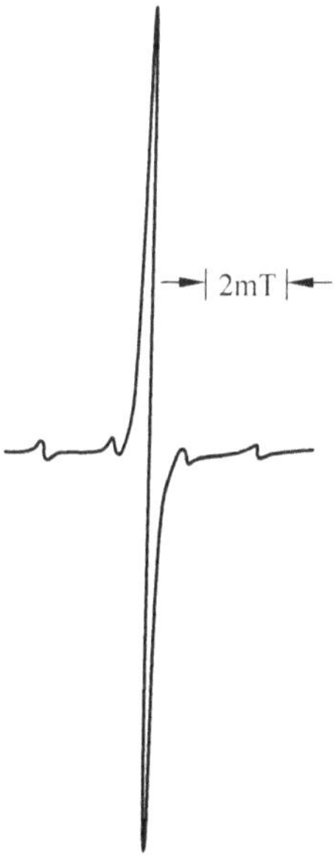

Figure 10.13: Room temperature EMR spectrum of Cr^{3+} ion in MgO.

Using $\Delta(^4T_{2g}) = 16900\ \text{cm}^{-1}$ and $\lambda = 91\ \text{cm}^{-1}$ of the Cr^{3+} ions in the MgO crystal, we can calculate $g = 1{\cdot}96$ using Eq. (10.50), which is very close to the experimental value. Similar calculations can also be applied to V^{2+} and Mn^{4+} ions having same electron numbers. When the ionic charge increases, the deviation between the calculated g value and the measured g value also increases. A more detailed theoretical calculation shows that the charge on the ligand is partially transferred to the central ion, and with the increase in central ion charge, the dynamics of the charge transfer also increases. It explains that the reason why the calculated value deviates from the experimental value is that there is an existence of covalent bond between the central ion and the ligand [27]. In the theory of crystal field, the ligand is also regarded as a point charge, which is not correct in many cases. It should use the theory of ligand field to deal with it. Taking the electron and energy level of ligand together into account is more realistic. It can be seen from Eq. (10.50) that when Δ is larger, the g value is closer to g_e. The experimental results show that the g value of no one $3d^3$ ion is less than 1.95. This proves that all the Δ values are very large, that the interval of energy level between excited state and ground state is quite large, and that the spin–orbit coupling action is rather small. Since the spin–lattice relaxation of the transition metal ions is carried out via the spin–orbit coupling action mainly, if the spin–orbit coupling interaction is very small, τ_1 should be very large, and the width of spectral line will be narrow relatively; the EMR spectrum can be observed at room temperature.

10.6.2.6 The $3d^7$ (high spin) ions in tetrahedral field

Its energy level and spin are same with $3d^3$ ion basically. Therefore, we can apply the expression of the g factor and the zero-field splitting parameter of the $3d^3$ ion, but it should be noted that the $3d^7$ ion is over half full, $\lambda < 0$, $g > g_e$. The behavior of the $3d^7$ ion in the cubic field is same as in the tetrahedral field.

The Co^{2+} ions in the ZnS or ZnTe of cubic crystal system, as well as the Co^{2+} ion [28] in cubic coordination in the CaF_2 or CdF_2 crystal system, all belong to the $3d^7$ ions (high spin) in the tetrahedral field. Although the ground state is orbit nondegenerate, the energy-level interval of T_{2g} state with the ground-state A_{2g} is not large. For the Co^{2+} ions in CdF_2, the energy interval of the T_{2g} state with the A_{2g} state is $4200\,cm^{-1}$. Therefore, the interaction of spin–orbit coupling is rather strong and τ_1 is small relatively, so the EMR spectrum is at a temperature lower than 20 K; and $g = 2{\cdot}278$.

10.6.2.7 $3d^3$ Ions in the tetragonal distorted octahedral field

The $3d^3$ ion is in the oblique field or other field with lower symmetry because $S = 3/2$; the optional wave function are $|\frac{3}{2}\rangle, |\frac{1}{2}\rangle, |-\frac{1}{2}\rangle, |-\frac{3}{2}\rangle$, and in the case of $H \parallel Z$, the matrix representation of spin Hamiltonian is as follows:

$$\hat{\mathscr{H}} = \begin{array}{c} \\ \langle \frac{3}{2}| \\ \langle -\frac{1}{2}| \\ \langle \frac{1}{2}| \\ \langle -\frac{3}{2}| \end{array} \begin{array}{cccc} |\frac{3}{2}\rangle & |-\frac{1}{2}\rangle & |\frac{1}{2}\rangle & |-\frac{3}{2}\rangle \\ \left[\begin{matrix} \frac{3}{2}g\beta H_z + D & \sqrt{3}E & 0 & 0 \\ \sqrt{3}E & -\frac{1}{2}g\beta H_z - D & 0 & 0 \\ 0 & 0 & \frac{1}{2}g\beta H_z - D & \sqrt{3}E \\ 0 & 0 & \sqrt{3}E & -\frac{3}{2}g\beta H_z + D \end{matrix}\right] \end{array} \quad (10.58)$$

Solving secular equation, we have

$$E_{3/2} = \frac{1}{2}g\beta H_z + [(g\beta H_z + D)^2 + 3E^2]^{1/2} \quad (10.59a)$$

$$E_{-1/2} = \frac{1}{2}g\beta H_z - [(g\beta H_z + D)^2 + 3E^2]^{1/2} \quad (10.59b)$$

$$E_{1/2} = -\frac{1}{2}g\beta H_z + [(g\beta H_z - D)^2 + 3E^2]^{1/2} \tag{10.59c}$$

$$E_{-3/2} = -\frac{1}{2}g\beta H_z - [(g\beta H_z - D)^2 + 3E^2]^{1/2} \tag{10.59d}$$

Since $3d^3$ ions have odd number of electrons, Kramer degeneracy should exist. When the magnetic field is zero, at $\pm(D^2+3E^2)^{1/2}$, there are two pair levels of double degenerate. The simplest case is the quarter field. When E = 0, the level can be simplified as follows:

$$E_{\pm 1/2} = \pm\frac{1}{2}g\beta H_z - D$$

$$E_{\pm 3/2} = \pm\frac{3}{2}g\beta H_z + D$$

Thus, when D compared with hν is not too large, there are three transition lines as shown in Figure 10.14. It must be noted that these three lines are not equal; according to the transition theory, their intensity ratio should be proportional to $\left|\langle M_S|\hat{S}_+|M_S-1\rangle\right|^2$. Then,

$$\left|\left\langle -\frac{1}{2}\middle|\hat{S}_+\middle|-\frac{3}{2}\right\rangle\right|^2 = 3, \quad \left|\left\langle \frac{1}{2}\middle|\hat{S}_+\middle|-\frac{1}{2}\right\rangle\right|^2 = 4$$

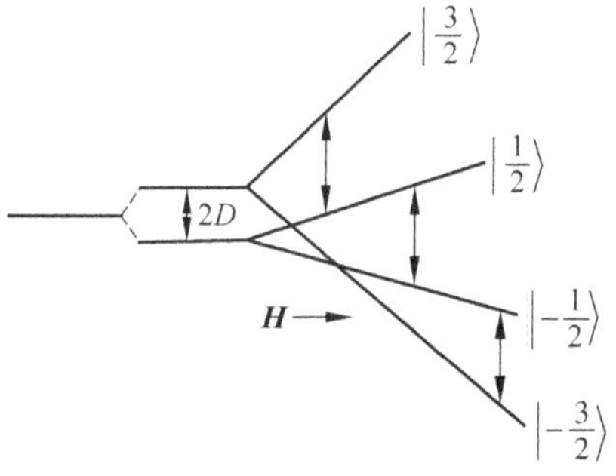

Figure 10.14: The energy level diagram of d^3 ion in tetragonal distortion octahedral field.

The relative intensity of these three lines should be 3: 4: 3.

All of above discussion is for $H \parallel Z$. For $H \parallel X$ and $H \parallel Y$, see Ref. [15].

10.6.3 *S*-state ions of the ground-state orbital nondegenerate

10.6.3.1 $3d^5$ (High spin) ions in the octahedral field

For the half-filled shell of the high-spin $3d^5$ ion, whose ground state should be the orbital singlet, that is, when $^6S_{5/2}$ is in the external magnetic field parallel to the main axis of octahedron, the relation between energy-level splitting and the magnetic field is shown in Figure 10.15. The EMR spectra of high-spin $3d^5$ ions in many matrices have been studied at a wide temperature range. Since the excited state is far from the ground-state $^6A_1$, the spin orbital coupling between the ground state and the excited state is quite small. Because it is an odd number electronic system, the energy level is at least a double degenerate (Kramer theorem). Thus, even if it has a large zero-field splitting, the spectrum can be observed with the usual X-band EMR spectrometer.

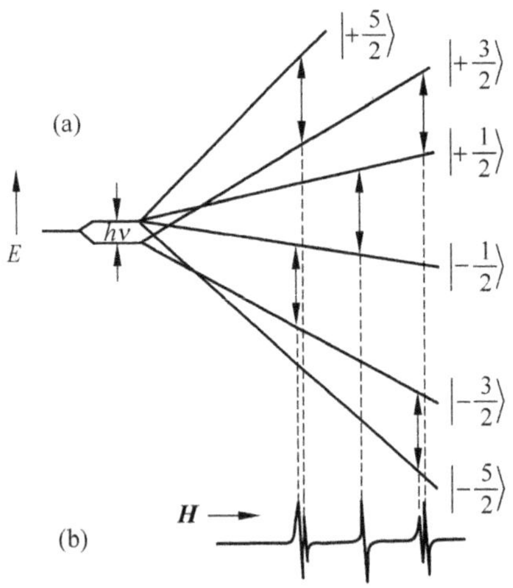

Figure 10.15: (a) The energy level of $3d^5$ ions in the octahedral field. (b) the EMR spectrum in the case.

The Fe^{3+} ions in $SrTiO_3$ are also $3d^5$ (high spin) ions in the octahedral field. Where Fe^{3+} replaces the Ti^{4+} ions in the lattice, the original Ti^{4+} ions are surrounded by eight octahedral coordination oxygen atoms. It should be noted that the Fe^{3+} ions have two crystal planes in the $SrTiO_3$ lattice: one is in the tetragonal symmetry field, and the other is in the octahedral symmetric field. Figure 10.16a shows the EMR spectrum when the extra magnetic field $\boldsymbol{H} \parallel [0, 0, 1]$[29], and its outer two lines are

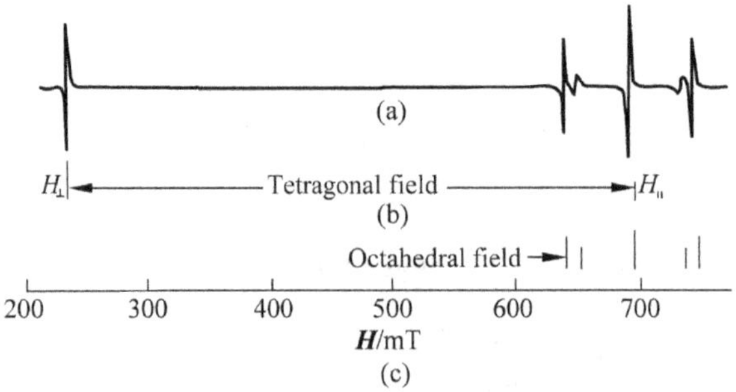

Figure 10.16: EMR spectra of Fe^{3+} ions in $SrTiO_3$ crystal.

transitions between $|\pm\frac{3}{2}\rangle \leftrightarrow |\pm\frac{1}{2}\rangle$, the inner two lines are the transition between $|\pm\frac{5}{2}\rangle \leftrightarrow |\pm\frac{3}{2}\rangle$, the center single line is the transition between $|+\frac{1}{2}\rangle \leftrightarrow |-\frac{1}{2}\rangle$, their intensity ratio should be 8:5:9:5:8. Figure 10.16b shows the magnetic field parallel to the octahedral main axis direction, the theoretical stick spectrum of Fe^{3+} ion in the tetragonal symmetry crystal. Figure 10.16c shows the theoretical stick spectrum of Fe^{3+} ion in octahedral symmetric field.

The Mn^{2+} ions in MgO are also $3d^5$ (high spin) ion in the octahedral field. Figure 10.17 is its EMR spectrum. The six strong lines are caused by the hyperfine interaction of $^{55}Mn(I = 5/2)$, five lines as a set containing one strong and one weak around a strong line, similar to the spectrum of Fe^{3+} ions in Figure 10.15. From left to right, there are six sets; while the fifth set overlaps on both sides, the sixth set is reversed, and the strong line moves from the outside to the inside. For Mn^{2+}, the parameter a' (0·001901cm^{-1}) is much smaller than the hyperfine coupling parameter a (0·00811cm^{-1}) and is much more smaller than the a' (0·02038cm^{-1}) of Fe^{3+}. So it is the same as the six lines of the $|-\frac{1}{2}\rangle \leftrightarrow |\frac{1}{2}\rangle$ transition and display only a small anisotropy. Therefore, even for the powder or solution samples, the hexalet of this set can still be seen. Because of a strong anisotropy of the other various sets of lines, it is not visible in the powder or solution. For Fe^{3+} ion in some octahedral field

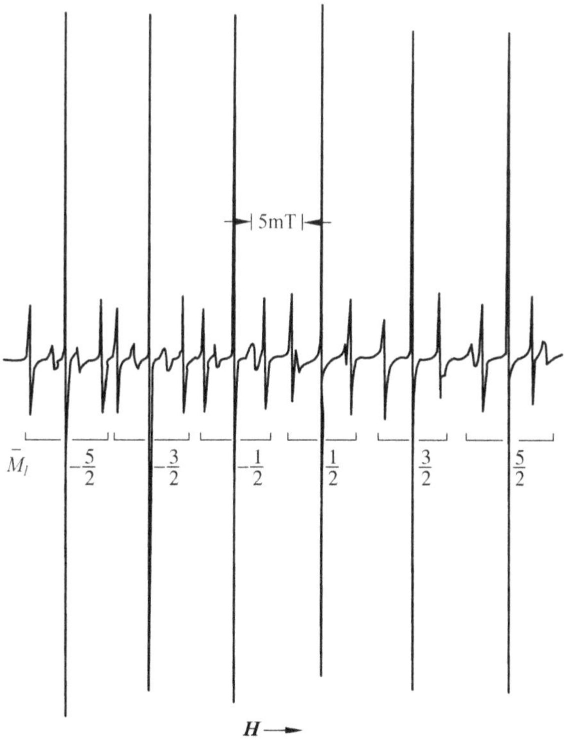

Figure 10.17: Room temperature EMR spectrum of Fe^{3+} ions in MgO.

matrices, $|-\frac{1}{2}\rangle \leftrightarrow |\frac{1}{2}\rangle$ transitions can also be seen, but only a wide line can be seen in the solution sample because its anisotropy is too large. The Fe^{3+} ion in solution, such as $[FeCl_4]^-$, may be tetrahedral coordination [30].

Since the energy-level interval between the excited state and ground state of $3d^5$ (high spin) ion is very large, its g value is always very close to g_e, such as Mn^{2+} ion in CaO [31], $g = 2{\cdot}0009$, Fe^{3+} ion in CaO, $g = 2{\cdot}0052$, Cr^+ ion in ZnS [32] $g = 1{\cdot}9995$.

10.6.3.2 $3d^5$ (High spin) ions in quadratic distortion octahedral field

In the low-symmetry crystal field, the spin Hamiltonian should be added:

$$D\left(\hat{S}_z^2 - \frac{1}{3}\hat{\boldsymbol{S}}^2\right) + E(\hat{S}_x^2 - \hat{S}_y^2) \tag{10.60}$$

Since D and E are usually much larger than a', so we can use Eq. (10.39). The g values are isotropic and very close to g_e. In the quadratic field, $E = 0$, if $\boldsymbol{H}$ is parallel to the square axis, the energy splitting is shown in Figure 10.18. Here $D \approx 0{\cdot}1\,\text{cm}^{-1}$. If $\boldsymbol{H}$ is not parallel square axis, the level representation is very complicated.

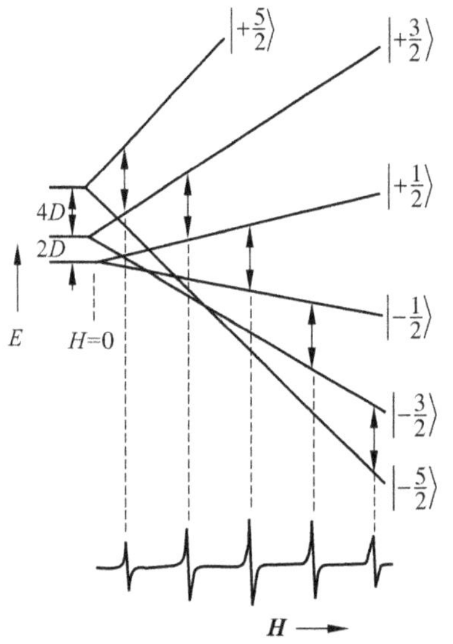

Figure 10.18: Energy level and allow transition of d^5 ions in the weak quadratic field.

If $D \gg h\nu$, only the $|-\frac{1}{2}\rangle \leftrightarrow |\frac{1}{2}\rangle$ transition can be observed. After proper treatment, the Fe^{3+} ions at octahedral coordination environment in $SrTiO_3$ can be converted into tetragonal system; $g_{\parallel} = 2{\cdot}0054$, $g_{\perp} = 5{\cdot}993$. Here, $D = 1{\cdot}42\,\text{cm}^{-1}$, which is really greater than $h\nu$.

Hemoglobin is an important biological example of Fe^{3+} ion (high spin) in the tetragonal field. Its structure is shown in Figure 10.19a. High-spin Fe^{3+} ions are in a

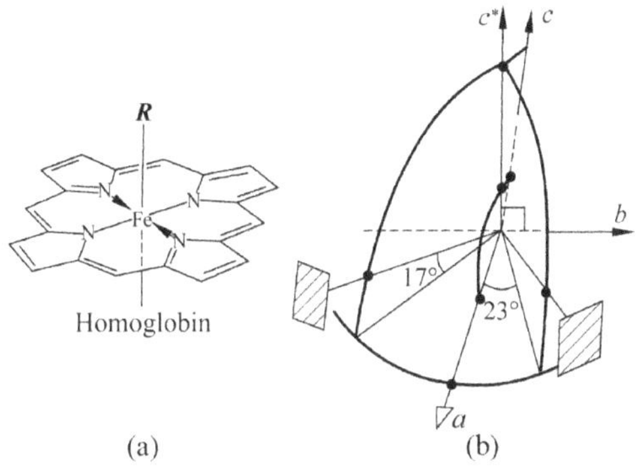

Figure 10.19: Hemoglobin structure.

very strong axisymmetric crystal field, and the D value is very large (20 – 30cm^{-1}), $g_{\parallel}=2$, $g_{\perp}=6$; the anisotropy of g tensor also is quite large, which provides a good test method for determining the plane of heme relative to the orientation of crystal axis. Using a typical method of measuring spectrum of single crystal, to determine the relationship between the g tensors in the three crystal axis planes with the angle, the plane of the heme and the orientation of its normal line could be determined. As shown in Figure 10.19b [33], the main value of the g tensor as well as the direction of the cosine could be obtained.

The EMR spectrum of hemoglobin can also be observed in a random system. The EMR spectrum of the acidic form of heme (pH 6.0) at 167 K is shown in Figure 10.20a [34]. This is an axisymmetric spectrum, strong line $g_{\perp}=6$, weak line $g_{\parallel}=2$. At this time, the sixth coordination of heme is H_2O. However, in the alkaline solution (pH10·5) at 77 K, the measured spectrum is shown in Figure 10.20b. The anisotropy of the g tensor at this time is quite small, indicating that the crystal field is no longer axisymmetric. At this time, the sixth coordination of heme is OH^-, and it combines with heme much stronger than H_2O. It seems that the Fe^{3+} ion has been changed from a high-spin state to a low-spin state. The magnetic susceptibility study also validates this point.

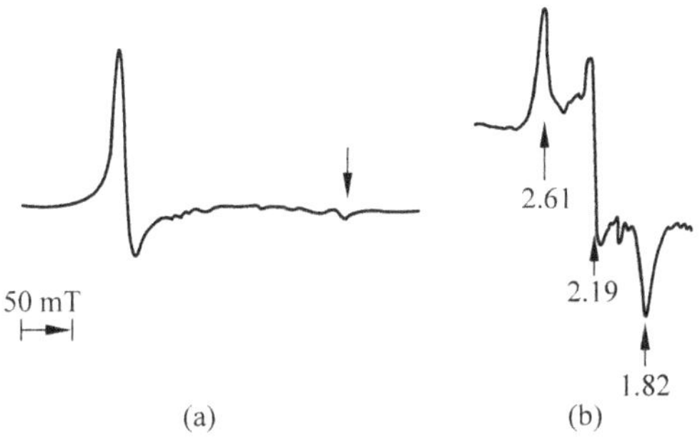

Figure 10.20: EMR spectrum of heme.

The EMR spectrum of Fe^{3+} ions in solution has a very broad (100 mT) line. The reason may be that the transient coordination distribution of the distortional octahedral symmetry brings about a large as well as zero-field splitting of variation with time. If pH > 5, complexing it with the F^- ion, a stable octahedral complex ion FeF_6^{3-} can be formed. Its EMR spectrum is a set of seven lines with a total width of only 1.1 mT [35] with intensity ratio of 1: 6: 15: 20: 15: 6: 1. Obviously, this is due to the contribution of hyperfine interactions of six equivalent F^- nuclei with unpaired electrons.

10.7 Ground-state ions with orbital degeneracy

When the crystal field cannot relieve completely the orbital degeneracy of free ions, the ground state still has pure orbital angular momentum. For this system, the spin-orbit coupling operator should be expressed by the total angular momentum of the ground state. Since the energy-level intervals after the splitting of energy levels are always much larger than kT, the lowest energy states that determine the degeneracy and eigenfunction of the energy states are considered mainly here because they determine the characteristics of its EMR spectrum.

10.7.1 *D*-state ions

10.7.1.1 Co^{2+} Ions in octahedral fields

In octahedral fields, the ^2D state of $3d^1$ free ions split into two and the lowest state is T_{2g}. It is orbital triplet degeneracy. Even if it is a zero-order approximation, the ground state still retains some orbital angular momentum. Here, taking the three degenerate T_{2g} states can be seen as three components of the state with "virtual" orbital angular momentum, and quantum number $L' = 1$. That is, we can use $|M_{L'}\rangle$ to characterize the three degenerate states; $M_{L'} = 1, 0, -1$. Strictly speaking, L can only be applied to the systems with spherical symmetry. If the ground state is regarded as a system of a state with $L' = 1$, then the Hamiltonian operator of Eq. (10.21) needs to be modified as follows [36]:

$$\hat{\mathscr{H}}' = \hat{\mathscr{H}}'_{\text{Zeeman}} + \hat{\mathscr{H}}'_{LS} = \beta(-\alpha\hat{L}' + g_e\hat{\mathbf{S}}) \cdot \boldsymbol{H} - \alpha\lambda\hat{L}' \cdot \hat{\mathbf{S}} \tag{10.61}$$

where α is a coefficient, $\alpha \cong 1$ for *D*-state ions, and $\alpha \cong 3/2$ for *F*-state ions.

For $S = 1/2$ of $3d^1$ free ions, there is at least one Kramer doublet degeneracy. If $|M_{L'}, M_S\rangle$ is the basic function of $\hat{\mathscr{H}}'_{LS}$, in these six basis functions, only two of $|\pm 1, \pm \frac{1}{2}\rangle$ are eigenfunctions of $\hat{\mathscr{H}}'_{LS}$. The other four are all not eigenfunctions of $\hat{\mathscr{H}}'_{LS}$.

Here introduce total angular momentum operator $\boldsymbol{J}' = \boldsymbol{L}' + \boldsymbol{S}$. Pay attention to $\hat{J}'_z$ and $\hat{L}'_z$, $\hat{S}_z$ is swappable. Therefore, $|M_{L'}, M_S\rangle$ is also the eigenfunction of $\hat{J}'_z$. Generally, they are not the eigenfunctions of $\hat{J}'^2_z$. For the eigenfunction of $|\pm 1, \pm \frac{1}{2}\rangle$, the eigenvalue of $\hat{J}'_z$ is $M_{J'} = \pm 3/2$. It is also the eigenfunction of $\hat{\mathscr{H}}'_{LS}$. Hence, the energy of state $|\pm 1, \pm \frac{1}{2}\rangle$ is

$$\langle \pm 1, \pm \frac{1}{2}|\hat{\mathscr{H}}'_{LS}|\pm 1, \pm \frac{1}{2}\rangle = -\alpha\lambda\langle \pm 1, \pm \frac{1}{2}|\hat{L}'_z\hat{S}_z + \frac{1}{2}(\hat{L}'_+\hat{S}_- + \hat{L}'_-\hat{S}_+)|\pm 1, \pm \frac{1}{2}\rangle = -\frac{\alpha\lambda}{2} \tag{10.62}$$

$M_{J'}$ of functions $|0, \pm \frac{1}{2}\rangle$ and $|\pm 1, \mp \frac{1}{2}\rangle$ are $M_{J'} = \pm \frac{1}{2}$; however, they are not the eigenfunctions of $\hat{\mathscr{H}}'_{LS}$. If we want to find out the energy, we should solve the secular equation. The matrix representation of $\hat{\mathscr{H}}'_{LS}$ is

$$\begin{array}{cc} & \left|\pm 1, \mp \frac{1}{2}\right\rangle \quad \left|0, \pm \frac{1}{2}\right\rangle \\ \hat{\mathscr{H}}'_{LS} = \begin{array}{c} \langle \pm 1, \mp \frac{1}{2}| \\ \langle 0, \pm \frac{1}{2}| \end{array} & \begin{pmatrix} \frac{1}{2}\alpha\lambda & -\frac{1}{\sqrt{2}}\alpha\lambda \\ -\frac{1}{\sqrt{2}}\alpha\lambda & 0 \end{pmatrix} \end{array} \tag{10.63}$$

By solving its secular equation, two energy levels can be obtained. One is $E_{1/2} = \alpha\lambda$ with doublet degeneracy, and the other is $E_{3/2} = -\frac{\alpha\lambda}{2}$ with quadruplet degeneracy. Their eigenfunctions are listed in Table 10.5.

Table 10.5: The eigenfunctions of $E_{1/2}$ and $E_{3/2}$.

Energy	Coupling representation function $\|J', M_{J'}\rangle$	Noncoupling representation function $\|M_{L'}, M_S\rangle$
$E_{1/2} = \alpha\lambda$	$\|\frac{1}{2}, \frac{1}{2}\rangle$	$\sqrt{\frac{2}{3}}\|1, -\frac{1}{2}\rangle - \sqrt{\frac{1}{3}}\|0, \frac{1}{2}\rangle$
	$\|\frac{1}{2}, -\frac{1}{2}\rangle$	$\sqrt{\frac{2}{3}}\|-1, \frac{1}{2}\rangle - \sqrt{\frac{1}{3}}\|0, -\frac{1}{2}\rangle$
$E_{3/2} = -\frac{\alpha\lambda}{2}$	$\|\frac{3}{2}, \frac{3}{2}\rangle$	$\|1, \frac{1}{2}\rangle$
	$\|\frac{3}{2}, \frac{1}{2}\rangle$	$\sqrt{\frac{1}{3}}\|1, -\frac{1}{2}\rangle + \sqrt{\frac{2}{3}}\|0, \frac{1}{2}\rangle$
	$\|\frac{3}{2}, -\frac{1}{2}\rangle$	$\sqrt{\frac{1}{3}}\|-1, \frac{1}{2}\rangle + \sqrt{\frac{2}{3}}\|0, -\frac{1}{2}\rangle$
	$\|\frac{3}{2}, -\frac{3}{2}\rangle$	$\|-1, -\frac{1}{2}\rangle$

For $3d^1$ ions, $\lambda > 0$, so, $E_{3/2}$ is the lowest energy level. If $\alpha\lambda \gg kT$, then only this energy level is occupied by electrons.

For the case of $H \parallel Z$:

$$\hat{\mathscr{H}}'_{\text{Zeeman}} = \beta(-\alpha L'_z + g_e S_z)H_Z \tag{10.64}$$

$$E_{3/2, \pm 3/2} = \pm(-\alpha + 1)\beta H_Z = \frac{2}{3}(-\alpha + 1)\beta H_z\left(\pm \frac{3}{2}\right) \tag{10.65a}$$

$$E_{3/2, \pm 1/2} = 2\left[\frac{1}{3}(-\alpha - 1) + \frac{2}{3}(+1)\right]\beta H_z\left(\pm \frac{1}{2}\right) = \frac{2}{3}(-\alpha + 1)\beta H_z\left(\pm \frac{1}{2}\right) \tag{10.65b}$$

Since for the D-state ions $\alpha \approx 1$, so, $g = 0$. In fact, the EMR spectrum of a $3d^1$ ion in octahedral field cannot be observed.

10.7.1.2 $3d^1$ Ions in a tetragonal distorted octahedral field $\Delta \gg \delta \gg \lambda$

For $3d^1$ ions in a tetragonal field without spin-orbit coupling interaction, the energy splitting is shown as the dotted line in Figure 10.21. Its lowest level depends on the sign of δ, which can be the orbital singlet or an orbital doublet degeneracy. If the ground state is orbital singlet, and there is no spin orbit coupling interaction, then $g_{\parallel} = g_{\perp} = g_e$, because this single state is still a spin-doublet degeneracy, namely, Kramer doublet degeneracy.

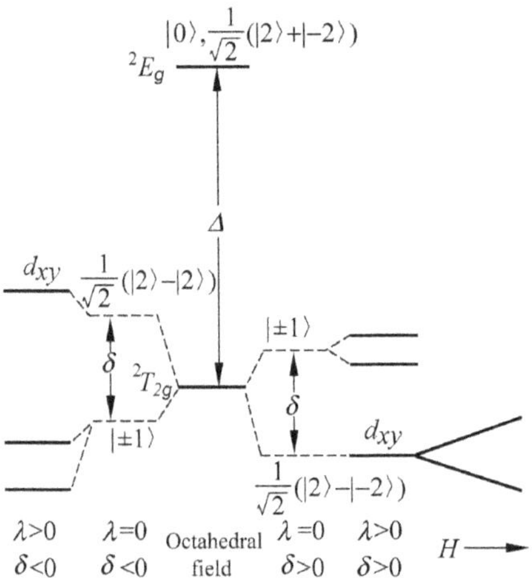

Figure 10.21: The energy-level splitting of $3d^1$ ion in tetragonal distorted octahedron field.

If there is spin-orbit coupling interaction, even small, orbital degeneracy will be relieved. For $0 < \delta \gg \lambda$, g value can be found from $\mathbf{\Lambda}$ tensor. Assume that the tetragonal axis is parallel to the Z-axis, then,

$$\Lambda_{zz} = -\frac{\left[\frac{1}{\sqrt{2}}(\langle 2| - \langle -2|)\hat{L}_z\left(\frac{1}{\sqrt{2}}\right)(|2\rangle + |-2\rangle)\right]^2}{\Delta} = -\frac{4}{\Delta} \tag{10.66}$$

So,

$$g_{\parallel} = g_{zz} = g_e - \frac{8\lambda}{\Delta} \tag{10.67}$$

$$\Lambda_{xx} = -\frac{\left[\frac{1}{\sqrt{2}}(\langle 2| - \langle -2|)\frac{1}{2}\hat{L}_+|1\rangle\right]^2 + \left[\frac{1}{\sqrt{2}}(\langle 2| - \langle -2|)\frac{1}{2}\hat{L}_-|-1\rangle\right]^2}{\delta} = -\frac{1}{\delta} \tag{10.68}$$

Hence,

$$g_\perp = g_{xx} = g_{yy} = g_e - \frac{2\lambda}{\delta} \tag{10.69}$$

For the $3d^1$ ion in tetragonal field with $\Delta \gg \delta$, $0 < \delta \gg \lambda$, *theoretical prediction*: $g_\perp < g_\parallel < g_e$

The experimental EMR spectra of $3d^1$ ions such as Sc^{2+}, Ti^{3+}, V^{4+}, Cr^{5+}, and Mn^{6+}, were exactly observed. Although all of them there have only one d electron, their covalent bond characteristics enhanced with the increase of the positive charge valence state, and the probability of the charge transfer to the ligand are also increased, which will decrease the effective value of the spin-orbit coupling parameter. Therefore, in some cases, even the $S = \frac{3}{2}$ can appear. It can be seen that the ion of lower charge using crystal field approximation agrees with the experimental result relatively, because its characteristics of covalent bond is the least.

There are two spectral lines of $VO(CN)_5^{3+}$ ion in KBr single crystal [37] as shown in Figure 10.22: one is ***H*** parallel to the axis of V—O and the other is ***H*** vertical to the axis of V—O; $g_\parallel = 1{\cdot}9711$, $g_\perp = 1{\cdot}9844$. This means $g_\perp > g_\parallel$, since the (CN) ligand has strong covalent bond property.

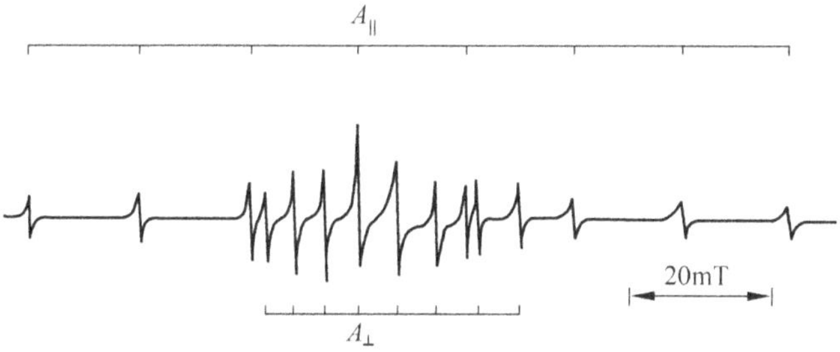

Figure 10.22: The EMR spectrum of $VO(CN)_5^{3+}$ in KBr single crystal.

10.7.1.3 $3d^1$ Ions in a tetragonal distorted octahedral field $\Delta \gg \lambda \approx \delta$

It is difficult to distinguish the influence between λ and δ when they are on the same order of magnitude. Let $\eta = \lambda/\delta$; $S = \left[\left(1 + \frac{1}{2}\eta\right)^2 + 2\eta^2\right]^{1/2}$, when $\delta \to 0$, $S \to 3\eta/2$; so, when $\lambda > 0$, all of the $g_\parallel$ and $g_\perp$ approach to zero. It is also to say, for the $3d^1$ ions in normal octahedral field, their EMR spectra cannot be observed. However, for $\lambda < 0$, $\delta = 0$, the ground state is corresponding to $J' = 1/2$. At this time, $\eta \to \infty$, then $g_\parallel \to -2$; $g_\perp \to +2$. It explains that the g value can be negative. When $\eta \to 0$, $S \to 1$, all of the $g_\parallel$ and $g_\perp$ approach to 2. Figure 10.23 shows the function relationship of the g value with η variation of the $3d^1$ ions in a tetragonal distorted octahedral field.

The EMR spectrum of Ti^{3+} ion in $CsTi(SO_4)_2 \cdot 12H_2O$ substrate shows that $g_\parallel = 1{\cdot}25$, $g_\perp = 1{\cdot}14$. Although it is a trilateral distortion, as a first-order approximation, the value $\eta \approx 0{\cdot}6$ could be estimated from Figure 10.23. If $\lambda = 154\ \text{cm}^{-1}$ and Ti^{3+} free ions would be used, then $\delta \approx 250\ \text{cm}^{-1}$.

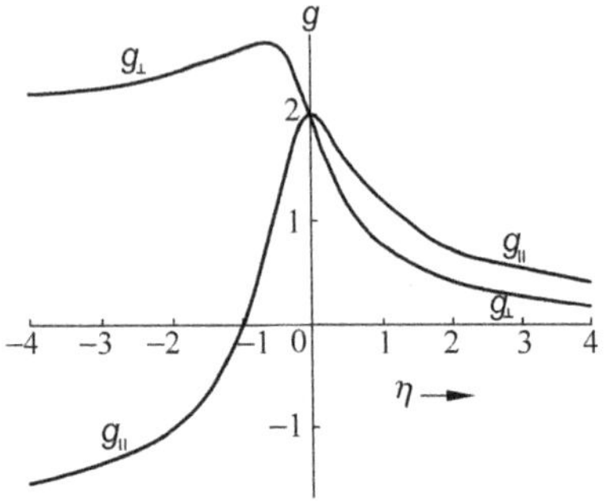

Figure 10.23: The relationship between g value and η of $3d^1$ ion in a tetragonal distorted octahedral field.

10.7.1.4 $3d^1$ Ions in a trilateral distorted octahedral field

Many 3d ions belong to this case; their major crystal field is an octahedral field with very small trilateral distortion. The diagonal line of the circumscribe cube of the octahedron is the symmetrical axis of the trilateral. Stretching or compressing along the direction of the axis forms a trilateral distortion. Usually, the trilateral axis is always chosen as the Z-axis. Thus, the operator and wave function of crystal field obtained will be the simplest. In the trilateral distorted octahedral field, the energy levels and wave functions of the $3d^1$ ion are shown in Figure 10.24. Its energy splitting and g factor are same as those of tetragonal distortion octahedral field of $3d^1$ ions: $g_\parallel \approx g_e$; $g_\perp \approx g_e - 2\lambda/\delta$ (the Δ^{-1} term to be neglected in equation).

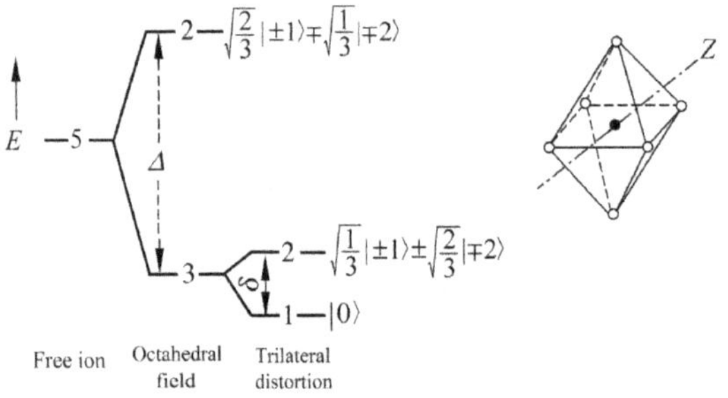

Figure 10.24: The energy level and wave function of $3d^1$ ions in a trilateral distorted octahedral field.

The titanium acetylacetonate in aluminum acetylacetonate is the $3d^1$ ion in the trilateral distorted octahedral field. Its spectral parameters [38] are $g_\parallel = 2{\cdot}000(2)$; $g_\perp = 1{\cdot}921(1)$; $A_\parallel = 0{\cdot}00063\,\mathrm{cm}^{-1}$; and $A_\perp = 0{\cdot}00175\,\mathrm{cm}^{-1}$. The estimated Δ value should be $2000-30,000\mathrm{cm}^{-1}$; however, the spectrum between 5000 and $14000\mathrm{cm}^{-1}$ is not absorption. According to T_1 dependent relation with temperature, the inferred Δ value should be $2000-5000\mathrm{cm}^{-1}$.

10.7.1.5 $3d^5$ (Low spin) ions in a tetragonal distorted octahedral field

The order of energy levels and the orbits of $3d^5$ (low spin) ions in a tetragonal distorted octahedral field are same was those of $3d^1$ ions in octahedral field. Since there are five electrons on the t_{2g} energy level, it can be seen that there is a positive hole in the t_{2g} orbit. Thus, $\lambda < 0$.

The ion with MX_6^{n-} form belongs to this case. It is generally assumed that it should have octahedron symmetry; however, in fact, it usually has some tetragonal or orthorhombic distortion. Since the magnitude of δ is similar to λ and $\delta > 0$, $\eta < 0$, the left part of Figure 10.25 should be used. For example, for Fe^{3+} ion in $K_3Co(CN)_6$, experimental determinate data [39] are $g_{\parallel} = 0{\cdot}915$, $g_{\perp} = 2{\cdot}2$. From Figure 10.25, $\eta \approx -0{\cdot}5$, and using its (free ion) $\lambda = -103\ \text{cm}^{-1}$, hence $\delta \approx 200\ \text{cm}^{-1}$.

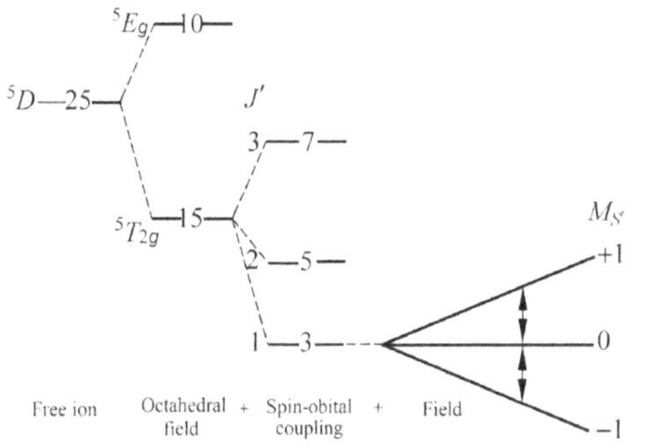

Figure 10.25: The energy splitting of $3d^6$ (high spin) in crystal field.

10.7.1.6 $3d^6$ (High spin) ions in octahedral field

In the octahedral field of moderate intensity, ^5D state of $3d^6$ ion will be split into two levels: ^5E_g and $^5T_{2g}$. The $^5T_{2g}$ is located in the lower energy level as shown in Figure 10.25. Since the strong field is diamagnetic, it is not considered here. In the weak field, the spin-orbit coupling interaction of the ground state, the $^5T_{2g}$ energy level will be split to $J' = 3, 2, 1$ three energy levels. Let operator

$$\mathscr{H}'_{LS} = -\alpha\lambda\hat{L}' \cdot \hat{\mathbf{S}} = -\frac{\alpha\lambda}{2}(\hat{J}'^2 - \hat{L}'^2 - \hat{S}^2)$$

applied to $|J', M_{J'}\rangle$, the $E_{J'}$ is obtained as

$$E_1 = 3\alpha\lambda \approx 3\lambda;\quad E_2 = \alpha\lambda \approx \lambda;\quad E_3 = -2\alpha\lambda \approx -2\lambda \tag{10.70}$$

If the spin–orbit coupling interaction with the ^5E_g energy state also is considered, to the above equation should be added a term $-\frac{18\lambda^2}{5\Delta}$.

Since the electrons of $3d^6$ ion is more than half full, $\lambda < 0$. Thus, the lowest energy state is E_1. If the temperature is low enough, only this energy level will be occupied by electrons. Since $J' = 1$ of E_1, thus, $M_{J'} = 1, 0, -1$. So, this energy level is threefold degenerate, and it has three $|J', M_{J'}\rangle$ functions. The energy levels in the magnetic field are as follows:

$$E_{-1}=\beta H\left(-\tfrac{3}{2}g_e-\tfrac{k'}{2}+\tfrac{18\lambda}{5\Delta}\right);\quad E_0=0;\quad E_{+1}=\beta H\left(\tfrac{3}{2}g_e+\tfrac{k'}{2}-\tfrac{18\lambda}{5\Delta}\right) \tag{10.71}$$

Considering the electrons possibly delocalize to ligands, the so-called orbital attenuation factor $k'(<1)$ is introduced here. For the $3d^6$ (high spin) ion in octahedral field, $k' \approx 1$, so,

$$g=\frac{7}{2}-\frac{18\lambda}{5\Delta} \tag{10.72}$$

For the Fe^{2+} free ion, $\lambda=-103\ cm^{-1}$. In MgO crystal, $\Delta \simeq 10{\cdot}000103\ cm^{-1}$, $g=3.494$ is calculated and the experimental value is g = 3.428. If using the $k' \approx 0{\cdot}8$, the calculated value will be more in line with the experimental value.

The spin orbit coupling interaction between the ^5E_g energy states lead to the spin lattice relaxation time being very short. Therefore, the EMR spectrum only can be observed at 20 K or lower temperature. The Fe^{2+} ions couple with environment more than any other 3d ion. The strict octahedron symmetry with a slight deviation could lead to a large zero-field splitting. In MgO, the linewidth of other 3d ions, except $\pm 1/2$ ions, are only 0.05 mT, while the linewidth of Fe^{2+} ions is as high as 50 mT. Figure 10.26 is the EMR spectrum measured at 4.2 K and microwave frequency of 25 GHz. On the octahedral field are superposed random low-symmetric distortion fields, leading to a very wide spectral line. The narrow line of the center is a double-quantum transition. The g value of wide line is determined by the central narrow line, thus, $g=3{\cdot}4277$. In addition, the transition of $\Delta M_s=2$ can be observed at the half field. It is further proved that a low-symmetry distortion field is superposed on the octahedral field. For the strict octahedral symmetry field, the transition of $\Delta M_s=2$ is forbidden.

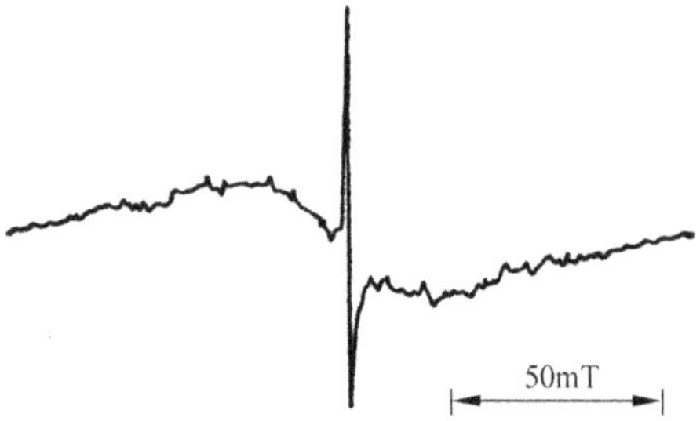

Figure 10.26: The EMR spectrum of Fe^{2+} ion in MgO octahedral field.

10.7.2 *F*-state ions

10.7.2.1 $3d^2$ Ions in octahedral field

The lowest energy level of the $3d^2$ ion in the octahedral field is T_{1g}, here S = 1. The results of spin–orbit coupling interaction arise further, splitting of energy level.

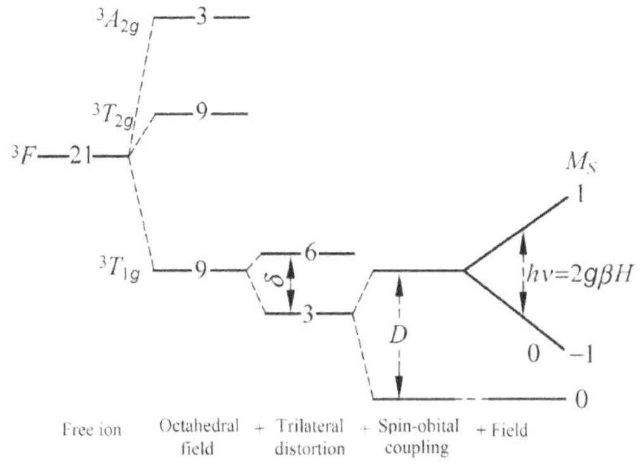

Figure 10.27: Energy-level splitting of ^3F ion (d^2) in octahedral field.

According to quantum number, $J' = 2, 1, 0$, split into three energy levels, as shown in Figure 10.27.

$$E_2 = -\alpha\lambda; \quad E_1 = \alpha\lambda; \quad E_0 = 2\alpha\lambda$$

Because $\lambda > 0$ of $3d^2$ ion, the lowest energy level is E_2. But so far, the experimental examples of $3d^2$ ions in the octahedral field still has not been observed in the text.

10.7.2.2 $3d^2$ Ions in a trilateral distorted octahedral field

In the known examples, the $3d^2$ ions in the octahedral field all have a strong trilateral distortion, and the energy level T_{1g} will be further split into two energy levels; the lowest energy level is an orbital singlet state, but has a triplet spin degeneracy. Since there is a strong spin–orbit coupling interaction in the upper two states, there arises a large zero-field splitting; the lowest energy level $|M_S\rangle = |0\rangle$ is nondegenerate state, and the higher energy level is doublet degenerate state $|\pm 1\rangle$ (see as Figure 10.27). Because of the external magnetic field, the transition of $h\nu = g\beta H$ between the states $|-1\rangle \leftrightarrow |+1\rangle$ can be seen, and the selection rule is $\Delta M_S = 2$.

From magnetic susceptibility test we known that D is about 5cm^{-1}. Because $|+1\rangle$ and $|-1\rangle$ states are far from the state $|0\rangle$, so the transition of $\Delta M_S = \pm 1$ cannot be observed in the ordinary magnetic field. In the trilateral distorted field, $E \neq 0$ and $|+1\rangle$ and $|-1\rangle$ are two states of mixing; there is a transition of $\Delta M_S = \pm 2$, but it should be observed when the microwave field is parallel to the external magnetic field.

The Al^{3+} in Ruby (Al_2O_3) is replaced by V^{3+} or Cr^{4+} ions, which is an example of $3d^2$ ions in a trilateral distorted octahedral field [40]. In ruby, all Al^{3+} ions are located on the trilateral axis of the distorted octahedral field surrounded by six oxygen ions. The trilateral field split the ground state T_{1g} of the V^{3+} ion into an orbital singlet state A_{2g} and an orbital doublet E_g. The energy of A_{2g} is the lowest, and E_g is higher than A_{2g} for 1200 cm^{-1}. Because of the strong spin–orbit coupling interaction between the ground state and the lowest excited state, the relaxation time V^{3+} is very short, and the EMR spectrum observed must be at a low temperature below 4 K.

The gyromagnetic ratio of ^{51}V nucleus ($I = 7/2$, natural abundance is 100%) is quite large, and all the eight lines of hyperfine structure usually can be observed. The S = 1 of V^{3+} ion, and its spin Hamiltonian is

$$\mathscr{H} = \beta[g_{\parallel} H_Z \hat{S}_z + g_{\perp}(H_X \hat{S}_x + H_Y \hat{S}_y) + D\left(\hat{S}_z^2 - \frac{2}{3}\right) + A_{\parallel}\hat{S}_z\hat{I}_z + \frac{1}{2}A_{\perp}(\hat{S}_+\hat{I}_- + \hat{S}_-\hat{I}_+) \quad (10.73)$$

Let the trilateral axis be Z-axis, when the external magnetic field $H \parallel$ is Z-axis; the above equation can be represented by an equivalent Hamiltonian operator:

$$\hat{\mathscr{H}}_{\text{eq}} = g_{\parallel}\beta H_Z \hat{S}_z + A_{\parallel}\hat{S}_z\hat{I}_z \quad (10.74)$$

Obviously, the x and y components are ignored here, and since $D\hat{S}_z^2$ has the same effect on the two states $|\pm 1\rangle$, they also are ignored. Then $\hat{H}_{\text{eq}}$ is applied to the state $|\pm 1\rangle$ and the following is obtained:

$$\Delta E = h\nu = 2(g_{\parallel}\beta H_Z + A_{\parallel}M_I) \quad (10.75)$$

The magnetic field position $H_r(M_I)$ corresponds to the hyperfine line of the M_I

$$H_r(M_I) = \left(\frac{1}{2}h\nu - A_{\parallel}M_I\right)/g_{\parallel}\beta \quad (10.76)$$

The EMR spectrum of V^{3+} ion in ruby (Al_2O_3) has been observed experimentally at 4.2 K. It is a hyperfine structure spectrum consisting of eight lines with nearly equal splitting distance (11.4 mT). It explains the fact that the spectrum should be attributed to the transition of $\Delta M_S = \pm 2$ between $|-1\rangle \leftrightarrow |+1\rangle$. Because the transition of $\Delta M_S = \pm 1$, in general, has a large hyperfine splitting, it is bound to produce considerable secondary shift and these eight lines cannot be equidistant. Conversely, the secondary shift is caused by the $\hat{S}_+\hat{I}_- + \hat{S}_-\hat{I}_+$ term of the Hamiltonian operator. These two states $|\pm 1\rangle$ with the state $|0\rangle$ could be coupled by this term of operator. Now, the transition takes place between $|-1\rangle \leftrightarrow |+1\rangle$, and it is impossible to couple these states by this term of operator; no secondary shift occurs, and the eight lines are exactly equidistant.

The $\Delta M_S = \pm 1$ transitions are not seen because the energy level difference of these two states $|\pm 1\rangle$ with the state $|0\rangle$ is too large. Using a strong pulsed magnetic field (10 T), the $\Delta M_S = 1$ transition between states $|-1\rangle$ and $|0\rangle$ is observed, thus determining $D = 7{\cdot}85 \pm 0{\cdot}4 \text{cm}^{-1}$.

The property of Cr^{4+} ion in ruby is similar to the V^{3+} ion, and the EMR spectrum parameter [41] is $g_{\parallel} = 1{\cdot}90$; $D = 7\ \text{cm}^{-1}$; $E < 0{\cdot}05\ \text{cm}^{-1}$.

10.7.2.3 3d^7 (High spin) ions in octahedral field

The lowest energy level after the energy level splitting of ^4F state of $3d^7$ (high spin) free ions in octahedral field is $^4T_{1g}$, similar to the $3d^1$ in the octahedral field. Here, the

$S = \frac{3}{2}$, the ground state can be considered as $L' = 1$, and $J' = \frac{5}{2}, \frac{3}{2}, \frac{1}{2}$; their energies are as follows:

$$E_{5/2} = -\frac{3}{2}\alpha\lambda = -\frac{9}{4}\lambda \tag{10.77a}$$

$$E_{3/2} = \alpha\lambda \approx \frac{3}{2}\lambda \tag{10.77b}$$

$$E_{1/2} = \frac{5}{2}\alpha\lambda \approx \frac{15}{4}\lambda \tag{10.77c}$$

The splitting of the energy levels is shown in Figure 10.28.

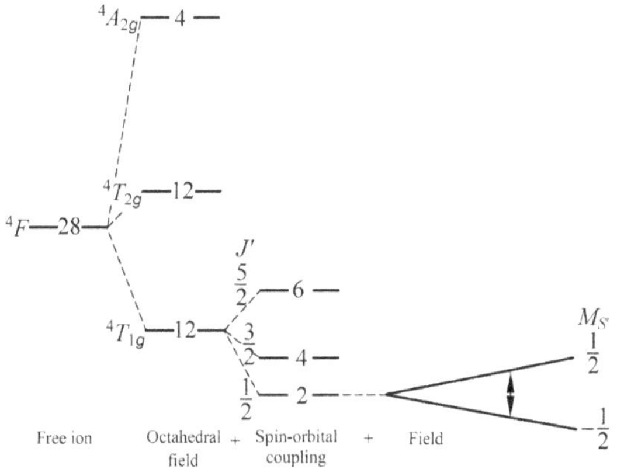

Figure 10.28: Energy-level splitting of $3d^7$ (high spin) ion in the Octahedral field.

The $\lambda < 0$ of $3d^7$ ion, the lowest energy level, should be $E_{1/2}$. If $\alpha\lambda \gg kT$, the electrons are only distributed at the energy level of $J' = \frac{1}{2}$. In the zero field, the ground state is doublet degenerate.

Because the ground state of the $3d^7$ (high spin) ion in the octahedral field is doublet degenerate, even though its true spin should be $S = \frac{3}{2}$, it still can be described by spin Hamiltonian of the equivalent spin $S' = \frac{1}{2}$.

The most common $3d^7$ (high spin) ion in the octahedral field is Co^{2+}, and its EMR spectrum [42] could be observed in many coordination environments. The eight lines hyperfine structure of ^{59}Co nucleus ($I = 7/2$, natural abundance 100%) can all be observed usually.

However, because the quadralet state of $J' = \frac{3}{2}$ and the hexalet state of $J' = \frac{5}{2}$ are only several hundred wave numbers higher than the ground state, the relaxation time T_1 is very short, and the EMR spectrum cannot be observed at normal temperature. It can only be seen at 20 K or even lower temperatures.

The calculated g value of the first approximation is 4.33. The experimental values [43] in the MgO are $g = 4.2785$ and $A = 0{\cdot}009779\ \text{cm}^{-1}$. The experimental values [44] in the CaO are $g = 4.372$ and $A = 0{\cdot}01322\ \text{cm}^{-1}$. In the CaO, the hyperfine splitting of the

Co nucleus is greater than in the MgO; it explains that the covalent properties of Co^{2+} ion in the CaO is less than in the MgO.

In the distorted octahedral field, the spin–orbit coupling and the nonoctahedral symmetry of the crystal field all can lead to the anisotropy of the g factor. For example, the g factor of Co^{2+} ion in TiO_2 is $g_{XX} = 2{\cdot}090$; $g_{YY} = 3{\cdot}725$; $g_{ZZ} = 5{\cdot}860$.

10.7.3 Jahn–Teller distortion

10.7.3.1 3d^9 Ions in octahedral field

The electronic configuration of the 3d^9 ion is $t_{2g}^6 e_g^3$, which can be considered as a positively charged hole in e_g orbit. In the octahedral field, the orbit is doublet degenerate. Even if the spin–orbit coupling interaction is introduced, the doublet degeneracy cannot be relieved, because the $\hat{H}_{LS} = \lambda L \cdot S$ operator cannot be coupled

$$|0\rangle = d_{z^2} \text{ with } \frac{1}{\sqrt{2}}(|2\rangle + |-2\rangle) = d_{x^2-y^2}$$

However, for the nonlinear system with residual orbital degeneracy, according to the Jahn–Teller theory, the system will be strongly coupled to the lattice vibration, the residual orbital degeneracy is relieved, and also the ground-state energy is reduced [45]. Tetragonal or orthorhombic distortion can relieve its degeneracy; however, it cannot be relieved by the crystal field of the trilateral distortion. Even the strong trilateral distorted crystal field, such as Al_2O_3, cannot relieve its degeneracy. For d_{z^2} and $d_{x^2-y^2}$, there are all possible locations at the lowest energy level; it depends on the sign of the distortion field. If the $|0\rangle = d_{z^2}$ is in the lowest energy level, then

$$g_{\parallel} = g_e g_{\perp} = g_e - 6\lambda/\Delta \tag{10.78}$$

If the $d_{x^2-y^2}$ is in the lowest energy level, then

$$g_{\parallel} = g_e - 8\lambda/\Delta_{\parallel} \quad g_{\perp} = g_e - 2\lambda/\Delta_{\perp} \tag{10.79}$$

If the Jahn–Teller distortion is very large, the g value of the tetragonal symmetry field should be satisfied with Eq. (10.78) or (10.79). If the Jahn–Teller distortion is comparatively small, the distortion axis can move toward the main axis of octahedron; at high temperatures, the time-averaged effect could occur and an isotropic g factor will be obtained, that is, $g = g_e - 4\lambda/\Delta$. The Cu^{2+} ion in MgO is the best example. At 77 K, observed g factor is isotropic [46] (g = 2.192; $A = 0{\cdot}0019$ cm^{-1}), while at 1.2 K, observed g factor is more anisotropic [47].

10.7.3.2 3d^7 (Low spin) ions in octahedral field

The electronic configuration of 3d^7 (low spin) ions is $t_{2g}^6 e_g^1$, and Ni^{3+} in Al_2O_3 is an example[48]. The above analysis is suitable for this kind of ions. Both the spin–orbit coupling and the trilateral field cannot relieve the residual orbital degeneracy, but the Jahn–Teller distortion can relieve it. Over 50 K, the isotropic g value is 2.146. The spectra exhibit a strong anisotropy when the temperature drops to 4.2 K. The reason is that each static Jahn–Teller distortion configuration contributes to each of the EMR respectively.

10.7.4 The palladium group (4d) and platinum group (5d) ions

In principle, 4d and 5d ions can be treated using the same method as for 3d ions. However, 4d and 5d ions have some complexities. Despite the experimental examples of $4d^3$ and $5d^3$ ions in intermediate fields (high spin) and strong fields (low spin), experimental examples of $4d^4$ to $4d^9$ and $5d^4$ to $5d^9$ are observed only in strong field.

Because the spin–orbital parameters of 4d and 5d group ions are larger than 3d group ions, the spin–lattice relaxation time τ is very short. Their EMR spectra could be observed only at very low temperatures, but in the special case of the orbital, singlet state energy is much more lower than the other states energy; it is thus possible that EMR spectrum is observed even at temperatures higher than 20 K. For Tc^{4+} ion in K_2PtCl_6 substance ($4d^3$ in octahedron), its EMR spectrum [49] can be observed at 77 K; the g value is 2.050.

For the d^1 ion in the Archimedean trans- (antiprism) tetragonal prism symmetric crystal field, the lowest energy level is orbital singlet. Thus, the EMR spectra of $Mo(CN)_8^{3-}$ and $W(CN)_8^{3-}$ ions can be observed in aqueous solutions at room temperature.

10.8 EMR spectra of rare earth Ions

Rare earth ions include lanthanide (the $4f^n$ electron shell is not full) and actinide (the $5f^n$ electron shell is not full). Since it is shielded by the outer electron shell, the crystal field seems to be a weak field.

10.8.1 Lanthanide ion

Since the shielding effect of outer electron shell, there is only very weak interaction between $4f^n$ electron with the environment. In some solutions or crystals, the spectra lines of lanthanide ions are very narrow and very close to the frequency of free atom transitions. Although the $4f^n$ electrons locate in the inner layer, the spin–orbit coupling interaction is very strong. λ value is in the range of 640–2940 cm^{-1}.

The angular momentum vectors $\boldsymbol{L}$ and $\boldsymbol{S}$ are coupled to $\boldsymbol{J}$, and the value of J is $\sqrt{J(J+1)}$. The ground state is $J = |L - S|$ for the ion that is less than half full ($n < 7$), and

Table 10.6: The ground-state g factor, spectral terms, and spin–orbit coupling parameters of rare earth free ions.

$4f^n$	1	2	3	4	5	6
Represent ion	Ce^{3+}	Pr^{3+}	Nd^{3+}	Pm^{3+}	Sm^{3+}	Eu^{3+}
Spectral terms	$^2F_{5/2}$	$^3H_4$	$^4I_{9/2}$	$^5I_4$	$^6H_{5/2}$	$^7F_0$
g factor	6/7	4/5	8/11	3/5	2/7	—
$\lambda(\mathrm{cm}^{-1})$	640	800	900	1070	1200	1410

the ground state is $J = L + S$ for the ion that is greater than half full ($n > 7$). Hamiltonian operator of spin–orbit coupling of ground state is

$$\hat{\mathscr{H}}_{LS} = \lambda L \cdot S = \frac{1}{2}\lambda[J(J+1) - L(L+1) - S(S+1)] \tag{10.80}$$

Table 10.6 shows the ***L, S, J*** and the ground-state spectral terms of ground states of $4f^n$ ions.

The splitting of the energy levels of the $4f^1$ ion is shown in Figure 10.29. For $J = 7/2$ and $5/2$, the energy levels of E_J are obtained by Eq. (10.80)

$$E_{7/2} = 3\lambda/2 \qquad E_{5/2} = -2\lambda \tag{10.81}$$

Because $\lambda > 0$, $E_{5/2}$ is the ground state. Its eigenfunctions are

$$\left|\frac{5}{2}, \pm\frac{5}{2}\right\rangle; \quad \left|\frac{5}{2}, \pm\frac{3}{2}\right\rangle; \quad \left|\frac{5}{2}, \pm\frac{1}{2}\right\rangle \tag{10.82}$$

Most rare earth ions are trilateral symmetry field; the trilateral symmetry crystal field operator is applied onto the eigenfunctions of Eq. (10.82), which leads the energy-level splitting into three Kramer doublet degenerate state, namely, $M_J = \pm 5/2$; $\pm 3/2$; $\pm 1/2$. $M_J = \pm 1/2$ is the lowest energy level. The operator $\hat{\mathscr{H}}_{\mathrm{Zeeman}} = \beta \boldsymbol{H} \cdot (L + g_e S)$ applied onto the function $|\frac{5}{2}, \pm\frac{1}{2}\rangle$, then the g value can be calculated. When $\boldsymbol{H}$ is parallel to the trilateral axis (Z), for the state $M_J = +1/2$, $E_{1/2} \simeq +\frac{3}{7}\beta H$; $E_{-1/2} \simeq -\frac{3}{7}\beta H$, then the $g_{\parallel} = \frac{6}{7}$. For $\boldsymbol{H}$ parallel to the X-axis, the operator $\hat{\mathscr{H}}_{\mathrm{Zeeman}} = \beta H_X(\hat{L}_X + g_e\hat{S}_X)$ applies onto the function of $|\frac{5}{2}, \pm\frac{1}{2}\rangle$; then $g_{\perp} = \frac{18}{7}$ can be obtained. For the rare earth ions containing an odd number of electrons, after splitting, the lowest level should be a doublet Kramer degenerate. But since its large spin–orbit coupling interaction, the EMR spectrum can usually be observed at a temperature below 20 K. For the rare earth ions containing an even number of electrons, with the C_{3v} or C_{3h} crystal field, each ***J*** state will be split into one singlet and one doublet degeneracy. If the external magnetic field is added parallel to the direction of the symmetry axis, this non-Kramer doublet degeneracy will produce a

Table 10.6 (continued)

7	8	9	10	11	12	13
Gd^{3+}	Tb^{3+}	Dy^{3+}	Ho^{3+}	Er^{3+}	Tm^{3+}	Yb^{3+}
$^8S_{7/2}$	$^7F_6$	$^6H_{15/2}$	$^5I_8$	$^4I_{15/2}$	$^3H_6$	$^2F_{7/2}$
2	3/2	4/3	5/4	6/5	7/6	8/7
1540	-1770	-1860	-2000	-2350	-2660	-2940

primary degree splitting. However, if it is added perpendicular to the symmetry axis, then this doublet state does not require to be split.

10.8.2 Actinide ions

There are many similar properties of actinide ions with the lanthanide ions, and the common valence is also +3. The difference is that the actinide 5f electrons do participate in the bonding process with ligands. Table 10.7 lists the common magnetic electron number of actinide ions and their spectral term of ground state. The spin–orbit coupling parameters of actinide ions are greater than those of the lanthanides. In actinide ions, $(O-M-O)^{n+}$ ion is linear complex, and its axial interaction with oxygen dominates, which makes these ions with some anomalous properties. Its ligand field is much stronger than the spin–orbit coupling interaction. The electronic configuration of U^{2+} ion is $5f6d7s^2$, and these four electrons are exactly composed covalent bond with the oxygen atoms, so, there is no magnetic electron of UO_2^{2+} ion. NpO_2^{2+}, PuO_2^{2+}, and AmO_2^{2+} have 1, 2, 3 magnetic electrons in order, respectively.

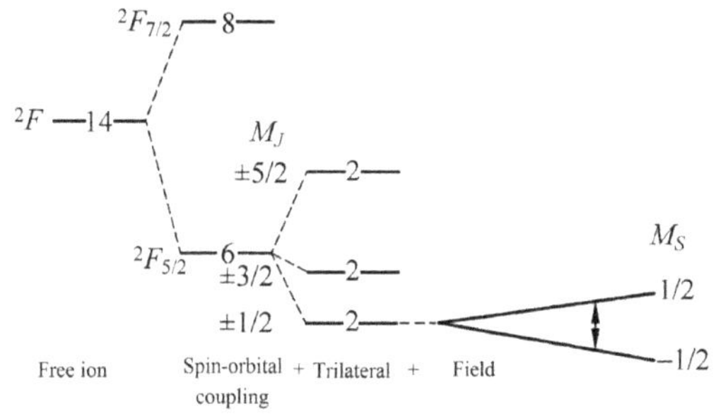

Figure 10.29: Energy-level splitting of rare earth $4f^1$ ion.

Table 10.7: Magnetic electron number of some actinide ions and their spectral terms of ground state.

Number of magnetic electrons	**0**	**1**	**2**	**3**	**4**	**5**	**6**	**7**
Representative ions	UO_2^{2+}	NpO_2^{2+}	PuO_2^{2+}	AmO_2^{2+}				
	Th^{4+}	Pa^{4+}	Pa^{3+}		Np^{3+}	Pu^{3+}	Am^{3+}	Cm^{3+}
		U^{5+}	U^{4+}	U^{3+}	Pu^{4+}	Am^{4+}		
		Np^{6+}	Np^{5+}	Np^{4+}				
			Pu^{6+}	Pu^{5+}				
				Am^{6+}				
Spectral terms of ground state		$^2F_{5/2}$	$^3H_4$	$^4I_{9/2}$	$^5I_4$	$^6H_{5/2}$	$^7F_0$	$^8S_{7/2}$

The ^2F ground-state sevenfold orbital degeneracy of NpO_2^{2+} ion are split into four energy levels $M_L = \pm 3,\ \pm 2,\ \pm 1,\ 0$, by strong axial symmetry field, and the lowest energy level is $M_L = \pm 3$. Spin orbital coupling makes this energy-level splitting into two Kramer doublet degeneracy: $M_J = \pm 7/2$ and $\pm 5/2$, and the lowest energy state is $M_J = \pm 5/2$. Because it is a Kramer doublet degeneracy, it can be described by $S' = 1/2$, and g factor is

$$g_{\parallel} = 2\left|\left\langle 3,\ -\frac{1}{2}\right|\hat{L}_z + g_e\hat{\mathbf{S}}_z\left|3,\ -\frac{1}{2}\right\rangle\right| \simeq 4$$

$$g_{\perp} = 2\left|\left\langle 3,\ -\frac{1}{2}\right|\hat{L}_x + g_e\hat{\mathbf{S}}_x\left|-3,\ \frac{1}{2}\right\rangle\right| \simeq 0$$

The experimental values [50] are $g_{\parallel} = 3{\cdot}405$ and $g_{\perp} = 0{\cdot}205$. Because the interaction with the excited state is not included, and electron delocalization to ligands is not considered, so, there is a large difference between the calculated value and the experimental value. For details on EMR descriptions of actinide ions, please refer to Ref. [1,7].

10.9 EMR spectra of transition metal complexes

The ligand field mentioned of the beginning of this chapter is generalized, in which the crystal electric field is considered as one of the ligand fields. Inorganic crystals of many transition metals are also not certainly pure crystal electric fields. As mentioned before, transition metal ions have some covalent characteristic. Even the unpaired electrons of the central transition metal ions are also delocalized into the surrounding ions. For example, EMR of $IrCl_6^{2-}$ appears to hyperfine splitting [51] of ^{35}Cl and ^{37}Cl. Above are all of the results dealt with the point charge crystal field model, there are certainly different from the experimental results usually.

For the transition metal complexes, their EMR spectrum should be treated by the "ligand field theory". In the study of EMR spectra of many transition metal complexes, the ligand field theory are all used by us. The Cu^{2+} complexes display very strong covalent characteristics [52–57]. The complexes of VO^{2+} ions appear quite strong ionic properties [58–67]. There are else some research of EMR spectra of transition metal complexes, including rare earth ions complexes, please refer to the literature [68–79].

References

[1] Low W. *Paramagnetic Resonance in Solids, Solid State Physics – Supplement 2*, Academic Press, New York, 1960.

[2] McGarvey B. R. "Electron spin resonance," in Carlin R. L., Ed., *Transition-Metal Chemistry*, Vol. 3, Marcel Dekker, New York, 1966, pp. 89–201.

[3] Orton J. W. *Electron Paramagnetic Resonance*, Iliffe, London, 1968.

[4] Abragam A., Bleaney B. *Electron Paramagnetic Resonance of Transition Metal Ions*, Oxford University Press, London, 1970, Chapter 7.

[5] Pake G. E., Estle T. L. "Ligand or crystal fields," *The Physical Principles of Electron Paramagnetic Resonance*, 2nd edn, Benjamin, Reading, 1973, Chapter 3.

[6] Al'tshuler S. A., Kozyrev B. M. *Electron Paramagnetic Resonance in Compounds of Transition Elements*, 2nd revised edn (Engl. transl.),John Wiley & Sons, Inc., New York, 1974.

[7] Pilbrow J. R. *Transition Ion Electron Paramagnetic Resonance*,: Oxford University Press, Oxford, 1990.

[8] Balhausen C. *Introduction to Ligand Field Theory*, McGraw-Hill, New York, 1962.

[9] Figgis B. N. *Introduction to Ligand Field Theory*, Interscience, New York, 1966 (1986 reprint).

[10] Gerloch M. *Magnetism and Ligand Field Analysis*, Cambridge University Press, Cambridge, 1983.

[11] Griffith J. S. *The Theory of Transition Metal Ions*, Cambridge University Press, Cambridge, 1961.

[12] Gordy W., Smith W. V., Trambarulo R. F. *Microwave Spectroscopy*, John Wiley & Sons, Inc., New York, 1953, 225.

[13] Cotton F. A. Chemical Applications of Group Theory. 1971. Liu C. W., You X. Z., Lai W. J.. Translate. Science Press, Beijing. 1975.

[14] Stevens K. W. H. *Proc. Phys. Soc.* 1952, **A65**: 209.

[15] Qiu Z. W. Electron Spin Resonance Spectroscopy. Science Press, Beijing. 1980: 265–268.

[16] Bleaney B. *Proc. Phys. Soc.* 1952, **A65**: 209.

[17] Koster G. F., et al. *Phys. Rev.* 1959, **113**: 445.

[18] Tanabe Y., Sugano S. *J. Phys. Soc. Jpn.* 1954, **9**: 753, 766.

[19] Mahootian N, Kikuchi C., Viehmann W. *J. Chem. Phys.* 1968, **48**: 1097.

[20] Assour J. M. *J. Am. Chem. Soc.* 1965, **87**: 4701.

[21] McMillan J. A., Smaller B. *J. Chem. Phys.* 1961, **35**: 1698.

[22] Garofano T., Palma-Vittorelli M. B., Palma M. U., Persico F., in Low W., Ed., *Paraqmagnetic Resonance*, Vol. 2, Academic Press Inc., New York, 1963, p. 582.

[23] Orton J. W., Wertz J. E., Auzins P. *Phys. Rev. Lett.* 1963, **6**: 339.

[24] Schneider J., Rauber A. *Phys. Lett.* 1966, **21**: 380.

[25] Ham F. S., Ludwig G. W., in Low W., Ed., *Paramagnetic Resonance*, Vol. 1. Academic Press Inc., New York, 1963.

[26] Carrington A., et al. *Proc.* R. Soc. (*Lond.*) 1960, **A254**: 101.

[27] McGarvey B. R. *J. Chem. Phys.* 1964, **41**: 3743.
[28] Hall T. P. P., Hayes W. *J. Chem. Phys.* 1960, **32**: 1871.
[29] Kirkpatrick E. S., Müller K. A., Rubins R. S. *Phys. Rev.* 1964, **135A**: 86.
[30] Hertel G. R., Clark H. M. *J. Phys. Chem.* 1961, **65**: 1930.
[31] Shuskus A. J. *Phys. Rev.* 1962, **127**: 1529.
[32] Title R. S. *Phys. Rev.* 1963, **131**: 623.
[33] Bennett J. E, Gibson J. F., Ingram D. J E. *Proc. R. Soc. (Lond)*. 1957, **A240**: 67.
[34] Ehrenberg A. *Arkiv Kemi*. 1962, **19**: 119.
[35] Levanon H., Stein G., Luz Z. *J. Am. Chem. Soc.* 1968, **90**: 5292.
[36] Abragam A., Pryce M. H. *Proc. R. Soc. (Lond.)*. 1951, **A205**: 135.
[37] Kuska H. A., et al. *Radical Ions*, Interscience Publishers, New York, 1968.
[38] McGarvey B. R. *J. Chem. Phys.* 1963, **38**: 388.
[39] Bleaney B., O'Brien M. C. M. *Proc. Phys. Soc.* 1956, **B69**: 1216.
[40] Lambe J., Kikuchi C. *Phys. Rev.* 1960, **118**: 71.
[41] Hoskins R. H., Soffer B. H. *Phys. Rev.* 1964, **133A**: 490.
[42] Thornley J. H. M., Windsor C. G., Owen J. *Proc.* R. Soc. (*Lond.*). 1965, **A284**: 252.
[43] Low W. *Phys. Rev.* 1958, **109**: 256.
[44] Low W., Rubins S. *Phys. Lett. (Netherlands)* 1962, **1**: 316.
[45] Ham F. S., Geschwind S. *Electron Paramagnetic Resonance*, Pienum Press, New York, 1969.
[46] Orton J. W., et al. *Proc. Phys. Soc.* 1961, **78**: 554.
[47] Coffman R. E. *J. Chem. Phys.* 1968, **48**: 609.
[48] Geschwind S., Remeika J P. *J. Appl. Phys.* 1962, **33**: 370.
[49] Low W., Llewellyn P M. *Phys. Rev.* 1958, **110**: 842.
[50] Bleaney B., et al. *Philos. Mag.* 1954, **45**: 992.
[51] Griffiths J. H. E., Owen J. *Proc. R. Soc. (Lond.)*. 1952, **A213**: 459; 1954, **A226**: 96.
[52] Xu Yuanzhi, Yu Linhua. *Sci. Sin. (B)* 1982, **25**(5): 453.
[53] Xu Yuanzhi, Chen Deyu, Li Xiaoping, Zhou Chengming, Chen Xing, Yuan Chengye. *Sci. Sin. (B)* 1982, **31**(1): 1.
[54] Xu Y. Z., Shi S. Acta Chimica Sinica. 1986, 44(4):336.
[55] Xu Y. Z., Yu L. H. Chin. J. Catal. 1981, 2(3):179.
[56] Deng Liqun, Zhao Ke, Xu Yuanzhi. *Chin. Sci. Bull.* 1989, **34**(6): 523.
[57] Xu Y. Z., Chen D. Y., Li X. P., Chen X., Yuan C. Y., Chin. Sci. Bull. 1986, 31(4): 275.
[58] Xu Y. Z., Li X. P., Yu L. H., Liu C. W., Lu J. X., Lin W. Z., Chin. J. Struc. Chem. 1982, 1(1):53.
[59] Li Xiaoping, Xu Yuanzhi. *Kexue Tongbao*. 1985, **30**(3): 340.
[60] Guo Jinliang, Chen Yidun, Xu Yuanzhi. *Sci. Sin. (B)*. 1985, **28**(5): 449.
[61] Xu Yuanzhi, Shi Shu, *Sci. Sin. (B)*. 1987, **30**(11): 1121.
[62] Xu Yuanzhi, Chen Deyu, Wang Shuyuan, Feng Yafei, Zhou Chengming, Yuan Chengye. *Kexue Tongbao* 1988, **33**: 111.
[63] Xu Yuanzhi, Shi Shu. *Appl. Magn. Reson.* 1996, **11**: 1.
[64] Wang L., Xu Y. Z. Acta Chimica Sinica. 1989, 47(11):1187.
[65] Chen D. Y., Huang L. B., Feng Y. F., Cheng C. R., Xu Y Z, Chin. J. Mag. Reson. 1991, 8(1):29.
[66] Chen D. Y., Xu Y. Z., Huang L. B., Feng Y. F., Cheng C. R., He L., Acta Chimica Sinica. 1992, 50(9): 793.
[67] Chen D. Y., Feng Y. F., Lu J. J., Xu Y. Z., He L., Chin. J. Mag. Reson. 1993, 10(3):287.
[68] Chen D. Y., Xu Y. Z., Lu J. J., Feng Y. F., He L., Chin. J. Appl. Chem. 1993, 10(3): 68.
[69] Xu Yuanzhi, Chen Yan, Ishizu K., Li Yong. *Appl. Magn. Reson.* 1990, **1**(2): 283.
[70] Xu Yuanzhi, Chen Deyu, Cheng Chaorong, Miyamoto R., Ohba Y., Iwaizumi M., Zhou Chengming, Chen Yaohuan. *Sci. China (B)*. 1993, **35**(9): 1052.
[71] Xu Yuanzhi, Chen Deyu. *Appl. Magn. Reson.* 1996, **10**: 103.

[72] Fukui K., Ohya-Nishiguchi H., Kamada H., Iwaizumi M., Xu Yuanzhi. *Bull. Chem. Soc. Japan.* 1998, **71**(12): 2787.
[73] Shi Weiliang, Chen Deyu, Wang Guoping, Xu Yuanzhi. *Appl. Magn. Reson.* 2001, **20**: 289.
[74] Xu Y. Z., Cheng C. R., Xu D. J., Chen D. Y., Chin. J. Chem. Phys. 1989, 2(6): 450.
[75] Zhuge X. M., Chen K., Chen Y., Feng Y. F., Chen J. S., Xu Y. Z., Chin. J. Appl. Chem. 1990, 7(2): 6.
[76] Liu J. G., Ma F. T., Xu Y. Z., Chem. J. Chin. Univ. 1990, 11(6): 628.
[77] Zhang L., Chen D. Y., Xu D. J., Xu Y. Z., Acta Phys-chim. Sin. 1996, 9(6): 522.
[78] Lu J. C., Cheng Y. X., Xu D. J., Xu Y. Z., Chin. J. Mag. Reson. 1999, 16(1):59.
[79] Cheng Y. X., Zhang Z. L., Xu Y. Z., Chem. J. Chin. Univ. 2001, 22(5): 796.

Further readings

[1] Carrington A., McLachlan A D. *Introduction to Magnetic Resonance*, Harper & Row, New York, 1967, Chapter 10.
[2] Ham F. S. "Jahn–Teller effects in EPR spectra," in Geschwind S, Ed., *Electron Paramagnetic Resonance*, Plenum Press, New York, 1972, Chapter 1.
[3] Sugano S., Tanabe Y., Kamimura H. *Multiplets of Transition-Metal Ions in Crystals*, Academic Press, New York, 1970.
[4] Wertz J. E., Bolton J. R. *Electron Spin Resonance*, McGraw-Hill, New York, 1972, Chapters 11 and 12.
[5] Pilbrow J. R. *Transition Ion Electron Paramagnetic Resonance*, Oxford University Press, Oxford, 1990.
[6] Cotton F. A. *Chemical Applications of Group Theory*, 3rd edn, John Wiley & Sons, Inc., 1990, Chapter 9.
[7] Sorin L. A., Vlasova M. V. *Electron Spin Resonance of Paramagnetic Crystals*. (Engl. transl.), Plenum Press, New York, 1973.
[8] Mabbs F. E., Collison D. *Electron Paramagnetic Resonance of d Transition Metal Compounds*, Plenum Press, New York, 1992.

Appendix 1: Extension and expansion of EMR

A-1 Pulse electron magnetic resonance

Unlike continuous-wave magnetic resonance spectroscopy (CW-EMR), this appendix introduces the microwave amplitude of exciting the electron transition, which is a time-dependent function $[H_1(t)]$, that is, the microwave excited field H_1 is pulsed. Much of the pioneering work of pulsed EMR has been carried out by W. B. Mims, beginning from 1961. Many advances have been made in the field of EMR research since the last 40 years. Although the development process of pulse EMR has been slow, the pace has been steady. One of the main reasons is that before the 1970s and 1980s, the technique of pulse microwave was mainly applied in military technology. The very high cost of spectrometer was another reason for not being used by civilians. The EMR spectrometer has been on the market since 1990s, but it is still very expensive, about three to four times that of CW-EMR spectrometer.

For $S = 1/2$ spin system, $\nu_{\alpha\beta} = (E_\alpha - K_\beta)/h = \nu_H$, the precession motion of applied angular momentum has relationship with the Heisenberg principle in quantum mechanics; the spin components perpendicular to the extra magnetic field **H** can be described by using "cone of uncertainty" (as shown in Figure A1-1.1) and only the axis direction of cone can be measured.

In practice, the selection of $H_1(t)$function obtained at least in principle can fit the requirements of experimentalists. The direction of H_1 relative to the extra magnetic field $\boldsymbol{H}$ and also to the orientation of sample (suppose the sample is an anisotropic single crystal) should be specified. If there are several microwave fields of differential frequencies, reasonable selection of phase relationship between their corresponding sinusoids can be done. The amplitude of $H_1(t)$ is very important for the decisive observation effects. Of course, when there is a change in H_1, the $-\hat{\boldsymbol{\mu}} \cdot \boldsymbol{H}_1$ term of spin-Hamiltonian should change simultaneously.

Now, if we want the $H_1(t)$ to be a monochromatic sinusoid, the amplitude must be kept constant for the entire duration. If it is switched on at extremely short interval, the microwave amplitude from zero will reach its threshold value ($\boldsymbol{H}_1$) instantaneously; likewise, when it is turned off at extremely short interval, the microwave amplitude from its threshold value ($\boldsymbol{H}_1$) falls down to zero immediately. Such microwave excitation magnetic fields constitute a square-wave pulse. Here we must focus on the rapid change of microwave field ($\boldsymbol{H}_1$) from zero to $\boldsymbol{H}_1$ as well as from $\boldsymbol{H}_1$ to zero that occurs in *frequencies-superimposed* distribution area, which is adjacent to the basic frequency ν of microwave magnetic field $\boldsymbol{H}_1$, which is a Fourier series [1,2] of various sinusoids.

https://doi.org/10.1515/9783110568578-011

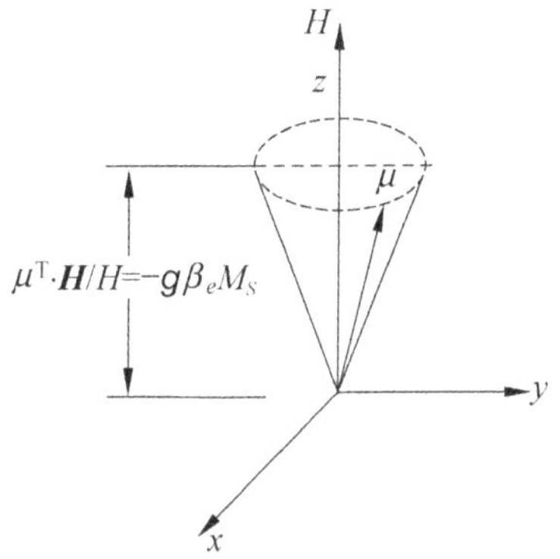

Figure A1-1.1: The precession model of spin moment in steady magnetic fields.

A1-1.1 The ideal switch-on of microwave field H_1

The spin magnetic moment $\hat{\boldsymbol{\mu}} = \alpha g\beta\hat{\boldsymbol{J}}$ in a frequency ν sinusoidal microwave excitation field H_1 is polarized linearly in the direction of static Zeeman field $\boldsymbol{H}\|\boldsymbol{z}$ present a 90 ° angle. Assuming these two magnetic fields are homogeneous, we can adopt a vector model to describe them. When $\boldsymbol{H}_1 = 0$, the spin vector $\hat{\boldsymbol{J}}$ acts near the extra field $\boldsymbol{H}$ with its natural frequency $\nu_H$and thus the precessing motion occur. Both states $M_J = +\ 1/2$ and $-1/2$ have the same rotational sense. When $\boldsymbol{H}_1 \neq 0$, two precessing motions occur simultaneously: one about the extra field $\boldsymbol{H}$ and the other about the microwave field $\boldsymbol{H}_1$ with a frequency lower than ν_H, and thus the precessing motion occurs (usually $\boldsymbol{H} \gg \boldsymbol{H}_1$). When $\nu = \nu_H$(ν is the basic frequency of $\boldsymbol{H}_1$), that is, during the production of magnetic resonance, the frequency and rotational direction of one component of $\boldsymbol{H}_1$ just matches with the frequency and rotational direction of the spin moment. It can be seen that this component of the $\boldsymbol{H}_1$, regarded as a polarized torque for applying on the magnetic moment, makes the moment reach to mean value $\langle\boldsymbol{\mu}\rangle$. When $\boldsymbol{H}$ is vertical to $\boldsymbol{H}_1$, it would reach to maximum. Then, the spin-moment vector between its J_z eigenstates is driven by $\boldsymbol{H}_1$ back and forth. A suitable energy exchange between spin and effective microwave field $\boldsymbol{H}_1$ with oscillating frequency $\nu_1 = g\beta H_1/h$ is observed.

For spin state M, the mean value of magnetic moment $\langle\boldsymbol{\mu}\rangle = \langle M|\hat{\boldsymbol{\mu}}|M\rangle$ obeys the differential equation:

$$\frac{\mathrm{d}\langle\boldsymbol{\mu}\rangle}{\mathrm{d}t} = \gamma\langle\boldsymbol{\mu}\rangle \wedge \boldsymbol{H} \qquad \text{(A-1.1)}$$

The external magnetic field ($\boldsymbol{H}$) consists of a static magnetic field $\boldsymbol{H}_o$ and a $\boldsymbol{H}_1$ around $\boldsymbol{H}_o$ rotation in the Cartesian coordinate system of the laboratory (as unit vector of x, y, z):

$$\boldsymbol{H} = \boldsymbol{x}H_1 \cos\omega t + \boldsymbol{y}H_1 \sin\omega t + \boldsymbol{z}H_o \qquad \text{(A-1.2)}$$

$\boldsymbol{H}_1 \perp \boldsymbol{H}_o$ and $\boldsymbol{H}_1 \ll \boldsymbol{H}_o$. In the case of resonance, $\omega = \omega_H = |\gamma|\, H_o$ can be transformed to the rotating coordinate system:

$$\boldsymbol{x}_\phi = \boldsymbol{x} \cos \omega t + \boldsymbol{y} \sin \omega t \tag{A-1.3}$$

$$\boldsymbol{y}_\phi = -\boldsymbol{x} \sin \omega t + \boldsymbol{y} \cos \omega t \tag{A-1.4}$$

This situation can be seen [3,4] more clearly from the so-called Rabi problem, that is, for the above-mentioned phenomena, the solution of the quantum mechanical dynamic equation of the amplitude probability is provided. For example, for an isolated $S = 1/2$ spin system, when the time is t_o, the probability [5–7] P from its ground state $M_J = -1/2$ transition to upper state $M_J = +1/2$ is

$$P = \left(\frac{\nu_1}{\nu_r}\right)^2 \sin^2[\pi\, \nu_r(t - t_o)] \tag{A-1.5}$$

where the frequency $\nu_r = [\nu_1^2 + (\nu - \nu_H)^2]^{1/2}$ is named *Rabi frequency*. ν_1 is the frequency of H_1. The probability P is maximum at $t - t_o = \ell/2\, \nu_r$, when ℓ is an odd integer. The probability of the system ground state is unity when resonance occurs and when $\nu = \nu_H$, ℓ is even. Between these timepoints, for each state of $M_J = \pm 1/2$, the probability of occurrence is not zero. The *cone axis* of magnetic moment about the extra field causes precessing motion, which looks like a screw motion back and forth between the top $+\boldsymbol{z}$ ($\boldsymbol{H}\|\boldsymbol{z}$) and bottom (-$\boldsymbol{z}$) of a sphere (shown as A1-1.2). The rate of spin "flipping" is slower than the rotation frequency ν_H when $\boldsymbol{H} \gg \boldsymbol{H}_1$. This flip corresponds to transfer of photon $h\nu$ between the spin and excitation field systems. Of course, the influence of spin–lattice relaxation and spin–spin relaxation have been ignored here.

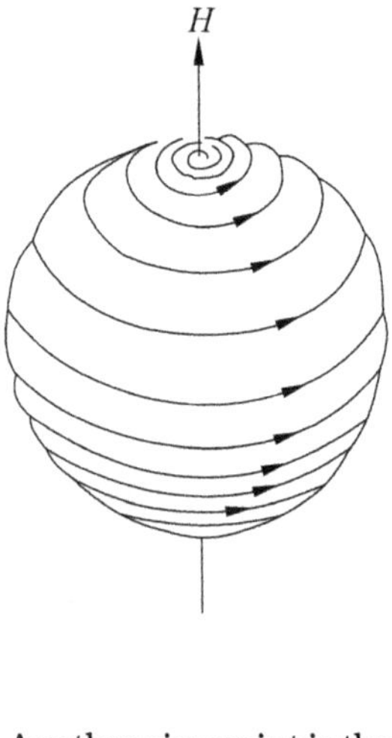

Figure A1-1.2: Diagram of magnetic moment about the extra field causing *precessing* motion.

Another viewpoint is that the "precession" motion of the magnetic dipole (the vector of magnetic moment) causes a detectable instantaneous energy transfer (added a frequency shift) in a resonator holding the spin system; for example, it can detect a voltage in a received coil of signal.

When $\nu \neq \nu_H$, and the frequency ν_r is higher than ν_1, the probability amplitudes occur as Rabi oscillations, and the probability of the upper lever cannot reach unity. The transition probability (roughly as $\nu_1/|\nu - \nu_{\text{Larmor}}|$) quite sharply goes down with ν departure from resonance, and the microwave field H_1relative to the spin becomes ineffective rapidly. However, the situation $\nu{\sim}\nu_H$ is very important. Also remember that the probability will decrease rapidly when the microwave field H_1 shifts away from the normal line of extra field $\boldsymbol{H}$ [8].

Clearly, if there is no irreversible energy transfer in or out of the total system (spin + radiation), then the average definite number of whole oscillation periods have no net change of any form. Some effective electromagnetic field other than $\boldsymbol{H}$ and H_1, that is, the spin–lattice relaxation (a mechanism of taking the energy coupling to the other reservoir), should be presented for observation of net EMR power absorption.

Next, we consider a set of independent spin particles in uniform fields ($\boldsymbol{H} \perp \boldsymbol{H}_1$), where there are identical magnetic moments in magnitude. They can be approached by density matrix, or statistically treated by using the Bloch theory. Using the net magnetization per unit volume $\boldsymbol{M}$ is most advantageous as it gives the expected value of spatially averaged macroscopic quantity. It is a vector of time that does not follow the uncertainty principle of quantum mechanics. The phase orientation (position of "orbits" along the z-axis and about the field $\boldsymbol{H}$) of the various individual spin vectors are random and remain as long as the microview field H_1 remains switched on. Therefore, in the isotropic material, the time average value of any component of magnetization $\boldsymbol{M}$, which is perpendicular to $\boldsymbol{H}$, is zero in the laboratory coordinate.

If the particle numbers of independent spins in states $M_J = -1/2$ and $M_J = +1/2$ are equal, then the net effect of the upward and downward transitions would keep the magnetization $\boldsymbol{M}$ constant. However, the common thermal equilibrium situation in a relaxing spin system would not be equivalent, the distributional number of down-level state is ΔN more than the up-level state. These excess spin numbers could adjust the value of magnetization $\boldsymbol{M}$; hence, if the $\boldsymbol{H}_1$ would drive the spin particles back and forth, transitioning between ±1/2 states harmoniously, then the M_z would oscillate between parallel and antiparallel out field [8] $\boldsymbol{H}$. An oscillating magnetic dipole association with an oscillating radiation field derive an alternate voltage. If H_1 did not turn off during the experiment, it is CW magnetic resonance spectroscopy (CW-EMR), as presented in the previous chapters. As already discussed, if some of the spin system energy goes to atomic motions rather than returning to $\boldsymbol{H}_1$, then ΔN is maintained and a net absorption signal is observable; unless the spin–lattice relaxation has long energy, ΔNcould approach to zero, so no energy absorption will occur and the spin system will become power-saturated (Section 7.1.5). Note that magnetization $\boldsymbol{M}$ approaches to zero when $\tau_1 \rightarrow \infty$.

To understand the physical interactions of the spin moments with the external magnetic fields and with each other, it is important to visualize the time-dependent behavior of the magnetization $\boldsymbol{M}$. This task can be simplified by choosing the most convenient coordinate system (CS). Very often, a "rotating frame" (Section 7.1.6), here it is $\boldsymbol{z}_\Phi(=\boldsymbol{z})$ axis, is used along the external magnetic field $\boldsymbol{H}$, and $\boldsymbol{x}_\Phi$ is the effective rotating component along $\boldsymbol{H}_1$ on the $\boldsymbol{x}$–$\boldsymbol{y}$ plane, since it seems that on the surface the rotation of $\boldsymbol{H}_1$ has no relationship [9] with the Larmor precession (when $\nu = \nu_H$).

Using the instantaneous solution [10] of the Bloch equation (Eq. (7.30)) can more completely analyze the pulse situation. As in the laboratory frame, magnetization $\boldsymbol{M}$ rotates about $\boldsymbol{H}(\| \boldsymbol{z})$ with frequency ν, where the motion is superimposed very slowly with H_1, nutation (change in angle between $\boldsymbol{M}$ and $\boldsymbol{H}$) with Rabi frequency ν_r occurs. When the microwave exciting field $\boldsymbol{H}_1 (\perp \boldsymbol{z})$ is switched on, this transient nutation has an initial amplitude because of the presence of magnetization $\boldsymbol{M}$, and the frequency ν is close to the resonance value. It is the result of damping with the spin–lattice and spin–spin interactions. When rotating about the z-axis with frequency ν, there is magnetization $\boldsymbol{M}$ about $\boldsymbol{H}_1$ in the direction of $+\boldsymbol{z}$ $(H - h\nu/g\beta_e)$ with Rabi frequency ν_r for precession to occur slowly. The pulse spectrometer can detect the modulation of the absorption (or dispersion) in the $\boldsymbol{x}$–$\boldsymbol{y}$ plane by using a very sensitive detector for the magnetization. This transient signal depends on the presence of a magnetization component $\boldsymbol{M}_\perp$ on the plane, and decays with $2(\tau_1^{-1} + \tau_2^{-1})$ giving a nutational relaxation time.

A1-1.2 The radiation field H_1 of Single Pulse

Even when the RF field H_1 is turned off, the nutation signal can be observed. Thus, the square-wave is used as excitation pulse; when the pulse is stopped, the temporal length τ appears as shown in Figure A1-1.3. The phenomenon of magnetization $\boldsymbol{M}$ variation with time observed at the end of the pulse is called "**f**ree **i**nduction **d**ecay" (FID). Here *free* means that $\boldsymbol{H}_1$ does not exist (while the external field $\boldsymbol{H}$ still remains switched on!) – in contrast to the situation of the spin transition being "driven" by $\boldsymbol{H}_1$ described above. Pulsed EPR (as well as NMR) focuses on the FID causing signal (time domain). Later, via Fourier transform, we can get a signal of frequency domain, as shown [12] in Figure A1-1.3.

If the excited field H_1 after being switched on, via time$\tau = \ell/4\nu_r$ is turned off instantaneously, then at that instant whether $\boldsymbol{M}$ is parallel or antiparallel to $\boldsymbol{H}$ will depend on whether ℓ is an even or odd integer [such as π pulse ($\ell = 2$), $\boldsymbol{M}$ from its initial direction turns 180°]. The $\boldsymbol{M}$ turning angle Ω is $2\pi\tau\nu_r$ (in radians). Note that the π pulse can invert the population of spin: the higher state population to be more than the lower one. Equivalently, the population of the spin system could be described by generating negative spin temperature. With time, the spin system resumes, and via radiationless relaxation and photons $h\nu$ emission, it moves toward the normal

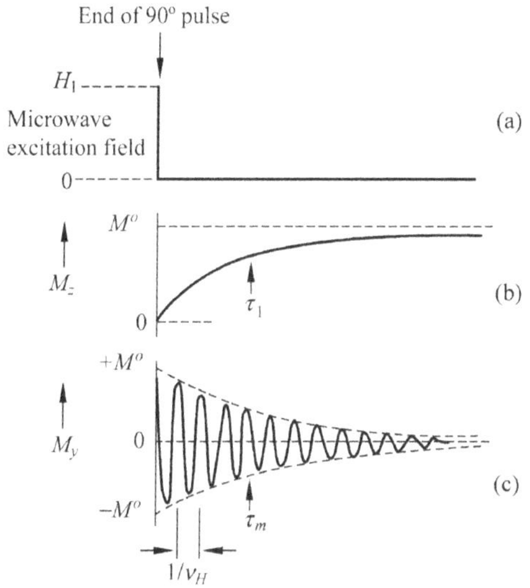

Figure A1-1.3: Behavior of the magnetization after $\pi/2$ pulse of H_1.

(positive spin temperature) Boltzmann population. When the $\boldsymbol{H}_1$ is switched off, the energy density ρ_p of photon is essentially zero, so, there is no induced transition. However, when the spin moment leads to superadiation system of affected FID, it will enhance spontaneous emission of photon. [There are no spin correlation in CW-EMR, hence most energy is lost from the spin system to the "lattice" of atoms and only a few (can be almost negligible) via spontaneous emission discontinuously go back to H_1.] The trace of vector $\boldsymbol{M}$ is visualized by retaining longitudinal (along the $\boldsymbol{H}$ direction) the inversion recovery, that is, along $-\boldsymbol{z}$ axis shrinking to zero, and then along $+\boldsymbol{z}$ regaining to its initial magnitude. The rate decay with exponential, as the typical relaxation time τ_1 (Chapter 7). The spin behavior can in many actual cases be described in terms of the Bloch equations for the time dependence of $\boldsymbol{M}$ (Section 7.2).

Now, let us discuss in a similar manner a very important case of 90°($\pi/2$) pulse ($\tau^{-1} = 4\nu_r$). After H_1 is stopped, $\boldsymbol{M}$ keeps constant its magnitude, and it turns immediately perpendicular to both $\boldsymbol{H}$ (if $\nu = \nu_H$) and $\boldsymbol{H}_1$, which is also perpendicular to $\boldsymbol{H}$ ($\boldsymbol{H}_1$ in the stopping state this time). Note, it is essentially different from the case of $\boldsymbol{M}_z = 0$, in which the spin vector are dispersed continuously transverse to the field $\boldsymbol{H}$ (i.e., in the $\boldsymbol{xy}$ plane), and is random orientation when the polarization field $\boldsymbol{H}_1$ is switched on or remains saturated completely. Under the situation of π pulse, $\boldsymbol{M}_z$ turns back toward its balance value (with spin–lattice relaxation time τ_1 exponentially), while each spin particles lose phase identity, so the transverse magnetization $M_\perp$ approaches to zero with relaxation time τ_m. This temporal behavior is shown in Figure 13.3. Moving phase can be realized in the following two types: reversible and irreversible (stochastics). The former is measured by τ_2 and the latter by τ_m.

The transverse magnetization of y_ϕ is in direction $M_\perp$ to precess continuously around $\boldsymbol{H}$ at the frequency $\nu = \nu_H$. The temporal behavior of $\boldsymbol{M}_y$ can be detected on the plane perpendicular to $\boldsymbol{H}$ with suitable apparatus. The coherency of the "phase," that detects signal relative to H_1 can be measured, since the sinusoidal (frequency ν) of voltage that supplies for field H_1 maintains H_1 until it is switched off. The detected signal is an "interferogram." The superposition of each spin particle and every instant from positive to negative contribute to the total signal (induced voltages), which varies with time rapidly. In practice, the FID signal should be long enough so that it can be recorded reliably.

Finally, we consider more realistic spin systems, where various interactions occur, so for the interpretation of the observed FID signal, precious chemical and structural information may be obtained. In other words, all the spin-Hamiltonian parameters discussed here affect FID; all are measurable in principle, although it is mainly used to measure the relaxation time. The usage of H_1 single pulses, as well as the design of pulse order that we shall see later, enables the measurement. All these considerations are valid equally for nuclei as well as electrons, but only the pulse EMR is discussed in this book.

A1-1.3 Analysis of FID curve and the EMR spectrum of fourier transform

The pulse detection gives two separate and complicated time functions, namely, signal of frequency ν as reference of continuous *phase-in* and *phase-out* signals (H_1 is turned off this time, but in the spin system it is still kept in the memory of original switch-on delaying). They can maintain the continuity of *phase* to be measured repetitively and separately, and can be stored in computer. These (start at the end of the pulse) signals will be subsequently analyzed with the time evolution. Usually it is carried out by means of modern computer technology using the mathematical technique called *Fourier transformation* [11], which has excellent program to be used. In essence, it is the data of digital time domain that are converted to digital frequency-domain data, including the line positions (ν_i or H_i) and relative intensities of peak absorption like those of the CW-EMR spectrum [12] (Figure A1-1.4). The top part of the figure shows the evolution process of the magnetization $\boldsymbol{M}$ during the FT-EMR experiment. The below part of the figure shows the FID and the corresponding FT-EMR spectrum. In order to obtain a best resolution of frequency-domain spectrum, it is important to follow the time expanding of the FID as long and completely as possible.

In any actual chemical system with unpaired electrons, various local fields (e.g., hyperfine effects, electron–electron interactions, and inhomogeneity of field $\boldsymbol{H}$) produce the collection of electron spin *precession* frequencies ν_H. Therefore, the given frequency ν of excitation field $\boldsymbol{H}_1$ rarely coincides with the actual ν_H. Thus, in the absence of resonance, the earlier frequency is a convention and not an exception. The result is that the total time τ of detecting EMR spectrum should be as short as possible.

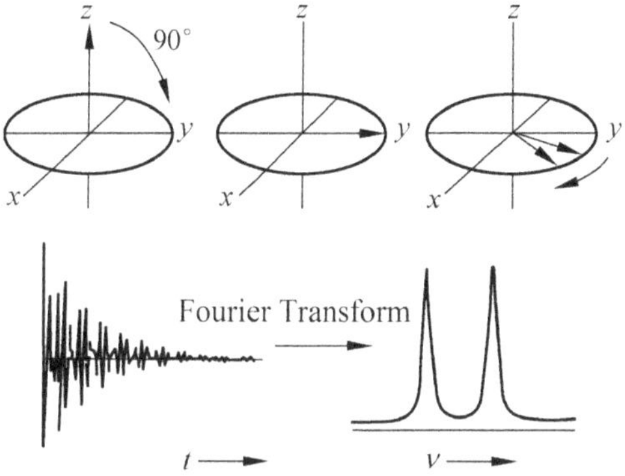

Figure A1-1.4: The top part shows the evolution process of the magnetization M during the FT-EMR experiment. The below part shows that the FID and the corresponding FT-EMR spectrum.

At the same time, the product of $H_1\tau$ is controlled by the condition of 90° pulse. Thus, H_1 must be as large as possible. In practice, the spectral region can be excited by FT-EMR just only $\sim$ 2 mT, and generally, τ is very short similar to τ_1 and τ_m.

Compared with CW-EPR, FT-EMR has the following advantages:

(1) *Efficient data collection:* All spectral lines can be detected simultaneously, and the required scan time does not depend on the coverage of scan; unlike CW-EMR, peak–peak scanning must be slow. It also means that by applying the pulse repetitively, scanning repetitively, and storing in computer, a clear spectrum can be obtained. Because the noise is random, through multitimes pile up, the +/– contributions can be counteracted. So the signal–noise ratio would be increased with the times n of pile up (the increase of $n^{1/2}$times can estimated roughly). It is very easy to be piled up 10^5 times.

(2) *Time resolution* A single FT-EMR spectrum can be recorded in ~1 μs. A spin system can be sampled easily in time. Thus, a pulse of given H_1 can be locked to a source of chemical energy, as a laser pulse, and after this first pulse, their reaction products in every 1 μs can be sampled. Many other types of dynamics (e.g., diffusion, phase changes, energy transfer between molecules) have also been studied [13]. Thus, the time-resolved EMR has become a powerful tool of kinetics for investigation.

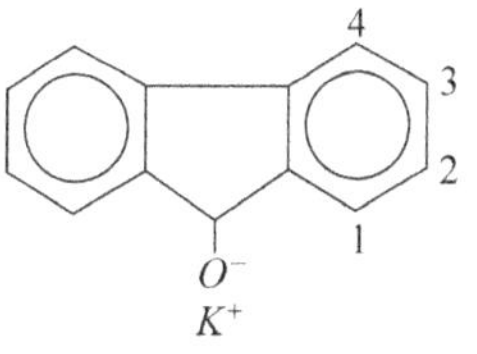

A-1-I: Fluorenone ketyl anion radical.

(3) *Efficient measurement of relaxation time* τ_1 and τ_m can be measured from the pulse responses directly, rather than via deconvolution of lineshapes or pass-through analysis of CW saturation behavior,

Probably the existing defects of FT-EMR are the limitations of scan spectral width not more than 2 mT and the relaxation time not too rapid.

A spectrum of fluorenone ketyl anion radicals is given here as an example. Taking fluorenone in tetrahydrofuran as solvent to be reduced by metal potassium in vacuum, the fluorenone ketyl anion radical (Figure **A-1-I**) is prepared. Figure A1-1.5 is the FID accumulative spectrum [14] of the fluorenone ketyl anion radical [14], which is detected in both phase orthogonal channels at 220 K. From the time-domain pass-through Fourier transformation, EMR spectrum of frequency domain is obtained; the central position of the spectrum is 25 MHz, which corresponds to the frequency of 9.271 GHz spectrometer, as shown in Figure A1-1.6b. The structural formula of fluorenone ketyl anion radical has four different sets of equivalent protons. The stick spectrum diagram of reconstruction based on a_1 = -5.77 MHz; a_2 = +0.27 MHz; a_3 = -8.80 MHz; a_4 = +1.84 MHz, respectively, is shown in Figure A1-1.6b.

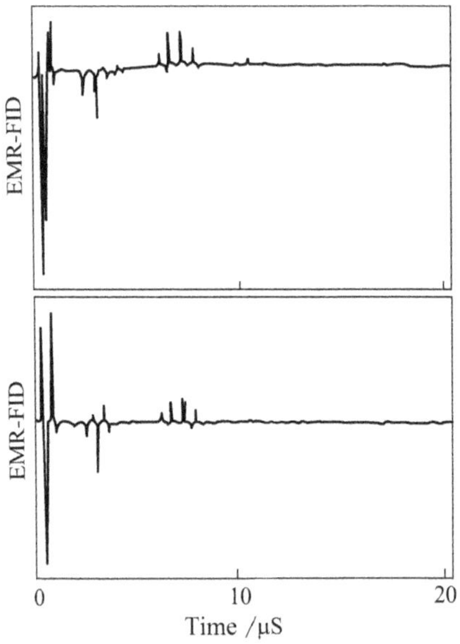

Figure A1-1.5: The FID accumulative spectrum of the fluorenone ketyl anion radical.

As a second example of the application of FT-EMR, we choose a short-lived organic free radical that is produced by photoinduced reversible electron-transfer reaction. Using a pulsed dye laser, excited zinc tetraphenylporphyrin (ZnTPP, formula as in Section A-1.II) and duroquinone (DQ, A-1.III), both diluted in liquid ethanol, have electron transfer between them, forming the corresponding positive and negative ion

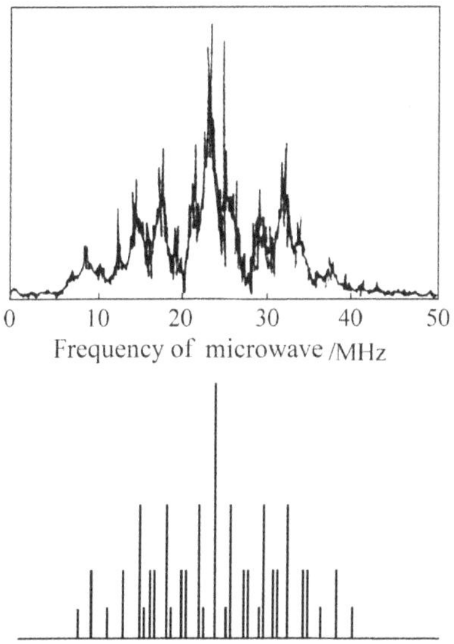

Figure A1-1.6: The top part is integral spectrum obtained from Fourier transformation; the below part is theoretical stick spectrum.

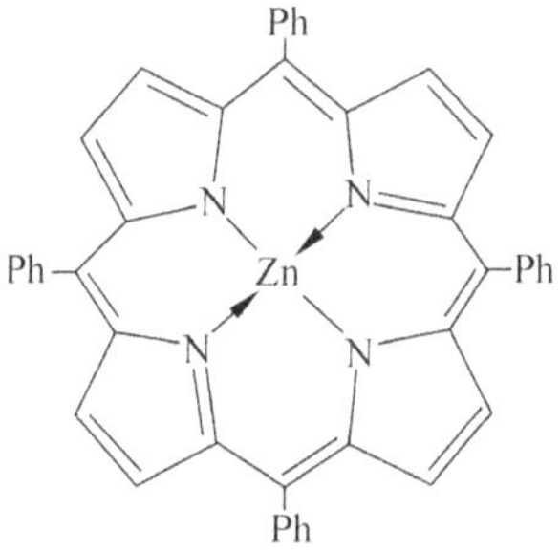

A-1.II: Zinc tetraphenylporphyrin.

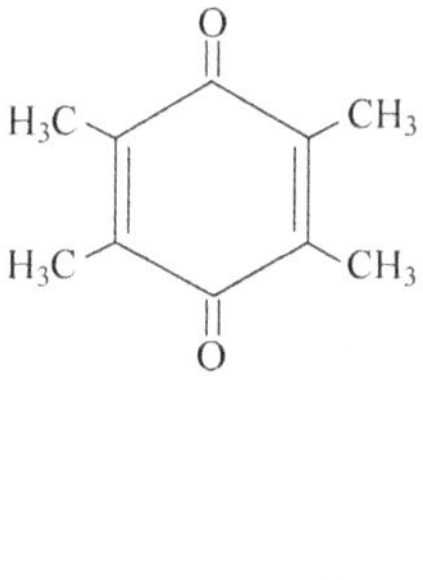

A-1.III: Tetramethyl-*p*-phenylquinone (DQ).

radicals [15–18]. The original EMR spectrum of DQ consists of 13 almost equally spaced narrow lines (0.19 mT), arising from the 12 equivalent methyl protons, with relative intensity ratios nominally of 1:12:66:220:495:792:924:792:495:220:66:12:1. *Note:* Only under the condition of 90° pulse, the FID signals of free radicals could be detected. The signal of $ZnTPP^+$ has not been observed here, because its time τ_m is too short. From the FT-EMR spectrum of DQ–, shown in Figure A1-1.7, and its

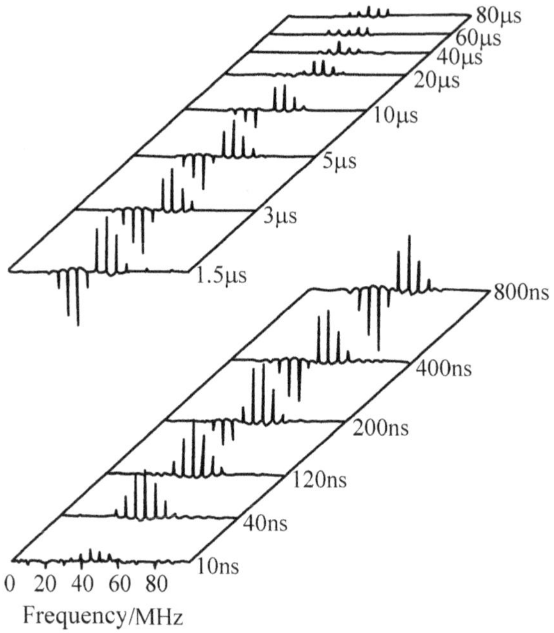

Figure A1-1.7: Integral FT-EMR spectra of DQ in ethanol solution at 245 K varying with time.

intensity ratio, we can see its 13 spectral lines could not be displayed at all. The variation of line intensity displays the electron transfer kinetics, which is same as the spin-polarization effects influencing the relative intensities of spectral line; for detailed discussion, refer to Refs [15–18].

A1-1.4 Multiple pulses

Using suitable pulse sequence, observer can obtain the selective spectral parameters, and remove the complexities of existing single-pulse in FID and in CW-EPR spectra. Thus, a multiline spectrum can be simplified usually, and both spectral and relaxation effects can be detected in the same experiment.

Now, we consider here n (labeled i, j = 1, 2, 3, . . ., n), square-wave H_1 pulses sequence of continued time τ_i and time intervals $\Delta_{ij}(j > i)$. The EMR work when $n < 4$ usually; external field $\boldsymbol{H}$ is kept constant.

Consider the two-pulse sequence $\pi - \Delta - \pi/2 - \Delta$. The first pulse makes the magnetization $\boldsymbol{M}$ inverse. The second pulse passes through an interval, placing M_z on the xy plane for the detection of FID. After a suitable interval (with a different Δ), repeat this sequence again. Clearly, the choice of this sequence provides an excellent method of measuring τ_1. Other choices can also do so.

On the contrary, consider the next sequence $\pi/2 - \Delta - \pi - \Delta$; the first pulse causes $\boldsymbol{M}$ from $z(=z_\Phi)$ invert to y_Φ. After an interval Δ, all individual spin moments undergo a moving phase. When τ_1 is relatively long, it makes $|M_z|$ remain rather small during this period, the π pulse makes $\boldsymbol{M}_y$ from $+y_\Phi$ turn to $-y_\Phi$. At this point, the continuous phase are all reversed suddenly; as seen from the surface, these spin particles move toward equal phase marvelously (the maximum in $M_\perp$). The radio frequency field H_1 is added by 90° pulse, after passing through an interval Δ, followed by 180° pulses to be added again. After passing through the interval Δ again, a signal can be established and then decay down, same as to produce an "echo." From the second pulse, it is switched off again and it passes through the interval Δ point, the amplitude of echo has a maximum value. It can be considered as the juxtaposition of two FIDs [19]. The entire process of is shown in Figure A1-1.8, and this order also is called "Hahn sequence" [20] sometimes. The first pulse is called the "excitation pulse" and the second pulse is called the "refocusing pulse."

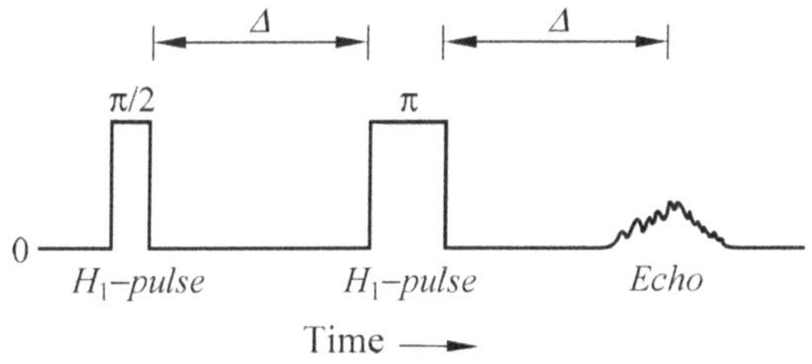

Figure A1-1.8: Hahn pulse sequence: $\pi/2 - \Delta - \pi - \Delta -$ echo.

A1-1.5 Spectrum of electron spin-echo envelope modulation (ESEEM)

The amplitude of spin-echo signal is a function of the pulse-interval time Δ. The decrease of any point height of the echo signal, as the function Δ increases, result in "phase-memory" relaxation time τ_m, which can be measured and stored in computer. Furthermore, in selected systems, the amplitude of echo versus interval Δ presents a periodic repeat, as shown in Figure A1-1.9a: that is, τ_m "modulation" added onto a decay curve (envelope). The echo signal is a function of time, and passing through Fourier transform, a signal of frequency domain can be obtained, as shown in Figure A1-1.9b. This is so called electron spin-echo envelope modulation (ESEEM) [21].

The magnitude of echo is recorded between second and third pulses of a pulse sequence in time function, which consists of a spectrum of a time domain. This spectrum is composed of two functions: one is an envelope of decay from

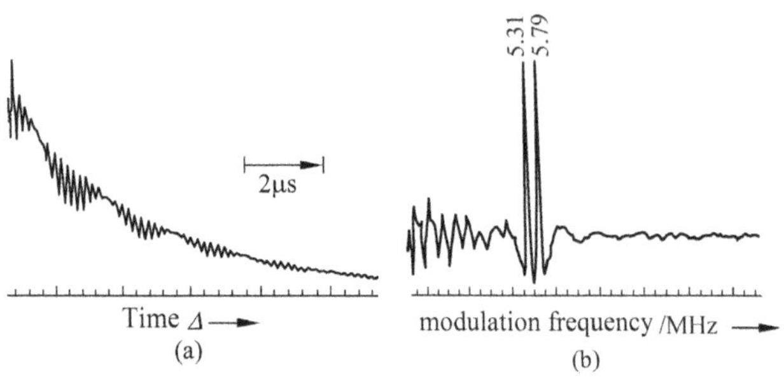

Figure A1-1.9: (a) The function relationship of the amplitude of echo (ε) with interval (Δ) in the experiment of a pulse order of $\pi/2$ - Δ - π - Δ - ε. (b) The ESEEM spectrum passing through Fourier transform.

beginning to end, and the other is a "modulation" pattern containing structural information on the attenuated envelope. The term "envelope modulation" may have originated from here. Owing to our interesting "modulation" pattern containing structural information, after treating the background of this envelope in the experiment, a polynomial function is derived, and the time-domain data and a frequency domain spectrum will be obtained through Fourier transform. The given formal data of frequency domain are similar to that of electron–nuclear double resonance (ENDOR) spectrum but different from the ENDOR spectrum. The intensity of the ESEEM spectrum can represent the number of nuclei interacting with it directly [22].

It is worth mentioning that one can measure the echo amplitudes at various different fixed fields ***H*** (or scan ***H*** slowly) and then assemble them to an *echo-modulated* EMR spectrum [23].

Figure A1-1.10 shows a fact example [24], that is, an ESEEM spectrum of similar ENDOR signal peaks that are obtained from the $\pi/2–\Delta–\pi–\Delta–\varepsilon$ experiment after being Fourier transformed. These signals arise from interstitial $Li^{+}(I = 3/2)$ ion interaction with unpaired electron of adjacent paramagnetic $Ti^{3+}(S = 1/2)$ ion that substitute the site of Si^{4+} in an α-quartz single crystal. The energy-level labeling of the CW-EMR hyperfine spectrum of the ^{7}Li nucleus as well as the theoretical stick spectrum is shown in Figure A1-1.11. Rotating the crystal, the ESEEM spectra of varied angles with the parameter matrices g, $\boldsymbol{A}(^{7}Li)$, and $\boldsymbol{Q}(^{7}Li)$ are measured, especially, the latter two are more accurate than those available from CW-EMR. The ESEEM spectra of different angles are measured and expressed as a stereodiagram of modulation frequency-dependent relation with angles, as shown in Figure A1-1.12.

The next example has several different frequencies, and hence considered as characteristic ESEEM spectra. The stable-free radical DPPH in frozen solution exhibits envelope modulation spectrum, which is attributed to contribution of the ^{14}N nucleus ($I = 1$) of the nitro groups. Using loop-gap resonator at C-band frequencies (4–7 GHz) combined

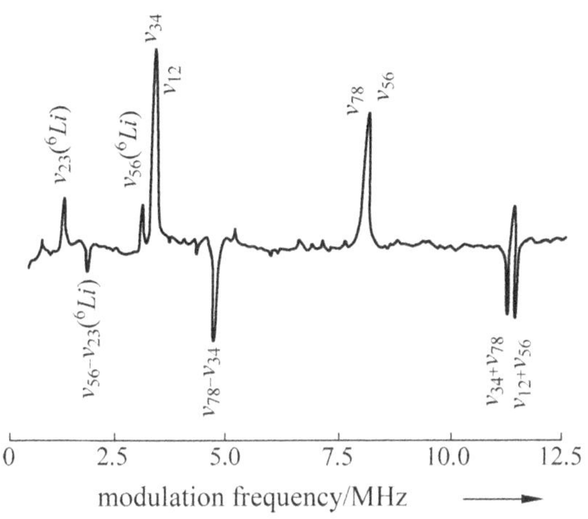

Figure A1-1.10: The FT spectrum of ESEEM signal of [TiO/Li] center in quartz.

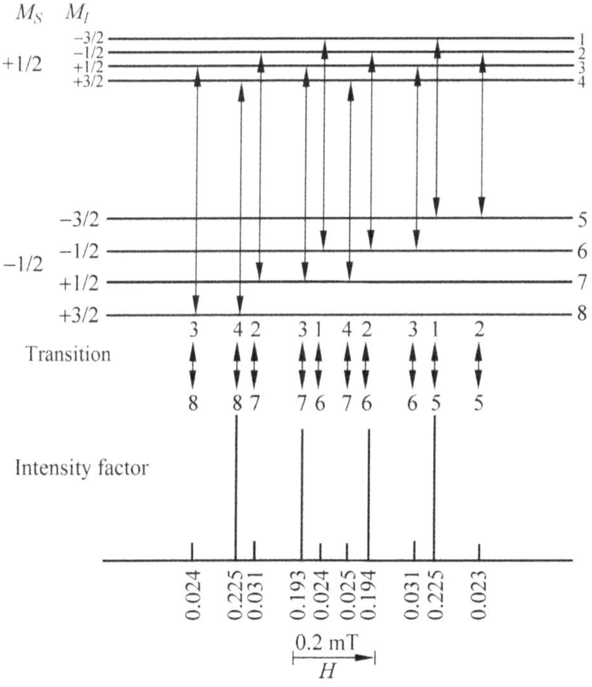

Figure A1-1.11: The level transition of hyperfine splitting of ^7Li nucleus in Quartz and its theoretical stick spectrum.

with X-band (9–10 GHz) detecting the data, the isotropic part [$A_{iso}/h = -1.12(8)$ MHz] of the hyperfine coupling parameters and accurate nuclear-quadrupole parameters [$Q/h = 0.280(2)$ MHz and $\eta = 0.37(3)$MHz] of ^{14}N nucleus on the *ortho*-nitro groups

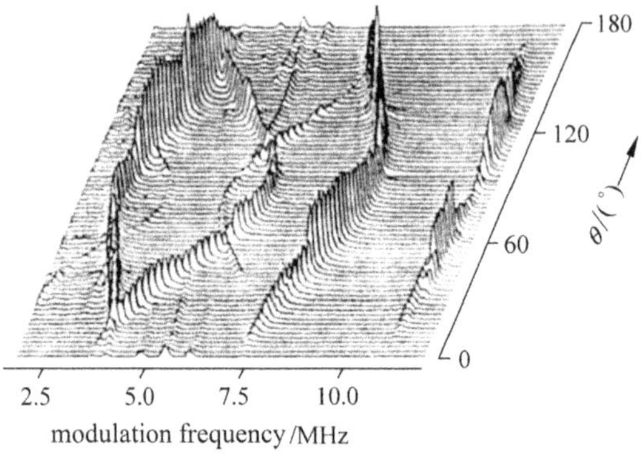

Figure A1-1.12: The stereodiagram of ESEEM spectrum measured from different angles.

are found. The ^{14}N nucleus on the *para*-nitro group is not observed, as being remote from the primary spin-density region. The nitrogen nucleus of hydrazyl did not undergo modulation, the reason being their coupling is too strong; the Zeeman and quadrupole energies of nuclei are smaller than the hyperfine interaction term.

As the last example of the ESEEM technique, the single crystal of perdeuteropyridine, which is prepared using perdeuterobenzene as solvent, has the structure of its lowest triplet state [25]. This molecule is planar in its ground state, but deduced from EMR hyperfine coupling data of ^{14}N nucleus that is detected by using ESE at 1.2 K; in the triplet state, this molecule is distorted to a *boat-type* structure, The ESEEM spectrum gives the matrices $\boldsymbol{A}(^2H)$ and $\boldsymbol{Q}(^2H)$ and obtains the spin-density distribution from them.

In Reference [26] it has been discussed that the relative advantages of measuring ESEEM spectrum at S-band (1.5–3.9 GHz) compared with X-band (9–10 GHz) is that the former provides contributions arising from weak Zeeman interaction of nucleus and quadrupolar coupling and enhances the depth of modulation.

A1-1.6 Advances in pulse EMR techniques

The introduction of pulse technique has led to the exciting development of advanced EMR technique. Like NMR spectroscopy, there are also two-dimensional correlation spectroscopy (COSY and HETEROCOSY) and its spin-echo variant (SECSY) as well as exchange spectroscopy (EXSY) [27–29] in EMR. An another kind of pulse EMR experiment is the so-called "hole burning" FID detection, which can obtain spectrum of excellent sensitivity and resolution [30].

The simultaneous application of CW and pulse microwaves produces coherent Raman beats (i.e., signal of oscillation absorption and scattering), on which detailed hyperfine information [31,32] can be achieved.

Various pulse ENDOR and electron–electron double resonance (ELDOR) techniques are gaining attention increasingly. Thus, the whole pulse EMR technique is being considered at par with CW-EMR spectroscopy.

References

[1] Farrar T. C., Beeker E. D. *Pulse and Fourier Transform NMR*, Academic Press, New York, NY, 1971.

[2] Starzak M. E. *Mathematical Methods in Physics and Chemistry*, Plenum Press, New York, NY, 1941.

[3] Allen L., Eberly J. H. *Optical Resonance and Two-Level Atoms*, Dover Publications, New York, NY, 1987.

[4] Louisell W. H. *Quantum Statistical Properties of Radiation*, John Wiley & Sons, Inc., New York, NY, 1973.

[5] Rabi I. I. *Phys. Rev.* 1937, **51**: 652.

[6] Weil J. A. "On the intensity of magnetic resonance absorption by anisotropic spin systems," in Weil J A, Ed., *Electronic Magnetic Resonance of the Solid State*, Canadian Society for Chemistry, Ottawa, ON, 1987, Chapter 1.

[7] Archibald W. J. *Am. J. Phys.* 1952, **20**: 368.

[8] Harris R. K. *Nuclear Magnetic Resonance Spectroscopy*, Pitman, London, U K, 1983.

[9] Pake G. E, Estle T. L. *The Physical Principles of Electron Paramagnetic Resonance*, 2nd ed., Benjamin, New York, NY, 1973, Chapter 2.

[10] Torrey H. C. *Phys. Rev.* 1949, **76**: 1059.

[11] Slichter C. P. *Principles of Magnetic Resonance*, 3rd edn, Springer, Berlin, 1990, Chapter 5.

[12] Schweiger A. *Angew. Chem., Int. Ed. Engl.* 1991, **30**: 265.

[13] Bowman M. K. "Fourier transform electron spin resonance," in *Modern Pulsed and Continuous-Wave Electron Resonance*, Kevan L, Bowman M K, Eds, John Wiley & Sons, Inc., New York, NY, 1990, Chapter 1.

[14] Dobbert O., Prisner T., Dinse K. P. *J. Magn. Reson.* 1986, **70**: 173.

[15] Prisner T., Dobbert O., Dinse K P., van Willigen H. *J. Am. Chem. Soc.* 1988, **110**: 1622.

[16] Angerhofer A., Toporowicz M., Bowman M K., Norris R., Levanon H. *J. Phys. Chem.* 1988, **92**: 7164.

[17] van Willigen H., Vuolle M., Dinse K. P. *J. Phys. Chem.* 1989, **93**: 2441.

[18] Pluschau M., Zahl A., Dinse K. P., van Willigen. *J. Chem. Phys.* 1989, **90**: 3153.

[19] Slichter C. P. *Principles of Magnetic Resonance*, 3rd edn, Springer, Berlin, 1990, pp. 42–43.

[20] Hahn E. L. *Phys. Rev.* 1950, **77**: 297.

[21] Mims W. B. "Electron spin echo," in Geschwind S, Ed., *Electron Paramagnetic Resonance*, Plenum Press, New York, 1972, Chapter 4.

[22] Snetsinger P. A, Comelius J. B., Clarkson R. B., et al. *J. Phys. Chem.* 1988, **92**: 3696.

[23] Kevan L., Bowman M K., Eds. *Modern Pulsed and Continuous-Wave Electron Resonance*, John Wiley & Sons, Inc., New York, NY, 1990,

[24] Isoya J., Bowman M. K., Norris J. R., Weil J. A. *J. Chem. Phys.* 1983, **78**: 1735.

[25] Buma W. J., Groenen E. J. J., Schmidt J., de Beer R. *J. Chem. Phys.* 1989, **91**: 6549.

[26] Clarkson R. B., Brown D. R., Cornelius J. B., Crookham H. C., Shi W.-J., Belford R L. *Pure Appl. Chem.* 1992, **64**: 893.

[27] Gorcester J., Millhauser G. L., Freed J. H. "Two-dimensional and Fourier-transform EPR," in Hoff A J, Ed., *Advanced EPR-Application in Biology and Biochemistry*, Elsevier, Amsterdam, 1989, Chapter 5.

[28] Angerhofer A., Massoth R. J., Bowman M. K. *Isr. J. Chem.* 1988, **28**: 227.

[29] Fauth J-M., Kababya S., Goldfarb D. *J. Mag. Reson.* 1991, **92**: 203.
[30] Wacker T., Sierra G. A., Schweiger A. *Isr. J. Chem.* 1992, **32**: 305.
[31] Bowman M. K., Massoth R. J., Yannoni C. S. "Coherent Raman beats in electron paramagnetic resonance," in Bagguley D M S, Ed., *Pulsed Magnetic Resonance: NMR, ESR and Optics*, Clarendon, Oxford, UK, 1992, pp. 423–445.
[32] Bowman M. K. *Isr. J. Chem.* 1992, **32**: 339.

Further readings

[1] Champeney D. C. *Fourier Transforms in Physics*, Hilger, Bristol, UK, 1983.
[2] Brigham E. O. *The Fast Fourier Transform*, Prentice-Hall, Englewood Cliffs, NJ, 1974.
[3] Kevan L., Schwartz R. N., Eds. *Time-Domain Electron Spin Resonance*, John Wiley & Sons, Inc., New York, NY, 1979.
[4] Mims W. B. *ENDOR Spectroscopy by Fourier Transformation of the Electron Spin Echo Envelope*, in Marshall A G, Ed., *Fourier Hadamard and Hilbert Transforms in Chemistry*, Plenum Press, New York, NY, 1982, pp. 307–322.
[5] Lin T-S. "*Electron spin echo spectroscopy of organic triplets*," *Chem. Rev.* 1984, **84**: 1.
[6] Weissbluth M. *Photon–Atom Interactions*, Academic Press, Boston, MA, 1989, Chapter 3.
[7] Macomber J. D. *The Dynamics of Spectroscopic Transitions*, John Wiley & Sons, Inc., New York, NY, 1976.
[8] Keijzers C. P., Reijerse E. J., Schmidt J. *Pulsed EPR: A New Field of Applications*, North Holland, Amsterdam, 1989.
[9] Bagguley D. M. S., Ed. *Pulsed Magnetic Resonance: NMR, EMR and Optics*, Oxford University Press, Oxford, UK, 1992.

A-2 Double-resonance techniques

In this section, we introduce double resonance techniques, focusing only on *ENDOR* and ELDOR and their CW and *pulse* spectra. The experimental aspects of these techniques have been described in detail by Poole [1].

A2-2.1 CW-ENDOR

1. The Basic Principles of ENDOR

As already mentioned, for the multiple-set nonequivalent magnetic nuclei, their spectral lines of hyperfine splitting tend to overlap and broaden usually to distinguish difficultly. Those nuclei that are more remote from the unpaired electron and their hyperfine interaction are very weak, their EMR spectra cannot display hyperfine structure pattern and cannot provide any information. G. Feher [2] proposed and established an electron–nuclear double resonance (ENDOR) apparatus in 1956. By applying his brilliant contribution of ENDOR the missing information on hyperfine interaction could be obtained [3].

In EMR, a weak microwave electromagnetic field H_1 is added to the external magnetic field H_o in a perpendicular direction. Because $H_1 \ll H_o$, the number of particles distribution between the various energy levels basically remains near to their thermal equilibrium value, and the energy levels are also not modified. But in ENDOR, two electromagnetic radiation fields is added in the perpendicular direction to the external magnetic field H_o: one is the microwave field, which is used to excite the electron transition, its selection rule is $\Delta M_S = \pm 1$, $\Delta M_I = 0$, the other is the radiation field, which is used to stimulate the nuclear spin transition, its selection rule is $\Delta M_S = 0$, $\Delta M_I = \pm 1$, the function is initiation pumping transition. It is different from EMR; in ENDOR, electron is pumped by using strong microwave, which brings them in partial saturation state, and then the enhancement of intensity by NMR transition is observed. It is different from not only the usual EMR but also from NMR. The variation of EMR signal occurring in NMR is also observed.

A. Brief Introduction of CW-ENDOR Experimental Technique

Before providing a detailed description of *ENDOR* processes, a phenomenological introduction of a simple continuous-wave ENDOR experiment of a solid-state system with $S = I = 1/2$ is briefly given.

1. The sample is first placed in a special cavity (shown in Figure A2-2.1). Under low microwave power, the magnetic field is scanned and passed through a resonance

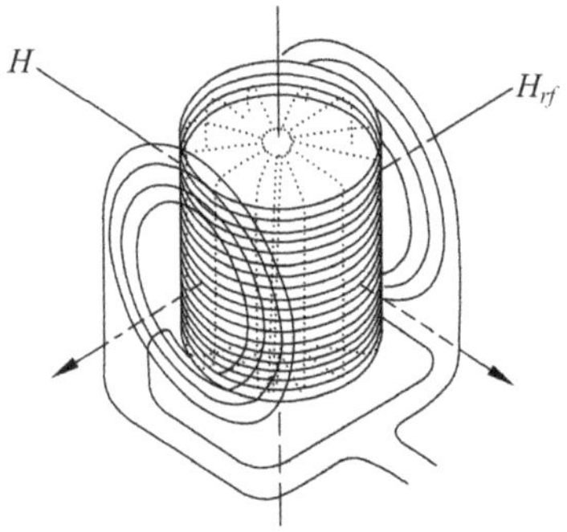

Figure A2-2.1: H_{011}-type cylindrical cavity for ENDOR [4].

point H_k of EMR. Then each parameter of spectrometer is readjusted enabling it to reach to optimum state and also the EMR signal to reach to maximum. Then the magnetic field in H_k is fixed.

2. With the increasing microwave power, the electrons from level 1 are pumped to level 2, until it reaches to $n_1 \approx n_2$. A new population diagram of particles in each level is obtained. In fact, the spin–lattice relaxation can prevent appearance of this extreme population case.
3. In the coils of the cavity, both sides input a radio frequency field (H_{rf}) of great power and wide frequency range scanning. The alternative radio frequency field H_{rf} acts on the sample this time, the scanning region is usually between 2 and 30 MHz; recording spectrum is shown in Figure A2-2.2. The base line expresses EPR absorption; when the radio frequency field scan through point ν_{n1} and ν_{n2}, the EMR signal increases, the two peaks in Figure A2-2.2 is the ENDOR signal.

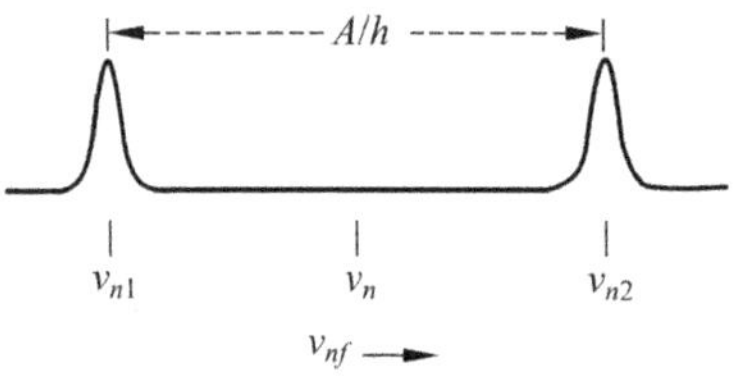

Figure A2-2.2: The spectrum of ENDOR.

B. Energy Splitting and ENDOR Transition

Let us begin with the simplest system: There is one unpaired electron only and the nucleus $I = 1/2$; its g and a all are isotropic. The spin-Hamiltonian of system is

$$\hat{\mathscr{H}} = g\beta \boldsymbol{H} \cdot \hat{\boldsymbol{S}} - g_n\beta_n \boldsymbol{H} \cdot \hat{\boldsymbol{I}} - a'\hat{\boldsymbol{I}} \cdot \hat{\boldsymbol{S}} \qquad \text{(A-2.1)}$$

If the first-order approximation is applied for the hyperfine interaction; $\hat{\boldsymbol{I}} \cdot \hat{\boldsymbol{S}}$ is replaced by $\hat{S}_z\hat{I}_z$, and the eigenvalue of the $\hat{\mathrm{H}}$ operator is

$$E(M_S,\ M_I) = g\beta \boldsymbol{H}\, M_S - g_n\beta_n H\, M_I + a'\, M_S\, M_I \tag{A-2.2}$$

Its corresponding eigenfunction is a simple product function $|M_S, M_I\rangle$. In order to make the unit of energy be expressed by the unit of frequency, let

$$g\beta H/h = \nu_e, \qquad g_n\beta_n H/h = \nu_n, \qquad a'/h = a$$

Then Eq. (A-2.2) can be rewritten as follows:

$$E(M_S,\ M_I) = \nu_e\, M_S - \nu_n\, M_I + a\, M_S\, M_I \tag{A-2.3}$$

$$E_1 = +\frac{1}{2}(\nu_e - \nu_n) + \frac{1}{4}a, \quad \text{corresponding eigenfunction}: \quad |+,+\rangle$$

$$E_2 = +\frac{1}{2}(\nu_e + \nu_n) - \frac{1}{4}a, \quad \text{corresponding eigenfunction}: \quad |+,-\rangle$$

$$E_3 - \frac{1}{2}(\nu_e + \nu_n) - \frac{1}{4}a, \quad \text{corresponding eigenfunction}: \quad |-,-\rangle$$

$$E_4 = -\frac{1}{2}(\nu_e - \nu_n) + \frac{1}{4}a, \quad \text{corresponding eigenfunction}: \quad |-,+\rangle$$

According to Eq. (A-2.3), draw the diagram of energy level as shown in Figure A2-2.3. The two cases should be distinguished so that one is $|a| < 2\nu_n$ and the other is $|a| > 2\nu_n$. For the X-band, $\nu_n = g_n\beta_n H/h = 13.5\,\text{MHz}$. This also implies that if $|a| < 0.965\,\text{mT}$, it belongs to the former; conversely, it belongs to the latter. In the thermal equilibrium, the number of each energy level obeys the Boltzmann distribution. However, because the nuclear Zeeman energy is much more smaller than the electronics, the population difference between E_1 and E_2 and between E_3 and E_4 can be ignored. Compared with kT, the electronic Zeeman energy is also very small, so with the exponential term of expansion, Boltzmann factor can be only retained to first-order term approximately. The population diagram of the thermal equilibrium is shown in Figure A2-2.3, and $\delta = \nu_e/kT$.

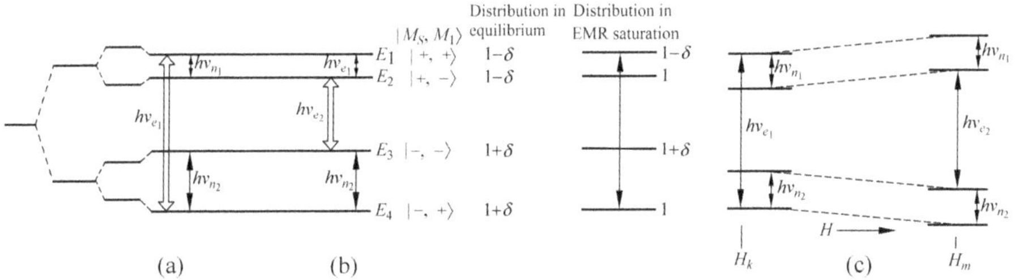

Figure A2-2.3: The diagram of level splitting of S = I = 1/2 system.

In general, the EMR transition corresponds to the selective rule of $\Delta M_S = \pm 1$; $\Delta M_I = 0$. The wide arrow of Figure A2-2.3a shows EMR transitions between E_4 and $E_1(h\nu_{e_1})$, this time $M_I = +\frac{1}{2}$. The frequency $\nu_{n_1}$of the radio electromagnetic field is as follows:

$$\nu_{n_1} = E_2 - E_1 = \nu_n - \frac{1}{2}a \tag{A-2.4}$$

Analogous to the selective rule, $\Delta M_S = 0$, $\Delta M_I = \pm 1$. It causes NMR transition between the levels E_2 and E_1, there are partial electrons of level E_2 that transit to the level E_1, enhancing the transition of EMR between the E_4 and E_1. The wide arrow of Figure A2-2.3b shows EMR transition $(h\nu_{e_2})$ between E_3 and E_2, this time $M_I = -\frac{1}{2}$. Thus, the frequency of radio electromagnetic field $\nu_{n_2}$is satisfied:

$$\nu_{n_2} = E_4 - E_3 = \nu_n + \frac{1}{2}a \tag{A-2.5}$$

NMR transition occurs between the levels E_4 and E_3, the partial electrons of energy level E_4 transit to the energy level E_3. The EMR transitions between E_3 and E_2 are enhanced. Figure A2-2.3c shows the transition of various levels at fixed microwave frequency. The transition of $h\nu_{n_1}$and $h\nu_{n_2}$ are the ENDOR lines.

From Eqs. (A-2.4) and (A-2.5), the following is obtained:

$$a = |\nu_{n_2} - \nu_{n_1}|, \; \ldots \nu_n = \frac{\nu_{n_1} + \nu_{n_2}}{2} \tag{A-2.6}$$

The prerequisite condition of the above discussion is $|a| < 2\nu_n$. If the case is $|a| > 2\nu_n$, then

$$a = |\nu_{n_1} + \nu_{n_2}|, \; \nu_n = |\nu_{n_1} - \nu_{n_2}|/2 \tag{A-2.7}$$

From here, we can see that by using ENDOR, we can measure the parameter $\boldsymbol{a}$ of hyperfine coupling:

$$a = |\nu_{n_1} \mp \nu_{n_2}| \tag{A-2.8}$$

The above formula is feasible only under the first-order approximation; if a is very small, the second-order effect should be considered generally. In addition, *the nuclear Landé factor* g_n can be detected by using ENDOR:

$$g_n = \frac{h|\nu_{n_1} \mp \nu_{n_2}|}{2\beta_n H} \tag{A-2.9}$$

In fact, there is a difference between the g_n value calculated in the above formula and the value from the handbook. It may have resulted from the Zeeman action of *pseudo*-nuclear.

In the EMR spectra of heterogeneous broadening, the ENDOR frequency to measure the hyperfine coupling parameter can enhance the accuracy greatly. Because the spectrum of ENDOR is very narrow, the average linewidth is about 10 kHz, and its range is from 3 to 1000 kHz. If the linewidth of the EMR spectrum is 0.1 mT, $g = 2.00$; it converts to frequency unit:

$$\Delta\nu = (g\beta\Delta H)/h = 2.8\,\text{MHz}$$

For the NMR, if the spectral linewidth is also 0.1 mT (if $\gamma_n = 6.3 \times 10^4\,\text{rad s}^{-1}\,\text{mT}^{-1}$), the corresponding frequency unit is $\Delta\nu = (2\pi)^{-1}\gamma_n\,\Delta H = 1.1\,\text{kHz}$. Therefore, the hyperfine coupling parameter a measured by ENDOR is much more accurate than that measured from the EMR spectrum directly.

C. Relaxation Process of Steady-State ENDOR

In the EMR experiment, only one spin–lattice relaxation time τ_1 is usually involved. However, in the ENDOR process, even the simplest four-level system has at least three spin–lattice relaxation times to adjust the distribution of population between various levels. These not only control the temperature range to enable ENDOR experiment to perform successfully but also control other experimental conditions and hence determine the properties of the observed spectrum. Its temperature sensitivity is also a disadvantage of the ENDOR technique.

Three spin–lattice relaxation times τ_{1e}, τ_{1n}, and τ_x of four-level system $(S = I = 1/2)$ are shown in Figure A2-2.4a. Here τ_{1e} is the electron spin–lattice relaxation time; τ_{1n} is the nuclear spin–lattice relaxation time; and τ_x is the "cross-relaxation" time arising between mutual "spin flips." They correspond to the process of $\Delta(M_S + M_I) = 0$. The τ_{xx} is the "cross-relaxation" time of $\Delta(M_S + M_I) = \pm 2$ process. Usually, $\tau_{1e} \ll \tau_x \ll \tau_{1n}$. The reciprocals of these times represent the rates of transition between these levels when the microwave or radio frequency field is absent (please refer to Figure A2-2.4a).

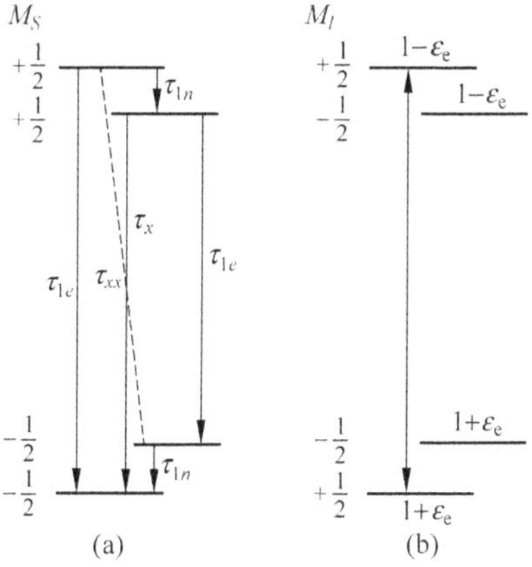

Figure A2-2.4: The relaxation path relation diagram of system S = I = ½.

When there is no microwave field or the microwave field is very weak, the relative population between each energy state is as shown in Figure A2-2.4b.

For most solid-state samples, at 4 K temperature, most ENDOR experiments can be performed successfully. At these temperatures, the microwave saturation only need a certain power, because τ_{1e} is relatively long. The long value of τ_{1e} makes the NMR transition ($\Delta M_I = \pm 1$) as compared with the EMR transition($\Delta M_S = \pm 1$). Under the extreme situation, for example, phosphorus-doped silicon, τ_{1e} is as long as an "hour" order; however, more common (usually) is a "second" order. If the ENDOR linewidth is 10 kHz, the corresponding $\tau_2 = 10^{-5}$ s. If the τ_{1e} does not contribute to the broadening of spin–lattice relaxation, the value of τ_{1e} is not much short. This also implies that both τ_{1e} and τ_2 are of same order. If τ_x is not too long, the steady-state ENDOR experiment could be done, that is, the ENDOR lines can be observed, in an arbitrarily slow speed sweep and with an indefinite times sweep back. In contrast, in the rapid-passage types of ENDOR experiment, only one spectral line of ENDOR could be observed during rapid sweep forward; at this time, there is a process of population equalization. So the ENDOR line cannot appear at scanning back immediately. In other words, it a couple of ENDOR lines cannot be observed in the rapid-passage type experiment. In order to determine hyperfine coupling parameters a, ν_n should be determined, which is the center of a pair of ENDOR spectral lines. The name of "steady state" is not accurate, if the relaxation time τ_{1n} is very long.

It makes the central intensity of EMR absorption line nearly reach the maximum value of homogeneous broadening curve with increased microwave power. T the optimum value [5] of the microwave magnetic field H_{1e} for the steady-state ENDOR is $\gamma_e^2 H_{1e}^2 \tau_1 \tau_2 \approx 3$. Here γ_e is the magnetogyric ratio of the electron, and H_{1e} is the amplitude of the microwave magnetic field. The power level of the radio frequency source is set sufficiently high so that the rate $(dN/dt)_\uparrow$ of induced upward transitions at frequency ν_{n_1} is large in comparison with τ_x^{-1}, that is, $(dN/dt)\tau_x \geq 1$. Another kind of setup requires very large value of H_{1n}, because $\Delta M_S = 0$; $\Delta M_I = \pm 1$ transitions should be comparable to the cross-relaxation $\Delta(M_S + M_I) = 0$ transitions of measurable τ_x. When the radiation frequency passes through the value ν_{n_2}, an ENDOR line can be observed. In many four-levels systems, the second ENDOR line can also be observed, when the radiation frequency pass through the value of ν_{n_2}.

Now let us discuss the relative population number of the various levels under different conditions. In the absence of out field, the population of each level in the four degenerate levels would almost be $N/4$, where N is the total number of unpaired electrons. When the out field is added and the hyperfine effects are not considered, then the populations are as follows:

$$M_S = +\frac{1}{2} \quad N_{+1/2} = \frac{1}{4} N \exp(-g_e \beta_e H / 2k_b T) \simeq \frac{1}{4} N(1 - \varepsilon_e) \tag{A-2.10a}$$

$$M_S = -\frac{1}{2} \quad N_{-1/2} = \frac{1}{4} N \exp(+g_e \beta_e H / 2k_b T) \simeq \frac{1}{4} N (1 + \varepsilon_e) \qquad \text{(A-2.10b)}$$

Here $\varepsilon_e = g_e \beta_e H / 2k_b T$. If $M_I = +\frac{1}{2}$, the effective relaxation path of microwave field excite EMR transition by τ_{1e} (as Figure A2-2.4). The path via τ_x is invalid, since τ_{1n} series will be much longer.

The steady-state ENDOR experiment discussed by us only involves partial saturation of the electron-spin transition, assuming the populations equalization of $M_I = +\frac{1}{2}$ states; this is shown in Figure A2-2.5a. The case of complete saturation of the electron-spin transition involving $M_I = -\frac{1}{2}$ state is shown in Figure A2-2.5b. It is still correct that for $\tau_{1e} \ll \tau_x$, the cross-relaxation almost does not occur. For the saturation transition of $M_I = +\frac{1}{2}$, using ε_e expresses the difference of the populationbetween the state$|+1/2, +1/2\rangle$ and state $|+1/2, -1/2\rangle$, whereas using $\varepsilon_n = g_n \beta_n H / 2k_b T$ expresses the population difference between them in case of absence of microwave saturation. If a short loop is provided between these two states, the population of the $|+1/2, +1/2\rangle$ state would be much more reduced compared with the population of the $|+1/2, -1/2\rangle$ state. The frequency $\nu_{n_1}$of intense rf field has provided such loop. The rate of transition between the state $|+1/2, +1/2\rangle$ and state $|+1/2, -1/2\rangle$ at least should be equal to τ_x^{-1}. If the $M_I = -\frac{1}{2}$ state is saturated, there is a very large population difference ε_e between the states of $|-1/2, -1/2\rangle$ and $|-1/2, +1/2\rangle$; hence, an intense rf field of frequency ν_{n_2} will arise in ENDOR transition line. The population numbers of the other various pair levels are all equal, if these relaxation processes are only operated, the ENDOR spectral line of frequency ν_{n_1}will not appear. Same operation makes the $M_I = +\frac{1}{2}$ state saturation, the ENDOR spectral line of frequency ν_{n_2} cannot be observed. So in order to observe a couple of ENDOR spectral lines, at least an additional relaxation path should be added in most systems.

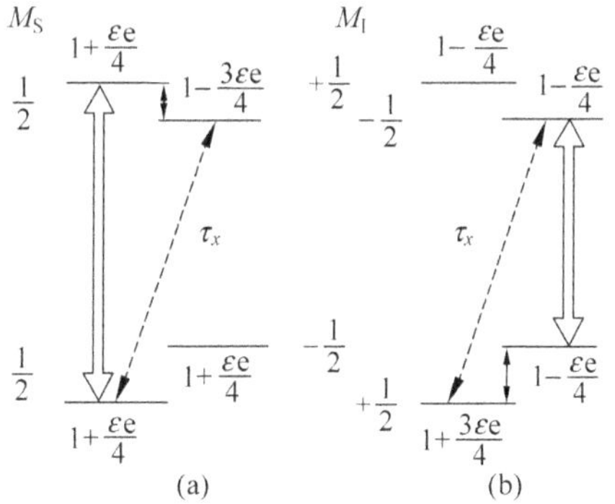

Figure A2-2.5: Diagram of relative population for S = I = 1/2 system and the relaxation of ENDOR.

Using operators $\hat{S}_+\hat{I}_- + \hat{S}_-\hat{I}_+$ expresses the third term on the right-hand side of Eq. A-2.1, that is, hyperfine interaction term, making the states $|M_S - 1, M_I + 1\rangle$ or

$|M_S+1, M_I-1\rangle$ admix into state $|M_S, M_I\rangle$. Because of such mixing, the $\Delta(M_S+M_I)=0$ transition of relation with the relaxation path τ_x is partially allowed. The alternative path τ_{xx} of Figure A2-2.4a involves $\Delta(M_S+M_I)=\pm 2$ transitions. There are two alternative relaxation paths (besides τ_{1e}) from the highest state $|+\frac{1}{2}, +\frac{1}{2}\rangle$ that reach to the lowest state $|-\frac{1}{2}, +\frac{1}{2}\rangle$: one is $\tau_{xx}+\tau_{1n}$, and the other is $\tau_{1n}+\tau_x$ (as shown in Figure A2-2.4a). In both cases, the relaxation rate is controlled by τ_{1n} since the time of τ_{xx} is much more shorter than τ_{1n}. Because of saturation, rf power at both nuclear frequency (ν_{n_1} or ν_{n_2}) can enhance the transition rate of the $\Delta M_S=0$, $\Delta M_I=\pm 1$ sufficiently. Because of the competition between relaxation paths, the effective value of τ_{1e} is reduced, it does not have relation with the saturation of microwave transition . It is indeed the essential characteristics of steady-state ENDOR.

If the magnitude of one or more cross-relaxation times commensurates with the nuclear spin–lattice relaxation time, the action of saturation making the rf power can sufficiently reduce the effective value of the electron spin–lattice relaxation time. This enables the observation of the ENDOR spectral lines continuously.

In most situations, the intensities of a pair of ENDOR lines are similar. However, these are also not equal, and even only one line could be detected. This phenomenon is marked particularly when the nuclear Zeeman interaction is comparable in magnitude with the hyperfine interaction [6–8]. For the system with nonaxial symmetry, the field arising from the hyperfine interaction of electron with nucleus can be calculated. The intensity difference of a couple of ENDOR lines, arising from different influence of rf field, coincides very well with the difference of observation. Compared to symmetrical axis, the orientation of the rf field is more important in determining the intensity of the anomalous ENDOR line, in spite of the rf field always being perpendicular to the outer field $\boldsymbol{H}$.

The above discussions are all limited to the simplest four-level system. For the system of general electron spin $\boldsymbol{S}$ and general nuclear spin $\boldsymbol{I}$, the maximum possible number of ENDOR lines is $16SI$; of course, some of them belong to forbidden transition. If spin $I\geq 1$ nucleus is present, the nuclear quadrupole interaction term should also be considered. It will increase additional relaxation paths, so it would because more difficulty in the prediction of the intensity of the ENDOR lines and for discussion of the relaxation mechanism of ENDOR. It will be discussed in Sections A-2.3 and A-2.4.

2. CW-ENDOR Spectra in Liquid Solution

The original observed ENDOR spectrum used single-crystal sample. Due to the difficulty of experimental technique, the ENDOR spectrum in the liquid was observed until 1964 [9]. The reason is that in order to excite an ENDOR transition, the driving rate of NMR transition should correspond to the electron-spin relaxation rate; since the latter is a very fast process in liquid, the RF power should be increased greatly. Let us first discuss the ENDOR spectrum of the liquid solution.

A. Determination of Hyperfine Coupling Parameters

There are n equivalent magnetic nuclei in the solution of free radical. The total nuclear spin are

$$\hat{\boldsymbol{I}} = \sum_i \hat{\boldsymbol{I}}_i$$

The component of $\hat{\boldsymbol{I}}$ in the Z-direction is

$$M_I = I,\ \ I-1,\ \ \ldots,\ \ -I+1,\ \ -I \quad (\text{there are } 2I+1 \text{ values of } M_I \text{ all together})$$

The first-order approximation energy is

$$E(M_S,\ M_I) = \nu_e M_S - \nu_n M_I + aM_S M_I \tag{A-2.11}$$

The selection rule of EMR $\Delta M_S = \pm 1$, $\Delta M_I = 0$. Then

$$\nu = \nu_e + aM_I \tag{A-2.12}$$

It has $2I+1$ hyperfine line of equal distance. While the selection rule of ENDOR is $\Delta M_S = 0$, $\Delta M_I = \pm 1$, then

$$\nu = \nu_n + aM_S \tag{A-2.13}$$

Since $M_S = \pm \frac{1}{2}$, there are only two lines of ENDOR spectrum, especially the larger the I, the much more hyperfine lines of EMR; the lines of ENDOR spectrum is not related with I. There are always only two lines.

Here taking the Coppinger radicals [10] as an example, the hyperfine coupling parameters are demonstrated in detail using ENDOR spectroscopy. The general formula of Coppinger radicals is shown in Table A2-2.1.

Table A2-2.1: Structural formula of Coppinger radical series.

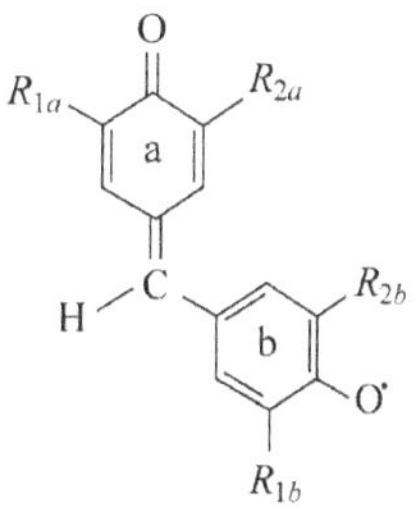

I	$R_{1a} = R_{1b} = R_{2a} = R_{2b} = C(CH_3)_3$
II	$R_{1a} = R_{1b} = R_{2a} = R_{2b} = OCH_3$
III	$R_{1a} = R_{2a} = C(CH_3)_3 = R_{1b} = R_{2b} = OCH_3$
IV	$R_{1a} = R_{2a} = C(CH_3)_3 = R_{1b} = R_{2b} = CH_3$
V	$R_{1a} = R_{2b} = C(CH_3)_3 = R_{2a} = R_{1b} = CH_3$
VI	$R_{1a} = R_{1b} = R_{2a} = R_{2b} = CH_3$

Hyde [4,9] was the first to use free radical I to determine the parameter of hyperfine coupling by ENDOR spectroscopy successfully. Radical I is a kind of natural free radicals. The hyperfine coupling parameter of a proton in methylidyne is 0.56 mT. The hyperfine coupling parameters of four protons on the ring is 0.14 mT. There are 12 methyl on the ring, and in total 36 protons on the methyl of ring; their hyperfine coupling parameters are very weak, only about 0.04 mT.

At that time, the ENDOR spectrum carried out by Hyde was discontinuous, as shown Figure A2-2.6. There are five peaks: 12,165 with 15.930 MHz is a pair; 12.065 with 16.032 MHz is an another pair; and the 21.850 MHz is a single peak. In order to seek the another single peak, the frequency of free proton $\nu_n$should be considered:

$$\nu_n = \frac{1}{2}(12.165 + 15.930) = 14.048\ (\text{MHz})$$
$$\nu_n = \frac{1}{2}(12.065 + 16.032) = 14.044\ (\text{MHz})$$

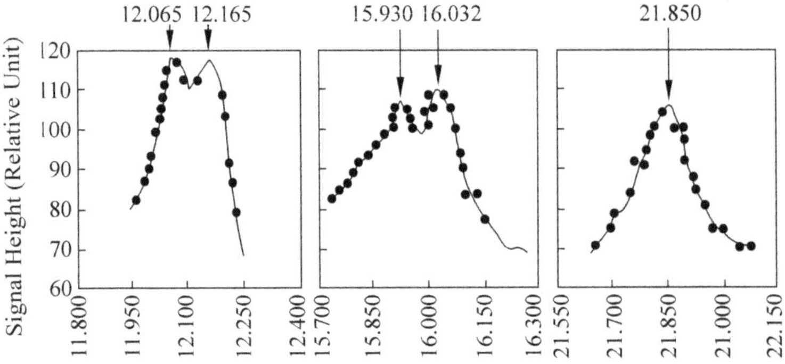

Figure A2-2.6: The ENDOR signal of Coppinger radical I.

Thus, the average value is $\nu_n = 14.046\,\text{MHz}$. Hyperfine coupling parameters are $a_a = \nu_{n_2} - \nu_{n_1} = 16.032 - 12.065 = 3.967\,\text{MHz}\ (0.142\,\text{mT})$ and $a_b = 15.930 - 12.165 = 3.765\,\text{MHz}\ (0.134\,\text{mT})$; the protons of methylidyne have only one peak because their scanning frequency range is too narrow, only from 11.80 to 22.15 MHz, but it can use the free proton frequency:

$$a = 2(21.850 - 14.046) = 15.608\,\text{MHz}\ (0.557\,\text{mT})$$

In 1968, Hyde et al.[11] made the ENDOR spectra of free radicals II, III, IV, and V at −80 °C, as shown in Figure A2-2.7.

There are three peaks for free radical II: the 21.5 MHz peak belongs to proton of methylidyne; the 15.3 MHz peak is for four protons on the ring and the 14.7 MHz peak is for 12 protons of methoxy group. Radical III has four peaks, the 21.2 peak belongs to the proton of methylidyne; the 15.2 and 15.6 MHz peaks belong to two pairs of protons of the ring, respectively; the 14.9 MHz peak is for six protons of

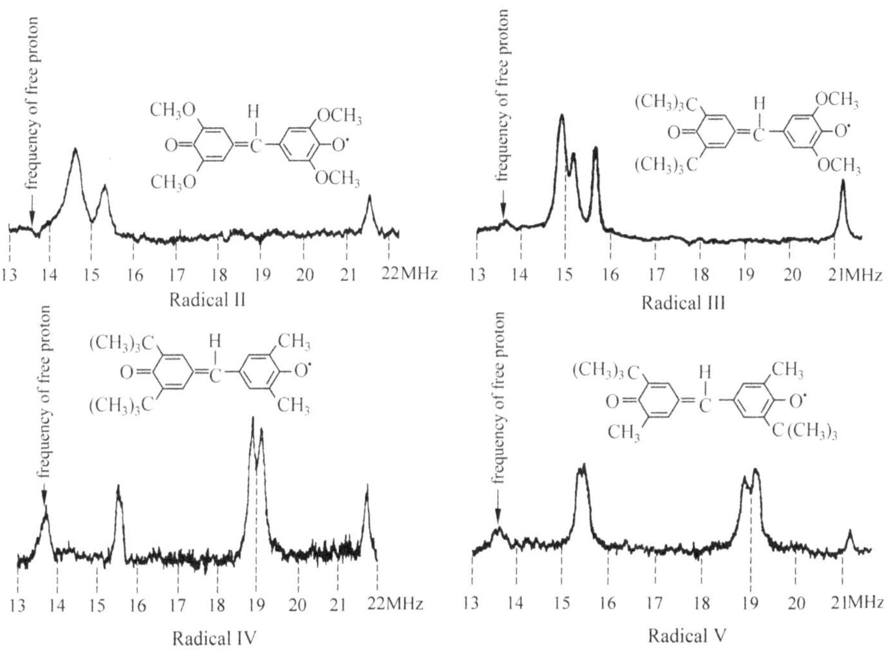

Figure A2-2.7: The ENDOR spectra [11] of free radical II, III, IV and V.

methoxyl group and 13.6 MHz peak is for 18 protons of tertbutyl group, respectively, which is an overlap with the frequency of free proton. Free radical IV also has four peaks: the 21.2 MHz peak is for the protons of methylidyne and the 15.6 MHz peak is for four protons on the ring; the 19 MHz peak is for six protons on the methyl group and the 13.6 MHz peak is for 18 protons of tertbutyl group, which is an overlap with free proton frequency. The hyperfine coupling parameters to be calculated from the ENDOR spectra are shown in Table 12.2. Concrete calculations are left to readers as exercises.

Table A2-2.2: The hyperfine coupling parameters obtained from ENDOR spectra (MHz)[11.].

Position	I[8,9]	II	III	IV	V	
Methylidyne	15.591	15.968	15.136	16.22	15.14	16.22
Ring (a)	3.869[a]	3.450[a]	4.108	3.76[a]	3.73[a]	
Ring (b)			3.120			
R_{1a}			0.110	0.20	0.28	
R_{2a}	0.08*	2.227			11.14	10.71
R_{1b}			2.558	10.48		
R_{2b}				10.94	0.28	

[a]The average value of both ring.

B. Determination of Nuclear Quadrupole Coupling Parameter

Start from the simplest situation: the system has only one isotropic hyperfine coupling parameter $a(a > 2\nu_n)$, the quadrupole coupling is axial symmetrical, and the Hamiltonian operator of first-order approximation is

$$\hat{\mathcal{H}} = g\beta \boldsymbol{H_z}\hat{\boldsymbol{S}}_{\boldsymbol{z}} - g_n\beta_n H_z\hat{I}_z + a\hat{S}_z\hat{I}_z + Q\left[\hat{I}_z^2 - \frac{1}{3}I(I+1)\right] \quad \text{(A-2.14)}$$

Its eigenvalue is

$$E(M_S, M_I) = \nu_e M_S - \nu_n M_I + aM_S M_I + Q\left[M_I^2 - \frac{1}{3}I(I+1)\right] \quad \text{(A-2.15)}$$

For the transition frequency of EMR($\Delta M_S = \pm 1,\ \Delta M_I = 0$) ν_{EMR}

$$\nu_{EMR} = \nu_e + aM_I \quad \text{(A-2.16)}$$

It has no relation with Q, so it cannot detect Q from EMR spectrum under the first-order approximation situation. However, for the transition frequency ν_{ENDOR} of ENDOR($\Delta M_S = 0,\ \Delta M_I = \pm 1$), there are four lines divided into two sets (please refer to Table A2-2.3): one is the ENDOR spectra of high frequency, its frequency is $\nu_n + a/2$ when the $Q = 0$; the another is the ENDOR spectra of low frequency, its frequency is $\nu_n - a/2$ when $Q = 0$. The intensity of low-frequency ENDOR spectrum is very weak when $Q = 0$, and when $a/2 \approx \nu_n$, the corresponding levels of low-frequency ENDOR transition admixes mutually, and then the common selection rule ($\Delta M_I = \pm 1$) could not be suitable again. The transition of high-frequency ENDOR ($M_S = -1/2$) can also be often observed in the experiment.

Table A2-2.3: The ENDOR transition of the $S = 1/2,\ I \geq 1$ system.

ENDOR transition	M_S	ν_{ENDOR}
$M_I \leftrightarrow M_I + 1$	−1/2	$\nu_+ = (a/2) + [\nu_n - Q(2M_I + 1)]$
	+1/2	$\nu_+ = (a/2) - [\nu_n - Q(2M_I + 1)]$
$M_I \leftrightarrow M_I - 1$	−1/2	$\nu_- = (a/2) + [\nu_n - Q(2M_I - 1)]$
	+1/2	$\nu_- = (a/2) - [\nu_n - Q(2M_I - 1)]$

When $Q \neq 0$, it would be divided into two lines; their interval distance is

$$\nu_- - \nu_+ = \delta = 2Q \quad \text{(A-2.17)}$$

The Q value could be detected from ENDOR spectrum quickly, and can also determine its size.

The sodium formate crystal is irradiated by X-ray; we discuss the ENDOR spectrum[12] of ^{23}Na($I = 3/2$) on the Na^+COO^- free radical as an example. The experiment is carried out at 77 K, 35 GHz. The high-frequency ENDOR line is in the range of 24–29 MHz, and $\nu_{Na} = (\gamma_{Na}/\gamma_e)\,\nu_e = 14.044$ MHz. The low-frequency ENDOR line is below 5 MHz. At low frequency, because of the level mixing, the selection rule ($\Delta M_I = 0$) is not suitable again. Figure A2-2.8 shows the ENDOR spectral line in 1.2–2.4 MHz range, $M_S = +1/2$. Due to the quadrupole coupling, the energy levels admix: the ENDOR spectrum ought to have six lines but not three lines. Figure A2-2.8 shows five lines of them. For high-frequency ENDOR, a first-order approximation is suitable. The experimental results show that the second-order corrective term has only several kHz. For $M_I = -3/2$, at high-frequency range, transition of $M_I \leftrightarrow M_I + 1$ does not exist. So one can only observe one ENDOR line of $(3/2 \leftrightarrow 1/2)$. The reason being same: There is also only one ENDOR line $(-3/2 \leftrightarrow -1/2)$. For the $M_I = \pm 1/2$, two lines of ENDOR spectrum could be observed, as shown in Figure A2-2.9 and Table A2-2.4.

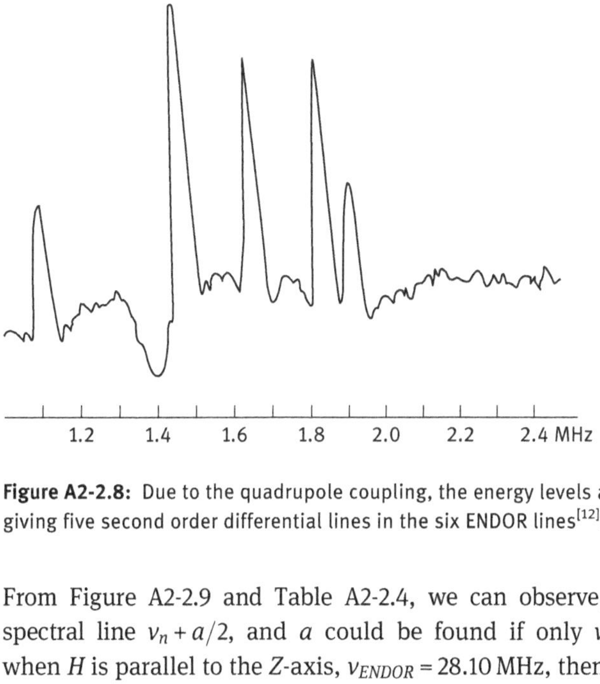

Figure A2-2.8: Due to the quadrupole coupling, the energy levels are mixed at $MS = +1/2$, which giving five second order differential lines in the six ENDOR lines[12].

From Figure A2-2.9 and Table A2-2.4, we can observe, for II and III, a common spectral line $\nu_n + a/2$, and a could be found if only ν_n is known. For example, when H is parallel to the Z-axis, $\nu_{ENDOR} = 28.10$ MHz, then

$$a_{zz} = 2(28.10 - 14.04) = 28.12\,(\text{MHz})$$

Similarly, a set of parameters could be obtained: $a_{zz} = 28.12$, $a_{yy} = 22.58$, $a_{xx} = 22.50$, $a_{xy} = 0.14$, $a_{yz} = a_{zx} = 0=$, $Q_{zz} = -0.340$, $Q_{yy} = 0.250$, $Q_{xx} = 0.065$, $Q_{xy} = -0.030$, $Q_{yz} = Q_{zx} = 0$. Using a set of these parameters and $\nu_{Na} = 14.044$, $\nu_e = 35\,000$ (the unit of all the above parameters are in MHz). The ENDOR transition frequency of various angles under the situation of magnetic field parallel to the different crystal planes can

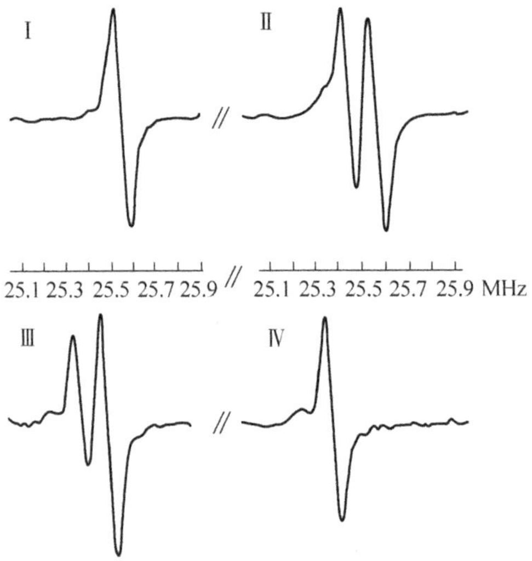

Figure A2-2.9: The first-order differential line of the ^{23}Na-ENDOR transition spectrum.

Table A2-2.4: Frequency of ENDOR high frequency.

No.	M_I	High frequency of ENDOR
I	−3/2	$\nu_+ = (a/2) + \nu_n + 2Q$
II	−1/2	$\nu_- = (a/2) + \nu_n + 2Q$
		$\nu_+ = (a/2) + \nu_n$
III	+1/2	$\nu_- = (a/2) + \nu_n$
		$\nu_+ = (a/2) + \nu_n - 2Q$
IV	+3/2	$\nu_- = (a/2) + \nu_n - 2Q$

be calculated by inputting the data into computer as shown in Figure A2-2.10 and Table A2-2.11. The sign of $\boldsymbol{Q}$ can be decided from the figures: If the curve II is on the top of curve III, the sign of $\boldsymbol{Q}$ should be positive; if the curve II is below the curve III, the sign of $\boldsymbol{Q}$ should be negative. From Figures A2-2.10 and A2-2.11, it can be decided that the Q_{zz} is positive and the Q_{yy} is negative.

Figure A2-2.10 and A-2.11 shows the variation of the high-frequency ENDOR transition with angles at various orientations. The $\boldsymbol{A}$ and $\boldsymbol{Q}$ tensors can thus be determined. After diagonalization of these tensors, the principal values and the principal axis directions of the g, $\boldsymbol{A}$, and $\boldsymbol{Q}$ tensors can be found, as shown in Table A2-2.5.

The ENDOR spectrum of ^{23}Na is only discussed above; however, the ENDOR spectrum of its proton can also be understood in the same manner. Because $\nu_H = (\gamma_H/\gamma_e)\,\nu_e = 52.96$ MHz, the experiment should be carried out in the range of 50–55MHz. The second-order differential spectrum of the proton ENDOR transition

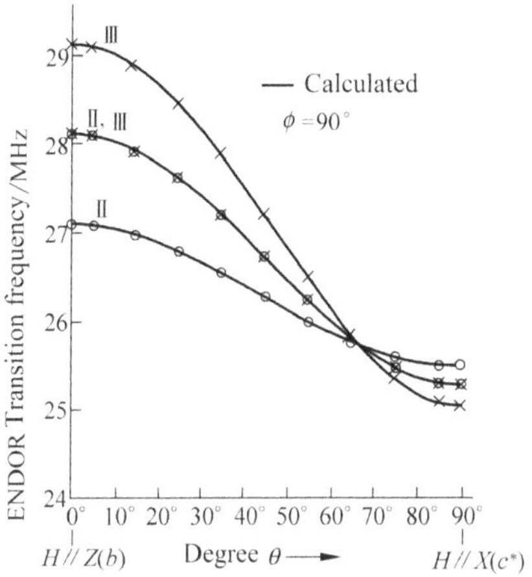

Figure A2-2.10: The values of ENDOR transition frequency at φ=90°, θ=0–90° by experiment and calculation.

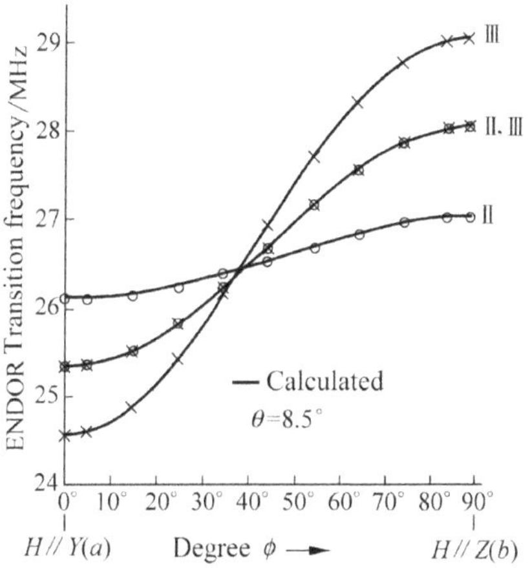

Figure A2-2.11: The value of ENDOR transition frequency at θ=8.5°, φ=0–90° by experiment and calculation.

Table A2-2.5: The principal values and directional cosines of the *g*, A, Q tensors in the coordination system of a(y), b(z), c*(x).

	Principal value (77 K)	Directional cosine		
		a	b	c*
	28.12±0.02	0	1	0
A (MHz)	22.69	0.798	0	0.602
	22.39	−0.602	0	0.798
	−0.33±0.01	0	1	0
Q (MHz)	0.07	0.156	0	0.988
	0.26	0.988	0	−0.156
	2.0019±0.0001	0	1	0
g	1.9980	0.968	0	0.250
	2.0034	−0.250	0	0.968

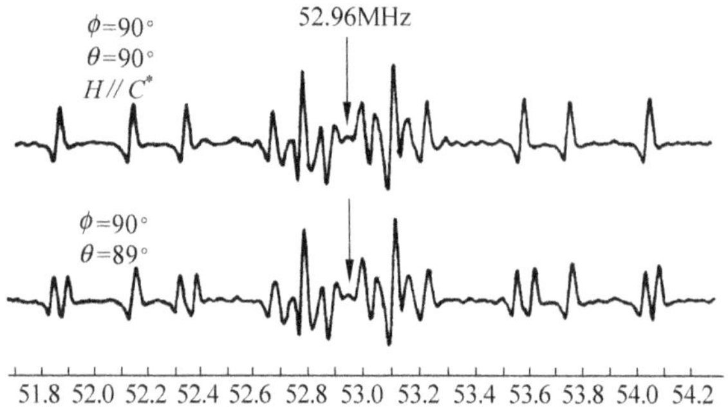

Figure A2-2.12: The second-order differential spectrum of the proton ENDOR transition.

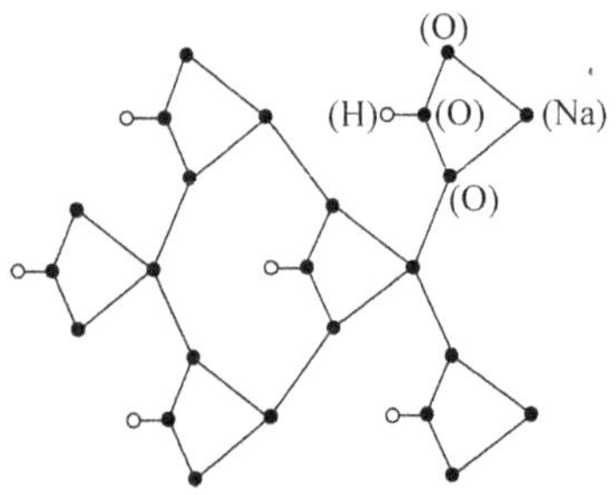

Figure A2-2.13: The orientation of free radical $Na^{+}COO-$ in single crystals.

is shown in Figure A2-2.12. The orientation of free radical $Na^{+}COO-$ in single crystals is shown in Figure A2-2.13.

3. ENDOR Spectrum of Crystal

In the beginning, ENDOR experiments were carried out in single-crystal samples. Some powder samples and liquid samples were used later, and since then tremendous progress has been made in ENDOR experiments . In solids, hyperfine coupling parameter $\boldsymbol{A}$ is a tensor that is strongly dependent on the orientation of the sample in the magnetic field. Because ν_{ENDOR} is dependent on $\boldsymbol{A}$, the variation of ν_{ENDOR} also change with the orientation of the sample. The principal axis coordinate system of $\boldsymbol{A}$ is X, Y, Z, the magnetic field is parallel to the Z-direction, and the direction cosine is $(l, m, 0)$, for the $S = I = 1/2$ system,

$$E = M_S \nu_e \pm \frac{1}{2}[\boldsymbol{l}^2(M_S A_{XX} - \nu_n)^2 + \boldsymbol{m}^2(M_S A_{YY} - \nu_n)^2]^{1/2} \tag{A-2.18}$$

According to the selection rule of ENDOR, $\Delta M_S = 0$, $\Delta M_I = \pm 1$, we obtain

$$\nu_{\text{ENDOR}} = [l^2(M_S A_{XX} - \nu_n)^2 + m^2(M_S A_{YY} - \nu_n)^2]^{1/2} \tag{A-2.19}$$

It is obvious that ν_{ENDOR} depends on the value of l, m. When $H//X$ and $H//Y$,

$$\nu_{\text{ENDOR}} = \nu_n \pm \frac{1}{2} A_{ii}(i = \boldsymbol{X}, \boldsymbol{Y}) \tag{A-2.20}$$

In the equation, A_{ii} is the matrix element of the $\boldsymbol{A}$ tensor. That is to say, the matrix elements of the $\boldsymbol{A}$ tensor can be obtained using the ν_{ENDOR}. But it should be pointed out that the ENDOR spectrum cannot be treated by the above first-order approximation, although it may be approximated to second correcting terms; even advanced approximation can be considered, by using digital diagonalization strictly. In the case of axial symmetry, applying the Bleaney formula by using second-order perturbation, we obtain

$$\begin{aligned} E\left(\left|\frac{1}{2}, M_I\right\rangle\right) &= \frac{1}{2} g\beta H + \frac{1}{2} K M_I - G_I M_I - \frac{AB^2}{4g\beta HK} M_I \\ &+ \frac{B^2(A^2+K^2)}{8g\beta HK^2}[I(I+1) - M_I^2] + \frac{(A^2-B^2)^2}{4g\beta H}\left\{\frac{g_{\parallel} g_{\perp} \sin 2\theta}{2g^2 K}\right\}^2 M_I^2 \end{aligned} \tag{A-2.21}$$

$$\begin{aligned} E\left(\left|-\frac{1}{2}, M_I\right\rangle\right) &= \frac{1}{2} g\beta H - \frac{1}{2} K M_I - G_I M_I - \frac{AB^2}{4g\beta HK} M_I \\ &- \frac{B^2(A^2+K^2)}{8g\beta HK^2}[I(I+1) - M_I^2] - \frac{(A^2-B^2)^2}{4g\beta H}\left\{\frac{g_{\parallel} g_{\perp} \sin 2\theta}{2g^2 K}\right\}^2 M_I^2 \end{aligned} \tag{A-2.22}$$

If it is the EMR saturated transition $\left|\frac{1}{2}, M_I\right\rangle \leftrightarrow \left|-\frac{1}{2}, M_I\right\rangle$, then it has four ENDOR transitions:

$$\begin{aligned} &\text{Transition } a: && \left|\tfrac{1}{2}, M_I+1\right\rangle \leftrightarrow \left|\tfrac{1}{2}, M_I\right\rangle \\ &\text{Transition } b: && \left|\tfrac{1}{2}, M_I\right\rangle \leftrightarrow \left|\tfrac{1}{2}, M_I-1\right\rangle \\ &\text{Transition } c: && \left|-\tfrac{1}{2}, M_I+1\right\rangle \leftrightarrow \left|-\tfrac{1}{2}, M_I\right\rangle \\ &\text{Transition } d: && \left|-\tfrac{1}{2}, M_I\right\rangle \leftrightarrow \left|-\tfrac{1}{2}, M_I-1\right\rangle \end{aligned}$$

$$\begin{aligned} \nu_{\text{ENDOR}}^{a,b} &= G_I - \frac{1}{2} K + \frac{AB^2}{4g\beta HK} \\ &+ \left\{\frac{B^2(A^2+K^2)}{8g\beta HK} - \frac{(A^2-B^2)}{4g\beta HK}\left[\frac{g_{\parallel} g_{\perp} \sin 2\theta}{2g^2 K}\right]^2\right\}(2M_I \pm 1) \end{aligned} \tag{A-2.23}$$

$$\nu_{\text{ENDOR}}^{c,d} = G_I + \frac{1}{2}K + \frac{AB^2}{4g\beta HK} - \left\{ \frac{B^2(A^2+K^2)}{8g\beta HK} - \frac{(A^2-B^2)^2}{4g\beta HK}\left[\frac{g_{\parallel}g_{\perp}\sin 2\theta}{2g^2K}\right]^2 \right\}(2M_I \pm 1) \tag{A-2.24}$$

The $2M_I \pm 1$ in the above two equations, for ν_{ENDOR}^{a} and ν_{ENDOR}^{c}, ought to take the sign of (+), and for the ν_{ENDOR}^{b} and ν_{ENDOR}^{d} it ought to take the sign of (−). Let us illustrate why the second-order approximation at least should be considered.

The ENDOR spectrum of the Co^{2+} in MgO at 4.2 K is shown in Figure A2-2.14. With $I = 7/2$ of ^{59}Co nucleus, the EMR spectral line of experimental saturation transition is carried out at $\Delta M_S = \pm 1$, and $M_I = +\frac{1}{2}$ with $\nu = 9.563\,GHz$, $g = 4.280$, $H = 156.1$ mT; there are four ENDOR lines in the figure, attributing line 1 to $|\frac{1}{2}, \frac{3}{2}\rangle \to |\frac{1}{2}, \frac{1}{2}\rangle$, line 2 to $|\frac{1}{2}, \frac{1}{2}\rangle \to |\frac{1}{2}, -\frac{1}{2}\rangle$, line 3 to $|-\frac{1}{2}, \frac{3}{2}\rangle \to |-\frac{1}{2}, \frac{1}{2}\rangle$, and line 4 to $|-\frac{1}{2}, \frac{1}{2}\rangle \to |-\frac{1}{2}, -\frac{1}{2}\rangle$. Experiments prove that the tensors g and A of this system are isotropic basically. For the isotropic g and A, there are

$$G_I = \nu_n, \quad K = A = B = a, \quad g\beta H = \nu_e$$

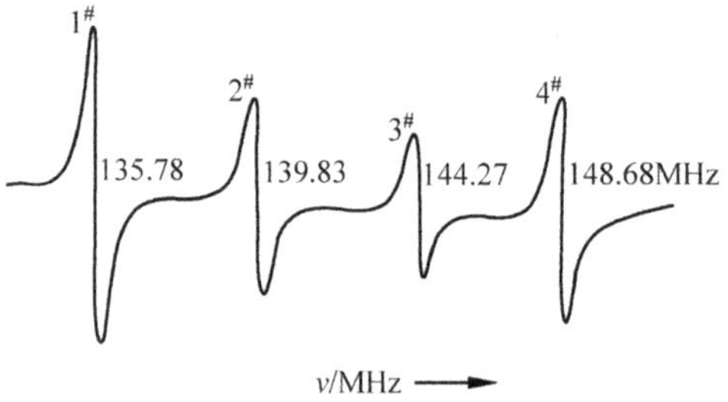

Figure A2-2.14: The ENDOR spectrum of Co^{2+} in the MgO under 4.2 K[13].

Equations A-2.23 and A-2.24 can be reduced to

$$\nu_{\text{ENDOR}}^{a} = \nu_n - \frac{1}{2}a + \frac{a^2}{4\nu_e}(2M_I + 2)$$

or

$$\nu_{\text{ENDOR}}^{a} = \frac{1}{2}a - \nu_n - \frac{a^2}{4\nu_e}(2M_I + 2) \tag{A-2.25a}$$

$$\nu_{\text{ENDOR}}^{b} = \nu_n - \frac{1}{2}a + \frac{a^2}{4\nu_e}(2M_I)$$

or

$$\nu_{\text{ENDOR}}^{a} = \frac{1}{2}a - \nu_n - \frac{a^2}{4\nu_e}(2M_I) \tag{A-2.25b}$$

$$\nu_{\text{ENDOR}}^{c} = \nu_n + \frac{1}{2}a - \frac{a^2}{4\nu_e}(2M_I) \tag{A-2.25c}$$

$$\nu_{\text{ENDOR}}^{d} = \nu_n + \frac{1}{2}a + \frac{a^2}{4\nu_e}(2 - 2M_I) \tag{A-2.25d}$$

We find that the frequency of ENDOR depends on M_I, that is,, on the saturated EMR spectral line:

$$\left|\frac{1}{2}, M_I\right\rangle \leftrightarrow \left|-\frac{1}{2}, M_I\right\rangle$$

In this experiment, that is transition of $\left|\frac{1}{2}, \frac{1}{2}\right\rangle \leftrightarrow \left|-\frac{1}{2}, -\frac{1}{2}\right\rangle$. Therefore, substituting $M_I = +\frac{1}{2}$ into Eq. (A-2.18), we obtain

$$\nu_{\text{ENDOR}}^{a} = \pm\left(\nu_n - \frac{1}{2}a + 3x\right) \tag{A-2.26a}$$

$$\nu_{\text{ENDOR}}^{b} = \pm\left(\nu_n - \frac{1}{2}a + x\right) \tag{A-2.26b}$$

$$\nu_{\text{ENDOR}}^{c} = \nu_n + \frac{1}{2}a - x \tag{A-2.26c}$$

$$\nu_{\text{ENDOR}}^{d} = \nu_n + \frac{1}{2}a + x \tag{A-2.26d}$$

Here $x = a^2/4\nu_e$. This is the result of taking into account the second-order approximation. If only the first-order approximation is considered, then

$$\nu_{\text{ENDOR}}^{a} = \nu_{\text{ENDOR}}^{b}, \quad \nu_{\text{ENDOR}}^{c} = \nu_{\text{ENDOR}}^{d}$$

This implies that there are only two ENDOR lines, but the experimental result is four lines. So it gives a corrective interpretation for the experimental results, the second-order approximation should be taken into account. The question arises: How to choose the sign of "±" in Eqs. (A-2.26a) and (A-2.26b)? When $\frac{1}{2}a < \nu_n$, it ought to take "+" sign. On the contrary, when $\frac{1}{2}a > \nu_n$, it should take the sign of "–".

According to Eq. (A-2.26), the attribution of these four lines of the ENDOR spectrum could be as shown in Figure A2-2.14. ν_n, a, and x could be determined from Eqs. A-2.26b, A-2.26c, and A-2.26d, and then ν_{ENDOR}^{a} will be calculated by substituting it into Eq. (A-2.26a), and the results are shown in Table A2-2.6. From the table it can be known that the experimental values of ν_{ENDOR}^{a} agree well with the calculated values, and the hyperfine coupling parameter is $a = 288.51$ MHz.

The ENDOR method is most successfully applied to the study of envelope spectral line that arises heterogeneously from the broadening of the EMR spectral lines that are formed by the overlapping of a large number of hyperfine spectral

Table A2-2.6: ν^i_{ENDOR} of the ENDOR spectral lines of Co^{2+} in MgO.

No.	ν^i_{ENDOR}	Experimental value (MHz)	Calculated value (MHz)[a]
1	ν^a_{ENDOR}	135.78	135.44
2	ν^b_{ENDOR}	139.83	139.84
3	ν^c_{ENDOR}	144.27	144.28
4	ν^d_{ENDOR}	148.68	148.68

[a]Calculated values obtained by using second-order approximation.

lines. For example, the F center of KBr, the total linewidth of its EMR spectral line (Gaussian type) is 12.5 mT. The six neighbors of first shell in the face-centered structure are either ^{39}K (natural abundance 93.08%) or ^{41}K (natural abundance 6.91%). These nuclei also exist in the third, fifth, and ninth shells. The second, fourth, sixth, and eighth shells also include the nuclei of ^{85}Br (abundance 50.69%) and ^{87}Br (abundance 49.31%). All of these nuclear spin are $I = 3/2$. The unpaired electrons carry out hyperfine interaction with all these nuclei and produce much more hyperfine structural lines. Assume that the differences of nuclear moment of the above-mentioned four nuclei as well as the differences of the anisotropics of hyperfine interaction are all neglected, and the hyperfine interaction of the six magnetic nuclei of the first shell and of the 12 magnetic nuclei of the second shell are only considered. Then, $\prod_i(2n_iI+1) = 19 \times 37 = 703$ lines of hyperfine interaction would be produced. Of course, the intensities of some of these lines are very weak. As an example, there are 19 hyperfine structural lines arising from the nuclei of first shell, and the intensity of the lowest line only has 1/580 of the central strongest line. However, the practical situation is much more complicated. Consequently, the EMR spectrum of the *F*-center of the KBr could only be an envelope line of 12.5 mT width, and could not provide any useful information for us.

With the method of ENDOR, the ENDOR spectral line of fine resolution could be obtained, as shown in Figure A2-2.15. It can be seen from Figure A2-2.15 that there are plentiful ENDOR spectral lines in the range of 0.5–28 MHz. Despite the considerable difference in their linewidth, the linewidths are very narrow, especially in the region of 3–4 MHz, the narrowest lines being only 10 kHz. Such narrow linewidth can resolve a pair of $\nu_{ENDOR} = \nu_n \mp a/2$ spectral line. In Figure A2-2.15,the numerals (I–VIII) of the nuclei express the nucleus on the various shells. The triplet line in the range of 8–12 MHz expresses the nuclear quadrupolar interaction: $Q = 0.054$ (^{39}K), 0.060 (^{41}K), 0.29 (^{79}Br), 0.27 (^{81}Br) $\times 10^{-28}$ m^2. The direction of the magnetic field in Figure A2-2.15 is parallel to the face of [1 0 0], and a set of ENDOR spectral lines can also be obtained from other directions; the hyperfine coupling parameters change with the change in the angle.

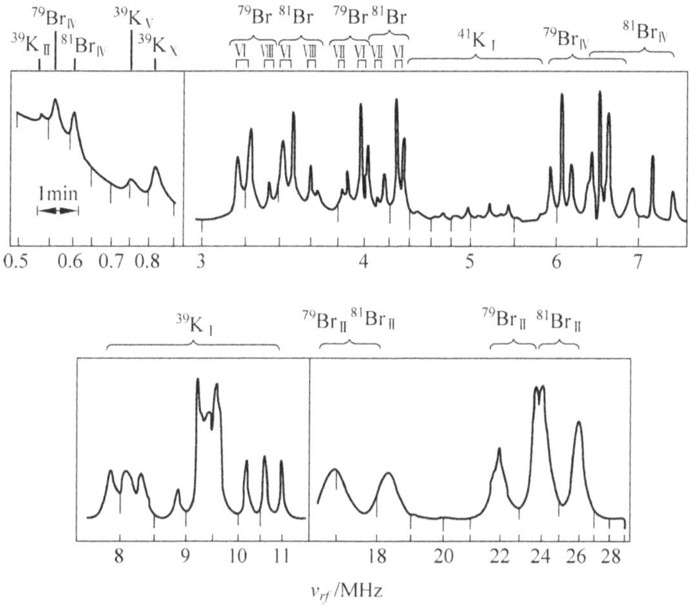

Figure A2-2.15: The ENDOR spectrum [14,15] of F-center in the KBr at 90 K, $H \parallel \langle 100 \rangle$.

The spin-Hamiltonian operator containing the nuclear quadrupole interaction [16,17] is

$$\hat{\mathscr{H}}_Q = \hat{\boldsymbol{I}} \cdot \boldsymbol{Q} \cdot \hat{\boldsymbol{I}} \tag{A-2.27}$$

This equation is correct only when $I \geq 1$; the $\boldsymbol{Q}$ matrix of the equation is

$$\boldsymbol{Q} = Q \begin{pmatrix} \eta - 1 & 0 & 0 \\ 0 & -\eta - 1 & 0 \\ 0 & 0 & 2 \end{pmatrix} \tag{A-2.28}$$

In case of axial symmetry$(\eta = 0)$, Eq. (A-2.27) can be written as follows:

$$\hat{\mathscr{H}}_Q = Q[3\hat{I}_z^2 - I(I+1)] \tag{A-2.29}$$

If both g and $\boldsymbol{A}$ are very close to isotropic, in the suitable external magnetic field, using first-order perturbation theory we can obtain

$$h\nu_Q = 3\left|Q[3\cos^2\theta - 1]\,(M_I - 1/2)\right| \tag{A-2.30}$$

where θ is the angle between $\boldsymbol{H}$ and $\boldsymbol{Z}$ and $\Delta M_I = \pm 1$, including the nuclear Zeeman term and the hyperfine interaction term; the ENDOR frequencies are given as follows:

$$\nu_{n1} = \left| + \frac{1}{2}[A_\parallel + A_\perp(3\cos^2\theta - 1)] - g_n\beta_n H + 3Q(3\cos^2\theta - 1)\left(M_I - \frac{1}{2}\right)\right| / h \tag{A-2.31a}$$

$$\nu_{n2} = \left| -\frac{1}{2}[A_{\|} + A_{\perp}(3\cos^2\theta - 1)] - g_n\beta_n H + 3Q(3\cos^2\theta - 1)\left(M_I - \frac{1}{2}\right)\right| / h \quad \text{(A-2.31b)}$$

Here A and Q are taken coaxial. The energy level $I = 3/2$ and the ENDOR spectrumare shown in Figure A2-2.16. The hyperfine and some quadrupole couplings of various shells for KBr are given in Table A2-2.7.

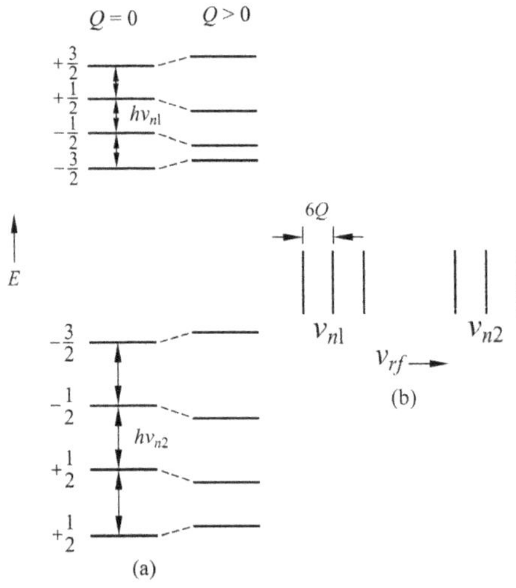

Figure A2-2.16: The level splitting of I = 3/2 system and the stick spectrum of ENDOR.

Table A2-2.7: The parameters of hyperfine and nuclear quadrupole interaction of KBr are detected by ENDOR at 90 K (MHz).

Shell	Nucleus	$A_{\|}/h$	$A_{\perp}/h$	Q/h
1	^{39}K	18.33	0.77	0.067
2	^{81}Br	42.85	2.81	0.077
3	^{39}K	0.27	0.022	
4	^{81}Br	5.70	0.41	0.035
5	^{39}K	0.16	0.021	
6	^{81}Br	0.84	0.086	
7	^{81}Br	0.54	0.07	

4. ENDOR in Powders and Noncrystalline Solids

As everyone knows, to get a single-crystal sample is not easy. In practical circumstances, the orientation of vast majority of solid samples is irregular (polycrystalline or powder). In some cases, the ENDOR spectra obtained from the powder samples are probably very similar to "single-crystal-type ENDOR spectra" obtained from the single-crystal samples. It is the ENDOR operation, fixed in some resonant magnetic field, increasing of the microwave power until saturation, and the sweep frequency of NMR that make the electron nuclear double resonance. If every small part of the powder sample corresponds to a unitary orientation of the paramagnetic molecule, the ENDOR spectrum of similar single-crystal samples can be obtained. The key is how to choose the saturation turning point of the EMR spectral line.

The Cu $(Pic)_2$ doped into $Zn(Pic)_2 \cdot 4H_2O$ had been prepared by Rist and Hyde[18] using coprecipitation method, and the second-order differential EMR spectrum is shown in Figure A2-2.17. It is a d^9planar molecule whose g and $\boldsymbol{A}$ tensors of Cu are axisymmetric, and the symmetrical axis is perpendicular to the plane of the complex molecule. The components of g and $\boldsymbol{A}$ tensors along the direction of symmetrical axis are much more larger than those on the complex plane, so the distribution of EMR spectrum is very wide at the low-field part, while at the high field $(g_\perp)$ part of the spectrum is very compact. If the fixed magnetic field is at the a point of Figure A2-2.17 and the microwave power increases to saturation, then the spectrum of Figure A2-2.18 would be the ENDOR spectrum of "single crystal type." The powder samples of the same series planar complexes are also prepared by them, and the parameters of hyperfine coupling and quadrupole coupling of N nuclei detected by using the method of ENDOR are listed in Table A2-2.8.

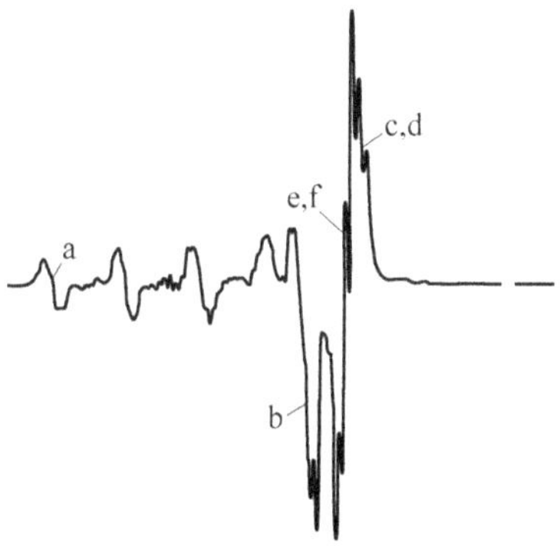

Figure A2-2.17: The second-order differential EMR spectrum of $Cu(Pic)_2$ doped in $Zn(Pic)_2 \cdot 4H_2O$ of powder sample.

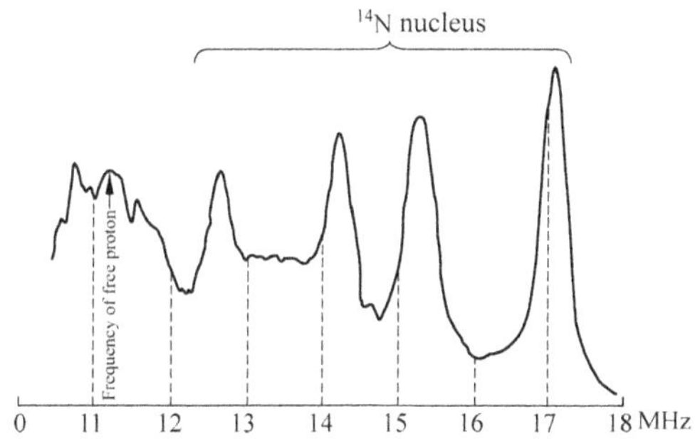

Figure A2-2.18: The single-crystal-type ENDOR spectrum of $Cu(Pic)_2$ doped in $Zn(Pic)_2 \cdot 4H_2O$.

Table A2-2.8: The hyperfine couple parameters of nitrogen nucleus (MHz)[a].

Complex	Basic substance	Parameters of hyperfine coupling	Parameters of quadrupole coupling	
			Measured value	Estimated value[b]
$Cu(Ox)_2$	Phthalimide	30.3	0.98	2.6
	$Zn(Ox)_2 \cdot 2H_2O$	27.5	1.34	3.6
$Cu(Pic)_2$	Phthalic acid	36.9	0.75	2.0
	$Zn(Pic)_2 \cdot 4H_2O$	29.7	1.35	3.6
$Cu(Qn)_2$	$Zn(Qn)_2 \cdot xH_2O$	29.0	0.95	2.5
$CuMe(Pic)_2$	$ZnMe(Pic)_2$	35.1	0.87	2.3
$Cu(Sal)_2$	$Pd(Sal)_2$	43.8	0.2	
$Cu(Dim)_2$	$Ni(Dim)_2$	47.2	0.57	
$Ag(Pic)_2$	$Zn(Pic)_2 \cdot 4H_2O$	50.1	1.2	
	Phthalic acid	~51		

a The field is perpendicular to the plane of complex.
b Assuming it is axial symmetrical, (Ox): 8-hydroxyquinoline; (Pic): picoline; (Qn): quinoline; Me(Pic): α-methyl picoline; (Dim):diketone oxime; (Sal): salicylaldoxime.

The typical ENDOR spectrum of the ^{14}N nucleus ($I = 1$) should be four lines, and the first-order approximation of the ENDOR frequency is

$$\nu_{ENDOR} = \left| \frac{A^N}{2} \pm \nu_n \pm Q^N \right| \tag{A-2.32}$$

Here $A^N$is the hyperfine coupling parameter of the ^{14}N; $\nu_n$is the Zeeman transition frequency of the ^{14}N nucleus; $Q^N$is the quadrupole coupling parameter of the ^{14}N. From Figure A2-2.18, the resonance frequency of the free proton is 11.2 MHz. Thus,

$$\nu_n \simeq (\gamma_n/\gamma_H)\nu_H = \frac{0.19324}{2.67510} \times 11.2 = 0.81\ (\text{MHz})$$

From Table A2-2.8, for the $Cu(Pic)_2$ in the $Zn(Pic)_2 \cdot 4H_2O$ substance, substitute $A^N = 29.7$ MHz; $Q^N = 1.35$ MHz into Eq. (A-2.32), and the frequencies of four ENDOR spectral lines of ^{14}N are calculated as follows:

$$\nu_{ENDOR} \frac{1}{2}(29.7) \pm 1.35 \pm 0.81 = 17.01; \quad 15.39; \quad 14.31, \quad \text{and} \quad 12.69\ \text{MHz}$$

These results are coincident with the experimental results (Figure A2-2.18) as well.

Let us look at Figure A2-2.17 again. Fixing the magnetic fields at *b*, *c*, *d*, *e*, and *f*, respectively, and increasing the microwave power up to saturation, a set of ENDOR lines can be obtained, as shown in Figure A2-2.19. *b*– *f* all are the ENDOR spectral

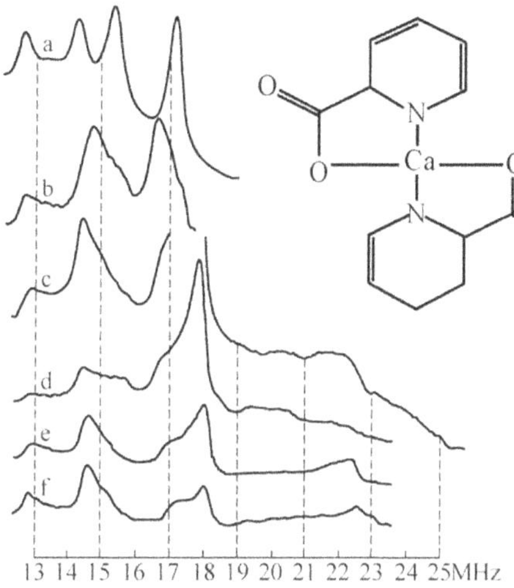

Figure A2-2.19: The ENDOR spectra of different magnetic field positions.

superposition of many different oriented molecules, which is called ENDOR spectrum of "powder type." Compared with the powder EMR spectrum, it still has the following advantages:

1. Powder ENDOR spectrum still contain information on the nuclear quadrupole coupling, which cannot be obtained from the EMR spectrum of the first-order approximation.
2. It is easy to distinguish the different nuclear coupling, such as the resonance of proton and nitrogen appear on the different frequency regions of the ENDOR spectrum.
3. There is no sum, difference, and linear combination of hyperfine lines in ENDOR spectrum, so it is more easy to analyze the ENDOR spectra, even for the powder type.

The ENDOR spectrum of another kind of solid sample of irregular orientation, such as bis-(2-hydroxyl -acytophenone-oxime)^{63}Cu complex (^{63}Cu-HAP) dissolved in DMSO/EtOH (5/1) solvent that freeze to 20K, mainly the ENDOR spectra [19–21] of ^{14}N and ^{15}N nuclei, has been studied to determine the superhyperfine coupling parameter caused by ^{14}N and ^{15}N nuclei and the quadrupole coupling parameter of ^{14}N nucleus. According to the selection rule $\Delta M_S = 0$, $\Delta M_I = \pm 1$, the first-order-approximated ENDOR spectrum will be composed of two groups of transition. Owing to the interaction of the ^{14}N nuclear quadrupole, splitting is caused again as shown in Figure A2-2.20. In the ENDOR spectrum of four lines, with $I = 1$, the central positions of four lines should be the half site of the hyperfine coupling parameter as shown in Figure A2-2.21. The transition frequency of first-order approximation can be simplified [22] to

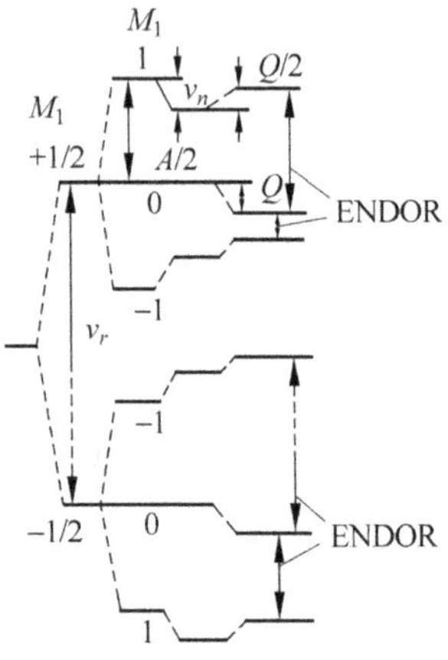

Figure A2-2.20: The level splitting diagram of $\Delta M_S = 0$, c = ± 1ENDOR spectrum of the first-order approximation.

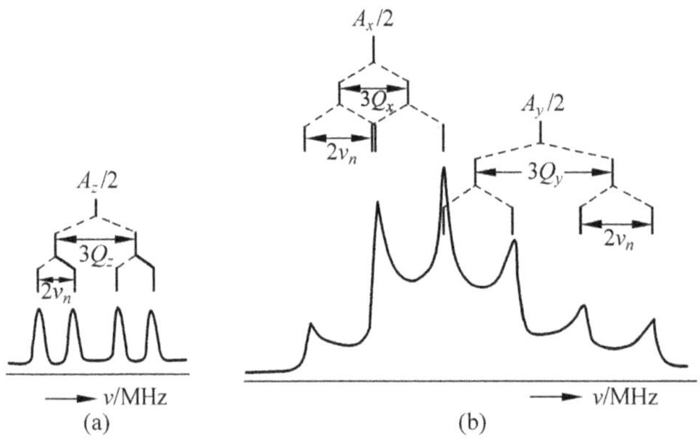

Figure A2-2.21: Typical ENDOR spectra: (a) single-crystal-type spectrum; (b) powder-type spectrum.

$$\nu_{\text{ENDOR}} = |(A_i/2) \pm (3Q_i/2) \pm \nu_n| \quad \text{(A-2.33)}$$

Here A_i, Q are the principal values of $\boldsymbol{A}$ and $\boldsymbol{Q}$ tensors along the i-axis ($i = x, y, z$); ν_n is the Zeeman transition frequency of the coordinate nucleus. When $(3Q/2) \approx \nu_n$, the two peaks in the middle may overlap, which can only observe three peaks, as shown in Figure A2-2.21b, and in some measurement, due to the broad spectral line, the line shape of the total signal will appear asymmetric.

For the nucleus of $I = 1/2$,

$$\nu_{\text{ENDOR}} = |(A_i/2) \pm \nu_n| \quad \text{(A-2.34)}$$

Assuming the proton ENDOR spectrum is as given above and $(A_i/2) \ll \nu_n$, the ENDOR peaks of proton on the spectrum are symmetrical disposition usually for the free proton peak. The improved method [18] of directional selective determination had been applied by authors [19]. The ENDOR spectrum is measured at the corresponding

g_ρ of EMR spectrum, in the direction parallel to external magnetic field $\boldsymbol{H}$, and with the g_ρ orientation corresponding to molecules. The measured spectrum is the "single-crystal-type" ENDOR spectrum shown in Figure A2-2.21a. The ENDOR spectrum is measured at the corresponding $g_\perp$ of EMR spectrum, in the direction perpendicular to the external magnetic field $\boldsymbol{H}$, and with the $g_\perp$ orientation corresponding to molecules. All the molecules on the x–y plane are contributions to the ENDOR spectrum, the measured spectrum is the "powder-type" ENDOR spectrum shown in Figure A2-2.21b.

The EMR spectrum of ^{14}N-^{63}Cu-HAP in DMSO/EtOH (5:1) solvent was detected at 20 K, as shown in Figure A2-2.22. The arrow in the figure indicates the position of the fixed magnetic field for the determination of the ENDOR spectrum – *a*: 257.6 mT, *b*: 276.4 mT, *c*: 304.8 mT, *d*: 311.1 mT, *e*: 315.2 mT, *f*: 319.5 mT. The corresponding ENDOR spectrum is shown in Figure A2-2.23.

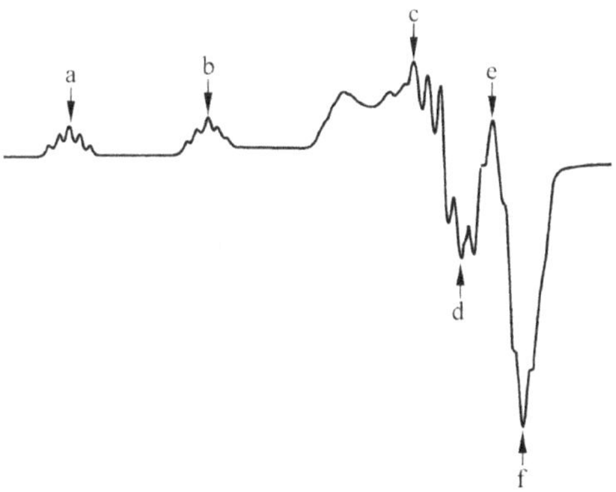

Figure A2-2.22: The EMR spectrum of ^{14}N-^{63}Cu-HAP at 20 K frozen solution.

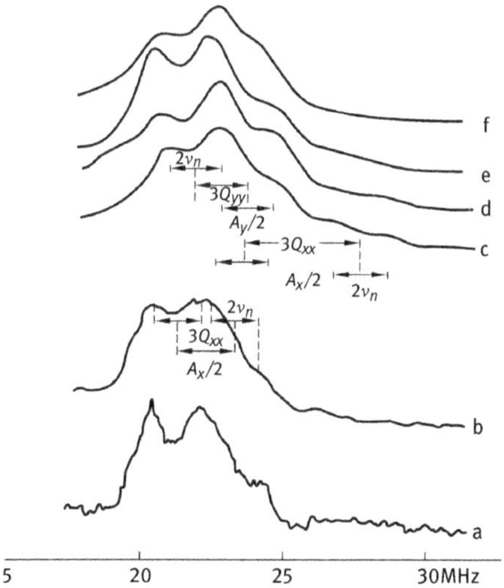

Figure A2-2.23: The ENDOR spectra of ^{14}N at 20 K frozen solution of ^{14}N-^{63}Cu-HAP.

Here we must pay attention to the concept of coordinate: When the external magnetic field $\boldsymbol{H}$ is parallel to the $\boldsymbol{z}$-axis, the hyperfine coupling parameters of Cu is ${}^{Cu}A_{\parallel}$, and superhyperfine coupling parameter for N nuclear is ${}^{N}A_{\perp}$; when the external magnetic field $\boldsymbol{H}$ is parallel to the $\boldsymbol{y}$-axis, the hyperfine coupling parameters of Cu is ${}^{Cu}A_{\perp}$, and the superhyperfine coupling parameter for N nuclear should be ${}^{N}A_{\perp}$; when the external magnetic field $\boldsymbol{H}$ is parallel to the $\boldsymbol{x}$-axis, the hyperfine coupling parameters of Cu is ${}^{Cu}A_{\perp}$, and for superhyperfine coupling parameter of N nuclear is ${}^{N}A_{\parallel}$. By using computer simulation, ${}^{\mathrm{Cu}}A_{\parallel} = -588\,\mathrm{MHz}$ and ${}^{\mathrm{Cu}}A_{\perp} = -74\,\mathrm{MHz}$ are obtained. From Figure 12.23, ${}^{\mathrm{N}}A_{\parallel} = 51\,\mathrm{MHz}$ and ${}^{\mathrm{N}}A_{\perp} = 44.8\,\mathrm{MHz}$.

Spin-Hamiltonian of the quadrupole interaction of the ^{14}N nucleus is

$$\hat{\mathscr{H}}_{\mathrm{Q}} = \frac{\mathrm{e}^2 qQ}{4I(I-1)h}\left[(3I_z^2 - I^2) + \eta(I_x^2 - I_y^2)\right] \tag{A-2.35}$$

Here $eq = V_{zz}$ is the electric field gradient tensor of the nucleus site; η is the asymmetry parameter, $\eta = (V_{xx} - V_{yy})/V_{zz}$, usually the $0 < \eta < 1$. In molecules, the electric field gradient (EFG) of the ^{14}N nucleus is defined as follows:

$$eq = -\mathrm{e}\sum_j \left\langle \psi_j \left| \frac{3\cos^2\theta - 1}{r^3} \right| \psi_j \right\rangle n_j + e\sum_{K \neq N} Z_K \left(\frac{3\cos^2\theta_K - 1}{R_K^3} \right) \tag{A-2.36}$$

Here the first term represents the electron contribution to the electric field gradient tensor, ψ_j is the occupied molecular orbital, n_j is the electron number of the j orbit; and the second term represents the K nucleus of the charged Z_K contribution to the electric field gradient tensor. R_K represents the distance from the electron to the center of the K nucleus, and $K \neq N$ denotes the other nuclei except the ^{14}N nucleus.

A nuclear quadrupole interaction is used to measure the extent of the deviation of the electric field gradient from the cubic symmetry of the neighboring nucleus, which is related to the distribution of all electrons (including valence electrons and atomic core electrons). The intensity of the Cu–N bond is directly represented by the quadrupole moment coupling parameter rather than the superhyperfine coupling parameter [18].

Assuming the principal axis of the quadrupole tensor of the ^{14}N nucleus agrees with the principal axis of the superhyperfine coupling tensor, the nuclear quadrupole moment is a trace tensor, that is, $Q_{xx} + Q_{yy} + Q_{zz} = 0$. As shown in Figure A2-2.23, ${}^{\mathrm{N}}Q_{xx} = -1.38$ MHz, ${}^{\mathrm{N}}Q_{yy} = 0.62$ MHz, ${}^{\mathrm{N}}Q_{zz} = 0.76$ MHz. It can be seen that the maximum principal value direction of the quadrupole moment tensor of the ^{14}N nucleus is directed to the Cu–N bond, indicating that the main axis of the electric field gradient is determined by the lone pair electron orbit of N, but the sign is opposite to the superhyperfine coupling parameter. According to the degree of deviation of the quadrupole tensor from the axial symmetry, it is estimated [23] as $\eta \approx 0.1$.

The proton ENDOR spectrum of the ^{14}N-^{63}Cu-HAP complex is shown in Figure A2-2.24. Free proton frequency $\nu_H = 12.8$ MHz is obtained from the figure. In the low-frequency region, there are three sets of peaks arranged symmetrically around the free proton peak. Their frequencies are 2.7, 5.8, and 8.0 MHz. This implies that there are three different equivalent protons in the ^{14}N-^{63}Cu-HAP complex. In fact, there should be four sets of different equivalent protons, and the other set of peak is overlapped by the peak of ^{14}N nucleus without identifying, which has been confirmed in ^{14}N-^{63}Cu-HAP complexes [20].

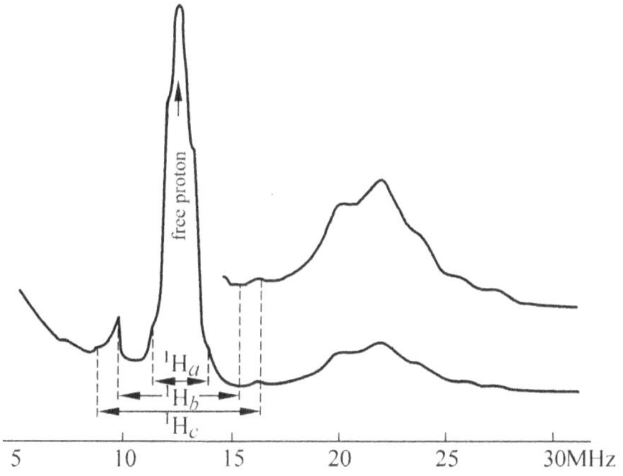

Figure A2.-2.24: The ENDOR spectra of ^1H at 20 K frozen solution of ^{14}N-^{63}Cu-HAP.

A2-2.2 Pulse ENDOR

Unlike with CW-ENDOR, the RF field is pulsed rather than continuous. In the pulse ENDOR spectrometer, all advantages of the time-resolved spectroscopy are applicable in ENDOR. The modern ENDOR technique can apply various radio frequency pulse sequences for understanding and recognizing the structure and relaxation properties of unpaired electron particles. These techniques include Polarization-Modulated ENDOR, "Double ENDOR" of two kinds of RF frequency, "Stochastic ENDOR," "Multiple-Quantum ENDOR," and "ENDOR-induced EMR. But these are beyond the scope of this book. For details, please to refer to Refs. [22, 24–31].

A2-2.3 Extension of ENDOR

Pulse-ENDOR is similar to CW-ENDOR. Besides "Double-ENDOR," and "ENDOR-Induced-EMR" (EI-EMR), "Circularly Polarized ENDOR" (CP-ENDOR), and "Polarization Modulated ENDOR" (PM-ENDOR) are also available [32]. The brief introduction is as follows:

1. Double-ENDOR

The paramagnetic system contains two equivalent magnetic nucleus such as $S = 1/2$, $I_1 = I_2 = 1/2$, which is the most simple one of this kind of system. To measure the hyperfine coupling constants of A_1 and A_2 simultaneously, two sheath mutually perpendicular RF field should be added to the perpendicular direction of the external magnetic field [33]. Two sets of doublet ENDOR spectrum can be obtained by using the method described above. This is called the Double-ENDOR spectrum.

The Hamiltonian operator of the system is described as follows:

$$\hat{\mathcal{H}} = g\beta \boldsymbol{HS} - g_n\beta_n \boldsymbol{HI} + \boldsymbol{A}_1\boldsymbol{I}_1\boldsymbol{S} + \boldsymbol{A}_2\boldsymbol{I}_2\boldsymbol{S} \tag{A-2.37}$$

Here, $I = I_1 + I_2$, according to selection rule; there are four ENDOR lines:

$$\begin{aligned} \Delta M_{I_2} &= 1, \quad h\nu_{n_1} = g_n\beta_n H - A_2/2 \\ \Delta M_{I_1} &= 1, \quad h\nu_{n_2} = g_n\beta_n H - A_2/2 \\ \Delta M_{I_1} &= 1, \quad h\nu_{n_3} = g_n\beta_n H + A_2/2 \\ \Delta M_{I_2} &= 1, \quad h\nu_{n_4} = g_n\beta_n H + A_2/2 \end{aligned}$$

2. ENDOR-Induced EMR (EI-EMR)

The ENDOR-induced EMR spectrum is a complement for the ENDOR spectra. From this, we can observe whole EMR spectrum with additional special ENDOR transition. Schweiger observed [34] that EI-EMR is a single ENDOR transition, expressed as a function of an external magnetic field rather than of the radio frequency field. Thus, it is an excellent method of peeling overlapped spectral lines. Figure A2-2.25a is the EMR spectrum [35] of Cu(acacen) doped in Ni(acacen) single crystal. Here, the

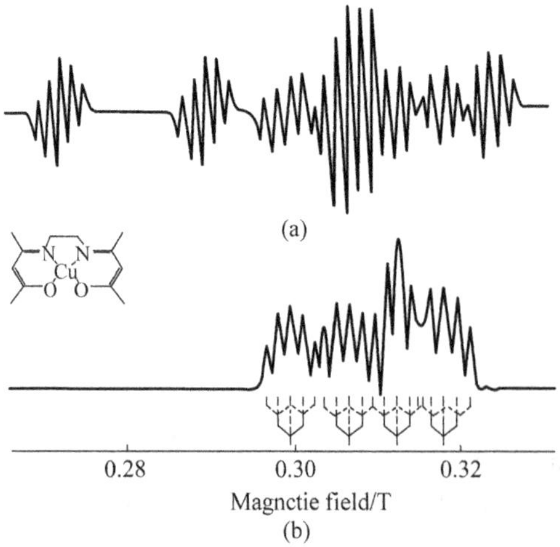

Figure A2-2.25: The EMR spectrum (a) and IE-EPR spectrum (b) of Cu(acacen) doped in Ni(acacen)single crystal [34].

(acacen) is ethylenediamine-2-acetylacetone. They are four sets of vertical and parallel, each having five super-hyperfine structural lines, but the vertical four sets overlap completely with the parallel third set and to distinguish them is very difficult. Figure A2-2.25b shows the vertical part of EI-EPR spectrum [34]; there are four sets of spectral lines, five superhyperfine structural lines of each set are resolved clearly. Figure A2-2.26a [35] is an EMR powder spectrum of $V(bz)_2$ diluted in ferrocene powder. Figure A2-2.26b –e is the single crystal EI-EMR spectrum of the powder sample. The value of $\Delta\nu$ is close to ${}^{H}A_{max}/2$, such as (b) $\Delta\nu = 8.50$ Mz, (c) $\Delta\nu = 8.70$ Mz, (d) $\Delta\nu = 8.75$ Mz, (e) $\Delta\nu = 8.80$ Mz. Figure A2-2.26c shows that the eight hyperfine structural lines of the ${}^{51}V$ nucleus ($I = 7/2$) are distinguishable clearly. This shows that not only the overlap spectral lines of the single crystal sample can be peeled off by EI-EMR spectrum but also could be done of the powder sample [35].

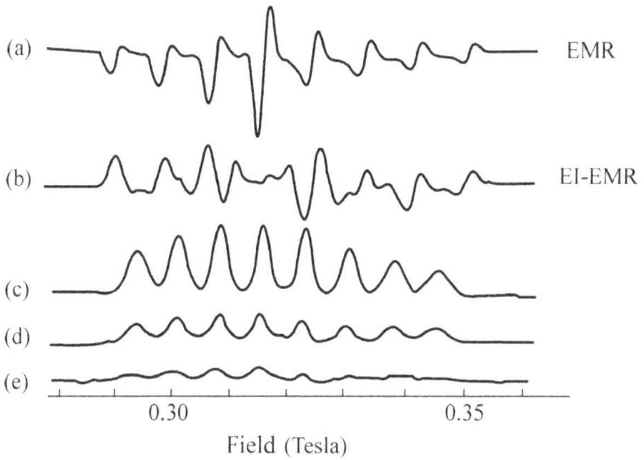

Figure A2-2.26: The EMR (a) and EI-EPR (b) ~ (e) spectra of $V(bz)_2$ diluted in ferrocene powder [35].

3. Circularly Polarized ENDOR (CP-ENDOR)

The RF field of circularly polarized ENDOR is the sum of two vectors of linearly polarized radio-frequency field:

$$\boldsymbol{H}_x(t) = \boldsymbol{iH}_{x0} \cos \omega t \tag{A-2.38a}$$

and

$$\boldsymbol{H}_y(t) = \boldsymbol{iH}_{y0} \cos(\omega t + \phi) \tag{A-2.38b}$$

$H_{x0} = H_{y0} = H_2$ corresponds to the left polarization, and $\phi = 90^\circ$ corresponds to the right polarization $\phi = -90^\circ$. These two linear polarization fields are produced by the RF current passing through two coils parallel to the external magnetic field $\boldsymbol{H}$, and these two pairs of coils form two half rings perpendicular to each other. The TE_{112} type cylindrical resonator is adopted. Figure A2-2.27a shows the ENDOR spectrum of Cu $(biyam)_2(ClO_4)_2$ doped in $Zn(biyam)_2(ClO_4)_2$ single crystals (sample is arbitrarily oriented at 20K), and measured by conventional linear polarized radio frequency fields; Figure A2-2.27b shows the CP-ENDOR spectrum under the same conditions. It can be seen that the latter is more simplified than the former.

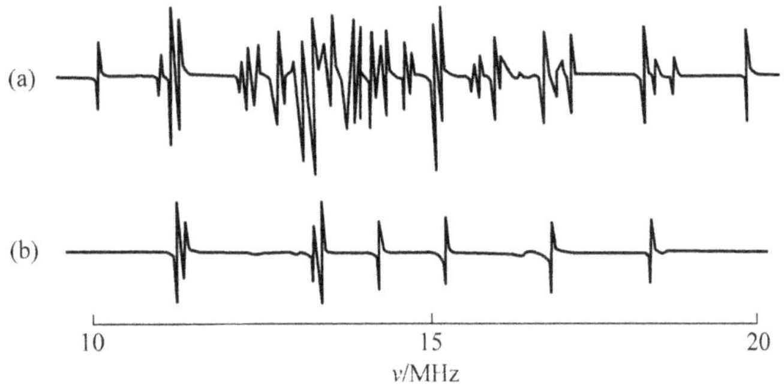

Figure A2-2.27: The conventional ENDOR spectrum of $Cu(biyam)_2(ClO_4)_2$ doped in $Zn(biyam)_2(ClO_4)_2$ single crystal (a); the CP-ENDOR spectrum of same sample (b) [36].

4. Polarization-Modulated ENDOR (PM-ENDOR)

In polarization modulation ENDOR spectroscopy, linear polarization RF field H_2 rotates with $\nu_r \ll \nu_m$ on the x–y plane of experimental coordinates. Here the ν_m is modulation frequency. The PM-ENDOR requires a smaller modulation frequency.

A2-2.4 Electron–electron double resonance (ELDOR)

The electron–electron double resonance (ELDOR) is different from electron–nuclear double resonance (ENDOR) and is: not added to the microwave field and an RF field to the direction perpendicular to the external magnetic field, while it is added to two microwave fields of different frequencies. Another difference is that the ENDOR, the "observing transition" and "drawing transition," has a share level, while the ELDOR does not have a share level. When the second hyperfine transition is saturated simultaneously, the intensity of the first hyperfine transition is observed to decrease [37]. For the same intensity of magnetic field (i.e., the intensity of magnetic field remains constant), in order to make two different transition frequencies, to produce electron spin resonance simultaneously, two different frequencies for microwave sources for irradiation should be required. To match two microwave sources of different frequencies, a "double model" cavity is required. The simplest situation is a single nucleus of spin 1/2, as illustrated in Figure A2-2.28. Although these two transitions have no common level, they can also be passed through the following two mechanisms:

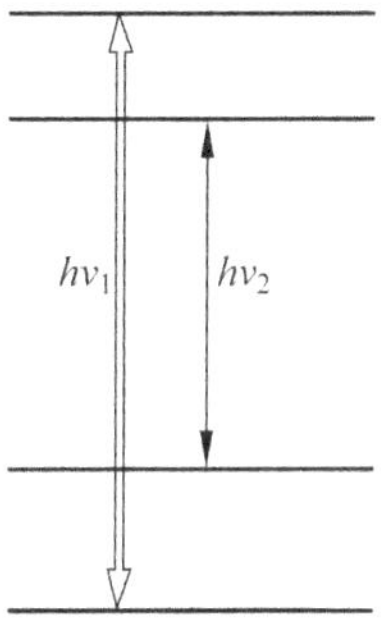

Fig. A2-2.28: The level transition.

(1) Electron dipolar coupling with nucleus-induced rapid nuclear relaxation. The spin flipping of electron can give rise to spin flipping of coupled nucleus at appropriate conditions. This mechanism is predominant at the case of the low concentration and low temperatures.

(2) Under the condition of high concentration or at sufficiently high temperature, spin exchange or chemical exchange tend to equalize the populations of all spin levels.

For the different relaxation mechanisms, the ELDOR technique is very sensitive. For example, it can distinguish the two neighboring nitrogen atoms, which have same magnitude of hyperfine splittings by using the difference of relaxation times of different seat of same nucleus in DPPH. The accurate value [38] of the hyperfine coupling parameter of ^{14}N nucleus can be obtained from this research. Both the spin relaxation time and the cross-relaxation time [39] between $M_I = \pm 1/2$ can be obtained by using the pulse ELDOR to study the CH fragment (arisen from irradiation malonic acid $CH(COOH)_2$)) ($S = I = 1/2$).

The two-dimensional ELDOR experiment can give valuable information about magnetization-transfer rates between nitroxide radicals [39], as well as about motional effects extending infinitively to these systems in the rigid limit region [40]. About the technique of two-dimensional ELDOR, please refer to Ref. [41].

A2-2.5 Optically detected magnetic resonance (ODMR)

The intensity of EMR transition is proportional to the difference ΔN of the two transition states $|M_S\rangle$ populations, that is, proportional to the spin polarization. Any other process of affecting ΔN can influence the intensity of EMR unless the spin–lattice relaxation is also very efficient. Under suitable circumstances, the intensity of optically induced transition is proportional to ΔN. There exists some link between electronic transition and EMR transition. Detection of this effect is not easy. To attain maximum value of ΔN, generally very low temperatures are required.

With the possibility of optically detected magnetic resonance (ODMR) experiment arise some of the specific unpaired electrons, which are characterized with broad optical bands. For reviews of ODMR, please refer to the Refs. [42–47].

Choose the naphthalene$^+$/naphthalene– in dilute solution (≥0.01M) at room temperature as an example[48] of ODMR. A fluorescence-dependent magnetic field (***H***) is detected from the free radical pairs, which are produced by electron irradiation and provide a very sensitive technique for observing the EMR spectrum of short life free radical triplet states.

A2-2.6 Fluorescence-detected magnetic resonance (FDMR)

In some circumstances, with higher sensitivity and time resolution, power can be obtained by using optical technique detection method than the ordinary (CW) EMR. It mainly uses the fluorescence that is produced from the recombination of short-lived free radicals in liquid or solid solution, to detect and display the EMR spectrum. For instance, the fluorescence-detected magnetic resonance (FDMR) can be utilized to observe original radical cations, [49] which is produced by the ion or photoion radiation (pulse sustain time 5–15 ns).

For example, the spin-correlated radical pair (cubane)$^+$ and (perdeuteroanthracene)$^-$ after 3MeV electron beam pulse ionization for some picoseconds recombination produce scintillant molecule [50] of excited singlet-state. At 190 K, the X-band FDMR spectrum is observed, displaying expected hyperfine structure of septet (1:6:15:20:15:6:1 with spacing 1.61 mT), which arise from the proton of the cubane$^+$ as shown in Figure A2-2.29.

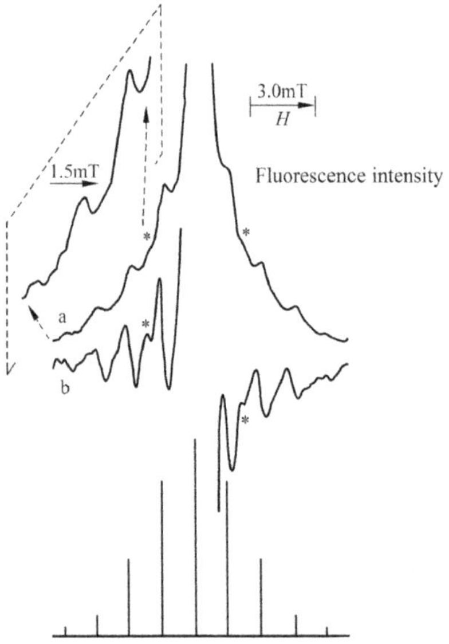

Figure A2-2.29: (a) The FDMR spectrum of (cubane)$^+$ in cyclopentane solvent (10^{-3} M) and (perdeuteroanthracene)$^-$ 10^{-4} M at 190 K.(b) is differential spectrum of (a) FDMR.

References

[1] Poole Jr.C P.*Electron Spin Resonance: A Comprehensive Treatise on Experimental Techniques*, 2nd edn, John Wiley & Sons, Inc., New York, NY, 1983.
[2] Feher G. *Phys. Rev.* 1956, **103**: 834.
[3] Feher G. *Phys. Rev.* 1959, **114**: 1219.
[4] Hyde J. S. *J. Chem. Phys.* 1965, **43**: 1806.
[5] Abragam A., Bleaney B. *Electron Paramagnetic Resonance of Transition Ions*, Oxford University Press, London, UK, 1970. p. 244.
[6] Davis E. R., Reddy T. Rs. *Phys. Lett.* 1970, **31A**: 398.
[7] Geschwind S. "Special topics in hyperfine structure in EPR," in Freeman A J, Frankel R B, Eds., *Hyperfine Interactions*, Academic Press, New York, NY, 1967, p. 225.
[8] Hyde J. S. *J. Chem. Phys.* 1965, **43**: 221.
[9] Hyde J. S., Maki A H. *J. Chem. Phys.* 1964, **40**: 3117.
[10] Coppinger G M. *J. Am. Chem. Soc.* 1957, **79**: 501.
[11] Steelink C., Fitzpatrick J. .D, Kispert L. D., Hyde J. S. *J. Am. Chem. Soc.* 1968, **90**: 4354.
[12] Cook R. J., Whiffen D. H. *J. Phys. Chem.* 1967, **71**: 93.
[13] Fry D. J. I., Llewellyn P M. *Proc. R. Soc. Lond.*, 1962, **A266**: 84.
[14] Seidel H. *Z. Phys.* 1961, **165**: 218, 239.
[15] Kevan L., Kispert L. D. *Electron Spin Double Resonance Spectroscopy*, John Wiley & Sons, Inc., New York, NY, 1976, Section 4.2.3.
[16] Drago R. S. *Physical Methods in Chemistry*, Saunders, Philadelphia, PA, 1977, Chapter 14.
[17] Slichter C. P. *Principles of Magnetic Resonance*, 3rd edn, Springer, New York, NY, 1990, Chapter 10.
[18] Rist G. H., Hyde J. S. *J. Chem. Phys.* 1970, **52**: 4633.
[19] Xu Y .Z., Chen D. Y., Cheng C. R., et al. Science in China. (B), 1992, 35(4): 337. Science in China. (B), 1992, 35(9): 1052.
[20] Chen D. Y., Xu Y. Z., et al. Acta Chimica Sinica., 1992, 50(8): 691.
[21] Yuanzhi, X., Deyu, C. *Appl. Magn. Reson.* 1996, **10**: 103.
[22] Schweiger A. *Struct. Bond.* 1982, **51**: 1.
[23] Rist G. H., Hyde J. S. *J. Chem. Phys.* 1969, **50**: 4532.
[24] Schweiger A. *Angew. Chem., Int. Ed. Engl.* 1991, **30**: 265.
[25] Gemperle C., Schweiger A. *Chem. Rev.* 1991, **91**: 1481.
[26] Grupp A., Mehring M. *Pulsed ENDOR Spectroscopy in Solids*, in, Kevan L, Bowman M K, Eds., *Modern Pulsed & Continuous-Wave Electron Spin Resonance*, John Wiley & Sons, Inc., New York, NY, 1990, Chapter 4, pp. 195ff.
[27] Thomann H., Bernardo M. *Spectrosc. Int. J.* 1990, **8**: 119.
[28] Thomann H., Mims W. B. *Pulsed Electron-Nuclear Spectroscopy and the Study of Metalloprotein Active Sites*, in Bagguley D M S, Ed. *Pulsed Magnetic Resonance: NMR, ESR and Optics*, Oxford University Press, Oxford, UK, 1992.
[29] Dinse K. P. "Pulsed ENDOR," in Hoff A J, Ed., *Advanced EPR: Applications in Biology & Biochemistry*, Elsevier, Amsterdam, 1989, Chapter 17, pp. 615–631.
[30] Hoffman B. M., Gurbiel R. J., Werst M. M., Sivaraja M. "Electron nuclear double resonance (ENDOR) of metalloenzymes," in Hoff A J, Ed., *Advanced EPR: Applications in Biology and Biochemistry*, Elsevier, Amsterdam, 1989, Chapter 15, pp. 541–591.
[31] Hoffman B. M. *Acc. Chem. Res.* 1991, **24**: 164.
[32] Pilbrow J. R. *Transition Ion Electron Paramagnetic Resonance*, Clarendon Press, Oxford, 1990, pp. 410–418.
[33] Forrer J, Schweiger A., Berchten N., Gunthard H. H. *J. Phys.* 1981, **E14**: 565.

[34] Schweiger A. *Struct. Bond.* 1982, **51**: 1.
[35] Schweiger A., Rudin M., Forrer J. *Chem. Phys. Lett.* 1981, **80**: 376.
[36] Schweiger A., Gunthard H. H. *Mol. Phys.* 1981, **42**: 283.
[37] Hyde J. S., Chien J. C. W., Freed J. H. *J. Chem. Phys.* 1968, **48**: 4211.
[38] Hyde J. S., Sneed R. C., Jr., Rist G. H. *J. Chem. Phys.* 1969, **51**: 1404.
[39] Nechtschein M., Hyde J. S. *Phys. Rev. Lett.* 1970, **24**: 672.
[40] Patyal B. R., Crepeau R. H., Gamliel D., Freed J. H. *Chem. Phys. Lett.* 1990, **175**: 445–453.
[41] Goreester J., Millhauser G. L., Freed J. H. "Two-dimensional electron spin resonance," in Kevan L, Bowman M H, Eds., *Modern Pulsed and Continuous-Wave Electron Spin Resonance*, John Wiley & Sons, Inc., New York, NY, 1990.
[42] Geschwind S. "Optical techniques in EPR in solids," in Geschwind S, Ed., *Electron Paramagnetic Resonance*, Plenum Press, New York, NY, 1972, Chapter 5.
[43] Cavenett B. C. *Adv. Phys.* 1981, **30**: 475.
[44] Clarke R. H. *Triplet-State ODMR Spectroscopy*, John Wiley & Sons, Inc., New York, NY, 1982.
[45] Spaeth J. M. "Application of magnetic multiple resonance techniques to the study of point defects in solids," in, Weil J A, Ed., *Electron Magnetic Resonance of the Solid State*, Canadian Society for Chemistry, Ottawa, ON, 1987, Chapter 34.
[46] Hoff A. J. "Optically detected magnetic resonance of triplet states," in Hoff A J, Ed., *Advanced EPR: Applications in Biology and Biochemistry*, Elsevier, Amsterdam, 1989, Chapter 18.
[47] Spaeth J. M., Lohse F. *J. Phys. Chem. Solids* 1990, **51**: 861.
[48] Molin Yu. N., Anisimov O. A., Grigoryants V. M., Molchanov V. K., Salikhov K. M. *J. Phys. Chem.* 1980, **84**: 1853.
[49] Werst D. W., Trifunac A. D. *J. Phys. Chem.* 1991, **95**: 3466.
[50] Qin X. Z., Trifunac A. D., Eaton P. E., Xiong Y. *J. Am. Chem. Soc.* 1991, **113**: 669.

Further readings

[1] Kevan L., Kispert L. D., *Electron Spin Double Resonance Spectroscopy*, Plenum Press, New York, 1976.
[2] Dorio M. M., Freed J. H., *Multiple Electron Resonance Spectroscopy*, Plenum Press, New York, 1979.
[3] Box H. C., *Radiation Effects: ESR and ENDOR Analysis*, Academic Press, New York, 1977.
[4] Deal R. M., Ingram D. J. E., Srinivasan R., *Electron Magnetic Resonance and Solid Dielectrics*, Proc. 12th Colloq. AMPERE, 1963, p. 239.
[5] Feher G., *Phys. Rev.* 1957, **105**: 1122.
[6] Blumberg W. E., Feher G., *Bull Am. Phys. Soc.* 1960, **5**: 183.
[7] Holton W. C., Blum H., Slichter C. P., *Phys. Rev. Lett.* 1960, **5**: 197.
[8] Holton W. C., Blum H., *Phys. Rev.* 1962, **125**: 89.
[9] Rist G. H., Hyde J. S. *J. Chem. Phys.* 1968, **49**: 2449.
[10] Pake G. E., Estle T. L. *The Physical Principles of Electron Paramagnetic Resonance*, Benjamin, Reading, MA, 1973, Chapter 11.
[11] Mobius K., Plato M., Lubitz W. "Radicals in solution studied by ENDOR & triple resonance spectroscopy," *Phys. Rep.* 1982, **87**: 171.
[12] Gemperle C., Schweiger A. "Pulsed electron nuclear double resonance technology," *Chem. Rev.*, 1991, **91**: 1481.
[13] Spaeth J. M., Niklas J. R., Bartram R. H. *Structural Analysis of point Defects in Solids – An Introduction to Multiple Magnetic Resonance Spectroscopy*, Springer, Berlin, Germany, 1982.

A-3 EMR imaging

Electronic magnetic resonance imaging (EMRI) was also developing at a fast pace in the 1980s, driven by nuclear magnetic resonance imaging (NMRI). They have found application in the fields of materials, catalysts, archaeology, biology, and medicine. Because free radicals associate with certain diseases (such as tumors), their potential applications in biology and medicine are of particular importance. However, most samples of biology and medicine fields are aqueous solutions, so that the wavelengths of the EMRI are the *L*-band (1–2 GHz, or even several hundred MHz). Their application in clinical medicine has not yet gone out from the laboratory, because the concentration of paramagnetic substances in the organism is very dilute. If the paramagnetic material to be used as spin labels is injected into the body, the consequences of adverse effect on health and personal security should be considered. These factors are the reasons of delaying the progress of EMRI development. A fairly comprehensive investigation has been made by G.R. Eaton and S.S. Eaton in 1996, and published in the *EMR Handbook* [1].

A3-3.1 Principle of electron magnetic resonance imaging

In general, the electron magnetic resonance spectrum is applied to sweep the field with uniform magnetic field, and the homogeneity of the magnetic field is quite high. The obtained spectrum at this time is the statistical average result of the EMR signal in the total sample tube, but it cannot get the information about the spatial distribution of paramagnetic particles. If we want to get the information about spatial distribution of paramagnetic particles (one or more), the heterogeneous field (gradient field) should be applied to scan, and realize one-dimensional or even multi-dimensional spatial spectral imaging [2].

1. Gradient Magnetic Field Imaging (EMR-CT)

Gradient magnetic field imaging in the common homogeneous magnetic field and a special linear gradient magnetic field is used to detect the concentration difference of paramagnetic materials in the different spatial positions. In the homogeneous field $\boldsymbol{H}$, it is impossible to distinguish the difference between EMR signals of spin particles between $\boldsymbol{z}_1$ and $\boldsymbol{z}_2$, both points at spectral diagram (as shown in Figure A3-3.1a). When an "anti-Helmholtz" coil (arrow indicates the direction of the current) is added to the homogeneous magnetic field producing an attach field, the entire fixed magnetic field becomes a gradient field, scanning field obtained from the EMR spectrum is discrete (shown as Figure A3-3.1b). The discrete distance and the space distance of

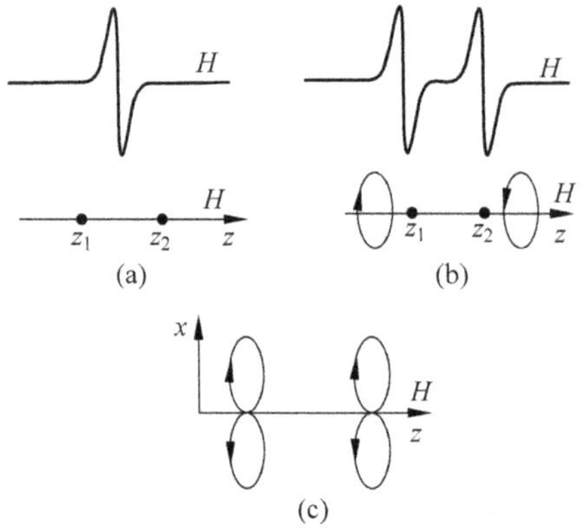

Figure A3-3.1: A diagram of producing gradient magnetic field.

particle spin is proportional to the intensity of gradient magnetic field. One-dimensional imaging in the *Z* direction can be obtained. If the "8" coil pair is added in the *X* direction or in the *Y* direction (shown as Figure A3-3.1c), a two-dimensional imaging map on the *X*–*Z* or *Y*–*Z* plane can be obtained. If there are three coils to be added simultaneously, a three-dimensional imaging map can be obtained.

The larger the gradient field, the higher the spatial resolution. These coils are installed on the outside of the cavity, because the mutual distance is far away; the gradient of the magnetic field is greatly limited (generally 0.1–0.5 T/m), and the resolution generally is not better than 0.1 mm. If the coil is placed in the inside of the cavity, the gradient of the magnetic field can be greatly improved. Smirnov et al. [3] applied the technique of superconducting (5T) and superhigh frequency (150 GHz); the magnetic field gradient reaches as high as 78 T/m, and the resolution can reach up to 1 μm. However, as high as this superhigh frequency, the volume of resonator will be too small, and bulk of sample can be less than 1 mm^3.

In addition, one-dimensional technique can be used multiple times to rotate the angle of the sample to collect a set of one-dimensional imaging data; after computer treatment, the reconstructing of a two-dimensional image can be done [4] (see Figures A3-3.2 and A3-3.3). This method can detect the distribution of paramagnetic substances in the sample but only if it does not destroy the specimen. However, the volume of the sample cannot exceed the size of the cavity. Moreover, it can only be applied to single paramagnetic species; the linewidth is wider and the resolution also is poorer.

In order to be suitable for imaging of biological and medical samples, Loop and Loop-Gap resonant cavities have been developed to study *L*-band and radio frequency (about 300 MHz) imaging. Reducing the frequency is for the sake of reducing the loss of microwave energy due to the nonresonant absorption of moisture in biological or medical samples. However, the sensitivity of signal detection is also decreased because of the lower frequency of the radio frequency field. Additionally, the paramagnetic particles content of the organism is very low, so, the animal

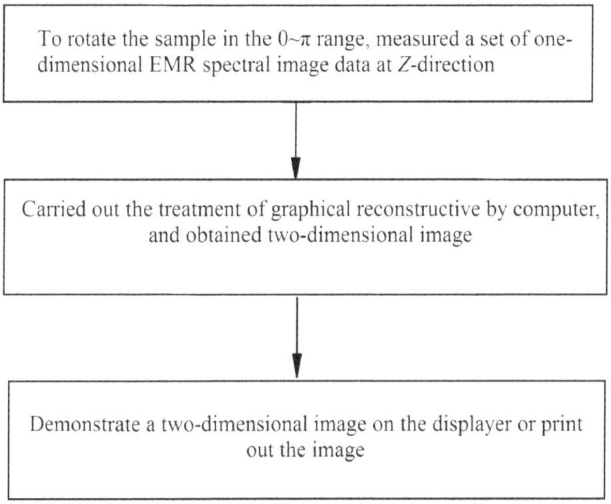

Figure A3-3.2: The block diagram of a 2D EMR imaged by using anti-Helmholtz coil causing 1D gradient field.

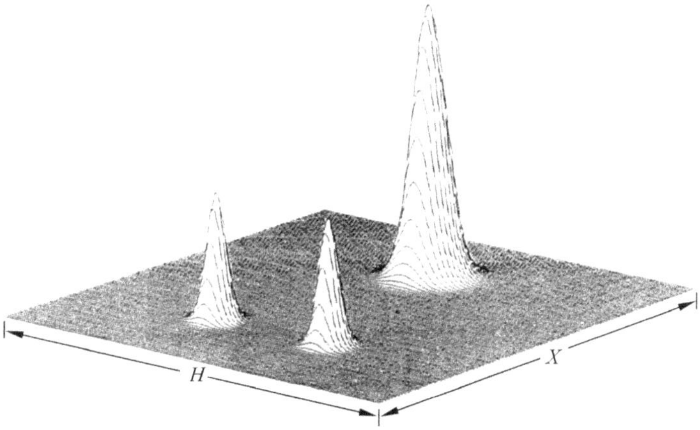

Figure A3-3.3: The diagram of three-spin signal simulation EMR imaging in the 2D space.

imaging experiments, at present, still remain in the stage of using spin label (spin graphic agent). Alecci et al. [5] reported a three-dimensional imaging technique with frequency 283 MHz, sample diameter 50 mm, and length 100 mm, which could be imaged for small animals. However, the concentration of nitroxide radicals could not exceed 10^{-4} mol L^{-1}, and the resolution will be about 1 mm. When using 300 MHz and Loop-Gap resonator, the sample diameter can extend to 100 mm and the length to 150 mm, but the S/N ratio will be 10 times lower than the *X*-band. If using the *L*-band concave cavity made of ceramic materials, in the 1.2 GHz of TEMPO aqueous solution (the concentration about 1 μmol L^{-1}) carry out measurement practically, the sensitivity can compare (almost) with X-band TE_{011} type resonant cavity. This type of resonator is more favorable to the in situ (in vivo) imaging of the biological organism.

2. Spectrogram of Linear Gradient Magnetic Field – Spatial Imaging [6]

When there exist multiple spin species in the sample, the simple EMR-CT imaging method cannot be used; however, we can use the spectrogram spatial imaging method. The spectrum to be obtained by the action of one-dimensional gradient magnetic field is integral of the signal in the direction of the θ angle intersection with Z-axis. The greater the field gradient, the greater the θ angle. The spectral signal integral of different θ angles can be obtained using multiple gradient magnetic field imaging. After computer treatment, one-dimensional spatial distribution and the imaging diagram of one-dimensional spectrum reconstruction can be obtained. After improvement, the measuring velocity of the spectrogram spatial imaging, which consists of 64×64 point sampled data, can be reduced to about 1 min [7]. This method can not only get the information about the spatial distribution of different spin species in one dimension but can also distinguish the variety of spin species from the imaging figure.

3. The EMR Microscope of Partial Modulated Magnetic Field Scanning Type [8]

The 100 kHz modulation field of the usual EMR spectrometer is action to the whole sample space. The local modulation magnetic field scanning technique is modulated by using a micro modulating coil close to the sample surface. Since the modulated magnetic field acts only on a small space, the measured paramagnetic signal is found in this tiny space. By scanning the surface of the sample, a two-dimensional distribution of spin species on the sample surface can be obtained.

This local magnetic field modulation method can apply to the surface imaging of sample with different paramagnetic particles and can simultaneously determine the distributing case of varied paramagnetic particles on the surface. The resolution of the imaging has no relation to the shape of the spectrum. Thus, signals for hyperfine structures depend on angles, and the signal of the linewidth can be distinguished.

An ordinary EMR spectrometer and a computer-controlled stepper motor scanner can be used in this method. The apparatus of this method is quite simple, and the spatial resolution is about 0.2 mm. This method can only detect the spin concentration distribution of the sample surface. If we want to know the depth distribution of the spin particles from the surface, the sample should be cut and scanned at a cross-section of the surface. Since only the information about the distribution of paramagnetic particles on the surface of the sample could be measured, which is similar to the ordinary microscope, so it is also called "The EMR microscope of partial modulated magnetic field scanning type." This method can be used to determine the dose distribution case of the alanine to be irradiated by radiation ray. The distributing map of alanine radical produced by the radiation ray is consistent with the position distribution of X-ray in space irradiation.

4. The EMR Microscope of Microlinear Array Scanning Field [9]

The EMR microscope of micro linear array scanning field is a typical application of the semiconductor technology, which is a micro linear array technique consisting of extremely fine parallel wires, or a micro linear array of computer flat cables. Each passes the opposite current through two adjacent parallel wires, which produce a local magnetic field between the two wires, to drive circuit and change switch wire position sequentially. It moves the position of the local magnetic field produced by the micro linear array. When the micro linear array is located in a homogeneous magnetic field, and the surface of the sample is very close to the micro linear array, it can carry out imaging for the surface of the sample. One- or two-dimensional imaging can be realized by using different linear arrays. The structure of this method is very simple but only one kind paramagnetic signal on the surface of the sample can be imaged. Owing to the limitation of the line diameter of micro linear array, too high resolution could not be obtained.

This method can be applied to determine the distribution of one-dimensional spin concentrations of shark's tooth fossil. The results showed that for the concentration of free radicals caused by natural radiation between the enamel part and the pulp part, there is remarkable difference. In an extinct cephalopod ancient animal belemnite fossil, from the imaging of fossil section, it can be seen that the distribution of Mn^{2+} ions is reducing from center to edge gradually. Presumably, it is related to the annual growth, or perhaps indicates climatic variety.

5. The EMR Microscope of Microwave Scanning Type [10]

The EMR microscope of microwave scanning type is application of a (TE_{102}) resonant cavity as shown in Figure A3-3.4. A hole is on the wall of cavity; the microwave after focusing on the resonant cavity inject out from the needle hole; the sample surface is close to the hole (distance is d); the microwave inject onto a tiny area of sample, and then through the computer control X–Y two-dimensional stepper motor movement occurs, the scanning of which determines the signal strength of the sample surface (spin concentration). The determination is expressed in two-dimensional graphics. The structure of these EMR microscope apparatus is rather simple (its block diagram

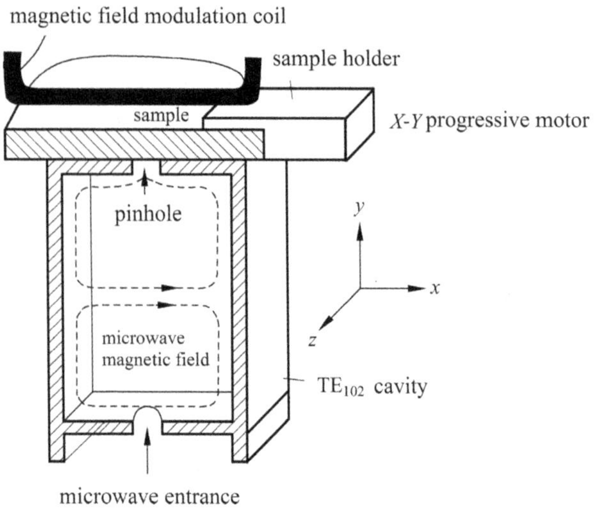

Figure A3-3.4: The diagram of TE_{102}-type pin hole cavity.

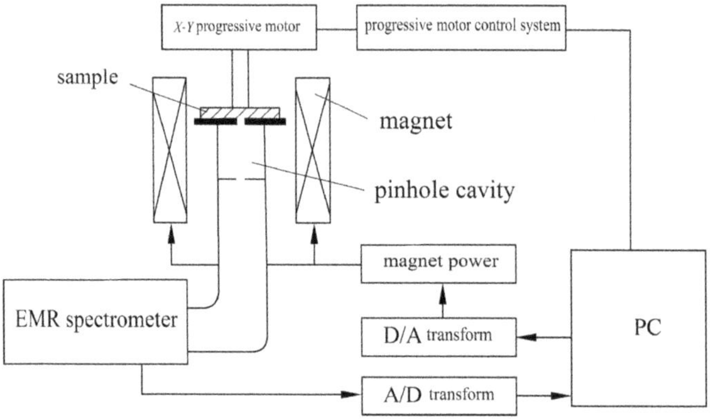

Figure A3-3.5: The block diagram of microwave scanning-type EMR microscope.

shown as Figure A3-3.5), although it can only determine the spin concentration distribution on the plane of sample, but because the sample is placed out of cavity, the bulk of sample is not limited by the volume of cavity. In addition, for the X-band (3 cm), the diameter of hole L is about 1 mm, the experimental data after deconvolution treatment, and the real resolution of the reconstructed graphic is only 1/5–1/10 of the hole diameter. The relation between the signal intensity with the pinhole diameter L and the distance d between the sample surface with the pin hole is shown in Figure A3-3.6.

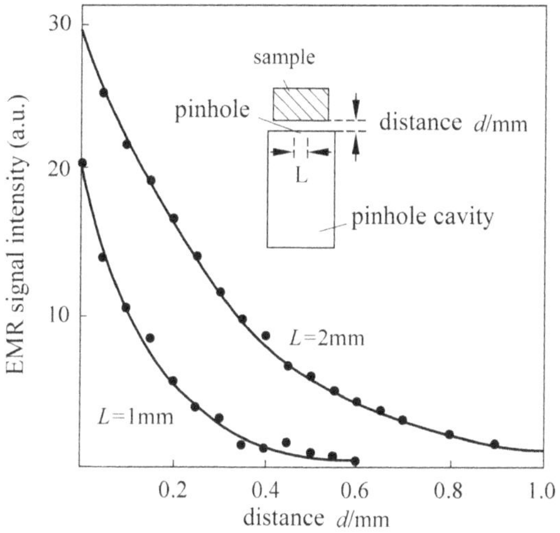

Figure A3-3.6: The relation of hole diameter (L) with the signal intensity and the distance (d) of hole with sample surface.

If encrinite fossil stem section and ammonite fossil are determined, the EMR signal of Mn^{2+} and the defects signal caused by radiation damage will be obtained. The content of paramagnetic Mn^{2+} ion is very rich; the encrinite fossils consist of calcium carbonate, whose EMR spectra is shown in Figure A3-3.7a. The graph of Mn^{2+} ions

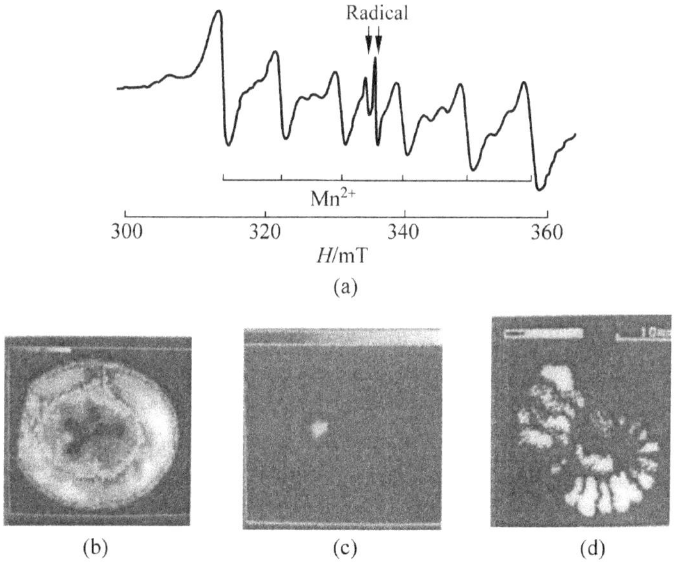

Figure A3-3.7: The EMR imaging of the Mn^{2+} ion in the encrinites and ammonite fossils.

concentration distribution is shown in Figure A3-3.7b; this difference of concentration distribution arises because of the crystallization process. It is thought that the Mn^{2+} ion diffuses toward inside from outside [11]. The concentration distribution graph of free radicals produced by radiation is shown in Figure A3-3.7c. The ammonite is also a cephalopod ancient soft body animal fossil that grew 400 million years ago, the content of Mn^{2+} ion as impurities is also very high. The concentration distribution graph of Mn^{2+} ion in the ammonite fossils is shown in Figure A3-3.7d. Interestingly, the shells had been partly FeS_2, and the middle parts are full of calcium carbonate ($CaCO_3$). Same with encrinite fossil, the high content of Mn^{2+} ions can also be observed in the part of $CaCO_3$ when its profile is measured using EMR imaging apparatus.

A3-3.2 Graphical display style of EMR imaging

One-Dimensional Graphical Display. It displays the signal intensity of one dimension and their corresponding position, so it is similar to the common EMR measurement, but abscissa represents the spatial position, not the intensity of the magnetic field. So, the graphical treatment of one-dimensional imaging is very simple. Practically, it is a two-dimensional graph: one is spatial position, and the other is signal intensity.

Two-dimensional Graphical Display. It displays the EMR signal on the two-dimensional space (plane), which require representation of both the two-dimensional space and one-dimensional signal intensities simultaneously. At present, map contour line method, gray or color graphical method, and top view graphical method are adopted mostly. In the contour line method, the point of height equality (i.e., the amplitude of the signal) link to a line, and using the amount of adjacent lines represents the intensity of the signal. The more crowded the contour line, the more stronger the signal intensity. Likewise, the more sparse the contour line, the more weaker the signal. The visual sense effect of contour lines is not better than the other methods. In the computer graphical display, most probably the gray and color graphical method is applied. These methods using the imaging plane represent 2D space, the gray degree or different colors represent different intensity of the signal. Color graphical method is especially suitable for the color computer display screen; the gray method although not as good as the color image display method, in the printing condition it is still adopted widely when there no color printer. In fact, a color image is displayed on a computer screen, and the black/white image is printed by common printer. Another one, called “vertical view” graphical method, looks very intuitive and is also often used. Its defect is that the small signal is hidden behind the large signal.

Three-dimensional Graphical Display. It is rarely seen in EMR imaging. If both three-dimensional spaces and one-dimensional signal intensities are to be displayed, it is quite difficult to use the usual three-dimensional graphical representation. But cross sectional representation can be made, that is, using the two-dimensional method of color imaging to represent the 3D local space, and then analyze the 3D graphical display of the .

At present, the graphical area of EMR imaging is usually not too large, due to the limitation of the spatial resolution of the instrument; for example, the graph of 2D imaging, most probably within 128 × 128 pixels. Obviously, the larger the data, the longer the time required for the sampling and treating the data. So, generally, the graphical software, color graphics (for screen display) and gray graphics (for printing), uses 10 colors and 10 gray levels, respectively. The data are all 128 × 128 pixels. The contour lines are displayed with 128 × 128 pixels – the "vertical view" graphical display using 200 × 200 pixels. In this general computer (586/133 compatible), the computation plus graphics display want about 20s, and for 100 × 100, need only 8s.

A3-3.3 The fundamental methods of graphical treatment

1. Numerical Graphical Method

In order to treat graphs conveniently by computer, at first we must transform the graphical analog data into digital data. An infinitesimal space has a slight change in the signal value and can only be represented by discrete values. This discrete sampling point is called "pixel." The approximate operation of this gray value is called "digitization," and the taking value range is called "gray range." If using the continuous function of x, y coordinates $f(x, y)$ represents the gray degree of the 2D graph, then the data graph of sampling obtained from discrete coordinate points (i, j) becomes a set of discrete concentration values f_{ij}. That is,

$$F = \{f_{ij}\} \tag{A-3.1}$$

where $i = 1, 2, 3, \ldots, M$; $j = 1, 2, 3, \ldots, N$.

This graph shows a two elementary determinant consisting of the matrix element f_{ij} (it means ith row and jth file), which can be treated by the computer. The larger value of the sampling point ($M \times N$) can be represented: the greater the digitization range of the f_{ij}, the more realistic the original graph. To get the high-resolution graph in space, it must enhance the pixels and increase the gray range. With the increasing data points, the data treating time is increased greatly. In the practical operation, it always adopts a compromise proposal. When using digital graphic to represent the photos, 512 × 512 pixels can perform fine shading part. The usual EMR imaging, using 128 × 128 pixels, can be enough.

2. Coordinate Transform Method

For the graphical treatment, the graphical magnification, reduction, rotation, translation, and other geometric transformation treatment are often used. If the original coordinate system (x, y) after transformation change to (u, v), then

$$u = h_1(x, y), \quad v = h_2(x, y), \tag{A-3.2}$$

There are primary transform, twice transform, projection transform [12], and so on for transformation. However, the most usual application is affine transformation. The relationships of coordinates before and after the transformation are as follows:

$$u = xT_{11} + yT_{12} + T_{13} \tag{A-3.3}$$

$$v = xT_{21} + yT_{22} + T_{23} \tag{A-3.4}$$

T_{ij} in the formula relationship with the specific transformation operation is as follows:

	T_{11}	T_{12}	T_{13}	T_{21}	T_{22}	T_{23}
Magnification or reduction	T_{11}	0	0	0	T_{22}	0
Rotation(θ)	$\cos\theta$	$-\sin\theta$	0	$\sin\theta$	$\cos\theta$	0
Translation	0	0	T_{13}	0	0	T_{23}

3. Algebraic Reconstruction Method

The method of CT (computer-assisted tomography) from multiple directions carry out projection for the detected object, and the obtained numerous projecting graphs need to be reduced to the profile of the concentrating distribution. To reconstruct the original graph from the projecting figure, there are many methods: algebraic reconstruction technique (ART) [13], simultaneous iterative reconstruction (SIRT) method [14], iterative least-square technique (ILST) [15], and others. We will introduce the ART method here only. In addition, the Fourier transform method, filtering correction inverse projection method, and others will also be introduced separately in the following.

Algebraic reconstruction method is carried out projection from various angles $(0 \le \theta \le \pi;\ \theta = \theta_o,\ \theta_1,\ \theta_2,\ \dots,\ \theta_m)$ obtained from a set of two-dimensional graphical data $f(i, j)$, and assumes the obtained projection data as $P(k, \theta)$. Giving the hypothesizing initial value $f_o(i, j)$ and then seeking the projection data $R_o(k, \theta_1)$ of the $f_o(i, j)$ in the $\theta_1$direction, and comparing it with the actual projection data $P(k, \theta_1)$ and correcting it make the least difference. Then, from the modified data $f_1(i, j)$, repeat the iterative operation until the difference between $f^n(k, \theta)$ and $f^{n-1}(k, \theta)$ is small enough, and then using the pixel value $f^n(i, j)$ of the least difference reconstruct the graph. Although the graph of high accuracy can be reconstructed by this method, the computating quantity is very huge, especially for the large pixels owing to the iterative operation. Therefore, the Fourier transform method and the filter correction inverse projection method are adopted most probably in the actual operation.

4. Fourier Transform Method [16]

For the functions $f(x)$, when the integral of real x is a continuous function, its one-dimensional Fourier transform is

$$F(\omega) = F[f(x)] = \int_{-\infty}^{\infty} f(x) \exp[-2\pi i\, \omega x] dx \tag{A-3.5}$$

Its inverse transformation is $f(x)$:

$$f(x) = F^{-1}[F(\omega)] = \int_{-\infty}^{\infty} F(\omega) \exp[2\pi i \omega x] d\omega \tag{A-3.6}$$

For the two-dimensional Fourier transform of the continuous functions $f(x, y)$,

$$F(\omega_1, \omega_2) = F[f(x, y)] = \iint_{-\infty}^{\infty} f(x, y) \exp[-2\pi i(\omega_1 x + \omega_2 y)] dxdy \tag{A-3.7}$$

Its inverse transformation is

$$f(x, y) = \mathrm{F}^{-1}[F(\omega_1, \omega_2)] = \iint_{-\infty}^{\infty} F(\omega_1, \omega_2) \exp[2\pi i(\omega_1 x + \omega_2 y)] d\omega_1 d\omega_2 \tag{A-3.8}$$

For the discrete function, one should adopt the discrete Fourier transform (DFT). The one-dimensional DFT is defined as follows:

$$F(m) = \frac{1}{N} \sum_{k=0}^{N-1} f(k) \exp[-2\pi i(mk/N)] \tag{A-3.9}$$

$$m = 0,\ 1,\ 2,\ \ldots,\ N-1$$

For the two-dimensional DFT,

$$F(m_1, m_2) = \frac{1}{MN} \sum_{k_1=0}^{M-1} \sum_{k_2=0}^{N-1} f(k_1,\ k_2) \exp[-2\pi i(m_1 k_1/M + m_2 k_2/N)] \tag{A-3.10}$$

$$m_1 = 0,\ 1,\ 2,\ \ldots,\ M-1$$
$$m_2 = 0,\ 1,\ 2,\ \ldots,\ N-1$$

Its inverse transformation is

$$f(k_1, k_2) = \frac{1}{MN} \sum_{m_1=0}^{M-1} \sum_{m_2=0}^{N-1} F(m_1, m_2) \exp[2\pi i(m_1 k_1/M + m_2 k_2/N)] \tag{A-3.11}$$

$$k_1 = 0, 1, 2, \ldots, M-1$$
$$k_2 = 0, 1, 2, \ldots, N-1$$

Assuming $f(x, y)$, after the (x, y) coordinate system rotate an angle θ, the system spin concentration distribution function is $f(x', y')$, its two-dimensional Fourier transform is $F(u', v')$. Both relationships are as follows:

$$F(u', v') = \iint_{-\infty}^{+\infty} f(x', y') \exp[-2\pi i(x'u' + y'v')] dx' dy' \tag{A-3.12}$$

When $v' = 0$

$$F(u', 0) = \int_{-\infty}^{+\infty} \left\{ \int_{-\infty}^{+\infty} f(x', y') dy' \right\} \exp[2\pi i\, x'u'] dx' \tag{A-3.13}$$

Let

$$P(x', \theta) = \int_{-\infty}^{+\infty} f(x', y') dy' \tag{A-3.14}$$

Substituting into above formula, we get

$$F(u', 0) = \int_{-\infty}^{+\infty} P(x', \theta) \exp[-2\pi i\, x'u'] dx' \tag{A-3.15}$$

Equation (A-3.15) is after the one-dimensional Fourier transformation of an original graph from some θ angle direction, it is equal to the incise central section of corresponding angle of the two-dimensional Fourier transform of the original graph. Thus, the plural radial sections of the original point of the two-dimensional Fourier transformation of the original graph are obtained by carrying out Fourier transformation on the one-dimensional projection graph, which is obtained through multidirectional determination, and the original graph is finally restored.

5. Filtered Modified Inverse Projection Method [17]

The simplest way of reconstructing an original graph is by means of the projection data combination with an inverse projection obtained, which is called "back projection" method. Simple inverse projection will produce dark spots around the graphic, and is inversely proportional to the distance of a high-concentration dot spin matter. In order to remove scotomata in the inverse projection graphic in the practical application of CT, the "filtered modified inverse projection" method is adopted. For the projection data, first carry out the filter modified method, and then the inverse projection is carried out. If the real space is represented by deconvolution, use Fourier transform in filter operation in the Fourier space, and then return to the real space by means of Fourier inverse transform, the effect of these filtering treatment is also good. When the projected graphics are much more, we can get almost the same with the original graphic, while the false image will appear when the projection graphics are quite less.

A3-3.4 Simulation of EMR imaging

The designed EMR imaging simulating program is installed on the EMR spectrometer, and the measured graph can be displayed directly on the spectrometer. In the study of EMR imaging, artificial samples are often used for imaging. Figure A3-3.8 displays a two-dimensional spatial contour simulation map of a planar sample of two different concentration of paramagnetic spots. Figure A3-3.9 is its grayscale simulation graphic. Figure A3-3.10 is its vertical view simulation graphic.

Figure A3-3.8: Simulation diagram of 2D contour line.

The one-dimensional map of one-dimensional space can distinguish not only the species of different paramagnetic materials but also their distribution in one-dimensional space at the same time. The simulated spectrography in Figure A3-3.3 shows that the sample can be seen from the magnetic field *H*-axis, the *g* values of these two paramagnetic substances are different, and one is two peaks of equal intensity, and the other is a single peak. In the direction of the *X*, the two paramagnetic substances are concentrated at two different points.

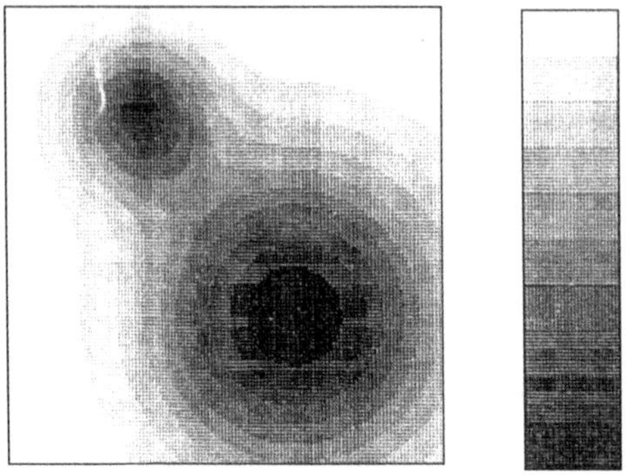

Figure A3-3.9: Simulation diagram of 2D grayscale.

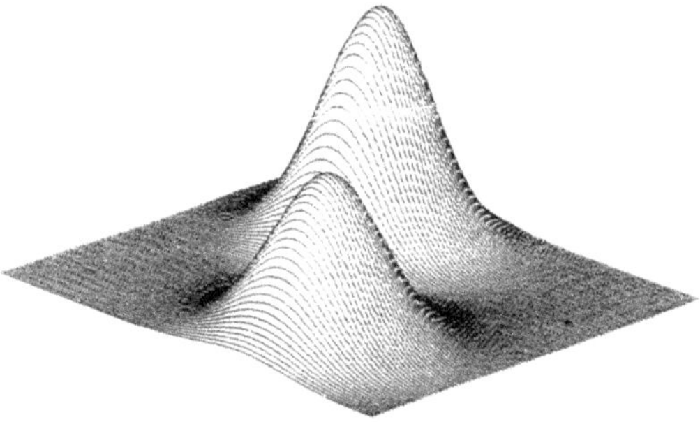

Figure A3-3.10: Vertical view simulation graphic of 2D space.

A3-3.5 Exemplification of EMR imaging technical application

EMR imaging was originally used to determine the consistency distribution of nitrogen radicals in two diamonds. Most materials are irradiated by rays, all can produce electron or hole trapping centers or free radicals. Through EMR imaging, one can understand the distribution of these "centers" or radicals in the body: quartz glass, calcium phosphate, sodium chloride, high-molecular antioxidant stabilizer HALS, alanine, and so on, after irradiation, will all produce color centers or free radicals.

In the heterogeneous catalysis system, the diffusion and mass transfer processes of molecules in porous catalysts are very important, which determine the activity and reaction rate of the catalysts. Yakimchenko et al.[18] used EMR radiographical agents (spin labels) to investigate its diffusion processes on alumina supports with different pore sizes and specific surfaces. Comparing two kinds (polar and nonpolar) of free radical, from solution toward alumina carrier that is saturated by the $(CCl)_4$ solvent, diffusing process carried out EMR imaging at different times. Figure A3-3.11a is a spatial distribution vertical view of the polar radical 2,2,6,6-tetramethyl-hydroxylpiperidine-1-oxyl in CCl_4 solution after

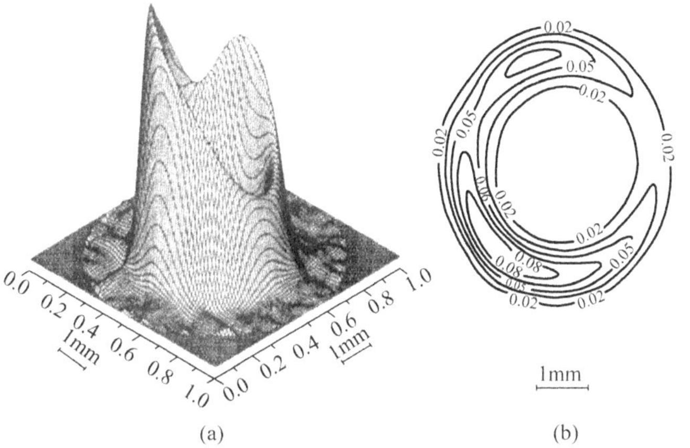

Figure A3-3.11: The radical solution diffusion on the Al_2O_3 carrier imaging: (a) vertical; (b) contour.

diffusion on the $\alpha - Al_2O_3$ support. Figure A3-3.11b is the corresponding contour map. From the actually measurement, diffusion coefficient can be seen: The diffusion rate of polar radicals perchlorotriphenylmethyl will be several decade times slower than the nonpolar radicals. The molecular diffusion coefficient of the former is much larger than the Knudsen diffusion coefficient, whereas the latter is just opposite. Therefore, this method can obtain various information about the mass transfer of organic molecules in porous media, and can be used as an experimental means of verifying the mass transfer theory.

The activities of transition metal/carrier catalysts should be reduced. Particles (flakes, spheres, or cylindrical) catalysts of the metal oxide are often reduced at shoal layer surface, and the deep layer (bulk) part is not reduced. Xu Yuanzhi et al.[19] found that $H_3PMo_{12}O_{40}$ was impregnated on plate-like silica gel and reduced at 450°C by propane. The bottom and periphery of the circular plate are EMR signals with uniform Mo^{5+} ions ($g = 1.904$), as shown in Figure A3-3.12a. The middle portions of the upper face are almost all of Mo^{6+} ions (without EMR signals) because they cannot contact with propane gas, while the edge portion has a very strong EMR signal of Mo^{5+} ions, as shown in Figure A3-3.12b. The imaging of the $A - A'$ section is shown in Figure A3-3.12c.

They also found that $H_4PVMo_{11}O_{40}$ is impregnated on a Y-zeolite and compressed as a plate. It is reduced with propane gas at 450°C. The EMR spectrum is obtained as shown in Figure A3-3.13a. It consists of the hyperfine structure of VO^{2+} and a strong free radical signal of $g = 2.0036$. The contour map of the concentration distribution of the free radical of $g = 2.0036$ on the catalyst surface is shown in Figure A3-3.13b.

The head of small animals (rats) by the L-band EMR imaging has been reported in Ref. [20]. The physiological salt water containing nitroxide radicals is injected into the body of small animal repeatedly, the free radicals enter the brain with the

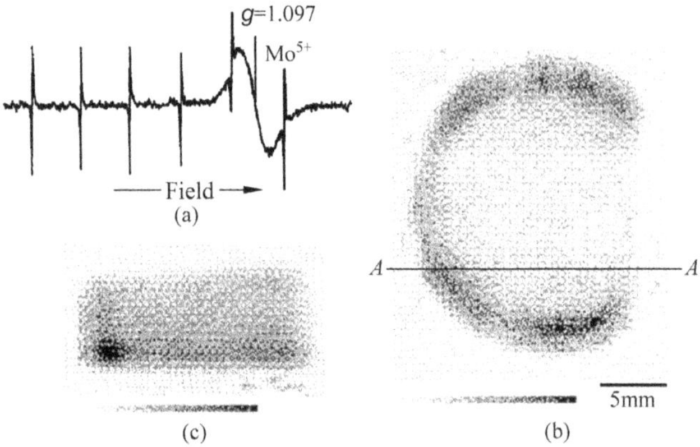

Figure A3-3.12: The EMR imaging graph of Mo^{5+} ions, after $H_3PMe_{12}O_{40}$ was impregnated on silica gel and reduced at 450 °C by propane.

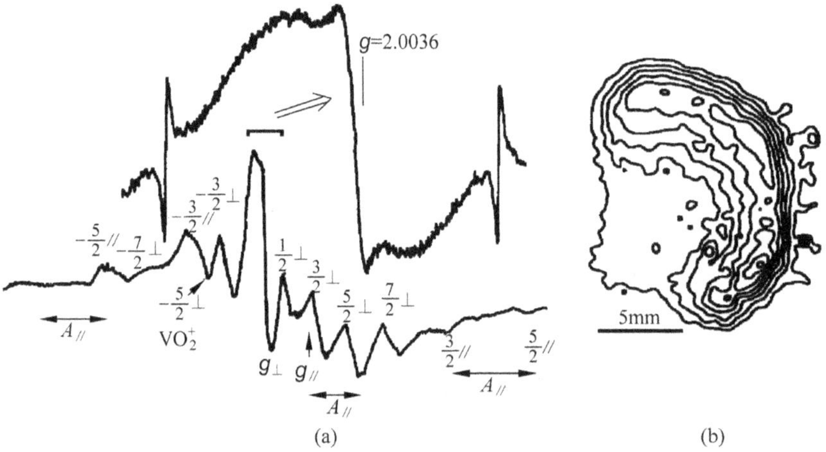

Figure A3-3.13: The EMR imaging graph of VO^{2+} ions and free radicals, after $H_4PVMe_{11}O_{40}$ was impregnated on a Y-zeolite and reduced at 450°C by propane.

increasing concentration. The sectional graphic of the rat head can be seen clearly by pass-through EMR imaging. Because the spin contrast agents usually have some toxicity, the in vivo application is limited greatly. There are also examples of determining free radical of small animal body surface directly: The melanoma of implanted rat tail is observed directly by EMR imaging with *L*-band cavity. It also uses X-band to determine the animal skin, which is injected the spin label into the skin and the diffusion dynamic behavior of the drug in the skin is studied [21]. Chlacicchi et al. [22] using the EMR imaging technique assess the reduction and elimination of nitroxides after injecting nitroxide radical of spin label, in order to understand the distribution and metabolic processes in vivo. The dynamic distribution of the drug in the head or whole body of rats was measured. The distribution image of the drug in the whole body of rats was obtained by using 280 MHz imaging apparatus after the injection of free radical label. A nitric oxide iron complex was injected into the white rat to carry out the *L*-band EMR imaging of the in vivo head. Using this imaging agent, clear imaging [23] of the white rat head is obtained.

References

[1] Eaton G. R., Eaton S. S. *Handb. Electron Spin Reson.*1999, **2**: 327–343.
[2] Xiang Z. M., Xu D. J., Xu Y. Z. *Chemistry (Huaxue Tongbao)*. 1997, (11): 48–53.
[3] Smirnov A. I, Poluectov O. G, Lebedev Y. S. *J. Magn. Reson.* 1992, **97**: 1–12.
[4] Zheng Y. G., Shen E. Z. *Chem. J. Chin. Univ.*, 1992, **13**(7): 981–984.
[5] Alecci M., Della P. S., Sotgiu A., et al. *Rev. Sci. Instrum.*1992, **63**(10 Pt. 1): 4263–4270.
[6] Eaton G. R., Eaton S. S., Maltempo M. M. *Appl. Radiat. Isot.* 1989, **40**(10–12): 1227–1231.
[7] Herring T., Thiessenhusen K., Ewert U. *J. Magn. Reson.* 1992, **100**: 123–131.
[8] Miki T. *Appl. Radiat. Isot.* 1989, **40**(10–12): 1243–1246.
[9] Ikeya M., Ishii H. *J. Magn. Reson.* 1990, **82**: 130–134.
[10] Furusawa M., Ikeya M. *Jpn. J. Appl. Phys.* 1990, **29**: 270–275.
[11] Furusawa M., Ikeya M. *Anal. Sci.* 1988, **4**: 649–651.
[12] Gonzalez R. C., Wintz. P. *Digital Image Processing*, Addison-Wesley, 1977.
[13] Gordon R., Bender R., Herman G. T. *J. Theor. Boil.* 1970, **29**: 471–481.
[14] Gilbert P. F. C. *J. Theor. Boil.* 1972, **36**: 105–117.
[15] Goitein M. *Nucl. Instum. Methods* 1972, **101**: 509–518.
[16] Gonzales R. C., Wintz P. *Digital Image Processing*, Addison-Wesley, 1977.
[17] Shepp L. A., Logan B. F. *IEEE Trans., Nucl. Sci.* 1974, **NS-21**: 21–43.
[18] Yakimchenko O. E., Degtyarev E. N., Parmon V. N., et al. *J. Phys. Chem.* 1995, **99**: 2038.
[19] Xu Yuanzhi., Furusawa M., Ikeya M., et al. *Chem. Lett.* 1991, 293.
[20] Ogata T., Kassei S. *Furi Rajikaru* 1992, **3**: 702.
[21] Fuchs J., Groth N., Herring T., et al. *J. Invest. Dermatol.* 1992, **98**: 713.
[22] Colacicchi S., Alecci M., Gualtieri G. *J. Chem. Soc. Perkin Trans.* 1993, **2**: 2077.
[23] Yoshimura T., Fujii Yokohama H., Kamada H. *Chem. Lett.* 1995, 309.

Further readings

[1] Ikeya M., Miki T. ESR Microscopy. Springer-Verlag Tokyo, (1992).
[2] Ohno K. ESR Imagination (1990).
[3] Miki T. High-Resolution ESR-CT ESR Application to Dosimetry (1990).
[4] Ohno K. "Development of high-resolution ESR imaging to investigate on paramagnetic species in thin samples," *Jpn. J. Appl. Phys.* 1984, **23**: L224.
[5] Ohno K. "Two-dimensional ESR imaging for paramagnetic species with anisotropic parameters," *J. Magn. Reson.* 1984, **64**: 109.
[6] Eaton G. R., Eaton S. S. "Electron spin echo detected EPR imaging," *J. Magn. Reson.* 1986, **67**: 73.

[7] Eaton G. R., Eaton S. S. "EPR Imaging using flip-angle gradients: a new approach to two dimensional imaging", *J. Magn. Reson.* 1986, **67**: 561.

[8] Janzen E. G., Kotake Y. "ENDOR imaging based on differences in viscosity or oxygen concentration in aqueous solutions of nitroxides," *J. Magn. Reson.* 1986, **69**: 567.

[9] Kotake Y., Oehler U. M., Janzen E. G. "Two-dimensional ENDOR imaging based on differences in oxygen concentration", *J. Chem. Soc. Faraday Trans.* 1988, **184**: 3275.

Appendix 2: Mathematical preparation

One would find many mathematical symbols and operation problems while reading this book. It is no problem for the reader who has excellent basic knowledge of mathematical physics. This appendix is meant for the readers who are not from the field of physics in order to acquaint them with the elementary mathematics, so that this book could be understood without any difficulty.

B-1 Complex number

A complex number can be expressed as follows:

$$u = x + iy = re^{+i\phi} \tag{B-1}$$

where x, y, r, and ϕ all are real numbers, $e^{i\phi} = \cos\phi + i\sin\phi$, $i^2 = -1$, x is the real component of the u, y is the imaginary component of u, r is the absolute magnitude of u, that is, $r = |u|$, and ϕ is the phase angle. u^* is the complex conjugate of u, $u^* = x - iy$. For the function of complex numbers, its conjugate complex function is to change a symbol in front of its i. The relation between the complex number and their conjugate, expressed in the complex plane (Argand diagram), is as shown in Figure B-1. Select the abscissa as the real axis (x) and the ordinate as the imaginary axis (y). *Attention:* $re(u)$ is a half of the sum of u and u^*; the product of the both is equal to r^2, that is,

$$u^*u = u\,u^* = |u|^2 = r\,e^{-i\phi}\,r\,e^{-i\phi} = r^2 \tag{B-2}$$

B-2 Properties of operators and their operation rules

An operator $\hat{\mathbf{A}}$ is a symbolic instruction to carry out a stipulated mathematical operation on some function (*operand*). The operator should be designated generally by a circumflex (ˆ), unless the operator itself has a clear meaning. One of the simplest operators is the product of constants, for example,

$$\hat{k}\alpha = k\alpha.$$

The operator $\hat{\Omega}$ is a linear operator, if it acts on a function that is summed by several function, the result is the sum of the operators acting on each function separately. If $\hat{\Omega}\alpha = \beta$, then

https://doi.org/10.1515/9783110568578-012

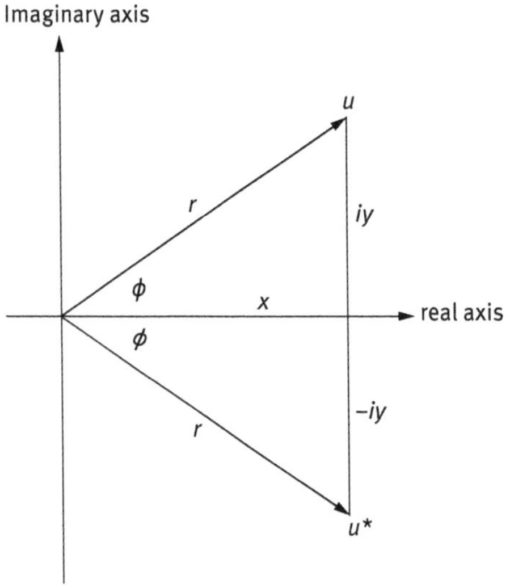

Figure B-1: Argand diagram.

$$\hat{\Omega}(\alpha_1+\alpha_2)=\hat{\Omega}\alpha_1+\hat{\Omega}\alpha_2=\beta_1+\beta_2 \tag{B-3}$$

If c is a constant, then

$$\hat{\Omega}(c\alpha)=c\hat{\Omega}\alpha=c\beta \tag{B-4}$$

For some continuous variable q_i, if $\alpha_i=f(q_i)$, then $\partial/\partial q_i$ is an example of a linear operator. An example of a nonlinear operator is $\sqrt{\ }$.

The reader may be familiar with Σ, as the summation operator, it means

$$\sum_{i=1}^{n} a_i=a_1+a_2+a_3+\cdots+a_n \tag{B-5a}$$

Similarly, production operator Π represents a simple series functions:

$$\prod_{i=1}^{n} a_i=a_1a_2a_3\cdots a_n \tag{B-5b}$$

It is often necessary to summarize a set of equations with constant coefficients as follows:

$$\psi_1=c_{11}\phi_1+c_{12}\phi_2+c_{13}\phi_3+\cdots+c_{1n}\phi_n \tag{B-6a}$$

$$\psi_2 = c_{21}\phi_1 + c_{22}\phi_2 + c_{23}\phi_3 + \cdots + c_{2n}\phi_n \tag{B-6b}$$

$$\psi_3 = c_{31}\phi_1 + c_{32}\phi_2 + c_{33}\phi_3 + \cdots + c_{3n}\phi_n \tag{B-6c}$$

$$\ldots = \ldots\ldots\ldots\ldots\ldots\ldots\ldots\ldots\ldots\ldots$$

The sum of the functions ψ_j can be regarded as the continuous action of function ϕ_k, and can be expressed by a double summation:

$$\sum_j \psi_j = \sum_j \sum_k c_{jk}\phi_k \tag{B-7}$$

Here we will encounter a juxtaposition of two operators, such as two parallel operators $\hat{\mathbf{A}}\hat{\mathbf{B}}$ should be understood as a result that lets the right-hand side of the operator $\hat{\mathbf{B}}$ act at first, then let the left-hand side of the operator $\hat{\mathbf{A}}$ act to the previous results, which is the correct approach; if two operators exchange order, it is possible to get different results, for example:

$$\hat{x}\frac{\hat{d}}{dx}(x^2) = 2x^2, \text{ however } \frac{\hat{d}}{dx}\hat{x}(x^2) = 3x^2$$

If $\hat{\mathbf{A}}\hat{\mathbf{B}} = \hat{\mathbf{B}}\hat{\mathbf{A}}$, then the operator $\hat{\mathbf{A}}$ and the operator $\hat{\mathbf{B}}$ are *commutative*. If they do not commute, then $\hat{\mathbf{A}}\hat{\mathbf{B}} - \hat{\mathbf{B}}\hat{\mathbf{A}}$ is called *Commutator*, shown as $[\hat{\mathbf{A}}, \hat{\mathbf{B}}]$. The commutator of the two operators is of great importance in the quantum mechanical system. We will discuss the commutator of angular momentum operator in Appendix III.

If the space operator $\hat{\Omega}$ submits to the following relation:

$$\int_\tau \psi_j^* \hat{\Omega} \psi_k d\tau = \int_\tau (\hat{\Omega}^* \psi_j^*) \psi_k d\tau \tag{B-8}$$

The operator $\hat{\Omega}$ is a Hermitian operator. ψ_j and ψ_k are continuous and well-behaved function of τ. Here τ stands for any position variable (length or angle) or multiple variables $(d\tau = dx\,dy\,dz)$, so $\int$ represents a multiple integral. A Hermitian conjugate operator is useful, the operator can "reverse" operate, when the operator is between the two conjugate functions, it can also act to the left function, as shown in Eq. (B-8). A very important property of Hermitian conjugate operators is that if the result of the operator acts on a function, it is the function itself multiplied by a constant, then the constant must be a real number.

Some of the most important operators in quantum mechanics are related to these compelling properties in the physical system. Some important linear operators are listed in Table B-1.

Table B-1: Some major variables in classical mechanics and quantum mechanics.

Variable	Classical mechanics	Quantum mechanics
Mass	M	m
Position	$q\ (= x, y, z)$	q
Time	T	t
Linear momentum	$p_q = m\frac{\partial q}{\partial t}$	$\hat{p}_q = -i\hbar\frac{\partial}{\partial t}$
Angular momentum (around the z-axis)$^{(a)}$	$\ell_z = xp_y - yp_x$	$\hat{l}_z = -i\hbar\left(x\frac{\partial}{\partial y} - y\frac{\partial}{\partial x}\right) = -i\hbar\frac{\partial}{\partial\phi}{}^{(b)}$
Kinetic energy (related to coordinate q)	$T = \frac{p_q^2}{2m}$	$\hat{H} = \frac{\hat{p}_q^2}{2m} = -\frac{\hbar^2}{2m}\frac{\partial^2}{\partial q^2}{}^{(c)}$
Potential energy$^{(d)}$	$V(r)$	$V(r)$

(*a*) See Eqs. (C-5)–(C-7). (*b*) The angle ϕ around the z-axis. (*c*) The kinetic energy of the Hamiltonian is valid only for the Cartesian coordinate. (*d*) Here $\boldsymbol{r} = x\boldsymbol{i} + y\boldsymbol{j} + z\boldsymbol{k}$ [in polar coordinates, $= r(\sin\theta\cos\phi\, \boldsymbol{i} + \sin\theta\sin\phi\, \boldsymbol{j} + \cos\theta\, \boldsymbol{k})$], see Section B-4.

B-3 Determinant

A determinant is a scalar that represents a linear combination of some product terms, and can be represented in square permutations as follows:

$$det\left|A^{(2)}\right| = \begin{vmatrix} A_{11} & A_{12} \\ A_{21} & A_{22} \end{vmatrix} = A_{11}A_{22} - A_{21}A_{12} \tag{B-9a}$$

The general formula is written as $det\left|A^{(\zeta)}\right|$, here $\zeta \geq 2$, and the general formula of k-order determinant is as follows:

$$det\left|A^{(k)}\right| = \begin{vmatrix} A_{11} & A_{12} & \cdots\cdots & A_{1k} \\ A_{21} & A_{22} & \cdots\cdots & A_{2k} \\ \vdots & \cdots\cdots & \cdots\cdots & \vdots \\ A_{k1} & A_{k2} & \cdots\cdots & A_{kk} \end{vmatrix} \tag{B-9b}$$

A determinant can also be represented by a minor determinant:

$$\begin{vmatrix} A_{11} & A_{12} & A_{13} \\ A_{21} & A_{22} & A_{23} \\ A_{31} & A_{32} & A_{33} \end{vmatrix} = A_{11}\begin{vmatrix} A_{22} & A_{23} \\ A_{32} & A_{33} \end{vmatrix} - A_{12}\begin{vmatrix} A_{21} & A_{23} \\ A_{31} & A_{33} \end{vmatrix} + A_{13}\begin{vmatrix} A_{21} & A_{23} \\ A_{31} & A_{32} \end{vmatrix} \tag{B-10}$$

$$= A_{11}A_{22}A_{33} - A_{11}A_{23}A_{32} - A_{12}A_{21}A_{33} + A_{12}A_{23}A_{31} + A_{13}A_{21}A_{32} - A_{13}A_{22}A_{31}$$

Generalized formula:

$$det|A^{(k)}| = \sum_{i-j} (-1)^{(i+j)} A_{ij} \left\{ det[A^{(k-1)}] \right\} \quad \text{(B-11)}$$

This method is very useful. It can simplify a higher order determinant into the sum of several minor determinants, and finally simplify it to the sum of many determinants with two orders. For example, the original is a four-order determinant that can be simplified as a linear combination of four determinants of order 3, and then simplify every three-order determinant as a linear combination of three two-order determinants, finally simplified to a polynomial.

The determinant has the following important properties:

1. If there are two rows (or two columns) equal in the determinant, the determinant is zero.
2. If the determinant switches two rows (or two columns), then the value of the determinant is changed.
3. If there is a row in the determinant (or column) multiplied by a constant, then the constant multiply the value of determinant.

A determinant is often used to solve simultaneous equations:

$$y_1 = c_{11}x_1 + c_{12}x_2 + c_{13}x_3 \quad \text{(F1.12a)}$$

$$y_2 = c_{21}x_1 + c_{22}x_2 + c_{23}x_3 \quad \text{(B-12b)}$$

$$y_3 = c_{31}x_1 + c_{32}x_2 + c_{33}x_3 \quad \text{(B-12c)}$$

The solution of the simultaneous equation is

$$x_1 = \frac{|\Delta_1|}{|\Delta|}; \qquad x_2 = \frac{|\Delta_2|}{|\Delta|}; \qquad x_3 = \frac{|\Delta_3|}{|\Delta|}; \quad \text{(B-13)}$$

Here

$$|\Delta| = \begin{vmatrix} c_{11} & c_{12} & c_{13} \\ c_{21} & c_{22} & c_{23} \\ c_{31} & c_{32} & c_{33} \end{vmatrix}; \ |\Delta_1| = \begin{vmatrix} y_1 & c_{12} & c_{13} \\ y_2 & c_{22} & c_{23} \\ y_3 & c_{32} & c_{33} \end{vmatrix}; \ |\Delta_2| = \begin{vmatrix} c_{11} & y_1 & c_{13} \\ c_{12} & y_2 & c_{23} \\ c_{13} & y_3 & c_{33} \end{vmatrix};$$

$$|\Delta_3| = \begin{vmatrix} c_{11} & c_{12} & y_1 \\ c_{21} & c_{22} & y_2 \\ c_{31} & c_{23} & y_3 \end{vmatrix} \quad \text{(B-14)}$$

The determinant is often used to represent the antisymmetric wave function, since the exchange of two electrons is equivalent to the exchange of two row of the determinants, and in accordance with the *property 2*, the symbol should be changed, which is exactly what the Pauli principle requires. For example, the two electron wave functions are

$$\Psi = \frac{1}{\sqrt{2!}}\begin{vmatrix} \psi(1)\alpha(1) & \psi(1)\beta(1) \\ \psi(2)\alpha(2) & \psi(2)\beta(2) \end{vmatrix} = \psi(1)\psi(2)\frac{1}{\sqrt{2}}[\alpha(1)\beta(2) - \alpha(2)\beta(1)] \tag{B-15a}$$

or expressed as

$$\Psi = \frac{1}{\sqrt{2!}}\|\psi\bar{\psi}\| \tag{B-15b}$$

where ψ represents spin α and $\bar{\psi}$ is spin β.

B-4 Vector algebra: Scalar, vector, and algebraic operations

The mathematical definition of a vector is to put a set of scalars into a single row or a single column. The scalar quantity that defines a vector is called a vector element. Vector $\boldsymbol{r}$ has three elements in orthogonal three-dimensional space: r_x, r_y, and r_z. Define $\boldsymbol{i}, \boldsymbol{j}$, and $\boldsymbol{k}$ as the unit vectors of the positive directions of x, y, and z, and scalar r represents the *magnitude* of the vector $\boldsymbol{r}$:

$$r = (r_x^2 + r_y^2 + r_z^2)^{1/2} \tag{B-16}$$

Assume the coordinate of a vector at a point in three-dimensional space x, y, z is 7, −3, 4, then

$$\boldsymbol{r} = 7\boldsymbol{i} - 3\boldsymbol{j} + 4\boldsymbol{k} \tag{B-17}$$

Thus, the vector can be represented as follows:

$$\boldsymbol{r} = \begin{bmatrix} 7 \\ -3 \\ 4 \end{bmatrix} \tag{B-18a}$$

or

$$\boldsymbol{r}^T = [7 \quad -3 \quad 4] \tag{B-18b}$$

where $\boldsymbol{r}$ represents column vector, $\boldsymbol{r}^T$ represents row vector, the superscript T denotes transposition, and turns the column vector into row vector.

Additive and subtractive methods of vectors:

Existing vectors **A** and **B:**

$$\mathbf{A} = a_x\boldsymbol{i} + a_y\boldsymbol{j} + a_z\boldsymbol{k} \tag{B-19a}$$

$$\mathbf{B} = b_x\boldsymbol{i} + b_y\boldsymbol{j} + b_z\boldsymbol{k} \tag{B-19b}$$

The additive and subtractive methods of vectors are as follows:

$$\mathbf{A} \pm \mathbf{B} = (a_x \pm b_x)\boldsymbol{i} + (a_y \pm b_y)\boldsymbol{j} + (a_z \pm b_z)\boldsymbol{k} \tag{B-20}$$

Multiplicative method of vector: There are three kinds of multiplication of vector.

1. **Scalar product**: Also called *inner product* or *dot product*, it is defined as follows:

$$\mathbf{A} \cdot \mathbf{B} = \mathrm{AB}\cos\theta_{AB} \tag{B-21}$$

Hence,

$$\boldsymbol{i} \cdot \boldsymbol{i} = \boldsymbol{j} \cdot \boldsymbol{j} = \boldsymbol{k} \cdot \boldsymbol{k} = 1 \tag{B-22a}$$

$$\boldsymbol{i} \cdot \boldsymbol{i} = \boldsymbol{j} \cdot \boldsymbol{k} = \boldsymbol{k} \cdot \boldsymbol{i} = 0 \tag{B-22b}$$

$$\mathbf{A} \cdot \mathbf{B} = a_x b_x + a_y b_y + a_z b_z \tag{B-23}$$

If **A** and **B** are complex numbers, then the definition of scalar product is $(\boldsymbol{A}^*)^{\mathrm{T}} \cdot \boldsymbol{B}$

2. **Vector product:** Also called cross-product, it is defined as follows:

$$\mathbf{C} = \mathbf{A} \times \mathbf{B} \quad \text{or} \quad \mathbf{C} = \mathbf{A} \wedge \mathbf{B} \tag{B-24}$$

C is also a vector, its direction is determined by the right-hand rule, and its magnitude is ***absin*** θ_{AB}. Thus,

$$\boldsymbol{I} \times \boldsymbol{I} = \boldsymbol{j} \times \boldsymbol{j} = \boldsymbol{k} \times \boldsymbol{k} = \boldsymbol{0} \tag{B-25a}$$

$$\boldsymbol{i} \times \boldsymbol{j} = \boldsymbol{k} \qquad \boldsymbol{j} \times \boldsymbol{k} = \boldsymbol{i} \qquad \boldsymbol{k} \times \boldsymbol{i} = \boldsymbol{j} \tag{B-25b}$$

$$\boldsymbol{j} \times \boldsymbol{i} = -\boldsymbol{k} \qquad \boldsymbol{k} \times \boldsymbol{j} = -\boldsymbol{i} \qquad \boldsymbol{i} \times \boldsymbol{k} = -\boldsymbol{j} \tag{B-25c}$$

$$\begin{aligned} \mathbf{C} = \mathbf{A} \times \mathbf{B} &= (a_x\boldsymbol{i} + a_y\boldsymbol{j} + a_z\boldsymbol{k}) \times (b_x\boldsymbol{i} + b_y\boldsymbol{j} + b_z\boldsymbol{k}) \\ &= (a_y b_z - a_z b_y)\boldsymbol{i} + (a_z b_x - a_x b_z)\boldsymbol{j} + (a_x b_y - a_y b_x)\boldsymbol{k} \\ &= \begin{vmatrix} i & j & k \\ a_x & a_y & a_z \\ b_x & b_y & b_z \end{vmatrix} \end{aligned} \tag{B-26}$$

3. **External product:** It is defined as follows::

$$\mathbf{AB} = \mathbf{C} \tag{B-27}$$

$$\begin{bmatrix} A_1 & A_2 & A_3 \end{bmatrix} \begin{bmatrix} B_1 & B_2 & B_3 \end{bmatrix} = \begin{bmatrix} C_{11} & C_{12} & C_{13} \\ C_{21} & C_{22} & C_{23} \\ C_{31} & C_{32} & C_{33} \end{bmatrix} \tag{B-27b}$$

$$C_{ij} = A_i B_j$$

B-5 Matrix

B-5.1 Definition of matrix and operation rules

Arrange $n \times m$ numbers (and/or operators) into a rectangular array with n rows and m columns. If there is only a set of numbers and no operators, it is the determinant det (A) of the matrix **A**. If $n = m = 1$, this matrix is a scalar. If $n = 1$, $m > 1$, it is called the row matrix, represented by **R**. If $n > 1$, $m = 1$, it is called the column matrix, represented by **C**.

$$\mathrm{C} = c = \begin{bmatrix} c_1 \\ c_2 \\ \vdots \\ c_n \end{bmatrix} \qquad \mathrm{R} = \mathrm{r}^{\mathrm{T}} = \begin{bmatrix} r_1 & r_2 & r_n \end{bmatrix} \tag{B-28}$$

When $n = m \geq 2$, it is called square matrix, or ***n***-order or ***n***-dimensional matrix:

$$\mathbf{B} = \begin{bmatrix} b_{11} & b_{12} & \cdots & b_{1n} \\ b_{21} & b_{22} & \cdots & b_{2n} \\ \vdots & \vdots & \vdots & \vdots \\ b_{n1} & b_{n2} & \cdots & b_{nn} \end{bmatrix} \tag{B-29}$$

Many books and documents often represent it in boldface **A** or a_{ij}. For matrices with the same dimension, the *addition and subtraction rule* is as follows:

$$\mathbf{D} = \mathbf{A} \pm \mathbf{B} \tag{B-30a}$$

or

$$d_{ij} = a_{ij} \pm b_{ij} \tag{B-31b}$$

For example,

$$\begin{bmatrix} 3 & -2 & 7 \\ -2 & 5 & -4 \\ 7 & -4 & 8 \end{bmatrix} + \begin{bmatrix} 6 & 4 & -2 \\ 4 & 2 & 3 \\ -2 & 3 & -5 \end{bmatrix} = \begin{bmatrix} 9 & 2 & 5 \\ 2 & 7 & -1 \\ 5 & -1 & 3 \end{bmatrix}$$

$$\begin{bmatrix} 3 & -2 & 7 \\ -2 & 5 & -4 \\ 7 & -4 & 8 \end{bmatrix} - \begin{bmatrix} 6 & 4 & -2 \\ 4 & 2 & 3 \\ -2 & 3 & -5 \end{bmatrix} = \begin{bmatrix} -3 & -6 & 9 \\ -6 & 3 & -7 \\ 9 & -7 & 13 \end{bmatrix}$$

If there are two matrices **A** and **B**, where the number of rows in **A** is equal to the number of columns in **B**, then the product of the matrices **B** and **A** can be defined as follows:

$$\mathbf{C} = \mathbf{BA} \tag{B-31a}$$

That is,

$$C_{ij} = \sum_{k=1}^{n} B_{ik} A_{kj} \tag{B-31b}$$

A row matrix dot multiplied by a column matrix is equal to the scalar product of two vectors, the result is a scalar.

For example, $\mathbf{B}^{\mathbf{T}} \cdot \hat{\mathbf{S}}$

$$\begin{bmatrix} B_x & B_y & B_z \end{bmatrix} \cdot \begin{bmatrix} \hat{S}_x \\ \hat{S}_y \\ \hat{S}_z \end{bmatrix} = B_x\hat{S}_x + B_y\hat{S}_y + B_z\hat{S}_z \tag{B-32a}$$

and

$$\begin{bmatrix} 3 & 5 & -4 \end{bmatrix} \cdot \begin{bmatrix} 2 \\ -1 \\ 1 \end{bmatrix} = 6 + (-5) + (-4) = -3 \tag{B-32b}$$

The result of a 1 × 3 matrix dot multiplied by a 3 × 3 matrix is a *row matrix*:

$$\begin{bmatrix} 3 & 5 & -4 \end{bmatrix} \cdot \begin{bmatrix} 3 & -2 & 7 \\ -2 & 5 & -4 \\ 7 & -4 & 8 \end{bmatrix} = \begin{bmatrix} -29 & 35 & -31 \end{bmatrix} \tag{B-33a}$$

The result of a 3 × 3 matrix dot multiplied by a 3 × 1 matrix is a *column matrix:*

$$\begin{bmatrix} 3 & -2 & 7 \\ -2 & 5 & -4 \\ 7 & -4 & 8 \end{bmatrix} \cdot \begin{bmatrix} -1 \\ -2 \\ 1 \end{bmatrix} = \begin{bmatrix} 8 \\ -12 \\ 9 \end{bmatrix} \tag{B-33b}$$

The multiplication of a matrix satisfies the union law; generally speaking, the commutative law is not satisfied, that is,

$$(\mathbf{AB})\mathbf{C} = \mathbf{A}(\mathbf{BC}) \tag{B-34}$$

but

$$\mathbf{AB} \neq \mathbf{BA} \tag{B-35}$$

if

$$\mathbf{AB} = \mathbf{BA} \tag{B-36}$$

Matrices **A** and **B** are called *commutative* (or *exchangeable*).

What is the result of a 1 × 3 matrix dot multiplied by a 3 × 3 matrix and dot multiplied by a 3 × 1 matrix?

$$\begin{bmatrix} a_1 & a_2 & a_3 \end{bmatrix} \cdot \begin{bmatrix} g_{11} & g_{12} & g_{13} \\ g_{21} & g_{22} & g_{23} \\ g_{31} & g_{32} & g_{33} \end{bmatrix} \cdot \begin{bmatrix} b_1 \\ b_2 \\ b_3 \end{bmatrix} = ? \tag{B-37}$$

The answer is to be calculated by the readers themselves. *No matter how to combine, the results are always*:

$$\begin{bmatrix} 3 & 5 & -4 \end{bmatrix} \cdot \begin{bmatrix} 3 & -2 & 7 \\ -2 & 5 & -4 \\ 7 & -4 & 8 \end{bmatrix} \cdot \begin{bmatrix} -1 \\ -2 \\ 1 \end{bmatrix} = 29 - 70 - 31 = -72 \tag{B-38}$$

A 3 × 3 matrix dot multiplied by a 3 × 3 matrix:

$$\begin{bmatrix} 3 & -2 & 7 \\ -2 & 5 & -4 \\ 7 & -4 & 8 \end{bmatrix} \cdot \begin{bmatrix} 6 & 4 & -2 \\ 4 & 2 & 3 \\ -2 & 3 & -5 \end{bmatrix} = \begin{bmatrix} -4 & 29 & -47 \\ 16 & -10 & 39 \\ 10 & 44 & -66 \end{bmatrix} \tag{B-39}$$

A further example of matrix multiplication is the operation of rotation about z coordinate axis in the xy plane, a point $p(x_1, y_1)$ through an arbitrary angle ϕ ; after counterclockwise rotation, the relationship between the point in the new coordinate system $p(x_2, y_2)$ and in the old coordinate system is as follows:

$$x_2 = +x_1 \cos\phi + y_1 \sin\phi \tag{B-40a}$$

$$y_2 = -x_1 \sin\phi + y_1 \cos\phi \tag{B-40b}$$

Represented by matrix:

$$\begin{bmatrix} x_2 \\ y_2 \end{bmatrix} = \begin{bmatrix} \cos\phi & \sin\phi \\ -\sin\phi & \cos\phi \end{bmatrix} \cdot \begin{bmatrix} x_1 \\ y_1 \end{bmatrix} = \begin{bmatrix} x_1 \cos\phi + y_1 \sin\phi \\ -x_1 \sin\phi + y_1 \cos\phi \end{bmatrix} \tag{B-41}$$

The square matrix in Eq. (B-41) is called the *coordinate rotation matrix*.

B-5.2 Some matrices related to matrix A

Some matrices are derived from a matrix whose definitions and names are listed in Table B-2.

Table B-2: Some matrices related to matrix **A**.

Matrix symbol	Matrix element	Example
A	a_{ij}	$\begin{bmatrix} 2 & 3+i \\ 4i & 5 \end{bmatrix}$
Transposed matrix $\mathbf{A}^T$	$(A^T)_{ij} = a_{ji}$	$\begin{bmatrix} 2 & 4i \\ 3+i & 5 \end{bmatrix}$
Complex conjugate matrix $\mathbf{A}^*$	$(A^*)_{ij} = a_{ij}^*$	$\begin{bmatrix} 2 & 3-i \\ -4i & 5 \end{bmatrix}$
Transposed conjugate matrix $\mathbf{A}^\dagger$	$(A^\dagger)_{ij} = a_{ji}^*$	$\begin{bmatrix} 2 & -4i \\ 3-i & 5 \end{bmatrix}$
Inverse matrix $\mathbf{A}^{-1}$	$(A^{-1})_{ij} = \frac{1}{\lvert A\rvert}\frac{\partial \lvert A\rvert}{\partial a_{ji}}$	$\frac{14+12i}{340}\begin{bmatrix} 5 & -3 \\ -4i & 2 \end{bmatrix}$

It is necessary to explain that the inverse matrix can only exist in a square matrix, and that its determinant cannot be equal to zero, and the 2 × 2 determinant, for example, has the following meaning about symbol $\partial|A|/\partial a_{ji}$:

$$|A| = \begin{vmatrix} a_{11} & a_{12} \\ a_{21} & a_{22} \end{vmatrix} = a_{11}a_{22} - a_{12}a_{21} \tag{B-42}$$

Then

$$\frac{\partial |A|}{\partial a_{21}} = -a_{12} \tag{B-43}$$

The following are some simple theorems about these matrices:

Theorem 1: $(AB)^T = (B)^T(A)^T$ (B-44)

Proof: $(AB)_{ij}^T = (AB)_{ji} = \sum_k A_{jk}B_{ki} = \sum_k (B)_{ik}^T (A)_{kj}^T = \left[(B)^T \quad (A)^T\right]_{ij}$

Theorem 2: $(AB)^* = (A^*)(B^*)$ (B-45)

Proof: $(AB)_{ij}^* = \sum_k A_{ik}^* B_{kj}^* = (A^*B^*)_{ij} = (A^*)_{ij}(B^*)_{ij}$

Theorem 3: $(AB)^\dagger = B^\dagger A^\dagger$ (B-46)

Proof: $(AB)^{\dagger}=[(AB)^{T}]^{*}=[(B)^{T}(A)^{T}]^{*}=B^{\dagger}A^{\dagger}$

Theorem 4: $(AB)^{-1}=B^{-1}A^{-1}$ (B-47)

Proof: $(B^{-1}A^{-1})(AB)=B^{-1}(A^{-1}A)B=B^{-1}B=E$

Table B-3: Some important special matrices and their properties.

Matrix name	Symbol and definition	Note
Unit matrix	E	$i{\neq}j,\ a_{ij}=0;\ a_{ij}=1$
Diagonal matrix	${}^{d}A$	$i{\neq}j,\ a_{ij}=0$
Symmetric matrix	$A^{T}=A$	$a_{ij}=a_{ji}$
Asymmetric matrix	$A^{T}=-A$	$a_{ij}=-a_{ji}$
Real matrix	$A^{*}=A$	$a_{ij}^{*}=a_{ij}$
Orthogonal matrix	$A^{-1}=A$	
Hermitian matrix	$A^{\dagger}=A$	$a_{ji}^{*}=a_{ij}$
Unitary matrix	$A^{-1}=A^{\dagger}$	

B-5.3 Some important special matrices

Some matrices have some special properties, and their names and definitions are shown in Table B-3.
Several theorems are also listed here:

Theorem 5: *The sum (or difference) of Hermitian matrices is still a Hermitian matrix*

Proof: $\boldsymbol{H}_1\pm\boldsymbol{H}_2=\boldsymbol{H}_1^{\dagger}\pm\boldsymbol{H}_2^{\dagger}=(\boldsymbol{H}_1\pm\boldsymbol{H}_2)^{\dagger}$

Theorem 6: *If the multiplication of the two Hermitian matrices* $\boldsymbol{H_1}$ *and* $\boldsymbol{H_2}$ *satisfies the commutative relation* $\boldsymbol{H_1H_2}=\boldsymbol{H_2H_1}$, then the matrix $\boldsymbol{H_1H_2}$ is also a Hermitian matrix.

Proof: $(\boldsymbol{H}_1\boldsymbol{H}_2)^{\dagger}=\boldsymbol{H}_2^{\dagger}\boldsymbol{H}_1^{\dagger}=\boldsymbol{H}_2\boldsymbol{H}_1=\boldsymbol{H}_1\boldsymbol{H}_2$

Theorem 7: *If a Hermitian matrix* $\boldsymbol{H}$ *is successively multiplied by unitary matrices* $\boldsymbol{U}^{-1}$ *and* $\boldsymbol{U}$ *on its left and right, then the resulting* $\boldsymbol{U}^{-1}\boldsymbol{HU}$ *matrix is also a Hermitian matrix.*

Proof: $(\boldsymbol{U}^{-1}\boldsymbol{HU})^{\dagger}=\boldsymbol{U}^{\dagger}\boldsymbol{H}^{\dagger}(\boldsymbol{U}^{-1})^{\dagger}=\boldsymbol{U}^{-1}\boldsymbol{HU}$

Theorem 8: *If* $\boldsymbol{H_1}$ *and* $\boldsymbol{H_2}$ *are two Hermitian matrices, then* $(\boldsymbol{H_1H_2}+\boldsymbol{H_2H_1})$ *and* $i(\boldsymbol{H_1H_2}-\boldsymbol{H_2H_1})$ *are also Hermitian matrices.*

Proof: $(\boldsymbol{H}_1\boldsymbol{H}_2+\boldsymbol{H}_2\boldsymbol{H}_1)^{\dagger}=(\boldsymbol{H}_2^{\dagger}\boldsymbol{H}_1^{\dagger}+\boldsymbol{H}_1^{\dagger}\boldsymbol{H}_2^{\dagger})=\boldsymbol{H}_2\boldsymbol{H}_1+\boldsymbol{H}_1\boldsymbol{H}_2=\boldsymbol{H}_1\boldsymbol{H}_2+\boldsymbol{H}_2\boldsymbol{H}_1$
Similarly, $[i(\boldsymbol{H}_1\boldsymbol{H}_2+\boldsymbol{H}_2\boldsymbol{H}_1)]^{\dagger}=(-\boldsymbol{i})((\boldsymbol{H_2^{\dagger}H_1^{\dagger}}-\boldsymbol{H_1^{\dagger}H_2^{\dagger}})=\boldsymbol{i}(\boldsymbol{H_1H_2}+\boldsymbol{H_2H_1})$

B-5.4 n-Dimensional linear space and n-dimensional vector

There are three-dimensional unit vectors ***i, j, k*** in three-dimensional space. Now expand it into n-dimensional space, there are n-dimensional base vectors $(\boldsymbol{e_1}, \boldsymbol{e_2}, \ldots, \boldsymbol{e_n})$, thus any vector can be represented as a linear combination of them:

$$\mathbf{X} = \sum x_i e_{i,} \qquad \mathbf{Y} = \sum y_i e_i \tag{B-48}$$

The n-dimensional vector has the following rules of operation:

$$\mathbf{X} \pm \mathbf{Y} = \sum_i (x_i \pm y_i) e_i \tag{B-49}$$

$$\lambda \mathbf{X} = \sum_i (\lambda x_i) e_i \tag{B-50}$$

where λ is a constant.

On multiplication: In the base vector $(\boldsymbol{e}_1, \boldsymbol{e}_2, \ldots, \boldsymbol{e}_n)$, the coordinates of **X** are $(x_1, x_2, \ldots, x_n)$, and the coordinates of **Y** are $(y_1, y_2, \ldots, y_n)$, then the inner product is defined as

$$(\mathbf{X}, \mathbf{Y}) = x_1^* y_1 + x_2^* y_2 + \ldots x_n^* y_n \equiv \sum_i x_i^* y_i$$

The inner product has the following properties:

$$(\mathbf{X}, \mathbf{Y}) = (\mathbf{Y}, \mathbf{X})^* \tag{B-51}$$

$$(\mathbf{Y}, \lambda \mathbf{X}) = \lambda (\mathbf{Y}, \mathbf{X}) \tag{B-52}$$

$$(\lambda \mathbf{Y}, \mathbf{X}) = \lambda^* (\mathbf{Y}, \mathbf{X}) \tag{B-53}$$

$$(\mathbf{Y}, \mathbf{X} + \mathbf{Z}) = (\mathbf{Y}, \mathbf{X}) + (\mathbf{Y}, \mathbf{Z}) \tag{B-54}$$

$(\mathbf{X}, \mathbf{X}) \geq 0$ Zero is only when **X** is a zero vector (B-55)

Orthonormal definition of base vectors:

$$(e_i, e_j) = \delta_{kj} \tag{B-56}$$

Several important theorems

Theorem 9: *The matrix that can makes the transformation between orthogonal normalized base vectors should be a unitary matrix.*

Proof: There is now a set of orthonormal base vectors $(\boldsymbol{e}_1, \boldsymbol{e}_2, \ldots, \boldsymbol{e}_n)$ that transform to another set of orthonormal base vectors $(\boldsymbol{e}'_1, \boldsymbol{e}'_2, \ldots, \boldsymbol{e}'_n)$. The new base vector can be represented as a linear combination of the primitive base vector:

$$e'_k = \sum_i a_{ik} e_i \tag{B-57}$$

Thus,

$$\begin{aligned}(\boldsymbol{e}'_k, \boldsymbol{e}'_j) &= \left(\sum_i a_{ik} e_i, \sum_m a_{mj} e_m\right) = \sum_i \sum_m a^*_{ik} a_{mj} (e_i, e_m) \\ &= \sum_i \sum_m a^*_{ik} a_{mj} \delta_{im} = \sum_i a^*_{ik} a_{ij}\end{aligned} \tag{B-58}$$

Because e'_i is a set of orthonormal base vector, so $(e'_k, e'_j) = \delta_{kj}$, that is to say,

$$\sum_i a^*_{ik} a_{ij} = \delta_{kj} \tag{B-59}$$

Equation (B-59) can be represented by matrix:

$$\begin{bmatrix} \boldsymbol{a}^*_{11} & \boldsymbol{a}^*_{21} & \cdots & \boldsymbol{a}^*_{n1} \\ \boldsymbol{a}^*_{12} & \boldsymbol{a}^*_{22} & \cdots & \boldsymbol{a}^*_{n2} \\ \vdots & \vdots & \vdots & \vdots \\ \boldsymbol{a}^*_{1n} & \boldsymbol{a}^*_{2n} & \cdots & \boldsymbol{a}^*_{nn} \end{bmatrix} \begin{bmatrix} \boldsymbol{a}_{11} & \boldsymbol{a}_{12} & \cdots & \boldsymbol{a}_{1n} \\ \boldsymbol{a}_{21} & \boldsymbol{a}_{22} & \cdots & \boldsymbol{a}_{2n} \\ \vdots & \vdots & \vdots & \vdots \\ \boldsymbol{a}_{n1} & \boldsymbol{a}_{n2} & \cdots & \boldsymbol{a}_{nn} \end{bmatrix} = \begin{bmatrix} 1 & 0 & \cdots & 0 \\ 0 & 1 & \cdots & 0 \\ \vdots & \vdots & \ddots & \vdots \\ 0 & 0 & \cdots & 1 \end{bmatrix} \tag{B-60}$$

Thus, $\boldsymbol{A}^\dagger \boldsymbol{A} = \boldsymbol{E}$; because $\boldsymbol{A}^{-1}\boldsymbol{A} = \boldsymbol{E}$, so $\boldsymbol{A}^\dagger = \boldsymbol{A}^{-1}$, hence $\boldsymbol{A}$ is a *unitary matrix.*

Theorem 10: *If the base vector* $(\boldsymbol{e}'_1, \boldsymbol{e}'_2, \ldots, \boldsymbol{e}'_n) = (\boldsymbol{e}_1, \boldsymbol{e}_2, \ldots, \boldsymbol{e}_n)\, \boldsymbol{A}$ *transform, then the coordinate of vector* **X** *has transformation as follows:*

$$\begin{bmatrix} x_1 \\ x_2 \\ \vdots \\ x_n \end{bmatrix} = \boldsymbol{A} \begin{bmatrix} x'_1 \\ x'_2 \\ \vdots \\ x'_n \end{bmatrix} \tag{B-61}$$

Proof: Because $\mathbf{X} = \sum_i x_i e_i = \sum_i x'_i e'_i$,

$$(\boldsymbol{e}_1, \boldsymbol{e}_2, \ldots, \boldsymbol{e}_n)\begin{bmatrix} x_1 \\ x_2 \\ \vdots \\ x_n \end{bmatrix} = (\boldsymbol{e}'_1, \boldsymbol{e}'_2, \ldots, \boldsymbol{e}'_n)\begin{bmatrix} x'_1 \\ x'_2 \\ \vdots \\ x'_n \end{bmatrix} = (\boldsymbol{e}_1, \boldsymbol{e}_2, \ldots, \boldsymbol{e}_n)\boldsymbol{A}\begin{bmatrix} x'_1 \\ x'_2 \\ \vdots \\ x'_n \end{bmatrix}$$

Thus, $\begin{bmatrix} x_1 \\ x_2 \\ \vdots \\ x_n \end{bmatrix} = \boldsymbol{A}\begin{bmatrix} x'_1 \\ x'_2 \\ \vdots \\ x'_n \end{bmatrix}$

B-5.5 Linear transformation

When a matrix $\boldsymbol{A}$ acts on the vector **X**, it is transformed into a new vector **Y**:

$$\mathbf{Y} = \boldsymbol{A}\mathbf{X} \tag{B-62}$$

If this transformation satisfies the following two conditions:

$$\boldsymbol{A}(\mathbf{X}_1 + \mathbf{X}_2) = \boldsymbol{A}\mathbf{X}_1 + \boldsymbol{A}\mathbf{X}_2 \tag{B-63}$$

$$\boldsymbol{A}(\lambda\boldsymbol{X}) = \lambda\boldsymbol{A}\boldsymbol{X} \tag{B-64}$$

then this transformation is called *linear transformation.* It can be represented by matrix:

$$\begin{bmatrix} y_1 \\ y_2 \\ \vdots \\ y_n \end{bmatrix} = \begin{bmatrix} a_{11} & a_{12} & \cdots & a_{1n} \\ a_{21} & a_{22} & \cdots & a_{2n} \\ \vdots & \vdots & \vdots & \vdots \\ a_{n1} & a_{n2} & \cdots & a_{nn} \end{bmatrix} \begin{bmatrix} x_1 \\ x_2 \\ \vdots \\ x_n \end{bmatrix} \tag{B-65}$$

It should be pointed out that the matrix representation of the linear transformation depends on the choice of base vector: If the base vector is changed, then the form of the matrix also changes.

Theorem 11: *Assume the matrix representation of the linear transformation* $\boldsymbol{T}$ *in the basis vector* $(\boldsymbol{e}_1, \boldsymbol{e}_2, \ldots, \boldsymbol{e}_n)$ *is* (a_{ik})*, and the matrix representation in the basis vector* $(\boldsymbol{e}'_1, \boldsymbol{e}'_2, \ldots, \boldsymbol{e}'_n)$ is (b_{ik}). If $(\boldsymbol{e}'_1, \boldsymbol{e}'_2, \ldots, \boldsymbol{e}'_n) = (\boldsymbol{e}_1, \boldsymbol{e}_2, \ldots, \boldsymbol{e}_n)C$, then $B = C^{-1}AC$

Proof: Because $\mathbf{Y} = \boldsymbol{T}\mathbf{X}$, the matrix representation in the basis vector $(\boldsymbol{e}_1, \boldsymbol{e}_2, \ldots, \boldsymbol{e}_n)$ is

$$\begin{bmatrix} y_1 \\ y_2 \\ \vdots \\ y_n \end{bmatrix} = \begin{bmatrix} a_{11} & a_{12} & \cdots & a_{1n} \\ a_{21} & a_{22} & \cdots & a_{2n} \\ \vdots & \vdots & \vdots & \vdots \\ a_{n1} & a_{n2} & \cdots & a_{nn} \end{bmatrix} \begin{bmatrix} x_1 \\ x_2 \\ \vdots \\ x_n \end{bmatrix} \tag{B-66}$$

and the matrix representation in the basis vector $(\boldsymbol{e}'_1, \boldsymbol{e}'_2, \ldots, \boldsymbol{e}'_n)$ is

$$\begin{bmatrix} y'_1 \\ y'_2 \\ \vdots \\ y'_n \end{bmatrix} = \begin{bmatrix} b_{11} & b_{12} & \cdots & b_{1n} \\ b_{21} & b_{22} & \cdots & b_{2n} \\ \vdots & \vdots & \vdots & \vdots \\ b_{n1} & b_{n2} & \cdots & b_{nn} \end{bmatrix} \begin{bmatrix} x'_1 \\ x'_2 \\ \vdots \\ x'_n \end{bmatrix} \tag{B-67}$$

because $(\boldsymbol{e}'_1, \boldsymbol{e}'_2, \ldots, \boldsymbol{e}'_n) = (\boldsymbol{e}_1, \boldsymbol{e}_2, \ldots, \boldsymbol{e}_n)C$. Equation (B-66) will be changed as

$$\boldsymbol{C} = \begin{bmatrix} y'_1 \\ y'_2 \\ \vdots \\ y'_n \end{bmatrix} = \boldsymbol{AC} \begin{bmatrix} x'_1 \\ x'_2 \\ \vdots \\ x'_n \end{bmatrix} \tag{B-68}$$

In comparison with Eq. (B-67), $\boldsymbol{B} = \boldsymbol{C}^{-1}\boldsymbol{AC}$.

Theorem 12: *If $(\boldsymbol{e}_1, \boldsymbol{e}_2, \ldots, \boldsymbol{e}_n)$ and $(\boldsymbol{e}'_1, \boldsymbol{e}'_2, \ldots, \boldsymbol{e}'_n)$ are all orthonormal base vectors, then*

$$\boldsymbol{B} = \boldsymbol{U}^{\dagger}\boldsymbol{AU}$$

Proof: In this case, C is an unitary matrix, and C can be represented by $\boldsymbol{U}$; thus,

$$\boldsymbol{U}^{-1} = \boldsymbol{U}^{\dagger}$$

B-5.6 Eigenvector and eigenvalue

If there are a linear transformation matrix $\boldsymbol{T}$ and a vector $\boldsymbol{X}$ satisfying $TX = \lambda X$, then $\boldsymbol{X}$ is the eigenvector of $\boldsymbol{T}$. λ is the eigenvalue of eigenvector $\boldsymbol{X}$. When the base vector $(\boldsymbol{e}_1, \boldsymbol{e}_2, \ldots, \boldsymbol{e}_n)$ is chosen, then according to Eq. (B-66) there is

$$\begin{bmatrix} a_{11} & a_{12} & \cdots & a_{1n} \\ a_{21} & a_{22} & \cdots & a_{2n} \\ \vdots & \vdots & \vdots & \vdots \\ a_{n1} & a_{n2} & \cdots & a_{nn} \end{bmatrix} \begin{bmatrix} x_1 \\ x_2 \\ \vdots \\ x_n \end{bmatrix} = \lambda \begin{bmatrix} x_1 \\ x_2 \\ \vdots \\ x_n \end{bmatrix} \tag{B-69}$$

This is a set of simultaneous equations:

$$\begin{cases} a_{11}x_1 + a_{12}x_2 + \cdots + a_{1n}x_n = \lambda x_1 \\ a_{21}x_1 + a_{22}x_2 + \cdots + a_{2n}x_n = \lambda x_2 \\ \vdots \cdots\cdots\cdots\cdots\cdots\cdots\cdots\cdots = \vdots \\ a_{n1}x_1 + a_{n2}x_2 + \cdots + a_{nn}x_n = \lambda x_n \end{cases} \tag{B-70}$$

The necessary condition for the nonzero solution of this set of simultaneous equations is

$$\begin{vmatrix} a_{11} - \lambda & a_{12} & \cdots & a_{1n} \\ a_{21} & a_{22} - \lambda & \cdots & a_{2n} \\ \vdots & \vdots & & \vdots \\ a_{n1} & a_{n2} & \cdots & a_{nn} - \lambda \end{vmatrix} = 0 \tag{B-71}$$

This is the secular determinant well known to us.

If the result of the operator $\hat{\boldsymbol{\Lambda}}$ acts on the function ψ_k,

$$\hat{\boldsymbol{\Lambda}}\psi_k = \lambda\psi_k \tag{B-72}$$

And if λ is a constant, then ψ_k is the eigenfunction of the operator $\hat{\boldsymbol{\Lambda}}$, and λ is the eigenvalue of the eigenfunction ψ_k. *Note:* If c is any nonzero scalar, then $c\psi_k$ also is an eigenfunction of the operator $\hat{\boldsymbol{\Lambda}}$. If ψ_k is a space function, it is also called a wave function. When the ψ_k only concerns the spin variables and has no function-dependent relationship, then the two terms of the "eigenfunction" and "eigenstate" can be exchanged mutually. That is, the "eigenfunction" ψ_k can also be called "eigenstate."

The set ψ_k of all *eigenfunctions* is often called "the basic set." In Chapter 3, we defined $\alpha(e) = \psi_{\alpha(e)}$, $\beta(e) = \psi_{\beta(e)}$. They are all electron spin functions, and all are eigenfunctions of the operator $\hat{S}_z$:

$$\hat{S}_z\,\alpha(e) = +\frac{1}{2}\alpha(e) \tag{B-73a}$$

$$\hat{S}_z\,\beta(e) = +\frac{1}{2}\beta(e) \tag{B-73b}$$

Given a set of eigenfunctions, they can be the eigenfunctions of several operators simultaneously; several different operators can have a set of identical eigenfunctions, and then these operators should be commutative. This property is very important and very useful. As for a particle of mass m to do a circular motion with a radius of r, the wave function is the common eigenfunction of the angular momentum operator $\hat{1}_z$ and the total energy operators $\hat{\mathscr{H}}$. The eigenequations

$$\hat{1}_z\psi = P\psi \tag{B-74}$$

and

$$\hat{\mathscr{H}}\psi = E\psi \tag{B-75}$$

In Table B-1, $\hat{1}_z = -ih\frac{d}{d\phi}$, here ϕ is the angular position where the particle is located. The kinetic energy of a classical particle with angular momentum P and rotational inertia I_o is

$$E = \frac{P^2}{2I_o} \tag{B-76}$$

For the system of potential energy V= 0, the Hamiltonian operator is

$$\hat{\mathscr{H}} = \frac{\hat{1}^2}{2I_o} = \frac{(-i\hbar)^2}{2I_o}\frac{\hat{d}^2}{d\phi^2} = \frac{-\hbar^2}{2I_o}\frac{\hat{d}^2}{d\phi^2} \tag{B-77}$$

Substituting the Hamiltonian operator of Eq. (B-77) into Eq. (B-75), we obtain

$$\frac{-\hbar^2}{2I_o}\frac{\hat{d}^2\psi}{d\phi^2} = E\psi \tag{B-78}$$

Rearranging Eq. (B-78),

$$\frac{\hat{d}^2\psi}{d\phi^2} = -\frac{2I_oE}{\hbar^2}\psi = -M^2\psi \tag{B-79}$$

Here, let $M^2 = 2I_oE/\hbar^2$, there are two solutions of Eq. (B-79):

$$\psi_1 = Ae^{+iM\phi} \tag{B-80a}$$

$$\psi_2 = Ae^{-iM\phi} \tag{B-80b}$$

Combining the normalized property of function ψ,

$$\int_0^{2\pi} \psi^*\psi d\phi = 1 \tag{B-81}$$

$A = (2\pi)^{-1/2}$ can be found, thus

$$\psi_1 = (2\pi)^{-1/2} e^{+iM\phi} \tag{B-82a}$$

$$\psi_2 = (2\pi)^{-1/2} e^{-iM\phi} \tag{B-82b}$$

Substituting ψ_1 into Eq.(B-78), we obtain

$$\frac{-\hbar^2}{2I_o}\frac{\hat{d}^2}{d\phi}[(2\pi)^{-1/2} e^{+iM\phi}] = \frac{M^2\hbar^2}{2I_o}[(2\pi)^{-1/2} e^{+iM\phi}] \tag{B-83}$$

The eigenvalue E of the Hamiltonian operator $\hat{\mathrm{H}}$ for ψ_1 and ψ_2 is $\frac{M^2\hbar^2}{2I_O}$, same for both.

The results of the operator $\hat{1}_z$ act on ψ_1 and ψ_2 as follows:

$$-i\hbar\frac{\hat{d}}{d\phi}[(2\pi)^{-1/2} e^{+iM\phi}] = +M\hbar[(2\pi)^{-1/2} e^{+iM\phi}] \tag{B-84a}$$

$$-i\hbar\frac{\hat{d}}{d\phi}[(2\pi)^{-1/2} e^{-iM\phi}] = -M\hbar[(2\pi)^{-1/2} e^{-iM\phi}] \tag{B-84b}$$

Corresponding to the eigenfunctions ψ_1 and ψ_2, the eigenvalues of operator $\hat{1}_z$ are $+M\hbar$ and $-M\hbar$. Corresponding two different rotational directions around the z-axis rotating.

Here we introduce the following theorems:

Theorem 13: *The eigenvalue is independent of the choice of base vectors.*

Proof: Assume the matrix representation of $\hat{\boldsymbol{A}}$ in the base vector $(\boldsymbol{e}_1, \boldsymbol{e}_2, \ldots, \boldsymbol{e}_n)$ is $\boldsymbol{A}$, and the matrix representation in the base vector $(\boldsymbol{e}'_1, \boldsymbol{e}'_2, \ldots, \boldsymbol{e}'_n)$ is $\boldsymbol{B}$,

$$\boldsymbol{B} = \boldsymbol{C}^{-1}\boldsymbol{A}\boldsymbol{C}$$

Then, the eigenvalues $\{\lambda_1, \lambda_2, \ldots, \lambda_n\}$ in the base vector $(e_1, e_2, \ldots, e_n)$ should be solved from

$$|\boldsymbol{A} - \lambda\boldsymbol{I}| = 0 \tag{B-85a}$$

and the eigenvalues in the base vector $(\boldsymbol{e}'_1, \boldsymbol{e}'_2, \ldots, \boldsymbol{e}'_n)$ should be solved from

$$|\boldsymbol{B} - \lambda\boldsymbol{I}| = 0 \tag{B-85b}$$

Thus,

$$|\boldsymbol{B} - \lambda\boldsymbol{I}| = |\boldsymbol{C}^{-1}\boldsymbol{A}\boldsymbol{C} - \lambda\boldsymbol{I}| = |\boldsymbol{C}^{-1}(\boldsymbol{A} - \lambda\boldsymbol{I})\boldsymbol{C}| = |\boldsymbol{C}^{-1}|\,|\boldsymbol{A} - \lambda\boldsymbol{I}|\,|\boldsymbol{C}| = |\boldsymbol{A} - \lambda\boldsymbol{I}|$$

So, it is independent of the choice of base vectors.

Theorem 14: *The eigenvalue of Hermitian matrix is real number.*

Proof: The eigenvalue λ of Hermitian matrix $\boldsymbol{H}$,

$$(X,\ HX) = (X,\ \lambda X) = \lambda(X,\ X) \tag{B-86a}$$

On the other hand,

$$(X,\ HX) = (\boldsymbol{H}^{\dagger}X,\ X) = (HX,\ X) = (\lambda X,\ X) = \lambda^*(X,\ X) \tag{B-86b}$$

Equation (B-86a) reduces Eq.(B-86b):

$$(\lambda - \lambda^*)(\boldsymbol{X}, \boldsymbol{X}) = 0 \tag{B-87}$$

because $(X, X) > 0$, so, $\lambda - \lambda^* = 0$, $\lambda = \lambda^*$, thus, λ is a real number.

Theorem 15: *For the Hermitian matrix* $\boldsymbol{H}$*, the eigenvectors belonging to different eigenvalues are orthogonal each other.*

Proof: Let $\boldsymbol{HX} = \lambda \boldsymbol{X}$, $\quad \boldsymbol{HY} = \lambda' \boldsymbol{Y}$
then,

$$(\boldsymbol{Y},\ \boldsymbol{HX}) = (\boldsymbol{Y}, \lambda \boldsymbol{X}) = \lambda(\boldsymbol{Y},\ \boldsymbol{X}) \tag{B-88}$$

On the other hand, $(\boldsymbol{Y},\ \boldsymbol{HX}) = (\boldsymbol{H}^{\dagger}\boldsymbol{Y},\ \boldsymbol{X}) = (\boldsymbol{HY},\ \boldsymbol{X}) = \lambda'^*(\boldsymbol{Y},\ \boldsymbol{X}) = \lambda'(\boldsymbol{Y},\ \boldsymbol{X})$ (B-89)
Comparing Eqs. (B-88) and (B-89),

$$(\lambda - \lambda')(\boldsymbol{Y},\ \boldsymbol{X}) = 0 \tag{B-90}$$

because $\lambda \neq \lambda'$, so, $(\boldsymbol{Y}, \boldsymbol{X}) = 0$

Theorem 16: *If the column vectors* $\boldsymbol{u}_1, \boldsymbol{u}_2, \ldots, \boldsymbol{u}_n$ *are the normalized eigenvectors of the Hermitian matrix* $\boldsymbol{H}$*, and the corresponding eigenvalues are* $\lambda_1, \lambda_2, \ldots, \lambda_n$*, then* $\boldsymbol{U}^{-1}\boldsymbol{H}\boldsymbol{U} = \boldsymbol{\Lambda}$*, where* $\boldsymbol{U}$ *is the unitary matrix, and the* $\boldsymbol{\Lambda}$ *should be diagonal matrix:*

$$\boldsymbol{U} = (\boldsymbol{u}_1, \boldsymbol{u}_2, \ldots, \boldsymbol{u}_n) = \begin{bmatrix} u_{11} & u_{12} & \cdots & u_{1n} \\ u_{21} & u_{22} & \cdots & u_{2n} \\ \vdots & \vdots & & \vdots \\ u_{n1} & u_{n2} & \cdots & u_{nn} \end{bmatrix}, \quad \boldsymbol{\Lambda} = \begin{bmatrix} \lambda_1 & 0 & \cdots & 0 \\ 0 & \lambda_2 & \cdots & 0 \\ \vdots & & \ddots & \vdots \\ 0 & 0 & \cdots & \lambda_n \end{bmatrix} \tag{B-91}$$

Proof: Because $H\boldsymbol{u_i} = \lambda\, \boldsymbol{u_i}$

Thus,

$$\boldsymbol{H} = \begin{bmatrix} u_{11} & u_{12} & \cdots & u_{1n} \\ u_{21} & u_{22} & \cdots & u_{2n} \\ \vdots & \vdots & & \vdots \\ u_{n1} & u_{n2} & \cdots & u_{nn} \end{bmatrix} = \begin{bmatrix} \lambda_1 u_{11} & \lambda_2 u_{12} & \cdots & \lambda_n u_{1n} \\ \lambda_1 u_{21} & \lambda_2 u_{22} & \cdots & \lambda_n u_{2n} \\ \vdots & \vdots & & \vdots \\ \lambda_1 u_{n1} & \lambda_2 u_{n2} & \cdots & \lambda_n u_{nn} \end{bmatrix} = \begin{bmatrix} u_{11} & u_{12} & \cdots & u_{1n} \\ u_{21} & u_{22} & \cdots & u_{2n} \\ \vdots & \vdots & & \vdots \\ u_{n1} & u_{n2} & \cdots & u_{nn} \end{bmatrix} \begin{bmatrix} \lambda_1 & 0 & \cdots & 0 \\ 0 & \lambda_2 & \cdots & 0 \\ \vdots & \vdots & \ddots & \vdots \\ 0 & 0 & \cdots & \lambda_n \end{bmatrix}$$

$\boldsymbol{u}_1, \boldsymbol{u}_2, \ldots, \boldsymbol{u}_n$ are known as normalized vectors that are orthogonal to each other, so the matrix $\boldsymbol{U}$ is a unitary matrix.

Theorem 17: *The sum (or difference) of both unitary matrices is no longer a unitary matrix, but the product of the two unitary matrices still is a unitary matrix.*

Proof: $(\boldsymbol{U}_1 + \boldsymbol{U}_2)^\dagger = \boldsymbol{U}_1^\dagger + \boldsymbol{U}_2^\dagger = \boldsymbol{U}_1^{-1} + \boldsymbol{U}_2^{-1} \neq (\boldsymbol{U}_1 + \boldsymbol{U}_2)^{-1}$

$$(\boldsymbol{U}_1\boldsymbol{U}_2)^\dagger = \boldsymbol{U}_2^\dagger \boldsymbol{U}_1^\dagger = \boldsymbol{U}_2^{-1}\boldsymbol{U}_1^{-1} = (\boldsymbol{U}_1\boldsymbol{U}_2)^{-1}$$

Theorem 18: *The eigenvalue of the unitary matrix has the form of $e^{i\theta}$.*

Proof: Let λ and $\boldsymbol{X}$ be the eigenvalue and eigenvector of a unitary matrix $\boldsymbol{U}$, that is, $\boldsymbol{U}\,\boldsymbol{X} = \lambda\,\boldsymbol{X}$, then,

$$(\boldsymbol{UX},\ \boldsymbol{UX}) = (\lambda\boldsymbol{X},\ \lambda\boldsymbol{X}) = \lambda^*\lambda\,(\boldsymbol{X}, \boldsymbol{X}) \tag{B-92}$$

On the other hand,

$$(\boldsymbol{UX},\ \boldsymbol{UX}) = (\boldsymbol{U}^\dagger\boldsymbol{UX},\ \boldsymbol{X}) = (\boldsymbol{U}^{-1}\boldsymbol{UX}, \boldsymbol{X}) = (\boldsymbol{X},\ \boldsymbol{X}) \tag{B-93}$$

Comparing Eqs. (B-92) and (B-93), we obtain

$$\lambda^*\lambda = 1,$$

Thus, $\lambda = e^{i\theta}$

Theorem 19: *For any unitary matrix $\boldsymbol{U}$, there should exist another unitary matrix $\boldsymbol{V}$, for example, $\boldsymbol{V}^{-1}\boldsymbol{UV} = \boldsymbol{\Lambda}$, then $\boldsymbol{\Lambda}$ must be a diagonal matrix.*

This theorem is left to prove by the readers themselves.

Further readings

[1] Anderson J. M. *Mathematics for Quantum Chemistry*, Benjamin, New York, NY, 1966.
[2] Atkins P. W. *Molecular Quantum Mechanics*, 2nd edn, Oxford University Press, Oxford, UK, 1983.
[3] Bak T., Lichtenberg J. *Mathematics for Scientists*, 3 Vols., Benjamin, New York, NY, 1966.
[4] Hamilton A. G. *Linear Algebra*, Cambridge University Press,, Cambridge, UK, 1989.
[5] Nye J. F. *Physical Properties of Crystals*, 2nd edn, Oxford University Press, Oxford, UK, 1985.
[6] Wooster W. A. *Tensors and Group Theory for the Physical Properties of Crystals*, Clarendon, Oxford, UK, 1972.
[7] Yariv A. *An Introduction to Theory and Application of Quantum Mechanics*, John Wiley & Sons, Inc., New York, NY, 1982.

Appendix 3: Angular momentum and stable-state perturbation theory in quantum mechanics

C-1 Angular momentum operator

In classical mechanics, angular momentum L is given by the following formula:

$$\boldsymbol{L} = \boldsymbol{r} \times \boldsymbol{P} = \begin{vmatrix} i & j & k \\ x & y & z \\ p_x & p_y & p_z \end{vmatrix} \tag{C-1}$$

Thus,

$$L_x = yp_z - zp_y \tag{C-2a}$$

$$L_y = zp_x - xp_z \tag{C-2b}$$

$$L_z = xp_y - yp_x \tag{C-2c}$$

According to the correspondence principle, in quantum mechanics, the orbital angular momentum has the same definition as classical mechanics. The difference between them is that the physical quantities of Eq. (C-2) are replaced by corresponding operators:

$$\hat{\boldsymbol{L}}_x = y\hat{p}_z - z\hat{p}_y = -i\hbar\left(y\frac{\partial}{\partial z} - z\frac{\partial}{\partial y}\right) \tag{C-3a}$$

$$\hat{\boldsymbol{L}}_y = z\hat{p}_x - x\hat{p}_z = -i\hbar\left(z\frac{\partial}{\partial x} - x\frac{\partial}{\partial z}\right) \tag{C-3b}$$

$$\hat{\boldsymbol{L}}_z = x\hat{p}_y - y\hat{p}_x = -i\hbar\left(x\frac{\partial}{\partial y} - y\frac{\partial}{\partial x}\right) \tag{C-3c}$$

The operator of the square angular momentum is

$$\hat{\boldsymbol{L}}^2 = \hat{L}_x^2 + \hat{L}_y^2 + \hat{L}_z^2 \tag{C-4}$$

The above is the case of Cartesian coordinates. In spherical coordinates, according to the relations of Cartesian coordinates with spherical coordinates,

$$x = r\sin\theta\cos\phi, \quad y = r\sin\theta\sin\phi, \quad z = r\cos\theta \tag{C-5a}$$

$$r^2 = x^2 + y^2 + z^2, \quad \cos\theta = \frac{z}{r}, \quad tg\phi = \frac{y}{x} \tag{C-5b}$$

$$\frac{\partial}{\partial x} = \frac{\partial r}{\partial x}\frac{\partial}{\partial r} + \frac{\partial \theta}{\partial x}\frac{\partial}{\partial \theta} + \frac{\partial \phi}{\partial x}\frac{\partial}{\partial \phi} = \sin\theta\cos\phi\frac{\partial}{\partial r} + \frac{1}{r}\cos\theta\cos\phi\frac{\partial}{\partial \theta} + \frac{1}{r}\frac{\sin\phi}{\sin\theta}\frac{\partial}{\partial \phi} \tag{C-6a}$$

https://doi.org/10.1515/9783110568578-013

$$\frac{\partial}{\partial y} = \frac{\partial r}{\partial y}\frac{\partial}{\partial r} + \frac{\partial \theta}{\partial y}\frac{\partial}{\partial \theta} + \frac{\partial \phi}{\partial y}\frac{\partial}{\partial \phi} = \sin\theta\sin\phi\frac{\partial}{\partial r} + \frac{1}{r}\cos\theta\sin\phi\frac{\partial}{\partial \theta} + \frac{1}{r}\frac{\cos\phi}{\sin\theta}\frac{\partial}{\partial \phi} \quad \text{(C-6b)}$$

$$\frac{\partial}{\partial z} = \frac{\partial r}{\partial z}\frac{\partial}{\partial r} + \frac{\partial \theta}{\partial z}\frac{\partial}{\partial \theta} + \frac{\partial \phi}{\partial z}\frac{\partial}{\partial \phi} = \cos\theta\frac{\partial}{\partial r} - \frac{1}{r}\sin\theta\frac{\partial}{\partial \theta} \quad \text{(C-6c)}$$

By substituting these relations into Eqs. (C-3) and (C-4), the operator expressed in spherical coordinates can be obtained:

$$\hat{L}_x = -i\text{h}\left(\sin\phi\frac{\partial}{\partial\theta} - ctg\theta\cos\phi\frac{\partial}{\partial\phi}\right) \quad \text{(C-7a)}$$

$$\hat{L}_y = -i\hbar\left(\cos\phi\frac{\partial}{\partial\theta} - ctg\theta\sin\phi\frac{\partial}{\partial\phi}\right) \quad \text{(C-7b)}$$

$$\hat{L}_z = -i\hbar\frac{\partial}{\partial\phi} \quad \text{(C-7c)}$$

$$\hat{L}^2 = -\hbar^2\left[\frac{1}{\sin\theta}\frac{\partial}{\partial\theta}\left(\sin\theta\frac{\partial}{\partial\theta}\right) + \frac{1}{\sin^2\theta}\frac{\partial^2}{\partial\phi^2}\right] \quad \text{(C-7d)}$$

C-2 Commutation relation of angular momentum operators

In quantum mechanics, the angular momentum operators have the following commutation relations:

Let symbol $[\hat{A}, \hat{B}] = \hat{A}\hat{B} - \hat{B}\hat{A}$, then

$$[\hat{L}_x, \hat{L}_y] = i\hbar\hat{L}_z \quad [\hat{L}_y, \hat{L}_z] = i\hbar\hat{L}_x \quad [\hat{L}_z, \hat{L}_x] = i\hbar\hat{L}_y \quad \text{(C-8)}$$

$$[\hat{L}^2, \hat{L}_x] = [\hat{L}^2, \hat{L}_y] = [\hat{L}^2, \hat{L}_z] = 0 \quad \text{(C-9a)}$$

$$[\hat{L}^2, \hat{B}] = [\hat{L}^2, \hat{L}] = 0 \quad \text{(C-9b)}$$

$$[\hat{L}_z, \hat{L}_-] = -\hbar\hat{L}_-, \qquad [\hat{L}_z, \hat{L}_+] = \hbar\hat{L}_+ \quad \text{(C-10a)}$$

$$[\hat{L}_+, \hat{L}_-] = 2\hbar\hat{L}_z \quad \text{(C-10b)}$$

In order to prove the above formula, the following three basic commutation relations should be proved first:

$$[q_i, q_j] = 0, \quad \text{where } q_i, q_j = x, y, z \quad \text{(C-11)}$$

$$[p_i, p_j] = 0 \quad \text{(C-12)}$$

$$[q_i, p_j] = i\hbar\,\delta_{ij} \quad \text{(C-13)}$$

It is enough to prove only one formula as follows:

$$[x, \hat{p}_x]\psi = (x\hat{p}_x - \hat{p}_x x)\psi = -i\hbar(x\tfrac{\partial}{\partial x} - \tfrac{\partial}{\partial x}x)\psi = -i\hbar\{x\tfrac{\partial}{\partial x} - \tfrac{\partial}{\partial x}(x\psi)\}$$
$$= -i\hbar\left\{x\tfrac{\partial\psi}{\partial x} - \psi\tfrac{\partial x}{\partial x}x\tfrac{\partial\psi}{\partial x}\right\} = i\hbar\psi$$

Equation (C-13), which is $[x, \hat{p}_x] = i\hbar$, is thus proved.

It is easy to prove Eqs. (C-8)–(C-10) according to the basic commutation relation:

$$[\hat{L}_x, \hat{L}_y] = [y\hat{p}_z - z\hat{p}_y, z\hat{p}_x - x\hat{p}_z] = [y\hat{p}_z, z\hat{p}_x] - [z\hat{p}_y, z\hat{p}_x] - [y\hat{p}_z, x\hat{p}_z] + [z\hat{p}_y, x\hat{p}_z]$$
$$= y[\hat{p}_z, z]\hat{p}_x + x[z, \hat{p}_z]\hat{p}_y = -i\hbar y\hat{p}_x + i\hbar x\hat{p}_y = i\hbar(x\hat{p}_y - y\hat{p}_x) = i\hbar\,\hat{L}_z$$

This is the proof of Eq. (C-8). Other formulas are left to be proved by the readers themselves.

The following should be pointed out here:

1. In the classical mechanics, the magnitude and direction of the angular momentum is completely determined, but in the quantum mechanics, only the total angular momentum and one component can be determined at the same time. The magnitudes of other two components cannot have determining value simultaneously. The reason is that the $\hat{L}_x$, $\hat{L}_y$, $\hat{L}_z$ in the operators $\hat{L}^2$, $\hat{L}_x$, $\hat{L}_y$, $\hat{L}_z$ cannot commute mutually.
2. The angular momentum operator has a deeper meaning for the commutation relation, and we will no longer use $\boldsymbol{L} = \boldsymbol{r} \times \boldsymbol{p}$ to define angular momentum, because it is only applicable to orbital angular momentum. In order to be applicable to the general angular momentum (including spin angular momentum), we use the commutation relations between angular momentum components as the definition of angular momentum. That is to say, where the mechanical quantity operator $\hat{\boldsymbol{J}}$ satisfies the commutative relation $\hat{\boldsymbol{J}} \times \hat{\boldsymbol{J}} = i\hbar\hat{\boldsymbol{J}}$, we call $\hat{\boldsymbol{J}}$ the angular momentum operator. Likewise, the spin operator $\hat{\boldsymbol{S}}$ satisfies the commutative relation $\hat{\boldsymbol{S}} \times \hat{\boldsymbol{S}} = i\hbar\,\hat{\boldsymbol{S}}$.

C-3 Eigenvalues of $\hat{J}^2$ and $\hat{J}_z$

Since $\hat{J}^2$ and $\hat{J}_z$ are commutative, they should have a common eigenfunction, $|j, m\rangle$. Then,

$$\hat{J}^2|j, m\rangle = \lambda_j|j, m\rangle \tag{C-14}$$

$$\hat{J}_z|j, m\rangle = \lambda_m|j, m\rangle \tag{C-15}$$

Here λ_j and λ_m are their corresponding eigenvalues; now it will be proved that

$$\lambda_j = j(j+1)$$
$$\lambda_m = m$$

First of all, for fixed λ_j and λ_m, there are an upper and a lower limit; it should satisfy

$$\lambda_j - \lambda_m^2 \geq 0 \tag{C-16}$$

Since

$$\langle j, m|\hat{J}_x^2 + \hat{J}_y^2|j, m\rangle = \langle j, m|\hat{J}^2 - \hat{J}_z^2|j, m\rangle = \lambda_j - \lambda_m^2$$
$$\langle j, m|\hat{J}_x^2|j, m\rangle = (\hat{J}_x|j, m\rangle, \hat{J}_x|j, m\rangle) \geq 0$$

Similarly,

$$\langle j, m|\hat{J}_y^2|j, m\rangle \geq 0$$

So $\lambda_j - \lambda_m^2 \geq 0$

Second, integer is the only difference that exist between the values λ_m, because $\hat{J}_+ = \hat{J}_x + i\hat{J}_y$ and $\hat{J}_- = \hat{J}_x - i\hat{J}_y$:

$$[\hat{J}^2, \hat{J}_+]_- = [\hat{J}^2, \hat{J}_-]_- = 0$$
$$[\hat{J}_z, \hat{J}_+]_- = \hat{J}_+$$
$$[\hat{J}_z, \hat{J}_-]_- = -\hat{J}_-$$
$$[\hat{J}_+, \hat{J}_-]_- = 2\hat{J}_z$$

Thus,

$$\langle j, m'|\hat{J}_z\hat{J}_+ - \hat{J}_+\hat{J}_z|j, m\rangle = \langle j, m'|\hat{J}_+|j, m\rangle \tag{C-17}$$

The left matrix element of this formula can be divided into two matrix elements:

$$\langle j, m'|\hat{J}_z\hat{J}_+ - \hat{J}_+\hat{J}_z|j, m\rangle = \langle j, m'|\hat{J}_z\hat{J}_+|j, m\rangle - \langle j, m'|\hat{J}_+\hat{J}_z|j, m\rangle$$

The latter can be obtained by Eq. (C-15):

$$\langle j, m'|\hat{J}_+\hat{J}_z|j, m\rangle = \lambda_m\langle j, m'|\hat{J}_+|j, m\rangle \tag{C-18a}$$

The former can be transformed as

$$\langle j, m'|\hat{J}_z\hat{J}_+|j, m\rangle = \lambda_{m'}^*\langle j, m'|\hat{J}_+|j, m\rangle \tag{C-18b}$$

Substituting Eqs. (C-18a) and (C-18*b*) into Eq. (C-17), we obtain

$$\langle j, m'|\hat{J}_+|j, m\rangle = (\lambda_{m'}^* - \lambda_m)\langle j, m'|\hat{J}_+|j, m\rangle \tag{C-19}$$

Because $\hat{J}_z$ is a Hermitian operator, so $\lambda_{m'}^* - \lambda_{m'}$; when it is put into Eq. (C-19) and reorganized, the following formula is obtained:

$$(\lambda_{m'} - \lambda_m - 1)\langle j, m'|\hat{J}_+|j, m\rangle = 0 \qquad \text{(C-20)}$$

From Eq. (C-20) we can know that if $(\lambda_{m'} - \lambda_m - 1) \neq 0$, then $\langle j, m'|\hat{J}_+|j, m\rangle = 0$; on the other hand, $\langle j, m'|\hat{J}_+|j, m\rangle \neq 0$, then $(\lambda_{m'} - \lambda_m - 1)$ should be equal to 0; hence, $\lambda_{m'}$ must be equal to $\lambda_m + 1$. This explains that

$$\hat{J}_+|j, m\rangle = x_m|j, m+1\rangle \qquad \text{(C-21)}$$

$$\hat{J}_-|j, m\rangle = y_m|j, m-1\rangle \qquad \text{(C-22)}$$

Here x_m and y_m can be complex numbers, because they can contain a phase factor$e^{i\phi}$. From Eqs. (C-21) and (C-22) it can be seen that $\hat{J}_+$ acting on $|j, m\rangle$ will result in $|j, m+1\rangle$, and so $\hat{J}_+$ is called the *raising operator*. Likewise, $\hat{J}_-$ acting on $|j, m\rangle$ will result in $|j, m-1\rangle$, so $\hat{J}_-$ is called the *reducing operator*. It explains that the eigenvalue of $|j, m\rangle$ should be

$$\ldots,\ \ldots,\ \lambda_{m-2}, \lambda_{m-1}, \lambda_m, \lambda_{m+1}, \lambda_{m+2}\ ,\ldots\ ,\ldots$$

Both terminals of this sequence should be limited, because they must satisfy $\lambda_m^2 \leq \lambda_j$, as the interval of the λ_m values can only differ in integer, and for a given value of j, the quantum number m is according to the integer increasing or decreasing. So, let $\lambda_m = m$, m has a maximum $(\overline{m})$ and also a minimum $(\underline{m})$. Thus,

$$\hat{J}_+|j, \overline{m}\rangle = 0 \qquad \text{(C-23)}$$

$$\hat{J}_-|j, \underline{m}\rangle = 0 \qquad \text{(C-24)}$$

Since

$$\hat{J}_-\hat{J}_+ = (\hat{J}_x - i\hat{J}_y)(\hat{J}_x + i\hat{J}_y) = \hat{J}_x^2 + \hat{J}_y^2 + i(\hat{J}_x\hat{J}_y - \hat{J}_y\hat{J}_x) = \hat{J}_x^2 + \hat{J}_y^2 - \hat{J}_z = \hat{J}^2 - \hat{J}_z^2 - \hat{J}_z$$

$$\hat{J}_-\hat{J}_+ = \hat{J}^2 - \hat{J}_z^2 - \hat{J}_z$$

So

$$\hat{J}_-\hat{J}_+|j, \overline{m}\rangle = (\hat{J}^2 - \hat{J}_z^2 - \hat{J}_z)|j, \overline{m}\rangle = (\lambda_j - \overline{m}^2 - \overline{m})|j, \overline{m}\rangle \qquad \text{(C-25)}$$

On the other hand,

$$\hat{J}_-\hat{J}_+|j, \overline{m}\rangle = 0$$

So

$$\lambda_j = \overline{m}(\overline{m} + 1) \qquad \text{(C-26)}$$

Similarly,

$$\hat{J}_+\hat{J}_-|j,\underline{m}\rangle = (\hat{J}^2 - \hat{J}_z^2 + \hat{J}_z)|j,\underline{m}\rangle = (\lambda_j - \underline{m}^2 - \underline{m})|j,\underline{m}\rangle = 0$$

So

$$\lambda_j = \underline{m}(\underline{m}+1) \tag{C-27}$$

To make Eqs. (C-26) and (C-27) valid simultaneously, the condition is $\overline{m} = -m$. Let the maximum value of m ($\overline{m}$) be j, then the sequence of m is

$$j, j-1, j-2, \ldots, \ldots, \quad -j+2, \ -j+1, \ -j$$

That is,

$$\lambda_j = j(j+1) \tag{C-28}$$

Thus,

$$\hat{J}^2|j,m\rangle = j(j+1)|j,m\rangle \tag{C-29}$$

$$\hat{J}_z|j,m\rangle = m|j,m\rangle \tag{C-30}$$

Look again

$$\begin{aligned}\langle j,m|\hat{J}_-\hat{J}_+|j,m\rangle &= x_m\langle j,m|\hat{J}_-|j,m+1\rangle = x_m\langle j,m+1|\hat{J}_+|j,m+1\rangle^* \\ &= x_m x_m^*\langle j,m+1|j,m+1\rangle = x_m x_m^*\end{aligned} \tag{C-31}$$

On the other hand,

$$\begin{aligned}\langle j,m|\hat{J}_-\hat{J}_+|j,m\rangle &= \langle j,m\left|\hat{J}^2 - \hat{J}_z^2 + \hat{J}_z\right|j,m\rangle = j(j+1) - m^2 - m \\ &= (j-m)(j+m+1)\end{aligned} \tag{C-32}$$

Hence,

$$x_m = \sqrt{(j-m)(j+m+1)} \tag{C-33}$$

Similarly, it can be proved

$$y_m = \sqrt{(j+m)(j-m+1)} \tag{C-34}$$

Thus, the result is

$$\hat{J}_+\left|j,m\right\rangle = \sqrt{(j-m)(j+m+1)}\left|j,m+1\right\rangle \tag{C-35}$$

$$\hat{J}_-\left|j,m\right\rangle = \sqrt{(j+m)(j-m+1)}\left|j,m-1\right\rangle \tag{C-36}$$

C-4 Matrix representation of angular momentum operator

If we choose the common eigenfunction $|j, m\rangle$ of $\hat{J}^2$ and $\hat{J}_z$ as the base function, then the matrix representation of $\hat{J}^2$ and $\hat{J}_z$ is a diagonal matrix, and the matrix elements of the diagonal line are

$$\langle j, m|\hat{J}^2|J, m\rangle = j(j+1) \tag{C-37}$$

$$\langle j, m|\hat{J}_z|J, m\rangle = m \tag{C-38}$$

For $\hat{J}_+$ and $\hat{J}_-$ or $\hat{J}_x$ matrices without diagonal elements, their nonzero matrix elements are as follows:

$$\langle j, m+1|\hat{J}_+|j, m\rangle = \sqrt{(j-m)(j+m+1)} \tag{C-39}$$

$$\langle j, m-1|\hat{J}_-|j, m\rangle = \sqrt{(j+m)(j-m+1)} \tag{C-40}$$

$$\langle j, m+1|\hat{J}_x|j, m\rangle = \frac{1}{2}\sqrt{(j-m)(j+m+1)} \tag{C-41}$$

$$\langle j, m-1|\hat{J}_x|j, m\rangle = \frac{1}{2}\sqrt{(j+m)(j-m+1)} \tag{C-42}$$

$$\langle j, m+1|\hat{J}_y|j, m\rangle = -\frac{i}{2}\sqrt{(j-m)(j+m+1)} \tag{C-43}$$

$$\langle j, m-1|\hat{J}_y|j, m\rangle = \frac{i}{2}\sqrt{(j+m)(j-m+1)} \tag{C-44}$$

Several common matrices are used. For $j = \frac{1}{2}$, there are only two base functions: $\alpha = |\frac{1}{2}, \frac{1}{2}\rangle$ and $\beta = |\frac{1}{2}, -\frac{1}{2}\rangle$, and the matrix is

$$J_i = \frac{1}{2}\boldsymbol{\sigma}_i \quad (i = x, y, z) \tag{C-45}$$

Here $\boldsymbol{\sigma}_i$ is the Pauli matrix, which is defined as

$$\boldsymbol{\sigma}_x = \begin{bmatrix} 0 & 1 \\ 1 & 0 \end{bmatrix}, \qquad \boldsymbol{\sigma}_y = \begin{bmatrix} 0 & -i \\ i & 0 \end{bmatrix}, \qquad \boldsymbol{\sigma}_z = \begin{bmatrix} 1 & 0 \\ 0 & -1 \end{bmatrix} \tag{C-46}$$

Their multiplication satisfies the following commutation relation:

$$[\boldsymbol{\sigma}_x, \boldsymbol{\sigma}_y] = 2i\boldsymbol{\sigma}_z \tag{C-47a}$$

$$[\boldsymbol{\sigma}_y, \boldsymbol{\sigma}_z] = 2i\boldsymbol{\sigma}_x \tag{C-47b}$$

$$[\boldsymbol{\sigma}_z, \boldsymbol{\sigma}_x] = 2i\boldsymbol{\sigma}_y \tag{C-47c}$$

$$\boldsymbol{\sigma}_x\boldsymbol{\sigma}_y + \boldsymbol{\sigma}_y\boldsymbol{\sigma}_x = 0 \tag{C-48a}$$

$$\boldsymbol{\sigma}_y\boldsymbol{\sigma}_z + \boldsymbol{\sigma}_z\boldsymbol{\sigma}_y = 0 \tag{C-48b}$$

$$\boldsymbol{\sigma}_z\boldsymbol{\sigma}_x + \boldsymbol{\sigma}_x\boldsymbol{\sigma}_z = 0 \tag{C-48c}$$

$$\boldsymbol{\sigma}_x\boldsymbol{\sigma}_y\boldsymbol{\sigma}_z = i \tag{C-49}$$

$$\boldsymbol{\sigma}_x^2 = \boldsymbol{\sigma}_y^2 = \boldsymbol{\sigma}_z^2 = 1 \tag{C-50}$$

$j = 1$ is usually used in the triplet, and its base functions are $|1,1\rangle$, $|1,0\rangle$, $|1,-1\rangle$, in accordance with Eqs. (C-41)– (C-44):

$$J_x = \frac{1}{\sqrt{2}}\begin{bmatrix} 0 & 1 & 0 \\ 1 & 0 & 1 \\ 0 & 1 & 0 \end{bmatrix}, \quad J_y = \frac{1}{\sqrt{2}}\begin{bmatrix} 0 & -i & 0 \\ i & 0 & -i \\ 0 & i & 0 \end{bmatrix}, \quad J_z = \begin{bmatrix} 1 & 0 & 0 \\ 0 & 0 & 0 \\ 0 & 0 & -1 \end{bmatrix} \tag{C-51}$$

C-5 Addition of two angular momenta

The two angular momenta coupling problem is often encountered. The two angular momenta coupling can be the orbital angular momentum and spin angular momentum of the same particle (electron), and can also be the angular momenta of two different particles (electron and nuclear). As generally considered, the two angular momentum operators of the system $\hat{\boldsymbol{J}}_1, \hat{\boldsymbol{J}}_2$ satisfy the following commutation relation:

$$\hat{\boldsymbol{J}}_1 \times \hat{\boldsymbol{J}}_1 = i\hat{\boldsymbol{J}}_1 \tag{C-52a}$$

$$\hat{\boldsymbol{J}}_2 \times \hat{\boldsymbol{J}}_2 = i\hat{\boldsymbol{J}}_2 \tag{C-52b}$$

$\hat{\boldsymbol{J}}_1$ and $\hat{\boldsymbol{J}}_2$ are commutative. Of course, this is no problem for the angular momenta of different particles. However, for the same particle (e.g., electron), as $\hat{\boldsymbol{J}}_1$ is the representative of orbital motion and $\hat{\boldsymbol{J}}_2$ is the representative of the spin motion, they represent two kinds of different motion, so they are also commutative.

$\hat{\boldsymbol{J}}$ is defined as the sum of $\hat{\boldsymbol{J}}_1$ and $\hat{\boldsymbol{J}}_2$:

$$\hat{\boldsymbol{J}} = \hat{\boldsymbol{J}}_1 + \hat{\boldsymbol{J}}_2 \tag{C-53}$$

Here one must first prove that $\hat{J}$ is also an angular momentum, and it must satisfy the commutation relation of $\hat{\boldsymbol{J}} \times \hat{\boldsymbol{J}} = i\hat{\boldsymbol{J}}$:

$$\begin{aligned} [\hat{J}_x, \hat{J}_y] &= [\hat{J}_{1x} + \hat{J}_{2x}, \hat{J}_{1y} + \hat{J}_{2y}] \\ &= [\hat{J}_{1x}, \hat{J}_{1y}] + [\hat{J}_{2x}, \hat{J}_{1y}] + [\hat{J}_{1x}, \hat{J}_{2y}] + [\hat{J}_{2x}, \hat{J}_{2y}] = i\hat{J}_{1z} + i\hat{J}_{2z} = i\hat{J}_z \end{aligned} \tag{C-54}$$

Since $\hat{\boldsymbol{J}}$ is an angular momentum, $\hat{J}^2$ can commute with any one of $\hat{J}_x, \hat{J}_y, \hat{J}_z$.

It should be mentioned here that $\hat{J}^2$ can also commute with $\hat{J}_1^2$ and $\hat{J}_2^2$, but not with $\hat{J}_{1x}, \hat{J}_{1y}, \hat{J}_{1z}$ and $\hat{J}_{2x}, \hat{J}_{2y}, \hat{J}_{2z}$, because

$$\hat{J}^2 = (\boldsymbol{J}_1 + \boldsymbol{J}_2)^2 = \hat{J}_1^2 + \hat{J}_2^2 + 2\hat{J}_1 \cdot \hat{J}_2 \tag{C-55}$$

So

$$[\hat{J}^2 + \hat{J}_1^2] = [\hat{J}_1^2 + \hat{J}_1^2] + [\hat{J}_2^2 + \hat{J}_1^2] + 2\left\{[\hat{J}_{1x} + \hat{J}_1^2]\hat{J}_{2z} + [\hat{J}_{1y} + \hat{J}_1^2]\hat{J}_{2y} + [\hat{J}_{1z} + \hat{J}_1^2]\hat{J}_{2z}\right\} = 0 \tag{C-56}$$

Of these 12 operators, 4 commutative operators can be chosen. There are two options: one is $(\hat{J}^2, \hat{J}_1^2, \hat{J}_2^2, \hat{J}_z)$ and the other is $(\hat{J}_1^2, \hat{J}_{1z}, \hat{J}_2^2, \hat{J}_{2z})$. For the latter, they have a set of common eigenfunctions:

$$|j_1, m_1\rangle |j_2, m_2\rangle \equiv |j_1, j_2, m_1, m_2\rangle$$

Moreover, they consist of complete orthonormal sets, the representation of these functions as the base functions is called "coupled representation":

$$\hat{J}_1^2 |j_1, j_2, m_1, m_2\rangle = j_1(j_1 + 1) |j_1, j_2, m_1, m_2\rangle \tag{C-57a}$$

$$\hat{J}_2^2 |j_1, j_2, m_1, m_2\rangle = j_2(j_2 + 1) |j_1, j_2, m_1, m_2\rangle \tag{C-57b}$$

$$\hat{J}_{1z} |j_1, j_2, m_1, m_2\rangle = m_1 |j_1, j_2, m_1, m_2\rangle \tag{C-57c}$$

$$\hat{J}_{2z} |j_1, j_2, m_1, m_2\rangle = m_2 |j_1, j_2, m_1, m_2\rangle \tag{C-57d}$$

Here

$$m_1 = j_1,\ j_1 - 1,\ j_1 - 2,\ \ldots,\ \ldots,\ -j_1 + 1,\ -j_1 \tag{C-58a}$$

$$m_2 = j_2,\ j_2 - 1,\ j_2 - 2,\ \ldots,\ \ldots,\ -j_2 + 1,\ -j_2 \tag{C-58b}$$

For the base functions $|j_1, j_2, m_1, m_2\rangle$ of the fixed j_1 and j_2, there are $2j_1 + 1$ and $2j_2 + 1$.

For a set of commutative operators $\hat{J}^2, \hat{J}_1^2, \hat{J}_2^2, \hat{J}_z$, they must also have a set of common eigenfunction $|j, m, j_1, j_2\rangle$. It is called the base function of "coupled representation":

$$\hat{J}^2 |j, m, j_1, j_2\rangle = j(j + 1) |j, m, j_1, j_2\rangle \tag{C-59a}$$

$$\hat{J}_z |j, m, j_1, j_2\rangle = m |j, m, j_1, j_2\rangle \tag{C-59b}$$

$$\hat{J}_1^2 |j, m, j_1, j_2\rangle = j_1(j_1 + 1) |j, m, j_1, j_2\rangle \tag{C-59c}$$

$$\hat{J}_2^2 |j, m, j_1, j_2\rangle = j_2(j_2 + 1) |j, m, j_1, j_2\rangle \tag{C-59d}$$

Here, the value of m can be anyone of the following $2j + 1$ values:

$$j,\ j - 1,\ j - 2,\ \ldots,\ -j + 1,\ -j \tag{C-60}$$

The values of j that can be chosen are as follows:

$$j_1+j_2,\ j_1+j_2-1,\ \ldots,\ |j_1-j_2| \tag{C-61}$$

Now let us prove this: Let the maximum and minimum values of j be j and j. Since the number of linear nonrelational base function should not be changed with the base vector transformation,

$$\sum_{j=j}^{j}(2j+1)=(2j_1+1)(2j_2+1) \tag{C-62}$$

and

$$\sum_{j=j}^{j}(2j+1)=2\sum_{j=j}^{j}j+\sum_{j=j}^{j}j=2(j-j+1)[\frac{1}{2}(j+j]+(j-j+1) \tag{C-63}$$
$$=(j-j+1)(j+j+1)$$

To establish Eqs. (C-62) and (C-63), it must be $j=j_1+j_2$, $j=j_1-j_2$, or j_1-j_2; thus, $j=|j_1-j_2|$. This proves Eq. (C-61).

There are two methods of choosing base function. After determining a base function (if $|j_1,m_1,j_2,m_2\rangle$ is chosen as the base vector), another base function can expand for it, that is,

$$|j_1,j_2,j,m\rangle=\sum_{m_1,m_2}\langle j_1,m_1,j_2,m_2|j_1,j_2,j,m\rangle|j_1,m_1,j_2,m_2\rangle \tag{C-64}$$

Here expanding coefficients are called *vector coupling coefficient*, or *Clebsch–Gordon coefficient*, or *Wigner coefficient*. This coefficient has an analytic closed expression formula, but very complicated, and here this will not be deduced. Equation (C-64) is usually written in a more concise form:

$$|j,m\rangle=\sum_{m_1,m_2}C(j_1j_2j\,;m_1m_2m)|m_1m_2\rangle \tag{C-65}$$

There are two important properties:

1. If $m\neq m_1+m_2$, then

$$C(j_1j_2j\,;m_1m_2m)=0 \tag{C-66}$$

Proof: Assuming $\hat{J}_z=\hat{J}_{1z}+\hat{J}_{2z}$ act on both sides of Eq. (C-65), we obtain

$$m|j,m\rangle=\sum_{m_1,m_2}(m_1+m_2)\,C(j_1j_2j\,;m_1m_2m)|m_1m_2\rangle$$

Thus,

$$\sum_{m_1, m_2} (m - m_1 - m_2)\, C(j_1 j_2 j\,; m_1 m_2 m)|m_1 m_2\rangle = 0 \tag{C-67}$$

Since $|m_1, m_2\rangle$ is linear nonrelation, Eq. (C-67) establishes that the coefficients must be equal to zero:

$$(m - m_1 - m_2)C(j_1 j_2 j\,; m_1 m_2 m) = 0$$

So if $m \neq m_1 + m_2$, then $C(j_1 j_2 j\,; m_1 m_2 m) = 0$ is necessary

According to this property, Eq. (C-65) can be written as follows:

$$|j, m\rangle = \sum_{m_1, m_2} C(j_1 j_2 j\,; m_1, m - m_1)|j_1, j_2, m_1, m - m_1\rangle \tag{C-68}$$

Here only summation form_1 is desired.

2. For $m_1 = j_1,\ \ m_2 = j_2$, then

$$|j_1 j_2, j_1 + j_2, j_1 + j_2\rangle = |j_1 j_2, j_1 j_2\rangle \tag{C-69a}$$

Similarly,

$$|j_1 j_2, j_1 + j_2, -j_1 - j_2\rangle = |j_1 j_2, -j_1 - j_2\rangle \tag{C-69b}$$

As for the relation between the coupling representation function and the noncoupling representation function, it can also be solved step by step using the raising and reducing operators. This method is tedious but not difficult. For example, $j_1 = 1,\ \ j_2 = 2$, from the property 2, we can know

$$|3, 3\rangle = |2, 1\rangle \tag{C-70}$$

Let J_- act on the both sides of Eq. (C-70), then the left-hand side

$$J_-\Big|3, 3\rangle = \sqrt{(3+3)(3-3+1)}\Big|3, 2\rangle = \sqrt{6}|3, 2\rangle \tag{C-71}$$

And the right-hand side

$$\begin{aligned} J_-|2, 1\rangle = (J_{1-} + J_{2-})|2, 1\rangle = \sqrt{(2+2)(2-2+1)}|1, 1\rangle = \sqrt{(1+1(1-1+1)}|2, 0\rangle \\ = 2|2, 1\rangle + \sqrt{2}|2, 0 \end{aligned} \tag{C-72}$$

Thus,

$$|3, 2\rangle \frac{1}{\sqrt{6}}(2|1, 1\rangle + \sqrt{2}|2, 0\rangle) \tag{C-73}$$

Let J_- act on both sides of Eq. (C-73), then

$$J_-\Big|3, 2\rangle = \sqrt{(3+2)(3-2+1)}\Big|3, 1\rangle = \sqrt{10}|3, 1\rangle \tag{C-74}$$

$$
\begin{aligned}
(J_{1-}+J_{2-}) &= \left\{\frac{2}{\sqrt{6}}|1,1\rangle + \frac{\sqrt{2}}{\sqrt{6}}|2,0\rangle\right\} \\
&= \frac{2}{\sqrt{6}}\left\{\sqrt{(2+1)(2-1+1)}|0,1\rangle + \sqrt{(1+1)(1-1+1)}|1,0\rangle\right\} \\
&+ \frac{\sqrt{2}}{\sqrt{6}}\left\{\sqrt{(2+2)(2-2+2)}|0,1\rangle + \sqrt{(1+0)(1-0+1)}|2-1\rangle\right\}
\end{aligned}
\tag{C-75}
$$

On regulation, we obtain

$$
|3,1\rangle = \frac{1}{\sqrt{30}}\left\{2\sqrt{3}\,|0,1\rangle + 4|1,0\rangle + \sqrt{2}|2,-1\rangle\right\} \tag{C-76}
$$

Thus, in this way, we can get entire functions $|3,m\rangle$ listed in Table C-1.

Table C-1: Entire functions j3,mi.

Coupling representation $\|j, m\rangle$	Noncoupling representation $\|m_1, m_2\rangle$
$\|3,\pm 3\rangle$	$\|\pm 2,\pm 1\rangle$
$\|3,\pm 2\rangle$	$\frac{1}{\sqrt{6}}\{2\|\pm 1,\pm 1\rangle + \sqrt{2}\|\pm 2,0\rangle\}$
$\|3,\pm 1\rangle$	$\frac{1}{\sqrt{30}}\{2\sqrt{3}\,\|0,\pm 1\rangle + 4\|\pm 1,0\rangle\sqrt{2}\,\|\pm 2,\mp 1\rangle\}$
$\|3,0\rangle$	$\frac{1}{\sqrt{15}}\{\sqrt{3}\,\|1,-1\rangle + 3\|0,0\rangle + \sqrt{3}\,\|-1,+1\rangle\}$
$\|2,\pm 2\rangle$	$\frac{1}{\sqrt{6}}\{\sqrt{2}\,\|\pm 1,\pm 1\rangle - 2\,\|\pm 2,0\rangle\}$
$\|2,\pm 1\rangle$	$\frac{1}{\sqrt{6}}\{\sqrt{3}\,\|0,\pm 1\rangle - \|\pm 1,0\rangle - \sqrt{2}\,\|\pm 2,\mp 1\rangle\}$
$\|2,0\rangle$	$\frac{1}{\sqrt{2}}\{\|-1,1\rangle - \|1,-1\rangle\}$
$\|1,\pm 1\rangle$	$\frac{1}{\sqrt{10}}\{\,\|10,\pm 1\rangle - \sqrt{3}\,\|\pm 1,0\rangle + \sqrt{6}\,\|\pm 2,\mp 1\rangle\,\}$
$\|1,0\rangle$	$\frac{1}{\sqrt{10}}\{\,\sqrt{3}\|1,-1\rangle - 2\|\,0,0\rangle + \sqrt{3}\,\|-1,1\rangle\,\}$

The solution method of $|2,2\rangle$ is as follows: Let

$$
|2,2\rangle = c_1\,|1,1\rangle + c_2|2,0\rangle \tag{C-77}
$$

Assuming that $\hat{J}_z$ acts on both sides, it can be proved that both sides of the formula are eigenfunctions of $\hat{J}_z$ and their eigenvalues are 2. According to the orthogonal normalization property of the function, we can know

$$
0 = \langle 3,2|2,2\rangle = c_1\frac{2}{\sqrt{6}} + c_2\sqrt{\frac{2}{6}} \tag{C-78a}
$$

$$
1 = \langle 2,2|2,2\rangle = {c_1}^2 + {c_2}^2 \tag{C-78b}
$$

Solve the simultaneous equations to obtain

$$
c_1 = \sqrt{\frac{1}{3}}, \quad c_2 = \sqrt{\frac{2}{3}}
$$

Thus,

$$|2,2\rangle = \frac{1}{\sqrt{6}}(\sqrt{2}\,|1,1\rangle - 2|2,0\rangle) \tag{C-79}$$

After having $|2,2\rangle$, use $\hat{J}_-$ to act on both sides so that all the functions of $|2,m\rangle$ are obtained, and finally only the function $|1,1\rangle$ is required. Let

$$|1,1\rangle = c_1\,|0,1\rangle + c_2\,|1,0\rangle + c_3\,|2,-1\rangle \tag{C-80}$$

Then,

$$\langle 3,1|1,1\rangle = \frac{1}{\sqrt{30}}(2\sqrt{3}\,c_1 + 4c_2 + \sqrt{2}\,c_3) = 0 \tag{C-81a}$$

$$\langle 2,1|1,1\rangle = \frac{1}{\sqrt{6}}(\sqrt{3}\,c_1 - c_2 - \sqrt{2}\,c_3) = 0 \tag{C-81b}$$

$$\langle 1,1|1,1\rangle = c_1^{\,2} + c_2^{\,2} + c_3^{\,2} = 1 \tag{C-81c}$$

Solving this equation, we obtain

$$c_1 = \tfrac{1}{\sqrt{10}}, \quad c_2 = \tfrac{-\sqrt{3}}{\sqrt{10}}, \quad c_3 = \sqrt{\tfrac{3}{5}} \tag{C-82}$$

After obtaining $|1,1\rangle$, we can seek $|1,0\rangle$ and $|1,-1\rangle$, and all the results of Table C-1 can be obtained by in this way.

C-6 Several useful tables

Orbital angular momentum operators are represented by $\hat{l}^2$ and $\hat{l}_z$, and their common eigenfunction is $|l,m\rangle$. The orbital functions of p and d are as follows:

$$p_x = (-\frac{1}{\sqrt{2}})\{|1,1\rangle - |1,1\rangle\}$$

$$p_y = \frac{i}{\sqrt{2}}\{\,|1,1\rangle + |1,1\rangle\,\}$$

$$p_z = 1,0\rangle$$

$$d_{x^2-y^2} = \frac{1}{\sqrt{2}}\{\,|2,2\rangle + |2,-2\rangle\,\}$$

$$d_{xy} = \frac{i}{\sqrt{2}}\{\,|2,2\rangle - |2,-2\rangle\,\}$$

$$d_{xz} = \frac{1}{\sqrt{2}}\{\,|2,1\rangle - |2,-1\rangle\,\}$$

$$d_{yz} = \frac{i}{\sqrt{2}}\{\,|2,1\rangle + |2,-1\rangle\,\}$$

$$d_{z^2} = |2,0\rangle\,\}$$

For example $\hat{l}_x$ acts on $|yz\rangle$:

$$\begin{aligned}\hat{L}_x|yz\rangle &= \frac{i}{\sqrt{2}}\frac{1}{2}\{\sqrt{(2-1)(2+1+1)}\ |\,2,2\rangle + 2\sqrt{(2+1)(2-1+1)}\ |2,0\rangle \\ &\quad + \sqrt{(2-1)(2+1+1)}|2,-2\rangle\ \} \\ &= \frac{i}{\sqrt{2}}\{|\,2,2\rangle\ +\ |\,2,2\rangle\ +\ \sqrt{3}i|\,2,0\rangle\} = i|x^2-y^2\rangle\ +\ i\sqrt{3}|z^2\rangle\end{aligned}$$

The new functions after the angular momentum operators $\hat{l}_x$, $\hat{l}_y$, $\hat{l}_z$ act on the orbitals p and d are listed in Table C-2. The matrix representation of $\hat{i}\cdot\hat{s}$ when using the p-orbital as base function is shown in Table C-3, and the matrix representation of $\hat{i}\cdot\hat{s}$ when using the d-orbital as base function is shown in Table C-4.

Table C-2: The obtained new functions after the operators $\hat{l}_x$, $\hat{l}_y$, $\hat{l}_z$, act on the orbitals p and d.

Function symbol	$\hat{l}_x$	$\hat{l}_y$	$\hat{l}_z$
$\|p_x\rangle \equiv \|x\rangle$	0	$-i\|z\rangle$	$i\|y\rangle$
$\|p_y\rangle \equiv \|y\rangle$	$i\|z\rangle$	0	$-i\|x\rangle$
$\|p_z\rangle \equiv \|z\rangle$	$-i\|y\rangle$	$i\|x\rangle$	0
$\|d_{x^2-y^2}\rangle \equiv \|x^2-y^2\rangle$	$-i\|yz\rangle$	$-i\|xz\rangle$	$2i\|xy\rangle$
$\|d_{xy}\rangle \equiv \|xy\rangle$	$i\|xz\rangle$	$-i\|yz\rangle$	$-2i\|x^2-y^2\rangle$
$\|d_{yz}\rangle \equiv \|yz\rangle$	$i\|x^2-y^2\rangle + \sqrt{3}i\|z^2$	$i\|xy\rangle$	$-i\|xz\rangle$
$\|d_{zx}\rangle \equiv \|zx\rangle$	$-i\|xy\rangle$	$i\|x^2-y^2\rangle - \sqrt{3}i\|z^2\rangle$	$i\|yz\rangle$
$\|d_{z^2}\rangle \equiv \|z^2\rangle$	$-\sqrt{3}i\|yz\rangle$	$\sqrt{3}i\|xz\rangle$	0

Table C-3: The matrix representation of $\hat{i}\cdot\hat{s}$ when using p-orbital as base function.

	$\|x,\alpha\rangle$	$\|y,\alpha\rangle$	$\|z,\alpha\rangle$	$\|x,\beta\rangle$	$\|y,\beta\rangle$	$\|z,\beta\rangle$
$\langle x,\alpha\|$	0	$-\frac{i}{2}$	0	0	0	$\frac{1}{2}$
$\langle y,\alpha\|$	$\frac{i}{2}$	0	0	0	0	$-\frac{i}{2}$
$\langle z,\alpha\|$	0	0	0	$-\frac{1}{2}$	$\frac{i}{2}$	0
$\langle x,\beta\|$	0	0	$-\frac{1}{2}$	0	$\frac{i}{2}$	0
$\langle y,\beta\|$	0	0	$-\frac{i}{2}$	$-\frac{i}{2}$	0	0
$\langle z,\beta\|$	$\frac{1}{2}$	$\frac{i}{2}$	0	0	0	0

Let us take another example of matrix element:

$$\begin{aligned}\langle x,\alpha|\hat{l}\cdot\hat{s}.|z,\beta\rangle &= \langle x,\alpha|\frac{1}{2}.(\hat{l}_+\hat{s}_- + \hat{l}_-\hat{s}_+)|\hat{l}_z\hat{s}_z|z,\beta\rangle \\ &= \frac{1}{2}\langle x|\hat{l}_-|z\rangle = \frac{1}{2}\{\langle x|\hat{l}_x|z\rangle - i\langle x|\hat{l}_y|z\rangle\} \\ &= \frac{1}{2}\{0-i^2\} = \frac{1}{2}\end{aligned}$$

C-7 Stable-state perturbation theory in quantum mechanics

After solving the concrete physical problem by using quantum mechanics, we seek to solve the eigenvalue and eigenfunction of Hamiltonian operator. However, as all the Hamiltonian operators are rather complex generally, we cannot seek strict solution, only a few very simple physical systems can seek strict solution. Only the approximate methods can be applied, under the most situations. For the stable-state problem, that is, the Hamiltonian operator is not an explicit function of time, the stable-state perturbation theory is commonly used as approximation method.

We need to solve the eigenvalue problem of the following equation:

$$(\hat{H}_0 + \hat{H}')\Psi = E\,\Psi \tag{C-83}$$

Here $\hat{H}_0 + \hat{H}' = \hat{H}$ does not contain time t obviously, so the system has definite energy. $\hat{H}_0$ is the expression of the Hamiltonian operator, when the system has not been influenced by the external perturbation, and its eigenfunction $|n,\alpha\rangle$ and eigenvalue E_n^o can be solved strictly, that is,

$$\hat{H}_0|n,\alpha\rangle = E_n^o|n,\alpha\rangle \tag{C-84}$$

The operator $\hat{H}'$ is a perturbation term, and its effect on the system is just only a small perturbation compared with the operator $\hat{H}_o$.

As already mentioned, Eq. (C-83) can neither be strictly solved nor can it be approximately solved. Equation (C-1) can be rewritten as

$$\hat{H}'\,\Psi = (E - \hat{H}_o)\Psi \tag{C-85}$$

$$(E - \hat{H}_o)^{-1}\hat{H}'\Psi = \Psi \tag{C-86}$$

If the eigenfunction $\{|n,\alpha\rangle\ ]$ of $\hat{H}_o$ is already a complete set of orthogonal normalization, then Ψ can spread out for it:

$$\Psi = \sum_{m,\beta}\langle m,\beta|\Psi\rangle|m,\beta\rangle \tag{C-87}$$

Table C-4: The matrix representation of $\hat{l}\cdot\hat{s}$ when using d-orbital as base function.

	$\vert x^2-y^2,\alpha\rangle$	$\vert xy,\alpha\rangle$	$\vert yz,\alpha\rangle$	$\vert xz,\alpha\rangle$	$\vert z^2,\alpha\rangle$
$\langle x^2-y^2,\alpha\vert$	0	$-i$	0	0	0
$\langle xy,\alpha\vert$	i	0	0	0	0
$\langle yz,\alpha\vert$	0	0	0	$\frac{i}{2}$	0
$\langle xz,\alpha\vert$	0	0	$-\frac{i}{2}$	0	0
$\langle z^2,\alpha\vert$	0	0	0	0	0
$\langle x^2-y^2,\beta\vert$	0	0	$\frac{i}{2}$	$-\frac{1}{2}$	0
$\langle xy,\beta\vert$	0	0	$-\frac{1}{2}$	$-\frac{i}{2}$	0
$\langle yz,\beta\vert$	$-\frac{i}{2}$	$\frac{1}{2}$	0	0	$-\frac{i\sqrt{3}}{2}$
$\langle xz,\beta\vert$	$\frac{1}{2}$	$-\frac{i}{2}$	0	0	$-\frac{\sqrt{3}}{2}$
$\langle z^2,\beta\vert$	0	0	$\frac{i\sqrt{3}}{2}$	$\frac{\sqrt{3}}{2}$	0

Similarly, function $\hat{H}'\Psi$ and $(E-\hat{H}_o)^{-1}\hat{H}'\Psi$ can also be spread out as follows:

$$\begin{aligned}\hat{H}'\Psi &= \sum_{m,\beta}\langle m,\beta|\Psi\rangle\,\hat{H}'|m,\beta\rangle \\ &= \sum_{m,\beta}\sum_{n,\alpha}\langle m,\beta|\Psi\rangle\,\langle n,\alpha|\hat{H}'|m,\beta\rangle\,|n,\alpha\rangle\end{aligned} \tag{C-88}$$

$$\begin{aligned}(\mathrm{E}-\hat{H}_o)^{-1}\hat{H}'\Psi &= \sum_{m,\beta}\sum_{n,\alpha}\langle m,\beta|\Psi\rangle\,\langle n,\alpha|\,\hat{H}'|m,\beta\rangle(E-\hat{H}_o)^{-1}|n,\alpha\rangle \\ &= \sum_{m,\beta}\sum_{n,\alpha}\langle m,\beta|\Psi\rangle\,\langle n,\alpha|\,\hat{H}'|m,\beta\rangle\,(E-E_n^o)^{-1}|n,\alpha\rangle\end{aligned} \tag{C-89}$$

Substituting Eqs. (C-89) and (C-87) into Eq. (C-86), we obtain:

$$\sum_{n,\alpha}\{\langle n,\alpha|\Psi\rangle\,(E-E_n^o)-\sum_{m,\beta}\left\langle n,\alpha|\hat{H}'m,\beta\rangle\langle m,\beta|\Psi\rangle\right\}|n,\alpha\rangle=0 \tag{C-90}$$

Since the already known t{$|n,\alpha\rangle$ } is a complete set of orthogonal normalization, they should be linear and nonrelative. If Eq. (C-9) is to be established, all of their coefficients should be zero. Thus,

$$\sum_{m,\beta}\left\langle n,\alpha|\hat{H}'|m,\beta\rangle\,\langle m,\beta|\Psi\rangle=\langle n,\alpha|\Psi\rangle\,(E-E_n^o)\right. \tag{C-91}$$

Now let us discuss two cases respectively: first, the stable-state perturbation theory in nondegenerate case; and second, the stable-state perturbation theory in degenerate case.

Table C-4 (continued)

$\lvert x^2-y^2,\beta\rangle$	$\lvert xy,\beta\rangle$	$\lvert yz,\beta\rangle$	$\lvert xz,\beta\rangle$	$\lvert z^2,\beta\rangle$
0	0	$\frac{i}{2}$	$\frac{1}{2}$	0
0	0	$\frac{1}{2}$	$-\frac{i}{2}$	0
$-\frac{i}{2}$	$-\frac{1}{2}$	0	0	$-\frac{i\sqrt{3}}{2}$
$-\frac{1}{2}$	$\frac{i}{2}$	0	0	$\frac{\sqrt{3}}{2}$
0	0	$\frac{i\sqrt{3}}{2}$	$-\frac{\sqrt{3}}{2}$	0
0	i	0	0	0
$-i$	0	0	0	0
0	0	0	$-\frac{i}{2}$	0
0	0	$\frac{i}{2}$	0	0
0	0	0	0	0

C-7.1 Stable-state perturbation theory in nondegenerate case

Since all the energy levels of $\hat{H}_o$ are nondegenerate, the eigenfunctions need only one marker. Equation (C-9) can be written as follows:

$$\sum_m \left\langle n|\hat{H}'|m\right\rangle \langle m|\Psi\rangle = \langle n|\Psi\rangle\ (E-E_n^o) \tag{C-92}$$

Now, let us consider the case of the *first-order approximation*. As the lth energy level, the energy E_l is E_l^o in the absence of the perturbation term $\hat{H}'$, and the corresponding eigenfunction is $|l\rangle$. When the perturbation term $\hat{H}'$ exist, the energy is E_l and the corresponding eigenfunction is Ψ_1. However, owing to the fact that $\hat{H}'$ is a very small perturbation term, it can be understood that E_l must be very close to E_l^o, and the Ψ_l must be very close to $|l\rangle$. Spreading Ψ_l for $\{|n\rangle\}$, we obtain

$$\Psi_l = \sum_m \langle n|\Psi_l\rangle|n\rangle = \langle l|\Psi_l\rangle|l\rangle + \sum_{n\neq l}\langle n|\Psi_l\rangle|n\rangle \tag{C-93}$$

As a first-order approximation, let

$\langle l|\Psi_l\rangle \approx 1,$ when $n\neq l$

时，$\langle n|\Psi_l\rangle \approx 0$ (C-94)

So

$$\Psi_l \approx |l\rangle \tag{C-95}$$

From Eq. (C-92), we can know

$$(E_l - E_l^o)\ \langle l|\Psi_l\rangle = \sum_m \langle l|\ \hat{H}'|m\rangle\ \langle m|\Psi_{l.}\rangle = \langle l|\hat{H}'|l\rangle\ \langle l|\Psi_l\rangle + \sum_{m\neq 1} \langle l|\hat{H}'|m\rangle\ \langle m|\Psi_l\rangle \quad \text{(C-96)}$$

Here $\langle l|\ \hat{\mathrm{H}}'|m\rangle$ is a perturbation small term; $\langle m|\Psi_l\rangle$ is also a perturbation small term, their product is a second-order minor term and can be ignored. So Eq. (C-95) can be written as

$$(E_l - E_l^o)\langle l|\Psi_l\rangle \approx\ \langle l|\hat{H}'|l\rangle = \langle l|\Psi_l\rangle$$

or

$$E_l \cong E_l^o +\ \langle l|\hat{H}'|l\rangle = \langle l.|\hat{H}_o|l\rangle + \langle l|\hat{H}'|l\rangle = \langle l|\hat{H}|l\rangle \quad \text{(C-97)}$$

Equations (C-97) and (C-95) are expressions of energy level and wave function in the first-order approximation.

Now consider the second-order approximation. Starting from Eq. (C-92), for $n \neq l$, there is

$$\langle n|\Psi_l\rangle\ (E - E_l^o) = \sum_m \langle n|\hat{H}'|m\rangle\ \langle m|\Psi_l\rangle = \langle n|\hat{H}'|l\rangle\ \langle l|\Psi_l\rangle + \sum_{m\neq l} \langle n|\hat{H}'|m\rangle\ \langle m|\Psi_l\rangle$$

Here $\langle l|\Psi_l\rangle \approx 1$; $\sum_{m\neq l} \langle n|\hat{H}'|m\rangle\ \langle m|\Psi_l\rangle$ is a second-order minor term, which can be neglected; moreover, $E \approx E_n^o$, for $n \neq l$:

$$\langle n|\Psi_l\rangle \approx\ (E_n^o - E_l^o)^{-1} = \langle n|\hat{H}'|l\rangle\ \langle l|\Psi_l\rangle \approx (E_n^o - E_l^o)^{-1}\langle n|\hat{H}'|l\rangle \quad \text{(C-98)}$$

Substituting into Eq. (C-93), we obtain

$$\Psi_l = \langle l|\Psi_l\rangle\ |l\rangle + \sum_{n\neq l} \langle n|\Psi_l\rangle|n\rangle\ \cong |l\rangle + \sum_{n\neq l} \frac{1}{(E_n^o - E_l^o)} \langle n|\hat{H}'|l\rangle|n\rangle \quad \text{(C-99)}$$

Substituting (C-16) into (C-14), we obtain

$$(E_l - E_l^o)\langle l|\Psi_l\rangle = \langle l|\hat{H}'|l\rangle\langle l|\Psi_l\rangle + \sum_{m\neq l} \langle l|\hat{H}'|m\rangle\ \langle m|\Psi_l\rangle$$

$$(E_l - E_l^o) = \langle l|\hat{H}'|l\rangle + \sum_{m\neq l} \frac{\langle l|\hat{\mathrm{H}}'|m\rangle\langle m|\hat{H}'|l\rangle}{E_m^o - E_l^o} \quad \text{(C-100)}$$

This is the result of nondegenerate case, and it is worth noting that under the case of first-order approximation, the energy adds a first-order correction term, and the wave function is still a zero-order wave function. In the case of second-order approximation, the energy adds a second-order correction term, and the wave function is added to a first-order correction term. That is to say, the approximate degree of the wave function is one order lower than the energy level.

C-7.2 Stable-state perturbation theory in degenerate case

Beginning from Eq. (C-91),

$$\sum_{m,\beta} \langle n,\alpha|\ \hat{H}'\ |m,\beta\rangle\ \langle m,\beta|\Psi\rangle = \langle n,\alpha|\Psi\rangle\ (E - E_n^o)$$

Considering the lth energy level,

$$(E - E_l^o)\ \langle l,\alpha|\Psi_l\rangle = \sum_{\beta} \langle l,\alpha|\hat{H}'\ |l,\beta\rangle\ \langle l,\beta|\Psi_l\rangle + \sum_{\substack{m,\beta \\ m\neq l}} \langle l,\alpha|\hat{H}'\ |m,\beta\rangle\ \langle m,\beta|\Psi_l\rangle \quad \text{(C-101)}$$

The latter is a second-order minor tern that can be ignored, so that the above formula can be written as

$$\sum_{\beta} \{\langle l,\alpha|\hat{H}'\ |l,\beta\rangle\ - (E - E_l^o)\ \delta_{\alpha\beta}\}\langle l,\beta|\Psi_l\rangle = 0 \quad \text{(C-102)}$$

If the above formula is to be established, $\langle l,\beta|\Psi_l\rangle$ should not be equal to zero and the coefficients must be equal to zero. Equation (C-102) is a group of level degenerate equation set. Assume the degeneracy of l level is s_l, then the index α and β can be 1, 2, . . .,s_l. Therefore, Eq. (C-102) is a set of linear algebraic equations. If we want to make the solutions of $\langle l,\beta|\Psi_l\rangle$ not equal to zero, the determinant consisting its coefficients must be equal to zero, that is,

$$|\langle l,\alpha|\hat{H}'\ |l,\beta\rangle - (E - E_l^o)\ \delta_{\alpha\beta}|\ = 0 \quad (\alpha,\beta = 1,\ 2,\ \ldots,\ s_l) \quad \text{(C-103)}$$

This is a s dimension secular determinant; expanding this determinant, $\Delta E_l \equiv (E - E_l^o)$ with s_l times algebraic equation can be obtained, and thus we can get s_l roots of ΔE_l, and the perturbation energy levels of s_l is also obtained.

$$E_l^{(k)} = E_l^o + \Delta E_l^{(k)} \quad (k = 1,\ 2,\ \ldots,\ s_l) \quad \text{(C-104)}$$

If there are no multiple roots in the s_l roots, then the original lth energy level E_l^o is s_l times degenerate. Due to the perturbation of $\hat{H}'$, this s_l times degeneracy is relieved completely. If there are also some multiple roots in the s_l roots, then the perturbation of $\hat{H}'$ only partially relieves the degeneracy of the lth energy level.

Let us look at wave functions. When there is no perturbation, E_l^o has s wave functions $|l,1\rangle,\ |l,2\rangle,\ \ldots,\ |l,s_l\rangle$. After being acted by the perturbation of $\hat{H}'$, it becomes $\Psi_l^{(1)},\ \Psi_l^{(2)},\ldots,\ \Psi_l^{(s_l)}$, and these wave functions should be the linear combination of the original s_l wave functions:

$$\Psi_l^{(k)} = \sum_{\alpha=1}^{s_l} \langle l,\alpha|\ \Psi_l\rangle^{(k)}|l,\alpha\rangle \quad (k = 1,\ 2,\ \ldots,\ s_l) \quad \text{(C-105)}$$

Substituting the kth root of ΔE_l into Eq. (C-102) and adding the normalization condition, a set of coefficients can be obtained:

$$\langle l,1|\Psi_l\rangle^{(k)},\ \langle l,2|\Psi_l\rangle^{(k)},\ \ldots,\ \langle l,s_l|\Psi_l\rangle^{(k)}$$

Substituting these coefficients into Eq. (C-105), $\Psi_l^{(k)}$ can be obtained, which is the result of the first-order approximation in the case of degeneracy.

Regarding the second-order approximation, and starting from Eq. (C-91), if $n \neq l$,

$$\langle n,\alpha|\Psi_l\rangle\,(E-E_n^o) = \sum_\beta \langle n,\alpha|\hat{H}'|l,\beta\rangle\,\langle l,\beta|\Psi_l\rangle + \sum_{m\neq l}\langle n,\alpha|\hat{H}'|m,\beta\rangle\,\langle m,\beta|\Psi_l\rangle$$

The last term can be ignored and $E \approx E_l^o$; therefore,

$$\langle n,\alpha|\Psi_l\rangle \cong (E-E_n^o)^{-1}\sum_\beta \langle n,\alpha|\hat{H}'|l,\beta\rangle\,\langle l,\beta|\Psi_l\rangle \tag{C-106}$$

For $n = l$, there is

$$\langle l,\alpha|\Psi\rangle(E_l - E_l^o) = \sum_\beta \langle l,\alpha|\hat{H}'|l,\beta\rangle\,\langle l,\beta|\Psi_l\rangle + \sum_{\substack{m\neq 1\\ m,\beta}}\langle l,\alpha|\hat{H}'|m,\beta\rangle\,\langle m,\beta|\Psi_l\rangle \tag{C-107}$$

Substituting (C-106) into (C-107), we obtain:

$$\begin{aligned}(E_l - E_l^o)\langle l,\alpha|\Psi\rangle &= \sum_\beta \langle l,\alpha|\hat{H}'|l,\beta\rangle\,\langle l,\beta|\Psi_l\rangle\\ &\quad + \sum_{\substack{m,\beta\\ m\neq 1}}\sum_\gamma \frac{\langle l,\alpha|\hat{H}'|m,\beta\rangle\langle m,\beta|\hat{H}'|l,\gamma\rangle}{E_l^o - E_m^o}\langle l,\gamma|\Psi_l\rangle\end{aligned} \tag{C-108}$$

This formula can be rewritten:

$$\sum_\beta\{\langle l,\alpha|\hat{H}'|l,\beta\rangle + \sum_{\substack{m,\gamma\\ m\neq 1}}\frac{\langle l,\alpha|\hat{H}'|m,\gamma\rangle\langle m,\gamma|\hat{H}'|l,\beta\rangle}{E_l^o - E_m^o}(\Delta E_l)\delta_{\alpha\beta}\}\langle l,\beta|\Psi_l\rangle = 0 \tag{C-109}$$

In order to make the above formula to have solutions of nonzero, the determinant of coefficients must be equal to zero. Then

$$|\langle l,\alpha|\hat{H}'|l,\beta\rangle + \sum_{\substack{m,\gamma\\ m\neq 1}}\frac{\langle l,\alpha|\hat{H}'|m,\gamma\rangle\langle m,\gamma|\hat{H}'|l,\beta\rangle}{E_l^o - E_m^o}(\Delta E_l)\delta_{\alpha\beta}| = 0 \tag{C-110}$$

Similarly, this is a secular determinant of s_l dimension, from which s_l roots of ΔE_l can be solved and a set of coefficients corresponding to each root can be obtained. Thus

$$\Psi_l^k = \sum_{n,\alpha} \langle n,\alpha|\Psi_l\rangle |n,\alpha\rangle = \sum_{\alpha} \langle l,\alpha|\Psi_l\rangle^k |l,\alpha\rangle + \sum_{\substack{n,\alpha \\ n\neq 1}} \langle n,\alpha|\Psi_l\rangle^k |n,\alpha\rangle \qquad \text{(C-111)}$$

Substituting Eq. (C-110) into Eq. (C-111), we obtain

$$\Psi_l^k = \sum_{\alpha} \langle l,\alpha|\Psi_l\rangle^k |l,\alpha\rangle + \sum_{\substack{n,\alpha \\ n\neq l}} \frac{1}{(E_l - E_n^o)} \Big(\sum_{\beta} \langle n,\alpha|\, \hat{H}' |l,\beta\rangle \langle l,\beta|\Psi_l\rangle^k\Big) |n,\alpha\rangle \qquad \text{(C-112)}$$

This is the approximate expression of the wave function.

Further readings

[1] Edmonds A. R. *Angular Momentum in Quantum Mechanics*, 2nd edn, Princeton University Press, Princeton, NJ, 1960.

[2] Rose M. E. *Elementary Theory of Angular Momentum*, John Wiley & Sons, Inc., New York, NY, 1974.

[3] Yariv A. *An Introduction to Theory and Application of Quantum Mechanics*, John Wiley & Sons, Inc., New York, NY, 1982.

[4] Zare R. N. *Angular Momentum*, John Wiley & Sons, Inc., New York, NY, 1988.

[5] Condon E. U., Odabasi H. *Atomic Structure*, Cambridge University Press, Cambridge, UK 1980.

Appendix 4: Fundamental constants and useful conversion actors

1. Speed of light in the electromagnetic vacuum:

$$c = 2.99792458 \times 10^8 \text{ m s}^{-1} (\text{defined})$$

2. Magnetic constant (permeability of the vacuum):

$$\mu_o = 4\pi \times 10^{-7} = 12.5663706147 \times 10^{-7} \text{ JC}^{-2} \text{ s}^2 \text{ m}^{-1} (= \text{T}^2 \text{ J}^{-1} \text{ m}^3)$$

3. Dielectric constant (permeability of the vacuum):

$$\varepsilon_o = \mu_o^{-1} \text{ c}^{-2} = 8.854187817 \times 10^{-12} \text{ J}^{-1} \text{ C}^2 \text{ m}^{-1}$$

4. Planck constant:

$$h = 6.6260693(11) \times 10^{-34} \text{J s}$$

$$\hbar = h/2\pi = 1.05457168(18) \times 10^{-34} \text{J s}$$

5. Electronic charge:

$$|e| = 1.60217653(14) \times 10^{-19} \text{ C}$$

$$e = (4.803250 \pm 0.000021) \times 10^{-10} \text{ esu}$$

6. Static mass of electron:

$$m_e = 9.1093826(16) \times 10^{-31} \text{ kg}$$

7. Static mass of proton:

$$m_p = 1.67262171(29) \times 10^{-27} \text{ kg}$$

8. Bohr magneton:

$$\beta_e = \frac{|e|\hbar}{2m_e} = 9.27400949(80) \times 10^{-24} \text{ J T}^{-1}$$

9. The g factor of free electrons:

$$g_e = 2.0023193043718(75)$$

10. Magnetic moment of electron:

$$\mu_e = -g_e \beta_e S = -9.28476412(80) \times 10^{-24} \text{ JT}^{-1} \qquad (S = 1/2)$$

https://doi.org/10.1515/9783110568578-014

11. Magnetogyric ratio of free electrons:

$$\gamma_e = \mu_e / S\hbar = -1.76085974(15) \times 10^{11}\ \mathrm{s}^{-1}\ \mathrm{T}^{-1}$$

12. Nuclear magneton:

$$\beta_n = \frac{|e|\hbar}{2m_p} = 5.05078343(43) \times 10^{-27}\ \mathrm{J\,T}^{-1}$$

13. The g factor of proton:

$$g_p = 5.585694701(56)$$

14. The g factor of proton（Corrected for diamagnetism in a spherical water sample at 298 K):

$$g'_p = 5.585551211(56)$$

15. Magnetic moment of proton:

$$\mu_p = g_p \beta_n I = 1.41060671(12) \times 10^{26}\ \mathrm{JT}^{-1} \qquad (I = 1/2)$$

16. Gyromagnetic ratio of proton:

$$\gamma_p = \mu_p / I\hbar = 2.\ 67522205(23) \times 10^{8}\ \mathrm{s}^{-1}\ \mathrm{T}^{-1}$$

17. Gyromagnetic ratio of proton (Corrected for diamagnetism in a spherical water sample at 298 K):

$$\gamma'_p = 2.67515333(23) \times 10^{8}\ \mathrm{s}^{-1}\ \mathrm{T}^{-1}$$

18. Bohr radius:

$$r_b = \frac{4\pi\varepsilon_o \hbar^2}{m_e e^2} = 5.291772108(18) \times 10^{-11}\ \mathrm{m}$$

19. Boltzmann constant:

$$k_b = 1.3806505(24) \times 10^{-23}\ \mathrm{J\,K}^{-1}$$

20. Avogadro constant:

$$N_{\mathrm{A}} = 6.0221415(10) \times 10^{-23}\ \mathrm{mol}^{-1}$$

Useful conversion factors

1. The conversion between the magnetic field H [mT] and the electron resonance frequency ν_e [MHz] and $\tilde{\nu}_e[\text{cm}^{-1}]$:

$$\nu_e[\text{MHz}] = \frac{g_e\beta_e \text{H}}{\text{h}}\frac{g}{g_e} = 28.02495\frac{g}{g_e}\text{H}[\text{m T}]$$

$$\text{H [mT]} = 0.03568249\frac{g_e}{g}\nu_e[\text{MHz}]$$

$$\nu_e\ [\text{MHz}] = \text{c}\ [\text{m s}^{-1}] \times 10^{-4}\ \tilde{\nu}_e\ [\text{cm}^{-1}] = 2.99792458 \times 10^{-4}\ \tilde{\nu}_e\ [\text{cm}^{-1}]$$

$$\tilde{\nu}_e\ [\text{cm}^{-1}] = 0.33356410 \times 10^{-4}\ \nu_e\ [\text{MHz}] = 9.3481139 \times 10^{-4}\ \frac{g}{g_e}\ \text{H}[\text{m T}]$$

2. The relation between magnetic field H [mT] and proton magnetic resonance frequency ν_p[MHz]:

$$\nu_p\ [\text{MHz}] = 0.04257748\ \text{H}[\text{mT}];\quad 0.04257639\ \text{H}[\text{m T}]\ (\text{in pure water})$$

$$\text{H}[\text{mT}] = 23.48659\ \ \nu_\text{p}[\text{MHz}];\ \ 23.48719\ \nu_\text{p}[\text{MHz}]\ (\text{in pure water})$$

3. The ratio of the resonance frequency of proton to electron:

$$\frac{\nu_p}{\nu_e} = 1.519270 \times 10^{-3}\ \frac{g_e}{g};1.519231 \times 10^{-3}\ \frac{g_e}{g}(\text{the proton in pure water})$$

4. Calculation of g factor:

$$g = \frac{h}{\beta_e}\frac{\nu_\text{e}}{\text{H}} = 0.07144773\frac{\nu_\text{e}}{\text{H}}\frac{[\text{MHz}]}{[\text{mT}]}$$

$$= \frac{g_n\beta_n}{\beta_e}\frac{\nu_e}{\nu_p} = 3.042064 \times 10^{-3}\frac{\nu_e}{\nu_p};\ 3.041987 \times 10^{-3}\ \frac{\nu_e}{\nu_p}(\text{the proton in pure water})$$

5. Hyperfine coupling and hyperfine splitting parameter:

$$\frac{A}{h}\ [\text{MHz}] = 28.02495\ a\ [\text{mT}]$$

$$a\ [\text{mT}] = 0.03568249\ \frac{A}{h}\ [\text{MHz}]$$

$$\frac{A}{hc}\ [\text{cm}^{-1}] = 0.33356410 \times 10^{-4}\frac{A}{h}\ [\text{MHz}]$$

$$= 9.3481182 \times 10^{-4}\frac{A}{g_e\beta_e}\ [\text{mT}]$$

Appendix 5: The natural abundance, nuclear spin, nuclear magnetogyric ratio of some magnetic nuclei and their hyperfine coupling parameters

https://doi.org/10.1515/9783110568578-015

Atomic order number	Nuclear name	Natural abundance[a] (%)	Nuclear spin	g_n[b]	$\frac{g_n\beta_n}{g_e\beta_e} \times 10^5$	Isotropic hyperfine coupling constant[c] a_o [mT]	Anisotropic hyperfine coupling parameter[d] b_o [mT]	ENDOR Frequency [MHz] (when the field of 350 mT)	Nuclear quadrupole moment $Q\|e\| \times 10^{-24}$ [cm^2]
1	^{1}H	99.9850	1/2	5.5856948	151.92704	50.6850 (50.68377)		14.90218	
	^{2}H	0.0148	1	0.8574388	23.32174	7.78027		2.287575	0.002875
2	^{3}He	0.000138	1/2	−4.255280	−115.7407	226.83		11.35266	
3	^{6}Li	7.5	1	0.8220575	22.35940	59.04908		2.193167	−0.000644
	^{7}Li	92.5	3/2	2.170977	59.04908	14.34 (13.02)		5.791950	−0.040
4	^{9}Be	100	3/2	−0.7850	−21.352	−16.11		2.094	0.053
5	^{10}B	19.8	3	0.600220	16.32557	30.43	0.760	1.60133	0.08608
	^{11}B	80.2	3/2	1.792437	48.75306	90.88	2.271	4.782043	0.040
6	^{13}C	1.11	1/2	1.40483	38.2104	134.77	3.832	3.74795	
7	^{14}N	99.63	1	0.4037637	10.98209	64.62	1.981	1.077201	0.0193
	^{15}N	0.366	1/2	−0.5663826	−15.40522	−90.65	−2.779	1.511052	
8	^{17}O	0.038	5/2	−0.757522	−20.60406	−187.80	−6.009	2.02099	−0.026
9	^{19}F	100	1/2	5.257771	143.0077	1886.53	62.82	14.02721	
10	^{21}Ne	0.27	3/2	−0.441200	−12.00034	−221.02	−7.536	1.17708	0.1029
11	^{23}Na	100	3/2	1.478402	40.21151	31.61 (33.08)		3.944228	0.108
12	^{25}Mg	10.00	5/2	−0.34218	−9.30713	17.338		0.91291	0.22
13	^{27}Al	100	5/2	1.456612	39.61883	139.55	2.965	3.886094	0.150

(continued)

(continued)

Atomic order number	Nuclear name	Natural abundance[a] (%)	Nuclear spin	g_n[b]	$\frac{g_n\beta_n}{g_e\beta_e} \times 10^5$	Isotropic hyperfine coupling constant[c] a_o [mT]	Anisotropic hyperfine coupling parameter[d] b_o [mT]	ENDOR Frequency [MHz] (when the field of 350 mT)	Nuclear quadrupole moment $Q\|e\| \times 10^{-24}$ [cm^2]
14	^{29}Si	4.67	1/2	−1.1106	−30.20778	−163.93	−4.075	2.96300	
15	^{31}P	100	1/2	2.26322	61.55793	474.79	13.088	6.03804	
16	^{33}S	0.75	3/2	0.42911	11.67158	123.57	3.587	1.1448	−0.064
17	^{35}Cl	75.77	3/2	0.5479198	14.90304	204.21	6.266	1.461795	−0.08249
	^{37}Cl	24.23	3/2	0.4560854	12.40521	169.98	5.216	1.216790	−0.06493
19	^{39}K	93.26	3/2	0.2609928	7.09882	8.238 (8.152)		0.6963030	0.054
	^{41}K	6.73	3/2	0.1432553	3.89644	4.525		0.382191	0.060
20	^{43}Ca	0.135	7/2	−0.376417	−10.23828	−22.862		1.00424	0.23
21	^{45}Sc	100	7/2	1.35962	36.9807	100.73	3.430	3.62586	−0.22
22	^{47}Ti	7.4	5/2	−0.31539	−8.5784	−27.904	−1.051	0.84144	0.29
	^{49}Ti	5.4	7/2	−0.315468	−8.58051	−27.910	−1.051	0.841667	0.24
23	^{50}V	0.250	6	0.556597	15.13906	56.335	2.368	1.48495	0.209
	^{51}V	99.750	7/2	1.46837	39.9387	148.62	6.246	3.91747	−0.0515
24	^{53}Cr	9.50	3/2	−0.3147	−8.560	−26.698	−1.470	0.8396	−0.0285/ +0.022
25	^{55}Mn	100	5/2	1.3819	37.587	179.70	−8.879	3.6868	0.33
26	^{57}Fe	2.15	1/2	0.1806	4.912	26.662	1.395	0.4818	
27	^{59}Co	100	7/2	1.318	35.849	212.20	12.065	3.516	0.42
28	^{61}Ni	1.13	3/2	−0.50001	−13.6000	−89.171	−5.360	1.3340	0.162
29	^{63}Cu	69.2	3/2	1.484	40.36	213.92	17.058	3.959	−0.222
	^{65}Cu	30.8	3/2	1.588	43.19	228.92	18.283	4.237	−0.195

30	^{67}Zn	4.10	5/2	0.350315	9.52831	74.470	5.021	0.934604	0.150
31	^{69}Ga	60.1	3/2	1.34440	36.5668	435.68	7.274	3.58673	0.168
	^{71}Ga	39.9	3/2	1.70819	46.4616	553.58	9.242	4.55729	0.106
32	^{73}Ge	7.8	9/2	−0.1954371	−5.315654	−84.32	−1.716	0.5214100	−0.19
33	^{75}As	100	3/2	0.959654	26.10193	523.11	11.905	2.56026	0.29
34	^{77}Se	7.6	1/2	1.0693	29.084	717.93	17.542	2.8528	
35	^{79}Br	50.69	3/2	1.404276	38.19535	1144.34	29.174	3.746469	0.293
	^{81}Br	49.31	3/2	1.513717	41.17206	1233.52	31.448	4.038446	0.27
36	^{83}Kr	11.5	9/2	−0.215706	−5.86704	−211.85	−5.515	0.575481	0.260
37	^{85}Rb	72.17	5/2	0.541257	14.72182	36.11 (37.00)		1.44402	0.273
	^{87}Rb	27.83	3/2	1.83428	49.8913	122.38		4.89369	0.130
38	^{87}Sr	7.0	9/2	−0.24291	−6.6070	−30.46		0.64806	0.15
39	^{89}Y	100	1/2	0.2748381	−7.475408	−44.60	−0.888	0.7332410	
40	^{91}Zr	11.2	5/2	−0.521452	−14.18313	−98.23	−2.221	1.39118	
41	^{93}Nb	100	9/2	1.3712	37.296	235.15	6.527	1.6583	−0.28
42	^{95}Mo	15.9	5/2	−0.3656	−9.944	−70.79	−2.151	0.9754	−0.019
	^{97}Mo	9.6	5/2	−0.3734	−10.56	−72.30	−2.197	0.9962	0.2
44	^{99}Ru	12.7	5/2	−0.249	−6.77	−62.94	−2.279	0.644	0.076
	^{101}Ru	17.0	5/2	−0.279	−7.59	−70.52	−2.554	0.744	0.44
45	^{103}Rh	100	1/2	−0.1768	−4.809	−43.85	−1.728	0.4717	
46	^{105}Pd	22.2	5/2	−0.256	−6.96		−2.683	0.683	0.66
47	^{107}Ag	51.83	1/2	−0.227251	−6.18106	−65.33	−2.924	0.606282	
	^{109}Ag	48.17	1/2	−0.261745	−7.11928	−75.25	−3.368	0.698309	
48	^{111}Cd	12.8	1/2	−1.19044	−32.3791	−487.07	−18.41	3.17597	
49	^{115}In	95.7	9/2	1.23130	33.4905	720.07	10.147	3.28498	0.861
50	^{117}Sn	7.75	1/2	−2.00209	−54.555	−1497.98	−24.98	5.34139	
	^{119}Sn	8.6	1/2	−2.09458	−56.971	−1567.18	−26.13	5.58812	

(continued)

(continued)

Atomic order number	Nuclear name	Natural abundance[a] (%)	Nuclear spin	g_n[b]	$\frac{g_n\beta_n}{g_e\beta_e} \times 10^5$	Isotropic hyperfine coupling constant[c] a_o [mT]	Anisotropic hyperfine coupling parameter[d] b_o [mT]	ENDOR Frequency [MHz] (when the field of 350 mT)	Nuclear quadrupole moment $Q\|e\| \times 10^{-24}$ [cm^2]
51	^{121}Sb	57.3	5/2	1.3455	36.597	1252.4	22.44	5.5897	0.33
	^{123}Sb	42.7	7/2	0.72877	19.8220	678.38	12.15	1.9443	0.68
52	^{125}Te	7.0	1/2	−1.7766	−48.322	−1983.6	−37.42	4.7398	
53	^{127}I	100	5/2	1.12531	30.6076	1484.40	28.989	3.00221	−0.789
54	^{129}Xe	26.4	1/2	−1.55595	−42.3211	−2418.92	−47.815	4.15115	
	^{131}Xe	21.2	3/2	0.461243	12.54550	717.06	14.143	1.23055	−0.120
55	^{133}Cs	100	7/2	0.7378532	20.06910	82.00 (88.03)		1.968518	−0.003
56	^{135}Ba	6.59	3/2	0.55884	15.2001	126.67		1.4909	0.20
	^{137}Ba	11.2	3/2	0.62515	17.0036	141.70		1.6679	0.34
57	^{139}La	99.911	7/2	0.79521	21.6292	214.35	3.384	2.1215	0.20
59	^{141}{Pr	100	5/2	1.6	43.5	445.7	12.62	4.3	−0.041
60	^{143}Nd	12.2	7/2	−0.3076	−8.367	−84.82	−2.268	0.8207	0.56
	^{145}Nd	8.3	7/2	−0.190	−5.17	−52.39	−1.401	0.507	0.29
62	^{147}Sm	15.1	7/2	−0.2322	−6.316	−71.86	−2.389	0.6195	−0.18
	^{149}Sm	13.9	7/2	0.1915	5.209	59.26	1.970	0.5109	0.056
63	^{151}Eu	47.9	5/2	1.389	37.78	462.33	15.606	3.607	1.53
	^{153}Eu	52.1	5/2	0.6134	16.684	204.17	6.892	1.637	3.92
64	^{155}Gd	14.8	3/2	−0.1723	−4.686	−69.48	−0.940	0.4597	1.30
	^{157}Gd	15.7	3/2	−0.2253	−6.128	−90.85	−1.229	0.6011	1.34
65	^{159}Tb	100	3/2	3.580	97.37	486.35	17.86	3.580	1.34

66	^{161}Dy	19.0	5/2	−0.189	−5.14	−75.12	−3.775	0.504	2.47
	^{163}Dy	24.9	5/2	0.266	7.24	105.73	3.905	0.710	2.51
67	^{165}Ho	100	7/2	1.192	32.42	483.86	18.340	3.180	2.73
68	^{167}Er	22.9	7/2	−0.1618	−4.401	−69.01	−2.705	0.4317	2.827
69	^{169}Tm	100	1/2	−0.466	−12.67	−208.21	−8.355	1.24	
70	^{171}Yb	14.4	1/2	0.9885	26.887	476.02	19.065	2.637	
	^{173}Yb	16.2	5/2	−0.27195	−7.3968	−130.96	−5.245	0.72554	2.8
71	^{175}Lu	97.39	7/2	0.63943	17.3921	379.31	3.985	1.7059	5.68
72	^{177}Hf	18.6	7/2	0.2267	6.166	157.36	1.771	0.6048	4.5
	^{179}Hf	13.7	9/2	−0.1424	−3.873	−98.84	−1.112	0.3799	5.1
73	^{181}Ta	99.9877	7/2	0.67730	18.4221	535.95	6.357	1.8070	3.44
74	^{183}W	14.3	1/2	0.2355711	6.407372	206.14	2.605	0.6284800	
75	^{185}Re	37.40	5/2	1.2748	34.674	1253.60	16.347	3.4011	2.33
	^{187}Re	62.60	5/2	1.2878	35.027	1266.38	16.514	3.4357	2.22
76	^{189}Os	16.1	3/2	0.488	13.27	471.0	6.637	1.30	0.8
77	^{191}Ir	37.3	3/2	0.097	2.64	112.96	0.907	0.295	0.78
	^{193}Ir	62.7	3/2	0.107	2.91	124.60	1.754	0.285	0.70
78	^{195}Pt	33.8	1/2	1.2190	33.156	1127.84	21.038	3.2522	
79	^{197}Au	100	3/2	0.097969	2.66469	102.62	1.884	0.261371	0.598
80	^{199}Hg	16.8	1/2	1.011778	27.51966	1494.4	22.99	2.688321	
	^{201}Hg	13.2	3/2	−0.373486	−10.15856	−551.6	−8.49	0.996423	0.42
81	^{203}Tl	29.5	1/2	3.244538	88.24919	6496.7	44.54	8.656103	
	^{205}Tl	70.5	1/2	3.2754	89.089	6558.5	44.96	8.7385	

(continued)

(continued)

Atomic order number	Nuclear name	Natural abundance[(a)] (%)	Nuclear spin	g_n[(b)]	$\frac{g_n\beta_n}{g_e\beta_e} \times 10^5$	Isotropic hyperfine coupling constant[(c)] a_o [mT]	Anisotropic hyperfine coupling parameter[(d)] b_o [mT]	ENDOR Frequency [MHz] (when the field of 350 mT)	Nuclear quadrupole moment $Q\|e\| \times 10^{-24}$ [cm^2]
82	^{207}Pb	22.1	1/2	1.1748	31.954	2908.49	23.208	3.1343	
83	^{209}Bi	100	9/2	0.938	25.51	2766.5	23.68	2.50	−0.46

a The data of this table are taken from Weil, J, A, and Rao P S 1985, compiled for the Bruker Instrument Company (which was checked with the 2006 data, has no new changes);

b The data were corrected by the Bohr magneton and nuclear magneton data in 1986;

c Isotropic hyperfine coupling parameters are obtained from experiments, except that the hydrogen atom and alkali metal atoms are calculated.

$$a_o = \frac{2\mu_o}{3} g_n\beta_n|\psi(0)|^2$$

d The anisotropic hyperfine coupling parameter is an unpaired electron in the p orbital. The expression is as follows:

$$b_o = \frac{2}{5}\frac{\mu_o}{4\pi} g_n\beta_n\langle r^{-3}\rangle_p$$

e The parenthesis in the formula is the representation of integral over the whole p orbital.

Index

https://doi.org/10.1515/9783110568578-016

www.ingramcontent.com/pod-product-compliance
Lightning Source LLC
Chambersburg PA
CBHW080130220326
41598CB00032B/5021